CO
HETEROCYCLI

CONTEMPORARY HETEROCYCLIC CHEMISTRY

Syntheses, Reactions, and Applications

George R. Newkome

Louisiana State University
Baton Rouge, Louisiana

William W. Paudler

Portland State University
Portland, Oregon

A Wiley-Interscience Publication

JOHN WILEY & SONS

New York Chichester Brisbane Toronto Singapore

To Our Children
and Their Children

Library of Congress Cataloging in Publication Data:

Newkome, George R. (George Richard)
Contemporary heterocyclic chemistry.

"A Wiley-Interscience publication."
Bibliography: p.
Includes index.
1. Heterocyclic compounds. I. Paudler, William W.,
1932- . II. Title

QD400.N49 547'.59 82-4795
ISBN 0-471-06279-0 AACR2

Printed in the United States of America

10 9 8 7 6 5

PREFACE

Many highly specialized topics in heterocyclic chemistry have recently been covered in book form. However, the general textbooks in organic chemistry have not treated heterocyclic chemistry in the important manner that it contributes to the general topic of organic chemistry. This becomes even more surprising when it is realized that about 40% of the papers published in organic chemistry deal with heterocyclic chemistry.

This book attempts to cover the area of heterocyclic chemistry, with special emphasis on heteroaromatic systems, in a manner that can be used by advanced undergraduates as well as graduate students. Additionally, there is sufficient information in the book so that it can be used as a reference text by researchers in the pharmaceutical industry and other areas where heterocyclic chemistry is an important aspect. The appendix of this book delineates the often needed information with respect to nuclear magnetic resonance spectroscopy, basicity of ring systems, ultraviolet spectral data, and molecular orbital calculations. The latter data are based on one identical set of criteria so that the various values are comparable among all the examined ring systems. The references at the end of each chapter are, hopefully, as up-to-date as is pertinent for the material covered in the chapter.

The authors would be most grateful to know of any errors which may have crept into the manuscript in spite of the proofreading that was done by many of their acquaintances, friends, and co-workers.

GEORGE R. NEWKOME
WILLIAM W. PAUDLER

Baton Rouge, Louisiana
Portland, Oregon
June 1982

ACKNOWLEDGMENTS

We wish to thank our research groups (in particular, Ms. V. K. Majestic and G. R. Baker, D. W. Evans, V. K. Gupta, D. C. Hager, G. E. Kieffer, D. K. Kohli, H. W. Lee, C. R. Marston, W. E. Puckett, and K. J. Theriot), who painstakingly proofed the text, drawings, and references. Special thanks go also to Hellen Rowland Taylor, who read and reread the text in an effort to keep our writing grammatically correct. All of the illustrations were elegantly drawn by Lisa Roper and the typing was masterfully completed by Jacqueline Terrebonne and Sandra Graham. Finally, I (GRN) wish to thank my friends at the Department of Chemistry at Emory University, who gave me the peace and quiet so necessary to permit completion of this book. We acknowledge the aid and guidance of these people who helped to make this work possible.

CONTENTS

Chapter 1

INTRODUCTION

More than half of the compounds produced by nature have heterocyclic rings incorporated in their structures. Nearly all the alkaloids are derived from heterocyclic molecules, as are a considerable number of the substances used as drugs. Heterocycles are also found in fossil fuels, much to the chagrin of environmentalists.

Despite this ubiquity, textbooks of organic chemistry frequently treat heterocyclic chemistry only as a one-chapter afterthought, if at all. Such sketchy coverage tends to leave students with the impression that "heterocyclic chemistry" is a discipline apart from "organic chemistry." Nothing could be further from the truth. As this book amply demonstrates, the chemistry of heterocyclic compounds encompasses most, if not all, of the general reactions of organic chemistry. The presence of heteroatoms merely endows heterocyclic systems with some additional, theoretically predictable properties. It is one of the principal aims of this book to instill in the student the realization that heterocyclic chemistry is simply an extension of aliphatic and aromatic chemistry.

Although it is true that the mechanistic and theoretical aspects of heterocyclic systems have in the past been neglected by physical organic chemists, this is no longer the case. In fact, mechanistic investigations of heterocyclic systems have, over the years, added greatly to a general understanding of the chemistry of organic systems. The application of nuclear magnetic resonance and mass spectrometric techniques to the study of heterocyclic systems has expanded understanding of the applicability of these methods to a great variety of related structural problems.

Advent of the worldwide energy shortage and recognition of the problem of air pollution have also led to an increased interest in heterocyclic chemistry.

For example, the effort expended on the production of solvent-refined coal has illuminated the necessity for techniques to remove nitrogen- and sulfur-containing compounds that occur as impurities. The carcinogenicity of many heteroaromatic compounds has prompted various industrial concerns to remove these chemicals from their smokestack effluents. In both cases, an understanding of the chemistry of heterocyclic systems is required.

One of the major problems associated with writing a book of this type is the vast extent of the topic and the consequent requirement to condense this immense amount of information. We describe most reactions in only the most general sense and hope that the interested reader will utilize the references given at the end of each chapter when detailed information is desired.

Basically, the book is divided into the chemistry of π-deficient (aza, etc., derivatives of benzene) and π-excessive (aza, etc., derivatives of the cyclopentadienyl carbanion) heteroaromatic compounds, that is, their syntheses and their chemical and spectroscopic properties. The final chapters deal with the chemistry of heteroaromatic rings of three, four, seven, and more members. An attempt has been made to include as many important recent developments as possible: more than half of the literature references are from the past two decades. It is interesting to note that the earliest reference cited in this text was published in 1776, more than 200 years ago. Hence our overview of this subject embraces primarily improvements of old procedures and the advances of modern methodology.

We hope that the inclusion of theoretical values (MINDO/3) of atomic electron populations; of proton, carbon, and nitrogen nuclear magnetic resonance data; and of mass spectral fragmentation patterns for various simple ring systems add to the general utility of this text.

Although several years have elapsed since the most recent text on heterocyclic chemistry was published, the information contained in these older books has continued to spark the imagination and research efforts of both experienced and new chemists. In the interest of providing other perspectives and, in many cases, more historical approaches to the classical reactions in this field, we offer the following list (by no means comprehensive) of other texts on the subject.

R. M. Acheson, *Heterocyclic Compounds*, 3rd ed., Interscience, New York, 1967

A. Albert, *Heterocyclic Chemistry*, 2nd ed., Oxford University Press, New York, 1968.

G. M. Badger, *The Chemistry of Heterocyclic Compounds*, Academic Press, New York, 1961.

J. A. Joule and G. F. Smith, *Heterocyclic Chemistry*, Van Nostrand–Reinhold, New York, 1972.

A. R. Katritzky and J. M. Lagowski, *Heterocyclic Chemistry*, Methuen, New York, 1960.

A. R. Katritzky and J. M. Lagowski, *The Principles of Heterocyclic Chemistry*, Academic Press, New York, 1968.

A. A. Morton, *The Chemistry of Heterocyclic Compounds*, McGraw-Hill, New York, 1946.

M. H. Palmer, *The Structure and Reactions of Heterocyclic Compounds*, E. Arnold, London, 1967.

L. A. Paquette, *Principles of Modern Heterocyclic Chemistry*, Benjamin, New York, 1968.

K. Schofield, (Ed.), *Heterocyclic Compounds*, Butterworths, London, 1975.

Chapter 2

NOMENCLATURE

2.1. GENERAL COMMENTS

Before a detailed study of heterocyclic chemistry is undertaken, an introduction to its nomenclature is appropriate. Since the IUPAC nomenclature rules have been presented[1] and expanded by McNaught,[2] only a brief overview is given here with the knowledge that these more comprehensive presentations are available.

2.2. MONOCYCLIC HETEROCYCLES

Heterocycles, derived from the related carbocycle through the replacement of one or more carbon atoms by heteroatoms, are named by appending "a" prefixes as shown in Table 2.1. This terminology is most easily envisioned if the

TABLE 2.1. Extended Version of the Replacement "a" Prefixes[a]

Element (Valence)	Prefix	Element (Valence)	Prefix
O (2)	Oxa	Sn (4)	Stanna
S (2)	Thia	Pb (4)	Plumba
Se (2)	Selena	B (3)	Bora
Te (2)	Tellura	Al (3)	Alumina
N (3)	Aza	Ga (3)	Galla
P (3)	Phospha	In (3)	Inda
As (3)	Arsa	Tl (3)	Thalla
Sb (3)	Stiba	Be (2)	Berylla
Bi (3)	Bismutha	Mg (2)	Magnesa
Si (4)	Sila	Zn (2)	Zinca
Ge (4)	Germa	Cd (2)	Cadma
		Hg (2)	Mercura

[a] See Reference 2.

parent carbocyclic IUPAC accepted name is available. Table 2.2 presents several examples of this adapted carbocyclic "a" nomenclature.

Reduction of heterocycles generally gives rise to several "hydro" isomers. The state of hydrogenation is indicated by either the appended stems given in Table 2.3 or additional prefixes, such as dihydro-, tetrahydro-, or perhydro- for the maximally reduced system, with the location given by locants (number prefixes). Table 2.4 presents several examples of these rules, known as the Hantzsch–Widman nomenclature.

TABLE 2.2. Selected Carbocyclic "a" Conversions

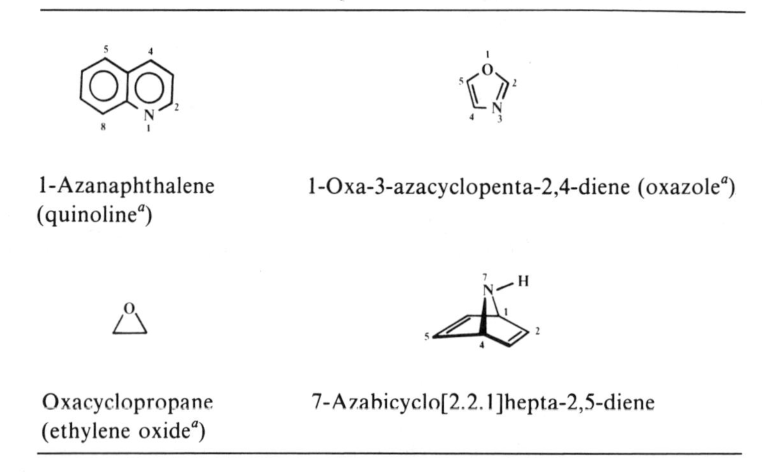

[a] Common name.

TABLE 2.3. IUPAC Unsaturation Suffixes for 3–10-Membered Heterocycles[a,b]

Number of Members in the Ring	Rings Containing Nitrogen		Rings Containing No Nitrogen	
	Unsaturated[c]	Saturated	Unsaturated[c]	Saturated
3	-irine	-iridene	-irene	-irane[e]
4	-ete (-etine)[g]	-eridine	-ete (-etene)[g]	-etane
5	-ole (-oline)[g]	-olidine	-ole (-olene)[g]	-olane
6	-ine[d]	[e]	-in[d]	-ane[f]
7	-epine	[e]	-epin	-epane
8	-ocine	[e]	-ocin	-ocane
9	-onine	[e]	-onin	-onane
10	-ecine	[e]	-ecin	-ecane

[a] See Reference 1.

[b] The syllables denoting the size of rings containing 3, 4, or 7–10 members are derived as follows: "ir" from *tri*, "et" from *tet*ra, "ep" from h*ep*ta, "oc" from *oc*ta, "on" from n*on*a, and "ec" from d*ec*a.

[c] Corresponding to the maximum number of noncumulative double bonds possible when the heteroatoms have the normal valences shown in Table 2.1.

[d] Preceding -ine or -in the special prefix names phosphor, arsen, and antimon are used rather than phospha, arsa, and stiba (see Table 2.1).

[e] Expressed by prefixing "perhydro" to the name of the corresponding unsaturated compound.

[f] Not applicable to silicon, germanium, tin, or lead. In this case, "perhydro" is prefixed to the name of the corresponding unsaturated compound.

[g] These suffixes indicate partial saturation.

Double bond positions between the atoms are designated by Δ^n, where n is the numerical position of the first atom of the unsaturated bond; for example, Δ^3 locates a 3,4-double bond. If a double bond can be arranged in more than one way, the precise location of an extra hydrogen is indicated by nH, where n indicates the position without reference to the double bond. Ring carbonyl groups are indicated by the suffix "-one" accompanied by a positional number, for example, -2-one. The term keto should be carefully used since many times lactam or lactone functionality can be misrepresented by the ketonic terminology. Multiplicity of the same heteroatom is designated by di-, tri-, etc., placed before the prefix; when more than one heteroatom is used per molecule, the order of citation is as presented in Table 2.1.

2.3. NOMENCLATURE OF POLYCYCLIC SYSTEMS

"Ortho-fused" or "benzo-fused" polycyclic hydrocarbons with the maximum number of noncumulative double bonds which contain at least two rings of

TABLE 2.4. Selected Examples of the Hantzsch–Widman Nomenclature System[a]

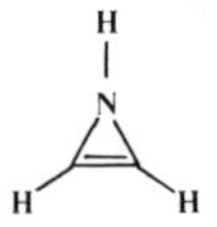 (N; 3; unsat.) *Az irine*	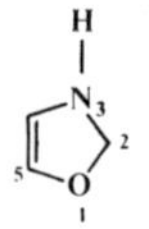 (O, N; 5; unsat.) Δ^4-1,3-*Ox az oline*	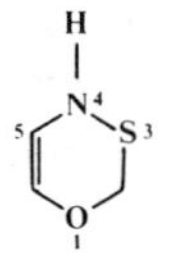 (O, S, N; 6; unsat.) (2*H*,4*H*)-1,3,4-*Ox thi az epin*
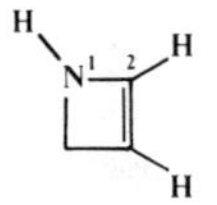 (N; 4; part sat.) Δ^2-*Az etine*	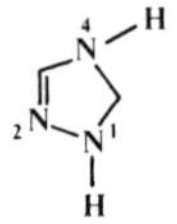 (N, N, N; 5; unsat.) Δ^2-1,2,4-*Tri az oline*	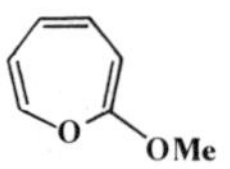 (O; 7; unsat.) 2-Methoxy *ox epin*
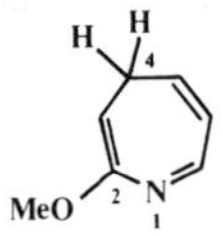 (N; 7; unsat.) 2-Methoxy-4*H*-*az epine*	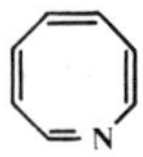 (N; 8; unsat.) *Az ocine*	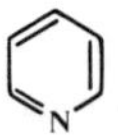 (N; 6; unsat.) *Azine* (pyridine[b])

[a] Analysis given in parentheses.

[b] Common and IUPAC name.

five or more members and which have no accepted trivial name are named by prefixing designations of the other components to the name of a component ring or ring system. The "base" component should be as large as possible and contain as many rings as possible yet still have an accepted[1] trivial name. The attached component should be as simple as possible. Prefixes are derived by dropping the "ene" ending and substituting "eno." Table 2.5 lists some of the common prefixes.

For other monocyclic prefixes, the simple names are used to represent the form with the maximum number of noncumulative double bonds, for example, cyclopenta—. The "base" component ending in "ene" signifies the maximum number of conjugated double bonds; it does not denote one double bond only.

Molecules are numbered as prescribed by the IUPAC rules, and the faces are then alphabetized with consecutive italic letters, beginning with *a* for the 1,2-face, *b* for the 2,3-face, and so on. If necessary, numbers of the positions of attachment of the other components are indicated. Both the numbers and

TABLE 2.5. Accepted, Shortened Prefixes for Some Common Polycyclic Systems

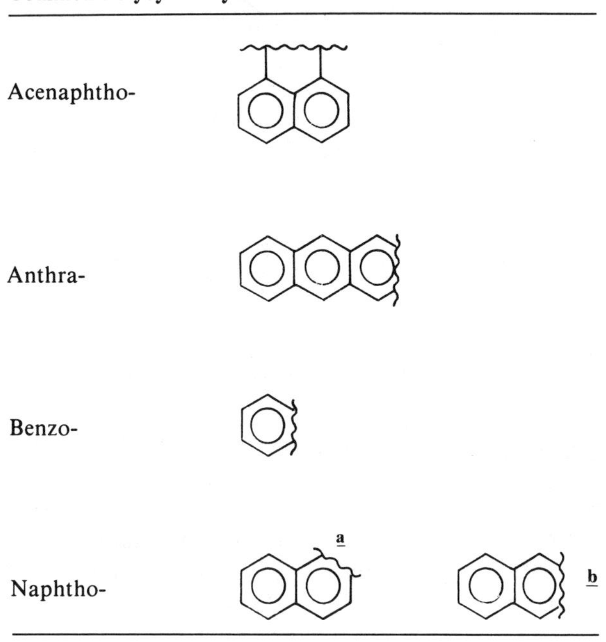

letters are enclosed in brackets immediately after the designated attachment. For the full application of this procedure see the references.

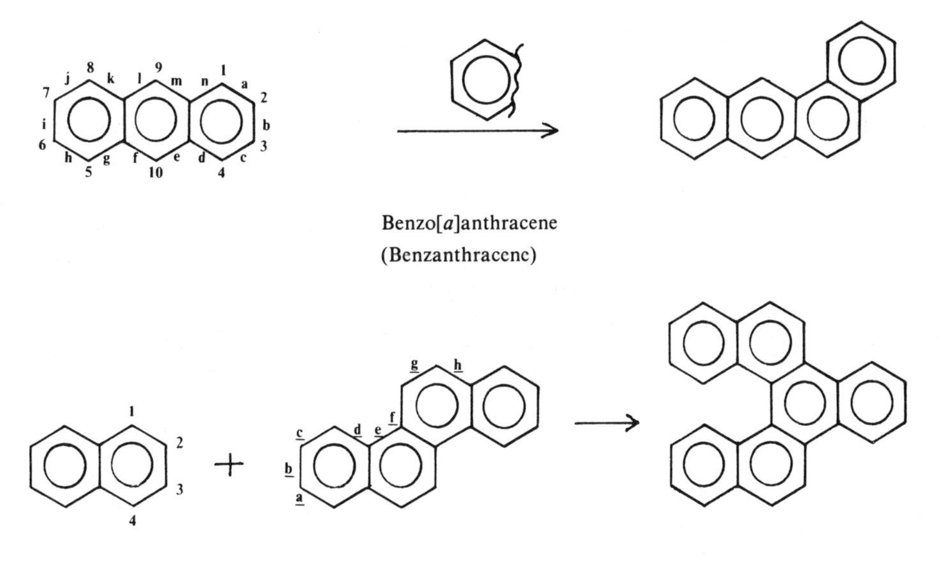

Benzo[*a*]anthracene
(Benzanthracene)

Naphtho[1,2-*g*]chrysene
via hydrocarbon base

7*H*-6-Oxa-1-thiacyclopenta[*a*]naphthalene
via heterocyclic base

1,2-Dihydro-1-oxa-thiopheno-[2,3-*f*]naphthalene.

2.4. BRIDGING NOMENCLATURE

The bridge is considered as an integral part of the ring system and is treated as a special example of the prefix ending in an "o." It is numbered as a continuation of the numbering of the parent fused system, starting at the higher-numbered bridgehead.

1,7-Ethano-1*H*-pyrano[3,2-*c*]pyridazine

1*H*-pyrano[3,2-*c*]pyridazine

In saturated bicyclic bridged hydrocarbons, the von Baeyer system[3] provides a naming procedure to cover skeletons containing any number of rings. (This number has been defined as the minimum number of bond scissions required to convert the complex skeleton into a monocycle.) The name obtained in the bicyclic case consists of the prefix "bicyclo" followed, in square brackets, by the number of carbon atoms separating bridgeheads by the three possible routes, in decreasing order, followed in turn by the name of the hydrocarbon containing the same number of carbon atoms as the *whole* skeletal system.

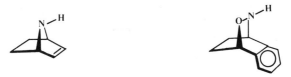

7-Azabicyclo[2.2.1]hept-2-ene

5,6-Benzo-2-oxa-3-azabicyclo[2.2.2]octane

REFERENCES

1. *IUPAC Nomenclature of Organic Chemistry*, Definitive Rules for Sections A and B (3rd ed.) and C (2nd ed.), Butterworths, London, 1971. "Definitive Rules for Nomenclature of Organic Chemistry," *J. Am. Chem. Soc.* **82,** 5545 (1960).
2. A. D. McNaught, *Adv. Heterocycl. Chem.* **20,** 175 (1976).
3. A. von Baeyer, *Chem. Ber.* **33,** 3771 (1900).
4. *Nomenclature of Organic Compounds*, J. H. Fletcher, O. C. Dermer, and R. B. Fox, Eds., *Advances in Chemistry Series 126*, American Chemical Society, Washington, D.C., 1974.

Chapter 3

ELECTRONIC DIVISION OF AROMATIC HETEROCYCLIC COMPOUNDS

3.1. INTRODUCTION

Heterocyclic chemistry in general and heteroaromatic chemistry in particular trace their roots to three discoveries: (1) the isolation of several furan derivatives from plant materials in 1780; (2) the discovery of pyrrole, obtained through dry distillation of horn and hoof, by Runge in 1834; and (3) the separation of picoline from coal tar by the Scottish chemist Anderson in 1846.[1-3] Since these discoveries, the study of heterocyclic chemistry has been the major effort of numerous chemists, physicists, and biologists.

Over the years it has become very clear that the chemical behavior of many *heteroaromatic* compounds can be explained on the basis of two different types of aromatic ring systems: those derived from the cyclopentadienyl carbanion (**1**) by replacement of one or more CH groups by a heteroatom such as O, S, Se, Te, N, or P (**2**) and those heteroaromatic compounds obtained by replacement of one or more CH groups (**3**) in benzene (**4**).

(X=O,S,Se,Te,NR,PR)

1 2

(Y=N, O^+, Sb,As)

4 3

3.2. DEFINITIONS AND PHYSICAL DATA

3.2.1. π-DEFICIENT HETEROAROMATIC COMPOUNDS

The ground-state π-electron distribution in benzene is such that one π-electron is associated with each carbon atom. The replacement of one of the CH groups by a heteroatom such as nitrogen, effecting pyridine (**5**), alters this ground-state electron distribution significantly. Since nitrogen is more electronegative than carbon, it tends to "withdraw" π-electron density from the remaining carbons in the molecule.

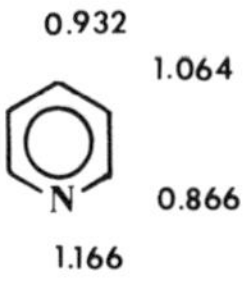

0.932 1.064 0.866 1.166

5

0.803 0.749 1.137 1.225

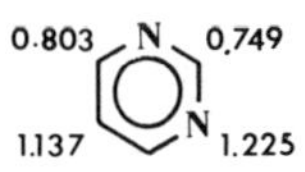

6

Molecular orbital calculations (see Appendix A for the results of the calculations on a large number of these heteroaromatic compounds) show this prediction to be true (**5**). The carbons that are α and γ to the nitrogen atom in pyridine have a ground-state π-electron density of less than one. Application of the concepts of resonance theory leads to the same conclusion, in qualitative terms. Thus, since two resonance-contributing structures (**5a** and **5b**) involving the α-carbons can be drawn, this position should be more positive than the γ-position, for which only one resonance-contributing (**5c**) structure can be written.

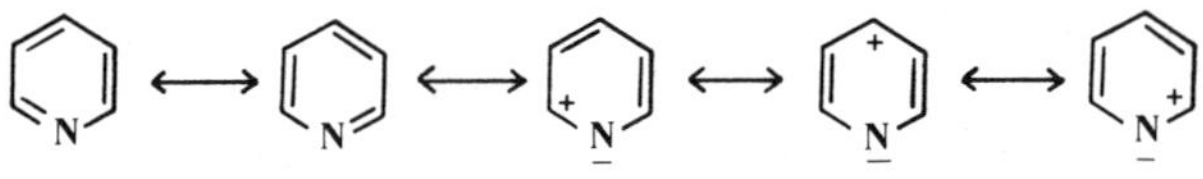

5 5a 5c 5b

The replacement of two *meta*-situated CH groups in benzene affords pyrimidine (**6**). Intuitively, the electron densities on the carbons α and γ to the nitrogen atoms should be even lower in this diaza analogue than on the corresponding positions in the monoazabenzene, pyridine (**5**). Molecular orbital calculations agree with this prediction as shown in structure **6**.

Again, application of the concepts of resonance theory to pyrimidine affords qualitatively similar conclusions. Though four resonance structures (**6c–6f**) can be drawn that place a positive charge on the carbons *ortho* to the nitrogen (positions 4 and 6), and only two (**6a** and **6b**) can be drawn that place a positive charge on the carbon sandwiched between the two nitrogens (positions 2), molecular orbital calculations show a greater charge depletion at C-2 than at C-4. Thus resonance-contributing structures **6a** and **6b** play a more important role than do the others in the ground–state structure of pyrimidine (**6**).

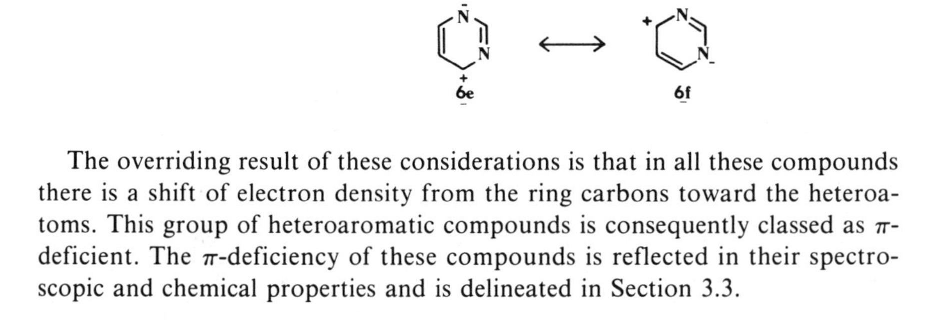

(1) (3)

6d 6c 6 6a 6b

6e 6f

The overriding result of these considerations is that in all these compounds there is a shift of electron density from the ring carbons toward the heteroatoms. This group of heteroaromatic compounds is consequently classed as π-deficient. The π-deficiency of these compounds is reflected in their spectroscopic and chemical properties and is delineated in Section 3.3.

3.2.2. π-EXCESSIVE HETEROAROMATIC COMPOUNDS

The cyclopentadienyl carbanion (**1**) is a five-membered carbocyclic aromatic species in which six π-electrons are distributed over five carbon atoms. Thus the π-electron density per carbon is 6/5, or 1.2. In comparison to benzene, which can be considered as a π-neutral system (1.0 π-electron per carbon), the cyclopentadienyl carbanion is a π-excessive carbocyclic aromatic compound.

The replacement of any one or more of the CH groups in this molecule by heteroatoms such as O, S, Se, Te, NR, or PR generates neutral heteroaromatic compounds **2**. For example, when X is S in structure **2,** the compound is thiophene. Molecular orbital calculations (see Appendix A) show that the ground–state electron distribution in thiophene is as given in structure **7**. When X = NH in the general structure **2**, pyrrole (**8**) is obtained. The molecular orbital calculations show the electron densities indicated in structure **8**.

1.036 0.912 S

7

1.064 1.030 N H 1.920

8

Although the electron density at each of the carbons is less than that of the carbons in the cyclopentadienyl carbanion (**1**), the general trend is still one of higher average electron densities than in benzene. Consequently, these five-membered ring heteroaromatic compounds are referred to as *π-excessive.* Resonance theory considerations are, again, in qualitative agreement with these molecular orbital calculations as shown by the resonance-contributing structures **2**.

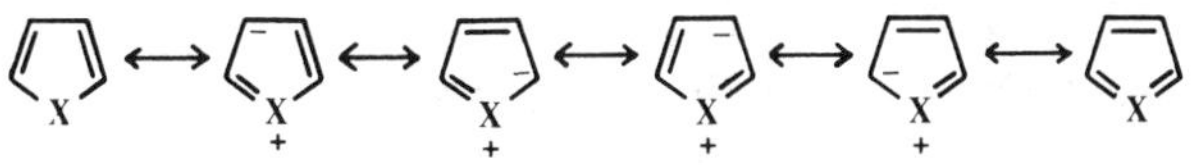

2 ⟷ 2c ⟷ 2a ⟷ 2d ⟷ 2b ⟷ 10 (X= S, Se)

3.2.3. SOME PHYSICAL PROPERTIES OF π-DEFICIENT AND π-EXCESSIVE HETEROAROMATIC SYSTEMS

3.2.3.1. Dipole Moment Studies. One of the earliest physical methods employed in the study of heteroaromatic systems was the determination of dipole moments. This technique offers significant information on the involvement of heteroatoms in the ground–state electronic distributions. Table 3.1 lists the dipole moments of a number of heteroaromatic systems along with those of their corresponding perhydro derivatives.

Based on the resonance-contributing structures, the dipole moments of the *π-excessive systems* should be *less* in the direction of the heteroatom than in the corresponding perhydro derivatives, since in the former case, structures such as **2a–2d** indicate withdrawal of electron density from the heteroatom. This is clearly the case for furan (**2,** X = O), thiophene (**2,** X = S), and selenophene (**2,** X = Se), and actually causes a *reversal* of the dipole moment of pyrrole (**2,** X = NH) as compared with that of its perhydro derivative (**9**). The

TABLE 3.1. Dipole Moments (D) of Selected Heterocycles[a]

Perhydro	Aromatic	ΔD[b]
1.68 (tetrahydrofuran, O)	0.71 (furan, O)	0.97
1.87 (tetrahydrothiophene, S)	0.52 (thiophene, S)	1.35
1.97 (tetrahydroselenophene, Se)	0.40 (selenophene, Se)	1.57
1.57 (pyrrolidine, N—H)	1.80 (pyrrole, N—H; reversed direction)	3.37
1.57 (piperidine, N—H)	2.20 (pyridine, N)	(−) 0.63

[a] See References 4 and 5.

[b] $\Delta D = D_{\text{perhydro}} - D_{\text{aromatic}}$. It should be noted that contrary to many textbook descriptions, these dipole directions are correct. See M. V. Sargent and T. M. Cresp, in *Comprehensive Organic Chemistry,* P. G. Sammes, Ed., Pergamon Press, New York, 1979, part 18.4, p. 624.

larger differences in the thiophene (**2,** X = S) and selenophene (**2,** X = Se) cases (ΔD 1.4) may well reflect the involvement of the *d*-orbitals of the heteroatom, as indicated in the resonance-contributing structure **10.**[6]

The vastly greater difference (ΔD 3.37) in the dipole moments of pyrrole (**8**) and perhydropyrrole (**11**) reflects the ionic character of the N—H bond, which also accounts for the acidic nature of the pyrrole ring. In stark contrast to this behavior, the *π-deficient systems,* as exemplified by pyridine (**5**), exhibit a *greater* dipole moment than do the corresponding perhydro derivatives (piperidine; **14**). This experimental observation confirms the importance of resonance-contributing structures such as **5a–5c** in pyridine and **6a–6f** in pyrimidine to the ground state of the π-deficient heteroaromatic systems.

3.2.3.2. Nuclear Magnetic Resonance Studies. The π-deficiency or π-excessiveness of heteroaromatic systems and the theoretically predicted ground-state electron densities are also experimentally demonstrable by the NMR chemical shifts of various nuclei.

In the π-excessive systems the carbon atom resonances appear at much more shielded positions [δ100–150 (ppm)] than do those of the carbons in π-deficient systems [δ150–170 (ppm)].[7]

Recently determined ^{15}N chemical shifts of a number of heteroaromatic systems[8] further demonstrate the electronic differences between π-deficient and π-excessive heteroaromatic compounds: for example, the pyridine nitrogen [δ ^{15}N 317 (ppm)] is much more deshielded than the nitrogen in pyrrole [δ ^{15}N 149 (ppm)]. Detailed proton, carbon, and nitrogen NMR data for a number of typical heteroaromatic systems can be found in Appendix B. In a *qualitative* sense, the relative π-deficiency of a number of mono-, di-, and triazabenzenes is reflected by the ^{15}N chemical shifts. The most deshielded nitrogen of this series is N_1 in 1,2,4-triazine ($\delta_N = 422$), whereas the most shielded ones are found in pyrimidine ($\delta_N = 298$). Thus the addition of nitrogen generally* causes an increase in the π-deficiency.

δ ^{15}N 298 < 312 < 338 < 400 < 422 (N1), 382 (N2), 338 (N4)

In the π-excessive aza analogues of pyrrole, a similar trend is observed for the ^{15}N chemical shifts. Again, the introduction of additional nitrogen atoms causes deshielding and thus *decreases* the π-excessiveness of the ring systems.

δ ^{15}N149 < 209 < 247

3.3. PREDICTIONS OF CHEMICAL REACTIVITIES

Based on this theoretical and experimental information, some predictions can be made with respect to the chemical reactivities of these heteroaromatic compounds. For example, it is anticipated that, because of the decreased electron density on the carbon atoms of *π-deficient systems*, electrophilic substitution will be more difficult than in either the π-neutral benzenoid or the π-excessive heteroaromatic systems. Conversely, nucleophilic substitution reactions in the π-deficient heteroaromatic systems are expected to be greatly facilitated by the decreased electron density on the carbon atoms, in comparison to the π-neutral aromatics.

Because of the increased electron density at the various carbon atoms in the *π-excessive* systems, electrophilic substitution should be more readily accom-

* Pyrimidine is an exception!

plished than in either the benzenoid or the π-deficient heteroaromatic systems. In contrast, nucleophilic substitution reactions on the π-excessive systems, barring the existence of any other effects, should be more difficult because of the increased charge density of some of the ring carbons.

REFERENCES

1. A. P. Dunlop and F. N. Peters, *The Furans*, Reinhold, New York, 1953.
2. H. Fischer and H. Orth, *Die Chemie des Pyrrols*, Vol. 1, Akademischer Verlag, Leipzig, 1934.
3. T. Anderson, *Trans. R. Soc. Edinb.* **16,** 123 (1849).
4. T. J. Barton, R. W. Roth, and J. G. Verkade, *J. Am. Chem. Soc.* **94,** 8854 (1972).
5. K. Schofield, *Heteroaromatic Nitrogen Compounds: Pyrroles and Pyridines*, Butterworths, London, 1967, p. 124.
6. Even though the S—C bond in thiophene has significant double bond character, the involvement of the sulfur *d*-orbitals has been questioned (D. T. Clark, in *Organic Compounds of Sulfur, Selenium, and Tellurium,* Vol. 1, The Chemical Society, London, 1970, Chap. 1).
7. G. C. Levy and G. L. Nelson, *Carbon-13 Nuclear Magnetic Resonance for Organic Chemists,* Wiley-Interscience, New York, 1972, p. 96.
8. G. C. Levy and R. L. Lichter, *Nitrogen-15 Nuclear Magnetic Resonance for Organic Chemists,* Wiley, New York, 1979, p. 28.

Chapter 4

SYNTHETIC ROUTES TO SELECTED MONOCYCLIC FIVE-MEMBERED π-EXCESSIVE HETEROAROMATIC COMPOUNDS

4.1. GENERAL SYNTHETIC ROUTES

Numerous superficially similar reactions have been modifed to give access to five-membered π-excessive heteroaromatic compounds, such as furan, thiophene, and pyrrole. For example, when α-diketones (**1**) are condensed with compounds that contain appropriately situated, activated methylene groups (**2**), the desired five-membered ring is generated. In the carbocyclic series, the cyclopentadienone **3a** ($R = R' = C_6H_5$) is readily prepared from benzil and 1,3-diphenylacetone under basic conditions.[1] The related furans (**3b**),[2] thiophenes (**3c**)[3] (the *Hinsberg thiophene synthesis*), pyrroles (**3d**),[4] and selenophenes (**3e**)[5] are easily prepared by this type of condensation reaction by substitution of **2a** by **2b–d,** respectively, where R′ is an electron-withdrawing group.

Base

1 2 3

a (X=C=O) d (X=NH)
b (X=O) e (X=Se)
c (X=S)

1,4-Diketones[6–8] (**4**), when treated with P_2O_5[9a] (or acids such as Ⓟ—SO_3H[9b]), P_2S_5 (*the Paal reaction*),[10] ammonia,[11] or P_2Se_5,[12] can be readily converted by an intramolecular cyclization into **5a, b, c,**[13] and **d,** respectively. In the case of pyrroles the reaction is known as the *Paal–Knorr synthesis* and is adaptable to the preparation of *N*-alkyl,[14a] *N*-hydroxy,[14b] or *N*-amino[14c] pyrroles through the use of 1° amines,[11] hydroxylamine, or hydrazine, respectively. The original synthesis of pyrrole (**3d**), in which ammonium mucate (**6**) is dry-distilled,[15] is a classical example of this synthesis.

P_2X_5

4

5a (X=O) c (X=Se)
b (X=S) d (X=NR¹)

6 3d

Although bis-acetylenes (**7**) have not yet been used to generate furans, treatment of **7** with $ArCH_2NH_2(CuCl)$,[16] Na_2S,[17] Na_2Se,[18] Na_2Te,[19] $ArAsLi_2$,[20] or $ArPLi_2$[21] provides an excellent route to the 2,5-substituted five-membered π-excessive heterocycles **5d** (X = NAr), **5b** (X = S), **5c** (X = Se), **5e** (X = Te), **5f** (X = ArAS), and **5g** (X = ArP), respectively. This synthetic procedure may well be related to the biogenetic synthesis of some of the naturally occurring thiophenes.[22] Phenylphosphine and phenylarsine in the presence of a catalytic amount of phenyllithium add smoothly to **7** even at 25°C. This facile, base-catalyzed cyclization of heavy atom dianions with bis-acetylenes (e.g., **7**) is an excellent route to these generally inaccessible compounds.

R–C≡C–C≡C–R $\xrightarrow{M_2X}$

7

5b (X=S) **c** (X=Se) **d** (X=N-Aryl) **e** (X=Te) **f** (X=AsAryl) **g** (X=PAryl)

In the instances of the five-membered π-excessive compounds, which possess heavier heteroatoms, the direct conversion of acetylenes (**8**) to a bis-organolithium intermediate (**9**) affords a general synthetic route to **5c,**[23] **5e,**[23-25] **5g,**[23] **5h** (X = SiR_2),[23] **5i** (X = ArSb),[23] and **5j** (X = R_2Ge)[26] by use of Se_2Br_2, $TeCl_4$, $ArPCl_2$, R_2SiCl_2, $ArSbCl_2$, and R_2GeBr_2, respectively. The related dibenzoplumbole (**10**) is prepared in an analogous manner,[27] whereas in several cases, a dihalodialkene (**11**) has been utilized to give **5e**[23,25] and **5g**[23,24] upon treatment with Li_2Te[19b] and $ArPNa_2$, respectively.

2 R–C≡C–R ⟶ [**9**] ⟶

8 **9**

5c (X=Se) **d** (X=NR′) **e** (X=Te) **g** (X=PAryl) **h** (X=SiR_2') **i** (X=SbAryl) **j** (X=GeR_2')

$\xrightarrow{(C_6H_5)_2PbCl_2}$

10

11 ⟶ **5e** (X=Te) **g** (X=PAryl)

Acetylenes, for example **12,** which are susceptible to nucleophilic additions, have been used to generate furans, pyrroles (**14**),[28] and thiophenes. In a representative synthesis, simple Michael addition of the amine to **12** generates an enamine intermediate **13,** which cyclizes to form the 3,4-carbon bond, then aromatizes via dehydration to give **14.**

X=SH,OH,NH$_2$ + $CH_3O_2CC{\equiv}CCO_2CH_3$ (12) —(X=NH$_2$)→ [13] → → 14

Müller and his co-workers[29] have used transition metal complexes of various diynes in the preparation of fused π-excessive heteroaromatic systems. Numerous diynes, exemplified by **15,** on treatment with tris(triphenylphosphine)rhodium(I) (**16**), generate the rhodium complex **17,** which reacts with oxygen,[30] sulfur,[30] selenium, and tellurium[31] to give varying yields of the corresponding fused heterocycles (**18**). The yields of furan products are generally better when hydrogen peroxide or organic peroxides are used instead of molecular oxygen, suggestive that singlet oxygen, formed through a catalytic effect of the transition metal, is the active species. Application to mixed heterocycles[32] can be demonstrated by the synthesis of **19.**

15 —$((C_6H_5)_3P)_3RhCl$ (16)→ 17 —X_2→ 18

18 (X=O; 10%) (X=S; 60%) (X=Se; 11%) (X=Te; —%)

—$((C_6H_5)_3P)_3RhCl$ (16)→ —O_2 (or S, Se)→ 19 (18%)

Transition metal complexes of cobalt are similarly used to generate non-fused heterocycles, but in better yields.[33] When cobaltacene **20** is allowed to react at 70–110°C with either sulfur, selenium or nitrosobenzene, the corresponding heterocyclic system is obtained.

Ar Ar $P(C_6H_5)_3$ Co Ar Ar **20** $\xrightarrow[70-110°C]{S/C_6H_6}$ Ar Ar S Ar Ar (75%)

Several bis-Wittig reagents[34] have been applied to the construction of these five-membered heterocycles. Although a potentially general route, only a few examples are available at this time. 3,4-Diphenylfuran (**21**)[35] has been synthesized by the reaction of benzil with dimethyl ether–α,α'-bis(triphenylphosphonium) dibromide. Both large and small fused-ring systems can be synthesized via this procedure, as is demonstrated by the preparation of **22.**[36] Unfortunately, the limited availability of macrocyclic α-diketones hinders some interesting applications.

Ar Ar O O + [Ar_3P O Ar_3P] ⟶ Ar Ar O **21** (30%)

O O + [Ar_3P S Ar_3P] ⟶ S **22** (~5%)

4.2. SPECIFIC ROUTES TO THE COMMON FIVE-MEMBERED π-EXCESSIVE HETEROAROMATIC COMPOUNDS

4.2.1. PYRROLES

The *Hantzsch synthesis*[27] is an "old" procedure which generates pyrroles by the condensation of α-haloketones (**23**) with β-ketoesters (**24**) in the presence of either ammonia or a primary amine. The reaction probably proceeds via an intermediate aminocrotonic ester (**25**), which undergoes *C*-alkylation, N—C bond formation, then dehydration. One of the numerous modifications that has been developed during the past 90 years is substitution of **23** with α-hydroxyaldehydes and ketones[38] or nitroalkenes.[39] The advantage of this procedure is that the initial reactants are, in general, readily accessible and the yields of the pyrroles are moderate to good.

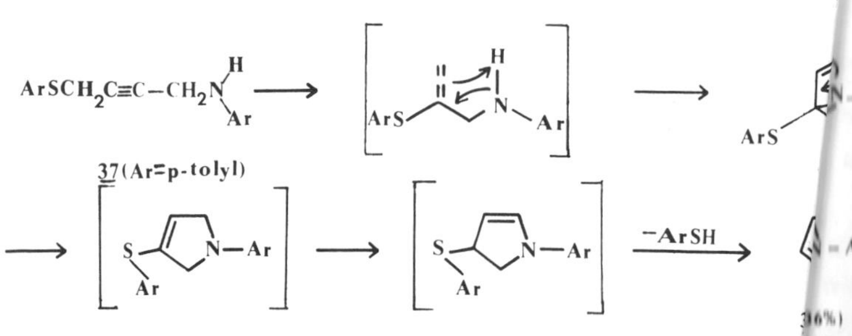

37 (Ar = p-tolyl)

Scheme 4.1

previously.[53] Vinylazirines **40** have also been observed to rearrange to the pyrrole nucleus **41** upon treatment with $Fe_2(CO)_9$ in dry benzene[54] or $Mo(CO)_6$ in anhydrous tetrahydrofuran.

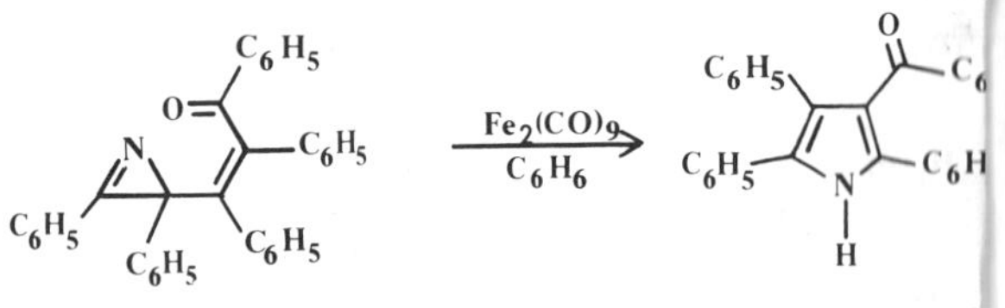

40 41 (75-85%)

With sufficiently activated methylenic ketones, such as **42**, reaction with 2*H*-azirine **43** proceeds at room temperature in the presence of a nickel(II) catalyst, such as $Ni(acac)_2$, to give pyrroles in nearly quantitative yields.[55] The only limit to the versatility of this procedure is the availability of the starting 2*H*-azirines.[56]

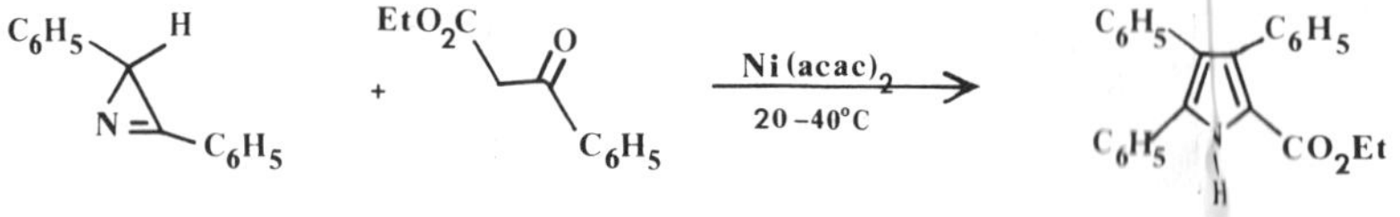

43 42

4.2.2. FURANS

Acidic reagents have been widely used to cyclize substituted 1,4-diketones; this reaction is known as the *Paal-Knorr synthesis* of furans and was considered previously among the general routes to five-membered π-excessive heteroaromatic compounds. Numerous alternative routes to the pivotal γ-dione starting materials have been described, such as those that employ β-chloroallyl ketones (**44**),[57] β-formylaldehyde acetals,[58] oxiranes[59] (an acid-catalyzed oxirane to carbonyl transformation), nitronic acids (via the Nef reaction),[60] and oxygenated cyclopropanes.[61]

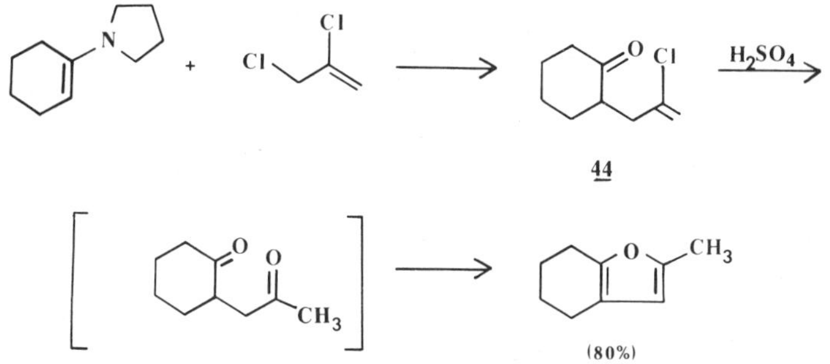

44

(80%)

The formation of furfural (**45**) from pentoses and of substituted furfurals from hexoses under acidic conditions is a simple, well-documented commercial procedure[62] from very inexpensive starting monosaccharides. The use of organic solvents and/or various chemical reagents, for example, iodine in *N,N*-dimethylformamide,[63,64] facilitates the transformation.

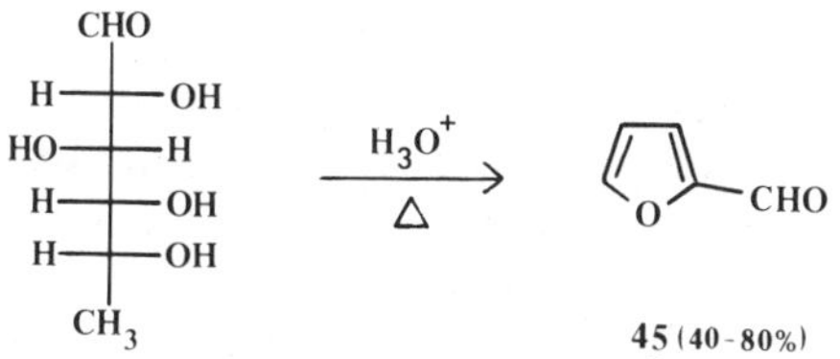

45 (40–80%)

The *Feist–Benary synthesis*[65] is used to prepare furans (e.g., **46**) which have an ester function in the β-position. The base-catalyzed condensation of β-ketoesters with either α-chloro aldehydes or ketones probably proceeds by an aldol condensation to give the intermediate β-hydroxydihydrofuran, which aromatizes by dehydration.

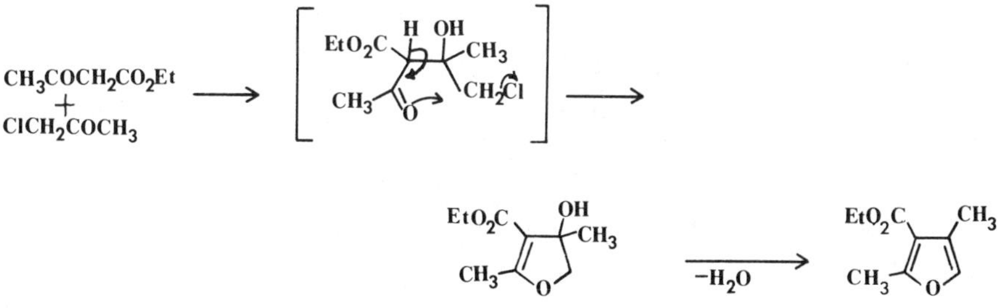

46

Gopalan and Magnus[66] have applied this procedure to the synthesis of **47**, which is the starting material in the total synthesis of linderalactone. The reagents necessary for the Feist–Benary synthesis can be conveniently generated *in situ* from lactones, for example **48**,[67] as well as from the appropriate pyrylium salts (**49**).[68]

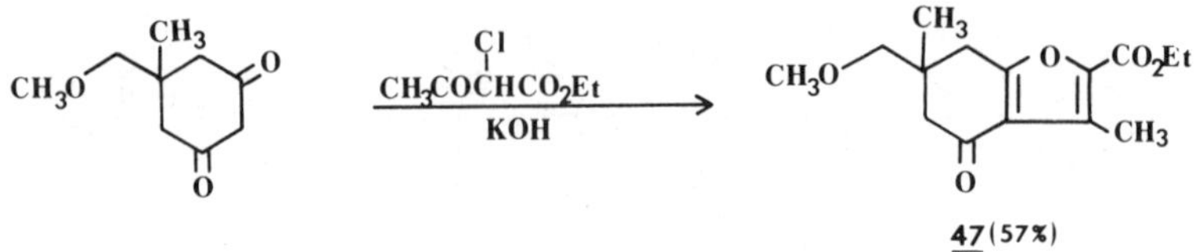

47 (57%)

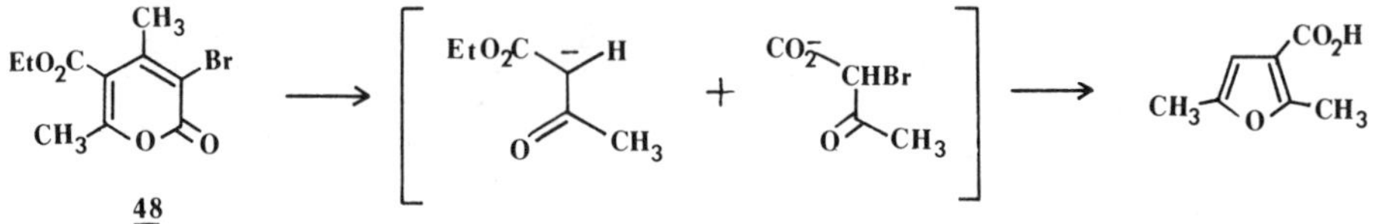

48

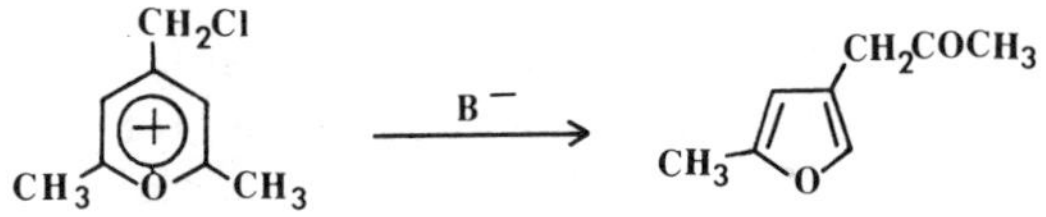

49

(80%)

Recent syntheses of several natural products have afforded the furan nucleus through the intermediacy of γ-hydroxy-α,β-unsaturated ketones or aldehydes.[69a] Acid-catalyzed cyclization of **50** gives methyl 8-(5-hexyl-2-furyl)octanoate (**51**),[69b] **52** gives perilla ketone (**53**),[70] isolated from *Perilla frutescens* Brit., whereas **54** affords **55**,[71] a compound isolated during the structure elucidation of genipic acid.

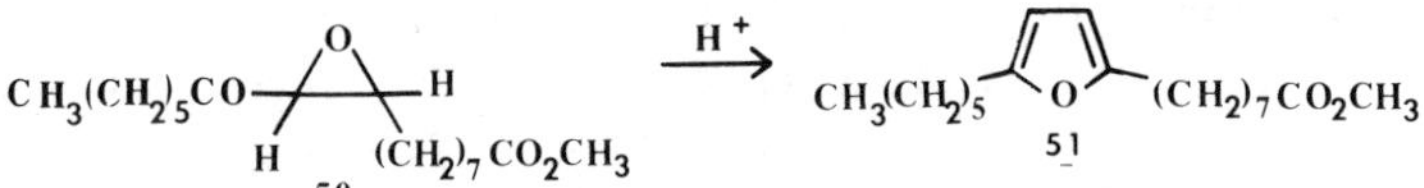

50 51

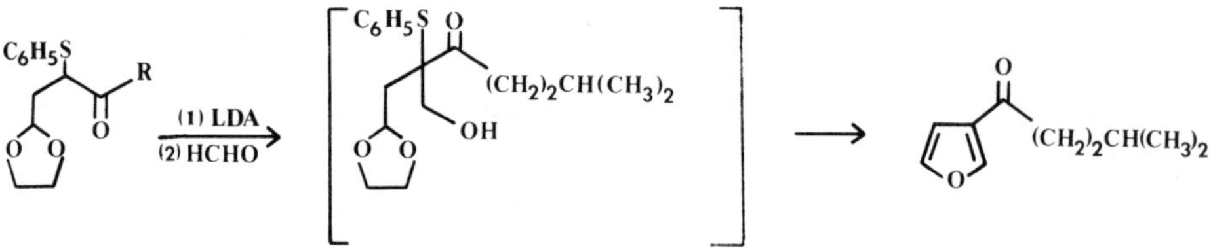

52 53 (62%)

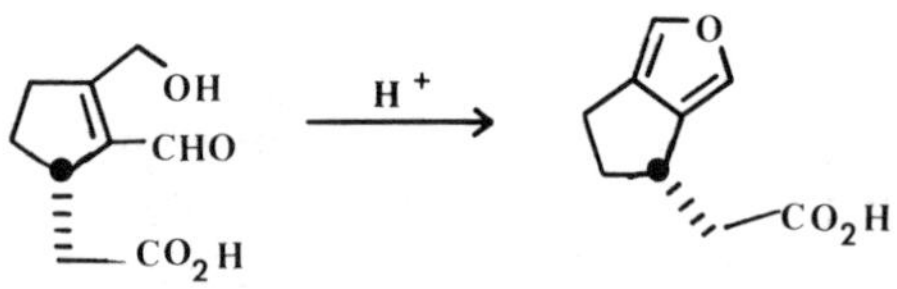

54 55

Numerous unsaturated heterocycles, such as oxazoles, undergo a Diels–Alder reaction with acetylenes to give addition compounds, which subsequently generate furans via a retro Diels–Alder procedure. Thus during the synthesis of vitamin B_6,[72] oxazole **56** is allowed to react with **57** to give the bicyclic intermediate **58,** which expels acetonitrile upon prolonged thermolysis; similarly **59** gives **60.**[73]

CH_3 H, N, O — 56 —[$ROCH_2C{\equiv}CCH_2OR$ (57) Δ]→ CH_3, O, N, OR, OR — 58 —[$-CH_3CN$]→ O, RO OR

CH_3, OEt, N, O — 59 —[$CH_3O_2CC{\equiv}CCO_2CH_3$ Δ]→ H_3CO_2C, CO_2CH_3, O, OEt — 60

A nucleophilic group located γ to a Lewis acid site directs the underlying orientation prior to the cyclization step in various furan syntheses. Therefore **61** on treatment with mercuric sulfate in the presence of acid gives **62.** Subsequent cyclization and dehydration affords the 3-substituted furan **63.**[74] The five-atom unit **64** can be generated in one step, as above, or from a ⟨3 + 2⟩ fragment combination as in the synthesis of **65.**[75] This more general synthetic approach to furans is an extension of known furan preparative routes.

[O] R, O, H — 61 —[$HgSO_4$ H^+]→ [R, OH, OH, H] 62 → R, O — 63

[O → O] $(CH_3)_2\overset{+}{S}CH_2C{\equiv}CH$ + H_3C, O, O, CH_3 —[NaOEt]→ $(CH_3)_2\overset{+}{S}$, $COCH_3$, O, CH_3 — 64

→ O, $COCH_3$, CH_3 → H_3C, $COCH_3$, O, CH_3 — 65 (86%)

4.2.3. THIOPHENES

General routes to thiophenes, which have already been considered, are the *Hinsberg synthesis* from α-diketones and activated methylene compounds (**2** → **3c**), the *Paal reaction* from γ-diketones and P_2S_4 (**4** → **5b**), bis-acetylenes with Na_2S (**7** → **5b**), nucleophilic addition to activated olefins or acetylenes (**66** → **67**),[76] transformation of diynes via a rhodium complex with sulfur, and action of sulfur-containing bis-Wittig reagents (to **22**). The *Socony vacuum process,*[77] which uses butane and sulfur at elevated temperatures, and the *Croda synthetic process,*[78] which employs functionalized hydrocarbons and carbon disulfide, are both highly successful industrial procedures.

$HSCH_2CO_2CH_3$ (68) + CH_3O_2C–CH=CH–CO_2CH_3 (66) $\xrightarrow[CH_3OH]{NaOCH_3}$ [3-oxo-tetrahydrothiophene with CO_2CH_3 groups] → → → [thiophene with CO_2CH_3 groups] (67)

In the *Fiesselmann synthesis,*[79] a combination of steps is involved: first, nucleophilic addition of a three-carbon unit, such as methyl thioglycolate (**68**), to an unsaturated center (or equivalent), and second, a base-catalyzed Dieckmann-type cyclization to generate substituted thiophenes (**69**).

$CH_3C{\equiv}C\text{-}CO_2CH_3$ + $HSCH_2CO_2CH_3$ (68) → [CO_2CH_3, CH_3, S, CO_2CH_3 adduct] $\xrightarrow{B^-}$ [OH, H_3C, S, CO_2CH_3 thiophene] (69)

The direct synthesis of substituted thiophenes has been reported by Marino and Kostusyk.[80] In this procedure, *S,S*-dimethyl-α-oxoketene dithioacetals undergo a regioselective deprotonation, followed by a cyclization. The diacetal **70,** prepared by methylation of the condensation product of the corresponding enolate with carbon disulfide, is treated with one equivalent of lithium diisopropylamide to generate anion **71,** which cyclizes to give the thiophene **72.**

70 [O, SCH_3, SCH_3, CH_2CH_3] $\xrightarrow{LDA}$ 71 [Li-CH_2, S, O, SCH_3, CH_2CH_3] → 72 (55%) [S, SCH_3, CH_2CH_3]

1,4-Dihalides (**73**) with the appropriate degree of unsaturation readily react with sulfide ion to form thiophenes.[81]

Cl CO$_2$Et Cl CO$_2$Et $\xrightarrow[\text{KOH/EtOH}]{H_2S}$ CH(CO$_2$Et)$_2$ S

73 (55%)

Despite numerous unsuccessful attempts,[82] a surprisingly simple synthesis of 3,4-di(*tert*-butyl)thiophene (**75**) was finally accomplished by addition of sulfur dichloride to 2,3-di(*tert*-butyl)butadiene (**74**).[83]

$ClCH_2C{\equiv}CCH_2Cl \xrightarrow[\text{THF/CuBr}]{t\text{-buMgCl}}$ 74 (>90%) $\xrightarrow[CH_2Cl_2]{SCl_2}$ 75

Finally, the first synthesis[84] of tetraphenylthiophene (**76**) should be mentioned. Laurent's preparation of **76** by thermolysis of thiobenzaldehyde indicated that most of the C_6H_5C units generated upon pyrolysis in the presence of sulfur go on to produce **76.**

CH$_3$ $\xrightarrow[\Delta]{S}$ S

76

4.2.4. SELENOPHENES

In addition to the general synthetic procedures of selenophenes, pyrolysis of substituted 1,2,3-selenodiazoles (**77**) affords predominantly **78.**[85,86] The expulsion of heteroatoms (**79**)[87] and carbon monoxide (**80**)[88] by thermolysis in the presence of hydrogen selenide and selenium, respectively, has also been described.

R N N Se $\xrightarrow{\Delta}$ R Se R + R Se R

77 78

O $\xrightarrow[\Delta]{H_2Se}$ Se

79

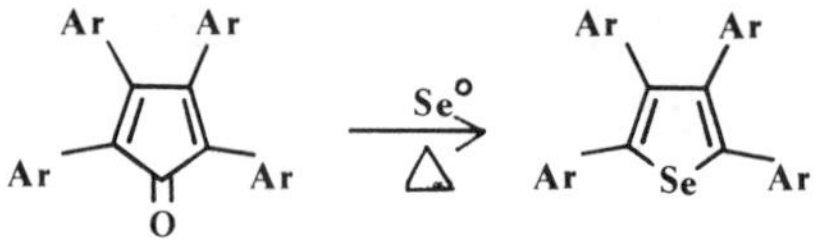

80

4.2.5. PYRAZOLES

Synthetic routes to the pyrazoles have been reviewed.[89] These reviews should be consulted for specific details and pertinent references prior to 1967, as well as for the more obscure procedures. One of the more utilized general routes to pyrazoles involves the reaction of β-diketones with hydrazine or monosubstituted hydrazine. Numerous synthons for β-dicarbonyl compounds can also be applied to this ring formation methodology. Thus acetylacetone (**81**) reacts with hydrazine to give 3,5-dimethylpyrazole (**82**)[90]; the reaction is generally exothermic but can be easily controlled by dilution or appropriate cooling. The transformation proceeds by initial hydrazone formation, followed by cyclization and acid-facilitated dehydration. As anticipated, unsymmetrical β-diketones usually afford, in most cases, an isomeric mixture.

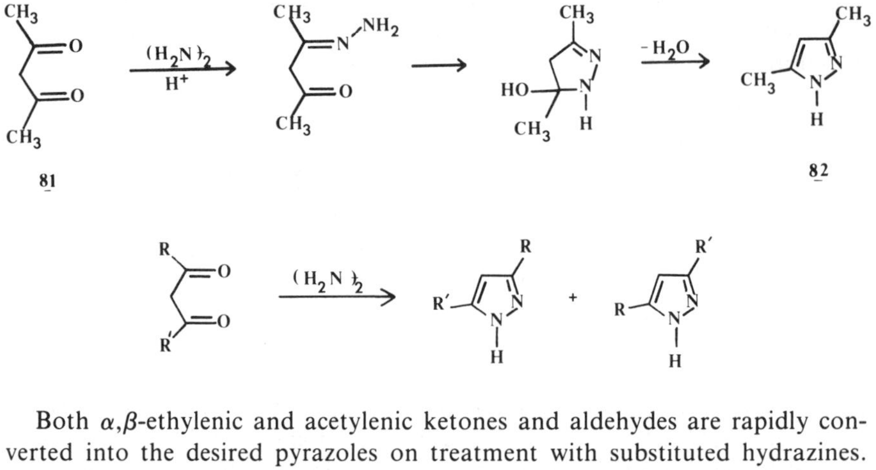

81 82

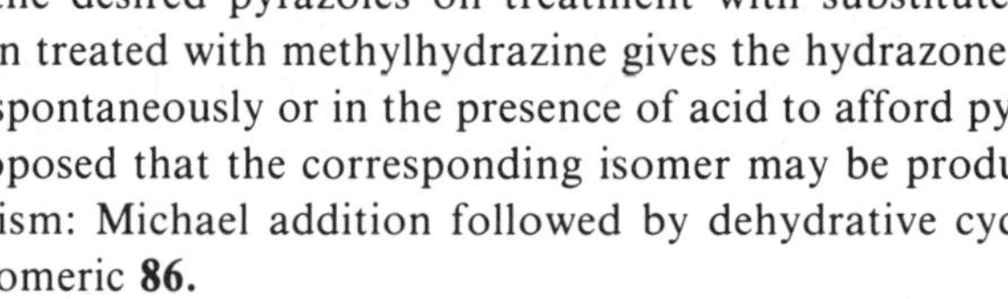

Both α,β-ethylenic and acetylenic ketones and aldehydes are rapidly converted into the desired pyrazoles on treatment with substituted hydrazines. Thus **83** when treated with methylhydrazine gives the hydrazone **84,** which cyclizes either spontaneously or in the presence of acid to afford pyrazole **85.**[91] It has been proposed that the corresponding isomer may be produced by a second mechanism: Michael addition followed by dehydrative cyclization gives rise to the isomeric **86.**

Cycloaddition of aliphatic diazo compounds to acetylenes gives the anticipated pyrazoles, usually as an isomeric mixture. In many cases, cycloaddition reactions may generate an exclusive regioisomer.[92] Regioselectivities of cycloadditions of diazoalkanes and acetylenes have been rationalized by frontier orbital theory. Diazoalkanes generally have elevated highest occupied molecular orbitals (HOMOs), and, as a consequence, are nucleophilic reagents.[92] For diazoalkanes, the carbon terminus is the more nucleophilic portion of the molecule.[93] One or two strongly electron-withdrawing groups on carbon decrease the nucleophilicity of the diazo moiety, and make the nitrogen terminus more nucleophilic.[92] In general, the site of largest HOMO coefficient (most nucleophilic site) on the diazoalkane becomes united to the site of largest LUMO coefficient (most electrophilic site) on the dipolarophile. Alkyl groups or other donors have relatively small effects on LUMO coefficients, although there is a preference for attack by the nucleophilic terminus of diazoalkanes at a site remote from the donor.[94] Conjugating or electron-withdrawing groups strongly polarize the LUMO, and attack by the diazoalkane carbon occurs at the remote site. High regioselectivity is observed unless similar groups are located at the two termini of an acetylene or an alkene. Thus for propiolic esters, there is a strong preference for formation of the 3-substituted pyrazole, whereas with phenylpropiolic esters, the phenyl and ester groups have an opposite antagonistic effect on the LUMO coefficients, so that a mixture of products (for example, **87** and **88**) is formed.[95] When methyl phenylpropiolate reacts with methyl diazoacetate, both regioisomers **87** and **88** are formed in approximately equal quantities.[95] Activated acetylenes can be substituted by acetylene equivalents (synthons): β-chlorovinyl ketones, enol ketones, cyanoolefins, and nitroolefins to mention but a few possibilities.

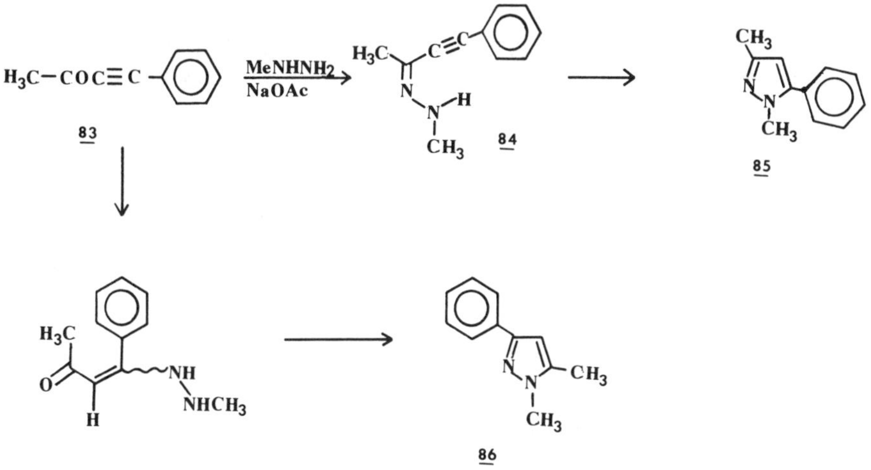

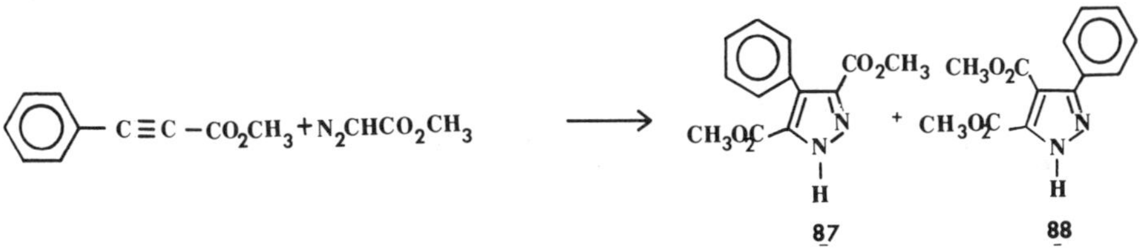

Norbornadiene (**89**) and ethyl diazoacetate in the presence of a catalytic amount of either $Fe(CO)_5$ or $Co_2(CO)_8$ give pyrazole (**90**) and cyclopentadiene; the intermediary 1,3-dipolar cycloaddition adduct undergoes a facile cycloreversion under the reaction conditions.[96]

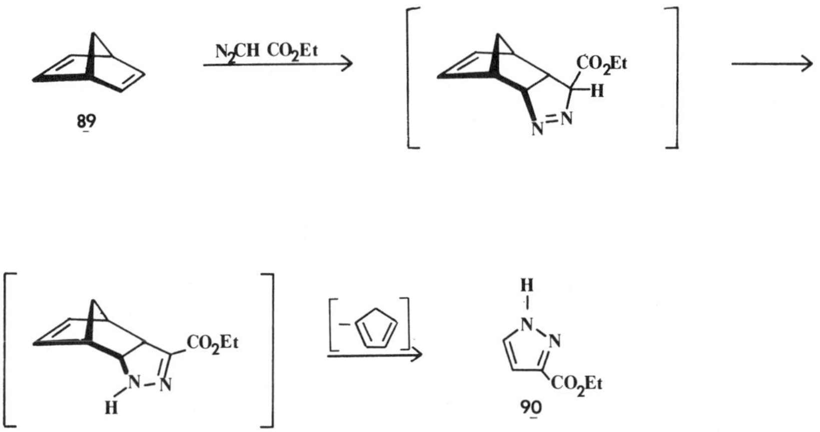

Pyrazolines (dihydropyrazoles; **91**) can be oxidized easily to the corresponding pyrazole by bromine,[97] permanganate, silver nitrate, mercuric oxide, chromic acid, or sulfur. If a good leaving group is present, it is often eliminated during the "aromatization" step.

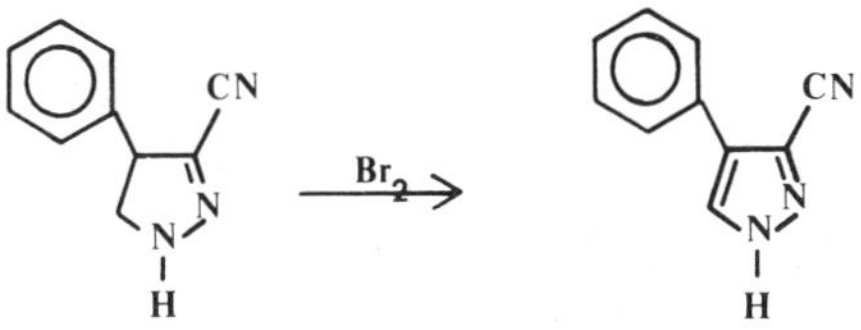

When disubstituted diazomethanes add to acetylenes the resultant isopyrazole cannot readily tautomerize. However, upon treatment with acid, base, or heat, these isopyrazoles (**92**) do isomerize to the pyrazole (**93**).[93a] Generally, one of the two-3-substituents rearranges to position 4, or is eliminated. If position 4 is occupied, migration to position 1 is observed (**94**).[98b]

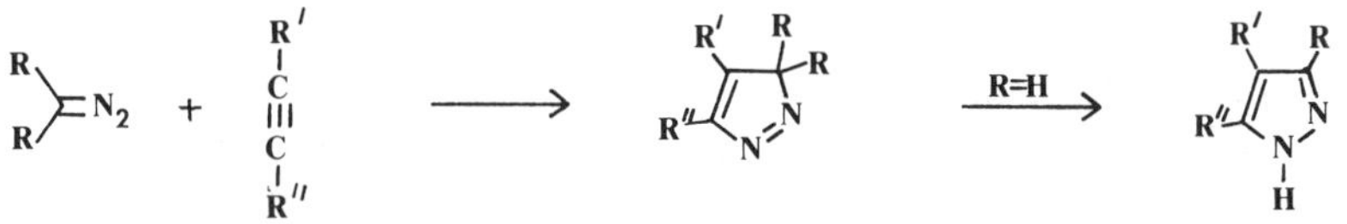

92 $\xrightarrow{R \neq H}$ 94

4.2.6. IMIDAZOLES

Grimmett[100] has reviewed imidazole chemistry and has delineated the important synthetic routes. The major, early routes to imidazoles generally included formation of the 1,5-, 2,3-, and 3,4(or 1,2)-bonds. The synthesis of 4,5-di-*tert*-butylimidazole (**95**) by Wynberg and DeGroot[101] represents an excellent example of this procedure; it utilizes a mediocyclic acryloin (**96**), ammonia, and formaldehyde. Acyloin starting materials are commonly used and are oxidized *in situ* to the corresponding α-diketone.

Bredereck and Theilig[102] developed the *formamide synthesis* of imidazoles (**97**) which involves the reaction of β-diketones, β-hydroxy-, β-halo-, or β-aminoketones, β-ketoesters, or β-oximinoketones with formamide. This procedure has wide application to the syntheses of this class of heterocycles. Imidazole itself can be prepared (60%) by the reaction of bromoacetaldehyde acetal, formamide, and ammonia at 175°C[103]; a reaction pathway has been suggested.[100]

96 $\xrightarrow{Pb(OAc)_4}$ (82%) $\xrightarrow{NH_3,\ H_2CO}$ $\xrightarrow{Ra/Ni}$ (76%) 95 (35%)

$\xrightarrow{HCONH_2}$ 97

Classically, the *Wallach synthesis* of imidazoles involves a cyclization via the formation of either the 1,2- or 1,5-bond.[104] Treatment of *N,N'*-disubstituted oxamide (**98**) with phosphorus oxychloride yields a chlorine-containing intermediate which, on reduction with hydrogen iodide, gives **99.** Similarly, imidates such as **100** condense with aminoacetaldehyde dimethylacetal **(101)** to give an amidine hydrochloride (**102**) that readily cyclizes to an imidazole (**103**).[105] This procedure has been applied to the synthesis of a wide range of substituted imidazoles.

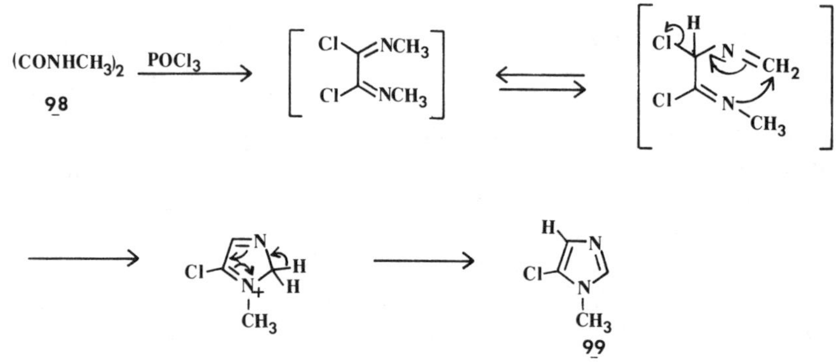

$(CONHCH_3)_2$ **98** $\xrightarrow{POCl_3}$ [...] ⇌ [...] ⟶ ⟶ **99**

100 + $NH_2CH_2CH(OCH_3)_2$ **101** ⟶ **102** $\xrightarrow{HCl}$ **103** (40%)

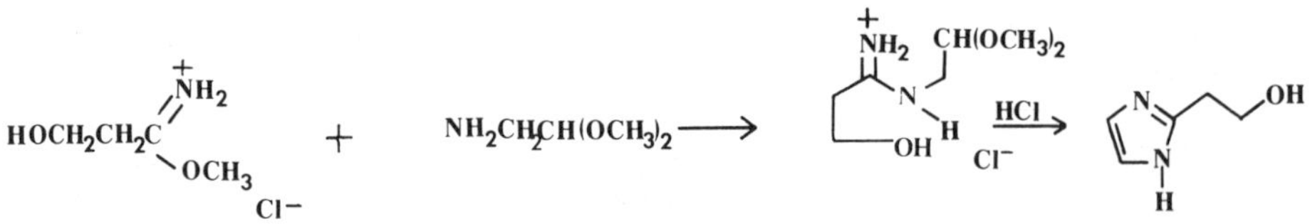

α-Hydroxy- and α-haloketones react with formamidine to give oxazoles (**104**) and imidazoles (**105**). In this synthesis, aliphatic acyloins preferentially yield imidazoles, whereas benzoins predominantly give oxazoles. When acetamidine and benzamidine react with either aliphatic acyloins or benzoins, imidazoles are formed exclusively.[106]

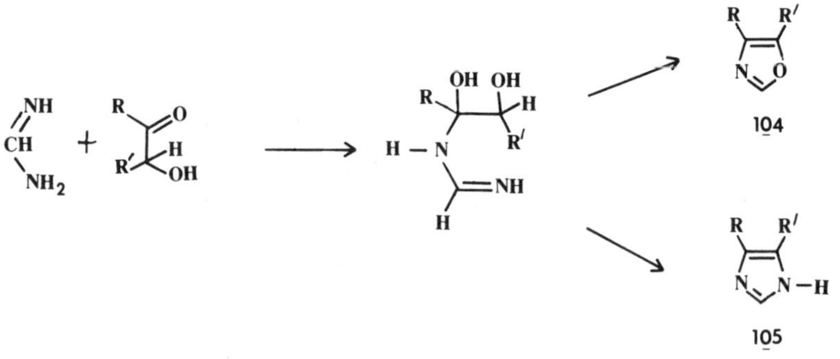

104

105

One of the simplest and most widely used routes to imidazoles depends on the generation of the 1,2- and 2,3-bonds. Diamines react with carboxylic acids to give the simple imidazoles (**106**).

$H_2NCH_2CH_2NH_2$ + $C_6H_5CO_2H$ $\xrightarrow[\Delta]{Pt/Al_2O_3}$ 2-phenylimidazole

106

Salts of α-aminoketones, for example, **107,** whose general syntheses and stability are severe limitations to the procedure, react with cyanates, thiocyanates, and isothiocyanates to give 3*H*-imidazol-2-ones or -2-thiones, which can be easily converted to the corresponding imidazoles.[107]

O, $\overset{+}{N}H_3$, Br^- — KCNS → SH, N, N, H → N, N, H

107

Oxazoles (**108**) react with ammonia or amines, as well as with formamide, to give imidazoles (**109**). The reaction is facilitated by electron-withdrawing substituents on the oxazole nucleus. The following mechanism has been proposed[106]:

R, R′, N, O, R″ $\xrightarrow[\Delta]{NH_3}$ R, R′, N, R″, NH_2, OH ⇌ R, R′, N, R″, $:NH_2$, O

108

$\xrightarrow[-H_2O]{NH_3}$ R, R′, N, R″, $:NH_2$, N, H $\xrightarrow{-NH_3}$ R, R′, N, N, R″, H

109

The irradiation of [9]-isoxazolophane **110** gives aziridine **111,** which rearranges in the presence of sodium carbonate to the [9]-oxazolophane **112.**

N-Alkylation of **112** with methyl tosylate affords **113**, which is ring-cleaved with aqueous ammonium hydroxide, then cyclized with ammonium acetate–acetic acid to give the [9]-imidazolophane **114.**[99]

CH_3 hν acetone CH_3 Na_2CO_3 MeOH Δ $(CH_2)_9$ CH_3 TosMe Δ

110 111 112

$(CH_2)_9$ CH_3 CH_3 Tos^- NH_4OH $(CH_2)_9$ CH_3 CH_3 NH_4OAc AcOH Δ $(CH_2)_9$ CH_3 CH_3

113 114

Numerous novel and less general routes to imidazoles have been devised and summarized.[100]

4.2.7. OXAZOLES AND THIAZOLES

As noted previously, mixed heteroaromatic π-excessive compounds are useful starting materials in the preparation of many related heterocyclic systems. Since the construction methodology is quite similar, they are treated together. Oxazoles are also excellent synthons and have tremendous synthetic potential if the desired substitution pattern is available.

Numerous routes to oxazoles have been described.[108] The disadvantages associated with the reaction of α-halo- or α-hydroxyketones with formamide can be circumvented by the use of α-metalated isocyanides with carboxylic acid derivatives. The reaction of **115** with acetyl chloride gives rise to **116** via the *p*-ketoisocyanide intermediate.[109] Use of the readily available tosylmethylisocyanide (TosMIC) **117** is very advantageous, since toluenesulfonic acid is eliminated, and causes facile aromatization.[110]

$N{\equiv}C{-}CH_2CO_2Et \xrightarrow{buLi} N{\equiv}C{-}\bar{C}HCO_2Et\ Li^+ \xrightarrow{CH_3COCl}$

115

CN CH_3 EtO_2C H O ⇌ $N{\equiv}C$ EtO_2C OH CH_3 $\xrightarrow{H_3O^+}$ CO_2Et CH_3

116 (66%)

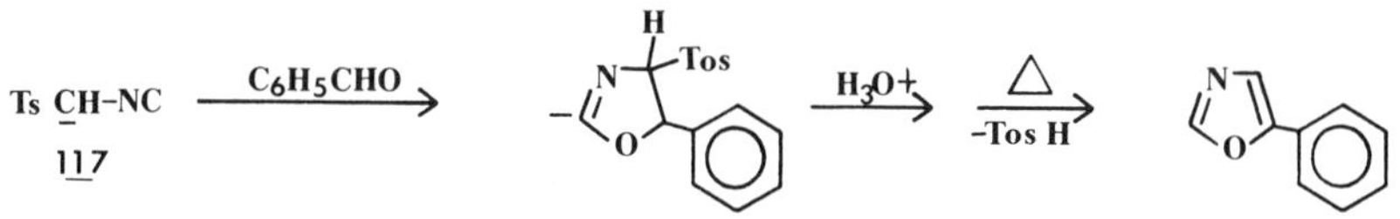

Ts CH-NC 117 $\xrightarrow{C_6H_5CHO}$ … $\xrightarrow{H_3O^+}$ $\xrightarrow[-\text{Tos H}]{\Delta}$ …

A common route to substituted oxazoles utilizes cycloaddition reactions; thus carbonylcarbenes (**118**) react with nitriles,[111] nitrilium methylides (**119**) with aldehydes,[112] α-carbonylnitrenes (**120**) with acetylenes,[113] and β-carbonylnitrenes (**121**) with acyl halides.[114]

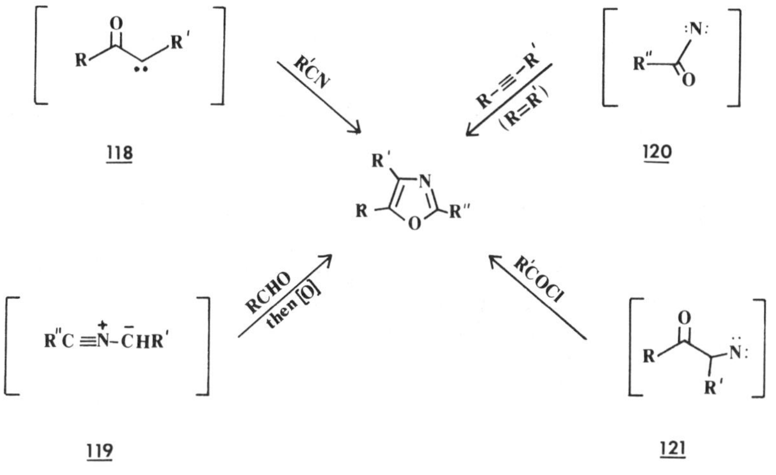

118 $\xrightarrow{R''CN}$ oxazole; 120 $\xrightarrow{R-\equiv-R'\ (R=R')}$ oxazole; 119 $\xrightarrow{RCHO\ \text{then}\ [O]}$ oxazole; 121 $\xrightarrow{R''COCl}$ oxazole

α-Acylaminocarbonyl compounds (**122**) readily cyclodehydrate to generate substituted oxazoles.[115] This procedure is known classically as the *Robinson–Gabriel synthesis,* and affords low yields of oxazoles when the standard dehydrating agents (PCl_5, $POCl_3$, $SOCl_2$, or H_2SO_4) are used. However, the use of polyphosphoric acid (PPA), anhydrous hydrogen fluoride, or phosgene in triethylamine greatly facilitates this process. The mechanistic aspects have recently been studied by Wasserman and Vinick[116] by means of ^{18}O-labeling techniques. These studies have established that the reaction is consistent with intermediate **123** rather than **124.** Because of the numerous modifications of this procedure, compounds of the general structure **125** can be prepared and subsequently converted to the related oxazole.

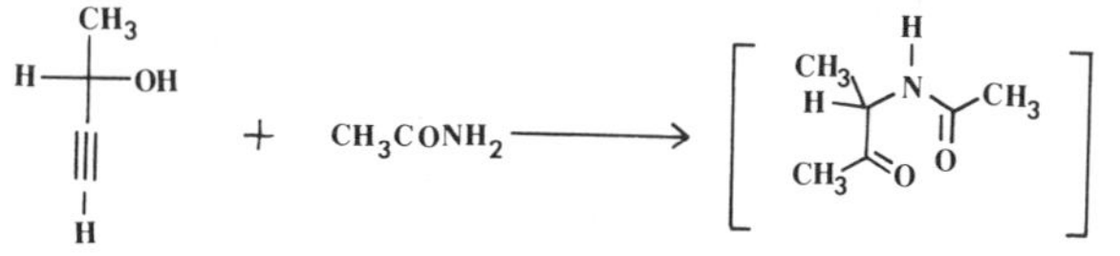

+ CH_3CONH_2 ⟶

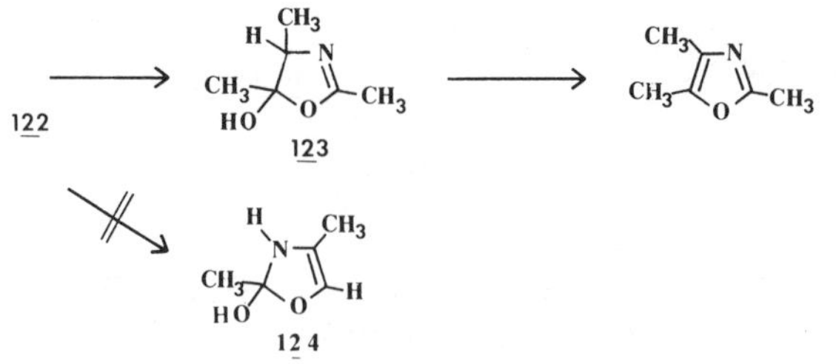

122 → 123 → ; 122 ⇸ 124

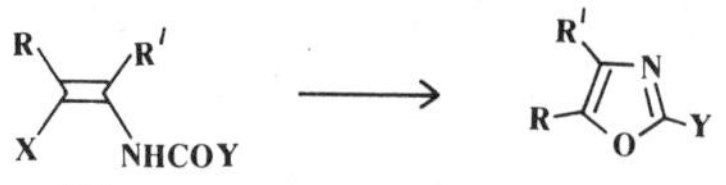

125 →

Acylaziridines undergo a thermal rearrangement at 220°C to afford substituted oxazoles. The reaction probably proceeds by the heterolytic cleavage of the aziridine C—C bond to give an azomethine ylide intermediate **126,** which cyclizes to the oxazoline **127.**[117] However, when 1*H*-azirine **128** is photolyzed (<3130 Å), the 2,5-diphenyloxazole **127** is isolated in 85% yield.[118] Use of light of wavelength >3340 Å results in a quantitative yield of the isomeric **129.** From sensitization experiments, it was suggested that triplet states of **128** and isoxazole **129** are intermediates in this interconversion and that **127** is derived from the singlet excited states of **128** and **129.** Thus the C—C bond is cleaved to give the nitrile ylide **130,** which smoothly cyclizes to oxazole **127,** whereas intersystem crossing of the excited singlet **128** to the triplet, CN cleavage, and subsequent cyclization affords the isomeric **129.**

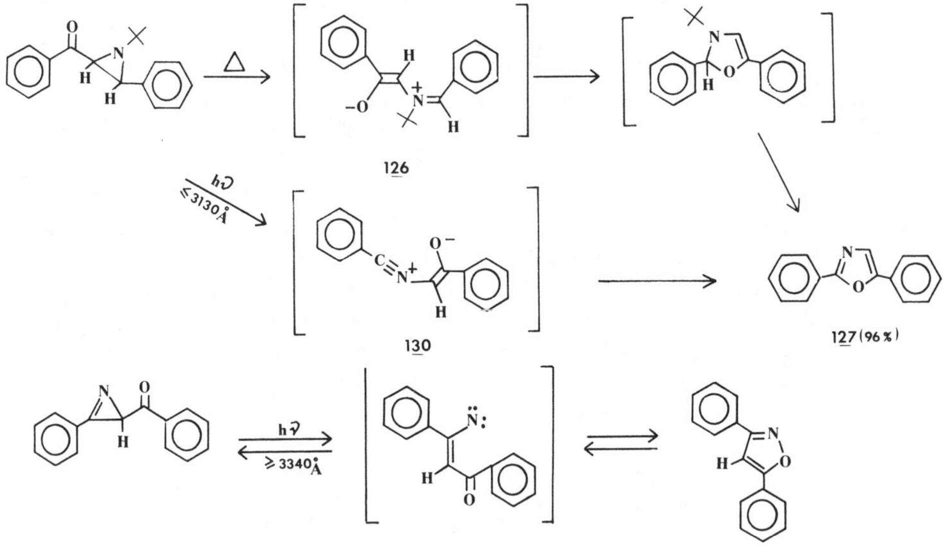

126

hν ≤3130 Å

130

127 (96%)

hν

≥ 3340 Å

128

129

In view of the facile photochemical bond reorganization of thiazoles (**131**) and isothiazoles (**132**),[119a–c] the isomeric interconversion is quite plausible. The distinct synthesis of a thiazole is possible through the application of many different synthetic schemes. The most common route to thiazoles is the *Hantzsch synthesis,* which constructs the ring system by a melding of C—C and S—C—N fragments: thioformamide (**133**) with chloroacetaldehyde gives the parent thiazole **134.** Although the procedure is straightforward in most cases, numerous modifications have been devised.

132 $\underset{}{\overset{h\nu}{\rightleftarrows}}$ [ ] $\rightleftarrows$ 131

$HCSNH_2 + ClCH_2CHO$ ⟶ [ ] $\xrightarrow{-H_2O}$ 134
133

As demonstrated in the synthesis of thiophenes from γ-dicarbonyl compounds, α-thioacylaminoketones (**135**) are transformed into the desired **136** in good yields[120] by a cyclization procedure similar to that used in the preparation of thiazoles, which, in this instance, effects the insertion of the *N*-heteroatom.[121] Thioacetic acid (**137**) reacts with α-haloketone **138** to give **139,** and subsequent treatment with ammonium acetate gives the thiazole **140.**

135 ⟶ 136

138 + $HSCOCH_3$ (137) $\xrightarrow{Et_3N}$ 139 $\xrightarrow{NH_4OAc}$ 140 (90%)

Related imines and enamines of dicarbonyl compound **139** have been generated specifically, then cyclized to the substituted thiazole by either catalyzed or noncatalyzed procedures. Alternative useful routes to 4,5-disubstituted thiazoles (**142**) are the cycloaddition of nitrile ylides with thiono compounds and the reaction of thionoesters with α-metalated isocyanides (**141**).[122]

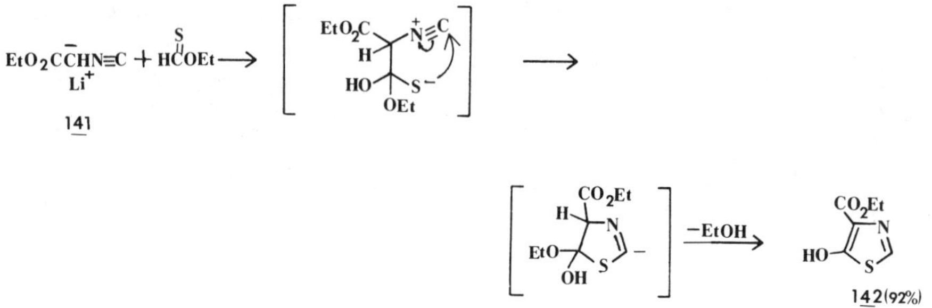

141

142 (92%)

4.2.8. ISOXAZOLES[123] AND ISOTHIAZOLES

One common route to the less important isoxazoles[118] and isothiazoles[119] is by photochemical isomerization[124] of the corresponding oxazole or thiazole derivatives. Irradiation of 2,5-diphenyloxazole (**127**) in ethanol gives **129** and **143.** However, a similar reaction conducted in benzene gives **129** and **144.** Since attempts to isolate 3-benzoylazirine (**128**) were unsuccessful, it was concluded that the reactions occur by the four-membered intermediates **145** and **146.** Photoreversion of **129** to **144** is proposed as proceeding via **128.**

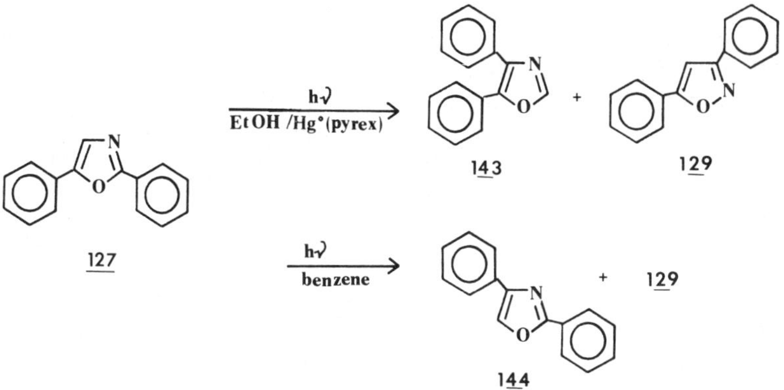

127

143 129

144

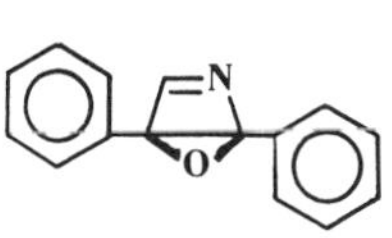

145 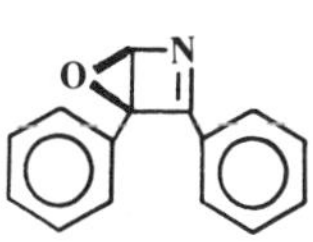 146

The condensation of γ-dicarbonyl compounds with hydroxylamines is the most common simple route to isoxazoles.[125] Acetoacetaldehyde sodium salt reacts with hydroxylamine hydrochloride to give a mixture of **147** and **148.** In certain cases the intermediate oxime is isolated, then cyclized separately. Since at times the isomer distribution depends on the acid concentration, the direction or extent of enolization of the intermediates may dictate the relative percentages of isoxazole formation. Other functional group equivalents to **149** are acetylenes **150** and enaminones or thioenaminone (**151**).[126]

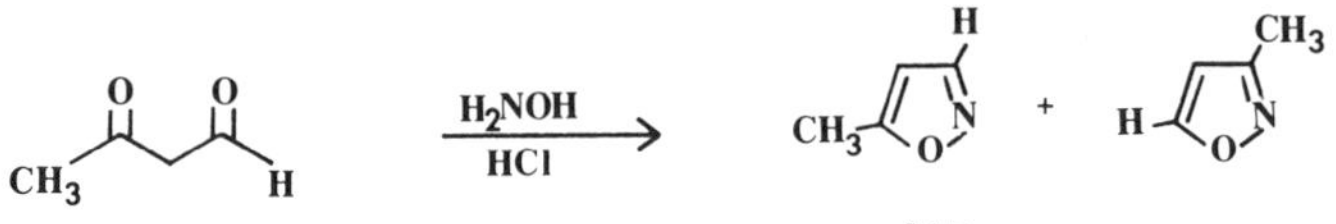

O O
CH_3 H
H_2NOH
HCl
H
CH_3 N O
147
+
CH_3
H N O
148

X O
R R'
149 (X = OH, halo)
R–C≡C– O R'
150
X
Ar $N(CH_3)_2$
151 (X = O or S)

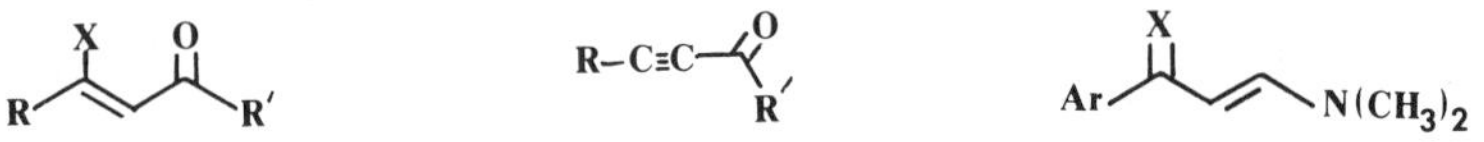

Dipolar cycloaddition reactions permit access to varied isoxazoles and isothiazoles, whereby addition of electron-poor acetylenes to nitrile oxide **152**[127] and nitrile sulfide **153**,[128] respectively, afford these compounds in poor yields.

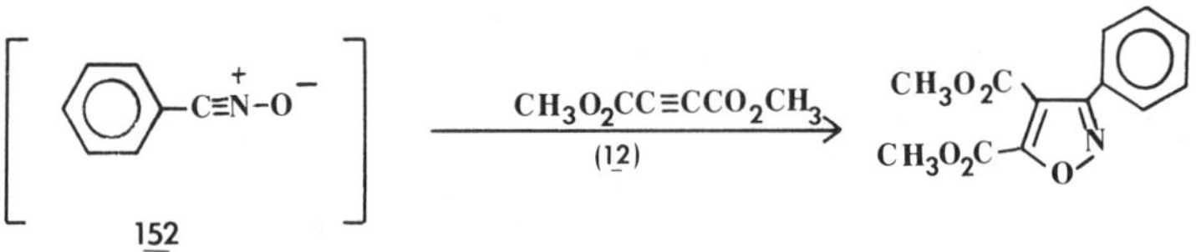

–C≡$\overset{+}{N}$–O^-
152 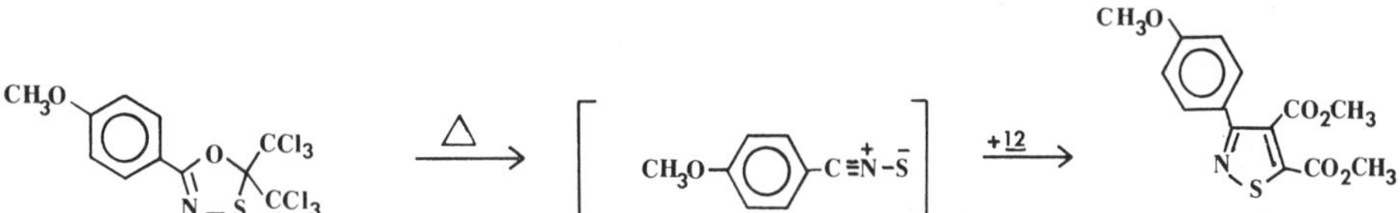
$CH_3O_2CC{\equiv}CCO_2CH_3$
(12)
CH_3O_2C
CH_3O_2C N O

CH_3O
O CCl_3
N – S CCl_3
△
CH_3O– –C≡$\overset{+}{N}$–$\bar{S}$
153
+12
CH_3O
CO_2CH_3
N S CO_2CH_3
(88%)

Isoxazoles (**154**) can be readily transformed into the corresponding isothiazole by initial catalytic reduction, followed by treatment with phosphorus pentasulfide and chloranil to induce recyclization.[129]

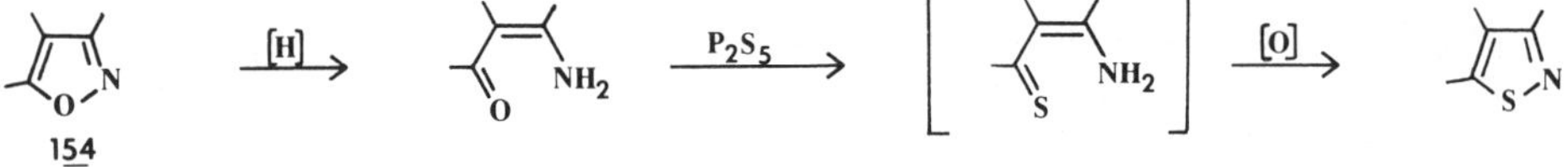

O N
154
[H]
O NH_2
P_2S_5
S NH_2
[O]
S N

REFERENCES

1. M. A. Ogliaruso, M. G. Romanelli, and E. I. Becker, *Chem. Rev.* **65,** 261 (1965).
2. O. Hinsberg, *Chem. Ber.* **43,** 901 (1910).
3. Y. Miyahara, T. Inazu, and T. Yoshino, *Bull. Chem. Soc. Japan* **53,** 1187 (1980); Y. Miyahara, *J. Heterocycl. Chem.* **16,** 1147 (1979).
4. K. Dimroth and U. Pintschorino, *Ann. Chem.* **639,** 102 (1961); M. Friedman, *J. Org. Chem.* **30,** 859 (1965).
5. H. J. Backer and W. Stevens, *Rec. Trav. Chim.* **59,** 423 (1940).
6. S. Watanabe, T. Fujita, K. Suga, and H. Abe, *J. Appl. Chem. Biotechnol.* **27,** 117 (1977).
7. D. C. Owsley, J. M. Nelke, and J. J. Bloomfield, *J. Org. Chem.* **38,** 901 (1973).
8. E. C. Kornfeld and R. G. Jones, *J. Org. Chem.* **19,** 1671 (1954).
9. (a) L. D. Krasnoslobodskaya and Ya. L. Gol'dfarb, *Russ. Chem. Rev.* **38,** 389 (1969); (b) C. U. Pittman, Jr. and Y. F. Liang, *J. Org. Chem.* **45,** 5048 (1980).
10. D. E. Wolf and K. Folkers, *Org. Reactions* **6,** 410 (1951).
11. H. S. Broadbent, W. S. Burnham, R. K. Olsen, and R. M. Sheeley, *J Heterocycl. Chem.* **5,** 757 (1968); A. Gossauer, *Die Chemie der Pyrrole,* Springer, Berlin, 1974, p. 240; also see R. A. Jones and G. P. Bean, in *The Chemistry of Pyrroles,* Academic Press, New York, 1977, Chap. 3.
12. C. Paal, *Chem. Ber.* **18,** 2255 (1885).
13. J. Hambrecht, *Synthesis* **1977,** 280.
14. (a) N. S. Kozlov, L. I. Moiseenok, and S. I. Kozintsev, *Khim. Geterosikl. Soedin.* **1979,** 1483; (b) R. Ramanasseul and A. Rassat, *Bull. Soc. Chem. Fr.* **1970,** 4330; (c) A. R. Katritzky and J. W. Suwinski, *Tetrahedron* **31,** 1549 (1975).
15. S. M. McElvain and K. M. Bolliger, *Org. Syn.* **9,** 78 (1929).
16. K. E. Schulte, J. Reisch, and H. Walker, *Chem. Ber.* **98,** 98 (1965).
17. K. E. Schulte, J. Reisch, W. Herrmann, and G. Bohn, *Arch. Pharm.* **296,** 456 (1963).
18. R. F. Curtis, S. N. Hasnain, and J. A. Taylor, *Chem. Commun.* **1968,** 365.
19. (a) F. Fringuelli and A. Taticchi, *J. Chem. Soc. Perkin I* **1972,** 199; (b) review of tellurophene: F. Fringuelli, G. Marino, and A. Taticchi, *Adv. Heterocycl. Chem.* **21,** 119 (1977).
20. G. Märkl and H. Hauptmann, *Tetrahedron Lett.* **1968,** 3257.
21. G. Märkl and R. Potthast, *Angew. Chem. Int. Ed.* **6,** 86 (1967).
22. F. Bohlmann, T. Burkhardt, and C. Zdero, in *Naturally Occurring Acetylenes,* Academic Press, New York, 1973.
23. E. H. Braye, W. Hubel, and J. Caplier, *J. Am. Chem. Soc.* **83,** 4406 (1961).
24. Review: A. N. Hughes, in *New Trends in Heterocyclic Chemistry,* R. B. Mitra, N. R. Ayyanger, V. N. Gogte, R. M. Acheson, and N. C. Cromwell, Eds., Elsevier, New York, 1979, pp. 216–249.
25. W. Mack, *Angew. Chem. Int. Ed.* **5,** 896 (1966).
26. I. M. Gverdtsiteli, T. P. Doksopulo, M. M. Mentshashvili, and I. I. Abhazabo, *Soobshch. Akad. Nauk. Gruz. SSR* **40,** 333 (1965) [*Chem. Abstr.* **64,** 11239b (1966)]; also see A. LaPorterie, G. Manuel, J. Dubac, P. Mazerolles, and H. Iloughmane, *J. Organometal. Chem.* **210,** C33 (1981).
27. D. C. van Beelen, J. Wolters, and A. van der Gen, *Rec. Trav. Chim.* **98,** 437 (1979).
28. J. B. Hendrickson, R. Rees, and J. F. Templeton, *J. Am. Chem. Soc.* **86,** 107 (1964).
29. E. Müller, *Synthesis* **1974,** 761.
30. E. Müller et al., *Ann. Chem.* **754,** 64 (1971).
31. E. Müller, E. Luppold, and W. Winter, *Synthesis* **1975,** 265.

32. E. Müller and W. Winter, *Chem. Ber.* **105,** 2523 (1972).
33. Y. Wakatsuki, T. Kuramitsu, and H. Yamazaki, *Tetrahedron Lett.* **1974,** 4549.
34. K. P. C. Vollhardt, *Synthesis* **1975,** 765.
35. K. Dimroth, G. Pohl, and H. Follmann, *Chem. Ber.* **99,** 634 (1966).
36. P. J. Garratt and D. N. Nicolaides, *J. Org. Chem.* **39,** 2222 (1974).
37. A. Hantzsch, *Chem. Ber.* **23,** 1474 (1890).
38. D. M. McKinnon, *Can. J. Chem.* **43,** 2628 (1965); H. George and H. J. Roth, *Arch. Pharm.* **307,** 699 (1974).
39. C. A. Grob and K. Camenisch, *Helv. Chim. Acta* **36,** 49 (1953).
40. H. Fischer and H. Orth, *Die Chemie des Pyrrols,* Vol. 1, Akademische Verlagsgesellschaft, Leipzig, 1934, pp. 3–5.
41. N. Vinot and J. Pinson, *Bull. Soc. Chim. Fr.* **1968,** 4970.
42. G. G. Kleinspehn, *J. Am. Chem. Soc.* **77,** 1546 (1955).
43. R. Huisgen, H. Stangl, H. J. Sturm, and H. Wagenhofer, *Angew. Chem. Int. Ed.* **1,** 50 (1962).
44. A. Padwa, J. Smolanoff, and A. Tremper, *Tetrahedron Lett.* **1974,** 29.
45. (a) R. Huisgen, H. Gotthardt, H. O. Bayer, and F. C. Schaefer, *Chem. Ber.* **103,** 2611 (1970); (b) H. Gotthardt, R. Huisgen, and H. O. Bayer, *J. Am. Chem. Soc.* **92,** 4340 (1970).
46. K. T. Potts and D. N. Roy, *Chem. Commun.* **1968,** 1061; K. T. Potts, J. Baum, E. Houghton, D. N. Roy, and U. P. Singh, *J. Org. Chem.* **39,** 3619 (1974).
47. K. T. Potts and S. Husain, *J. Org. Chem.* **36,** 3368 (1971).
48. S. -I. Murahashi, T. Shimamura, and I. Moritani, *J. Chem. Soc. Chem. Commun.* **1974,** 931.
49. K. Isomura, M. Okada, and H. Taniguchi, *Chem. Lett.* **1972,** 629.
50. J. P. Boukou-Poba, M. Farnier, and R. Guilard, *Tetrahedron Lett.* **1979,** 1717.
51. P. A. S. Smith, "Arylnitrenes and Formation of Nitrenes by Rupture of Heterocyclic Rings," in *Nitrenes,* W. Lwowski, Ed., Interscience, New York, 1970 and refs. cited therein.
52. B. S. Thyagarajan and P. E. Glaspy, *J. Chem. Soc. Chem. Commun.* **1979,** 515.
53. R. S. Atkinson and C. W. Rees, *J. Chem. Soc.* (*C*) **1969,** 778.
54. F. Bellamy, *J. Chem. Soc. Chem. Commun.* **1978,** 998.
55. P. F. dos Santos Filho and U. Schuchardt, *Angew. Chem. Int. Ed.* **16,** 647 (1977).
56. D. J. Anderson and A. Hassner, *Synthesis* **1975,** 483.
57. E. J. Nienhouse, R. M. Irwin, and G. R. Finni, *J. Am. Chem. Soc.* **89,** 4557 (1967).
58. C. Botteghi, L. Lardicci, and R. Menicagli, *J. Org. Chem.* **38,** 2361 (1973).
59. J. W. Cornforth, *J. Chem. Soc.* **1958,** 1310; R. A. Cormier and M. D. Francis, *Syn. Commun.* **11,** 365 (1981).
60. F. Boberg and A. Kieso, *Ann. Chem.* **626,** 71 (1959); F. Boberg and G. R. Schultze, *Chem. Ber.* **90,** 1215 (1957).
61. E. Wenkert, *Heterocycles* **14,** 1703 (1980); *Acc. Chem. Res.* **13,** 27 (1980).
62. See A. P. Dunlop and F. N. Peters, *The Furans,* Reinhold, New York, 1953.
63. T. G. Bonner, E. J. Bourne, and M. Ruszkiewicz, *J. Chem. Soc.* **1960,** 787.
64. C. H. Fawcett, D. M. Spencer, R. L. Wain, A. G. Fallis, E. R. H. Jones, M. LeQuan, C. B. Page, V. Thaller, D. C. Shubrook, and P. P. M. Whitham, *J. Chem. Soc.* (*C*) **1968,** 2455.
65. F. Feist, *Chem. Ber.* **35,** 1545 (1902); E. Benary, *Chem. Ber.* **44,** 493 (1911).
66. A. Gopalan and P. Magnus, *J. Am. Chem. Soc.* **102,** 1756 (1980).
67. J. Kagan and K. C. Mattes, *J. Org. Chem.* **45,** 1524 (1980).
68. V. I. Dulenko, N. N. Alekseev, V. M. Golyak, and L. V. Dulenko, *Khim. Geterotsikl. Soed.* **1977,** 1135.

69. (a) T. Mandai, S. Hashio, J. Goto, and M. Kawada, *Tetrahedron Lett.* **22,** 2187 (1981); (b) S. Ranganathan, D. Ranganathan, and M. M. Mehrotra, *Synthesis* **1977,** 838.
70. K. Inomata, M. Sumita, and H. Kotake, *Chem. Lett.* **1979,** 709.
71. S. W. Baldwin and M. T. Crimmins, *J. Am. Chem. Soc.* **102,** 1198 (1980).
72. W. Boll and H. König, *Ann. Chem.* **1979,** 1657; H. König, F. Graf, and V. Weberndörfer, *Ann. Chem.* **1981,** 668; also see J. Hutton, B. Potts, and P. F. Southern, *Syn. Commun.* **9,** 789 (1979).
73. R. Lakhan and B. Ternai, *Adv. Heterocycl. Chem.* **17,** 99 (1974); also see P. A. Jacobi, D. G. Walker, and I. M. A. Odeh, *J. Org. Chem.* **46,** 2065 (1981).
74. D. Miller, *J. Chem. Soc.* (*C*) **1969,** 12.
75. J. W. Batty, P. D. Howes, and C. J. M. Stirling, *J. Chem. Soc. Perkin I* **1973,** 65.
76. D. Binder and P. Stanetty, *Synthesis* **1977,** 200.
77. H. D. Hartough, *Thiophen and its Derivatives,* Interscience, New York, 1952.
78. N. R. Clark and W. E. Webster, Brit. Pat. 1,345,203 (1974) [*Chem. Abstr.* **78,** 11110 (1973)].
79. H. Fiesselmann *et al.,* Chem. Ber. **87,** 835, 841, 848 (1954); **89,** 1897, 1902, 1907 (1956).
80. J. P. Marino and J. L. Kostusyk, *Tetrahedron Lett.* **1979,** 2493.
81. J. P. Clayton, A. W. Guest, A. W. Taylor, and R. Ramage, *J. Chem. Soc. Chem. Commun.* **1979,** 500.
82. H. Wynberg, *Acc. Chem. Res.* **4,** 65 (1971).
83. L. Brandsma, J. Meijer, H. D. Verkruijsse, G. Bokkers, A. J. M. Duisenberg, and J. Kroon, *J. Chem. Soc. Chem. Commun.* **1980,** 922.
84. A. Laurent, *Ann. Chem.* **52,** 354 (1884).
85. I. Lalezari, A. Shafiee, and S. Sadeghi-Milani, *J. Heterocycl. Chem.* **16,** 1405 (1979).
86. I. Lalezari, A. Shafiee, F. Rabet, and M. Yalpani, *J. Heterocycl. Chem.* **10,** 953 (1973).
87. Yu. K. Yur'iev, *Zh. Obshch. Khim.* **16,** 851 (1946).
88. W. Dilthey, Ger. Pat. 628,954 (1936) [*Chem. Abstr.* **30,** 6009 (1936)].
89. (a) A. N. Kost and I. I. Grandberg, in *Advances in Heterocyclic Chemistry,* Vol. 6, A. R. Katritzky and A. J. Boulton, Eds., Academic Press, New York, 1966, pp. 347–429; (b) R. Fusco, in *Pyrazoles, Pyrazolines, Indazoles and Condensed Rings,* R. H. Wiley, Ed., Interscience, New York, 1967, Chap. 3, pp. 10–64.
90. R. H. Wiley and P. E. Hexner, *Org. Syn.* **31,** 43 (1951); C. M. Ashraf and F. K. N. Lugeniwa, *J. Prakt. Chem.* **322,** 816 (1980).
91. K. von Auwers and H. Stuhlmann, *Chem. Ber.* **59,** 1043 (1926).
92. K. N. Houk, *Acc. Chem. Res.,* **8,** 361 (1975); I. Fleming, *Frontier Orbitals and Organic Reactions,* Wiley, New York, 1976, and refs. cited therein.
93. K. N. Houk, *J. Am. Chem. Soc.* **94,** 8953 (1972).
94. K. N. Houk, L. N. Domelsmith, R. W. Strozier, and R. T. Patterson, *J. Am. Chem. Soc.* **100,** 6531 (1978).
95. K. von Auwers and O. Ungemach, *Chem. Ber.* **66,** 1205 (1933).
96. R. Paulissen, *J. Chem. Soc. Chem. Commun.* **1976,** 219.
97. C. S. Rondestvedt and P. K. Chang, *J. Am. Chem. Soc.* **77,** 6532 (1955).
98. (a) R. Huttel, K. Franke, H. Martin, and J. Riedel, *Chem. Ber.* **93,** 1433 (1960); (b) *Chem. Ber.* **93,** 1425 (1960).
99. E. M. Beccalli, L. Majori, A. Marchesini, and C. Torricelli, *Chem. Lett.* **1980,** 659.
100. M. R. Grimmett, *Adv. Heterocycl. Chem.* **12,** 103 (1970); also see A. F. Pozharskii, A. D. Garnovskii, and A. M. Simonov, *Russ. Chem. Rev.* **35,** 122 (1966).
101. H. Wynberg and A. DeGroot, *Chem. Commun.* **1965,** 171.

102. H. Bredereck and G. Theilig, *Chem. Ber.* **86,** 88 (1953).

103. H. Bredereck, R. Gompper, R. Baugert, and H. Herlinger, *Chem. Ber.* **97,** 827 (1964).

104. O. Wallach and E. Schulze, *Chem. Ber.* **14,** 420 (1881).

105. J. K. Lawson, *J. Am. Chem. Soc.* **75,** 3398 (1953).

106. H. Bredereck, R. Gompper, H. G. Schuh, and G. Theilig, in *Newer Methods of Preparative Organic Chemistry,* Vol. 3, W. Foerst, Ed., Academic Press, New York, 1964, p. 241.

107. H. Schubert, B. Ruhberg, and G. Fiedrich, *J. Prakt. Chem.* **32,** 249 (1966).

108. I. J. Turchi and M. J. S. Dewar, *Chem. Rev.* **75,** 389 (1975).

109. U. Schöllkopf and R. Schroder, *Angew. Chem. Int. Ed.* **10,** 3 (1971); *Synthesis* **1972,** 148.

110. H. A. Houwing, J. Wildeman, and A. M. van Leusen, *Tetrahedron Lett.* **1976,** 143.

111. R. Huisgen, *Angew. Chem. Int. Ed.* **2,** 565, 633 (1963).

112. K. Bunge, R. Huisgen, R. Raab, and H. Stangl, *Chem. Ber.* **105,** 1279 (1972).

113. (a) R. Huisgen and H. Blaschke, *Tetrahedron Lett.* **1964,** 1409; (b) *Chem. Ber.* **98,** 2985 (1965); (c) R. Huisgen and J. -P. Anselme, *Chem. Ber.* **98,** 2998 (1965).

114. E. Zbiral, E. Bauer, and J. Stroh, *Monatsh. Chem.* **102,** 168 (1971); *Synthesis* **1971,** 494.

115. W. Reppe and A. Magin, Fr. Pat. 1,340,996 (1963) [*Chem. Abstr.* **60,** 5506 (1964)].

116. H. H. Wasserman and F. J. Vinick, *J. Org. Chem.* **38,** 2407 (1973).

117. A. Padwa and W. Eisenhardt, *Chem. Commun.* **1968,** 380.

118. E. F. Ullman and B. Singh, *J. Am. Chem. Soc.* **88,** 1844 (1966); B. Singh and E. F. Ullman, *J. Am. Chem. Soc.* **89,** 6911 (1967); B. Singh, A. Zweig, and J. B. Gallivan, *J. Am. Chem. Soc.* **94,** 1199 (1972).

119. (a) J. Rokach and P. Hamel, *J. Chem. Soc. Chem. Commun.* **1979,** 786; (b) M. Ohashi, A. Iio, and T. Yonezawa, *Chem. Commun.* **1970,** 1148; (c) M. Kojima and M. Maeda, *Chem. Commun.* **1970,** 386.

120. R. H. Wiley, D. C. England, and L. C. Behr, *Org. Reactions* **6,** 367 (1951).

121. P. Dubs and R. Stuessi, *Synthesis* **1976,** 696.

122. G. D. Hartman and L. M. Weinstock, *Synthesis* **1976,** 681.

123. Review: C. Kashima, *Kagaku no Ryoiki, Zokan* **1980,** 71.

124. O. Buchardt, *Photochemistry of Heterocyclic Compounds,* Wiley-Interscience, New York, 1976, p. 162.

125. See A. Weissberger, Ed., *Five- and Six-Membered Compounds with Nitrogen and Oxygen,* Vol. 17 in *Chemistry of Heterocyclic Compounds,* Wiley-Interscience, New York, 1962, p. 6.

126. Y. -i. Lin and S. A. Lang, Jr., *J. Org. Chem.* **45,** 4857 (1980).

127. See C. Grundmann and P. Grunager, *The Nitrile Oxides,* Springer, Berlin, 1971.

128. R. M. Paton, F. M. Robertson, J. F. Ross, and J. Crosby, *J. Chem. Soc. Chem. Commun.* **1980,** 714.

130. K. R. H. Woolridge, *Adv. Heterocycl. Chem.* **14,** 1 (1972).

Chapter 5

SYNTHETIC ROUTES TO AND REACTIONS OF SOME COMMON BENZO-FUSED FIVE-MEMBERED π-EXCESSIVE HETEROAROMATIC COMPOUNDS

5.1. BENZOPYRROLES

Indole is the trivial name for the most common member of the benzopyrrole class and is represented by structure **1**. The accepted IUPAC name for **1** is 1*H*-benzo[*b*]pyrrole, although numerous other names have been used, including 1-azaindene, 1-benzazole, and α,β-benzopyrrole. Several isomers of **1** are also common: the double bond isomer, 3*H*-benzo[*b*]pyrrole (known as indolenine or 3*H*-indole) (**2**); the 2-aza isomer, 2*H*-benzo[*c*]pyrrole (1*H*-isoindole) (**3**); and the *a*-face benzo-fused isomer, indolizine (pyrrocoline) (**4**). Dibenzopyrrole is commonly known as carbazole and has structure **5**.

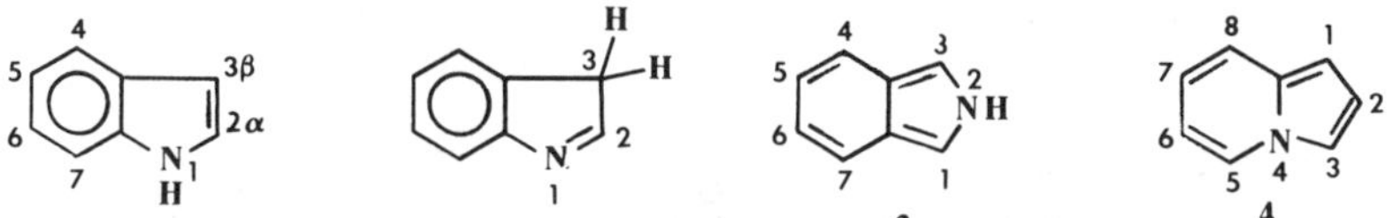

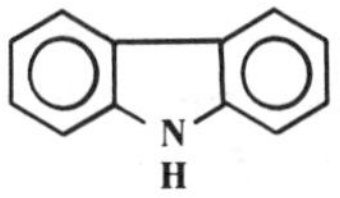

5.1.1. INDOLES AND CARBAZOLES

The most common and widely used route to indole (**6**) is the *Fischer indole synthesis*[1] (Scheme 5.1). By minor alteration in starting materials, carbazoles (**7**) are prepared; this procedure is known as the *Borsche synthesis.*[2] Among the numerous modifications to the Fischer indole synthesis reported over the past century is the *Fischer–Arbuzov reaction,*[3a] in which only catalytic amounts of a Lewis acid are used. Recently, excellent yields of 2,3-disubstituted indoles have been realized by use of phosphorus trichloride.[3b]

In the Fischer indole synthesis and its modifications, convenient syntheses of arylhydrazones are necessary. Although arylhydrazones are easily prepared by reduction of *N*-nitrosoarylamines[4,5] or of diazonium salts,[5,6] the *Japp–Klingemann reaction*[7] is an excellent alternate synthetic route to diverse arylhydrazones. The general procedure is demonstrated by the coupling of benzenediazonium salts with compounds that have activated methylene or methine

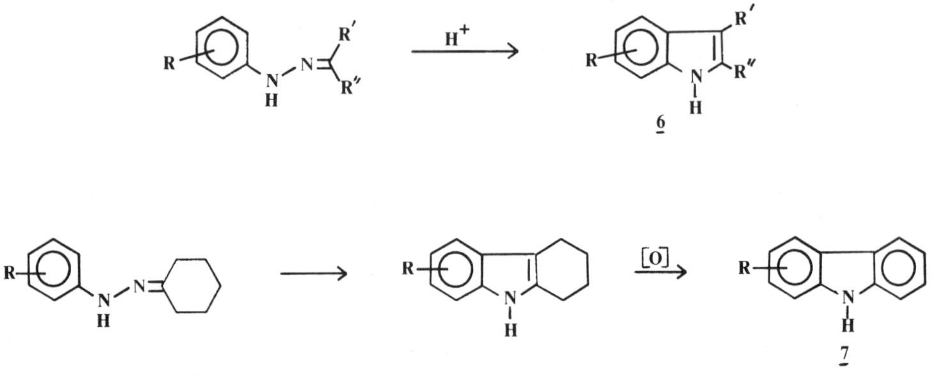

R N H N R′ R″ $\xrightarrow{H^+}$ R N H R′ R″ <u>6</u>

R N H N $\longrightarrow$ R N H $\xrightarrow{[O]}$ R N H <u>7</u>

Scheme 5.1

hydrogens. Thus reaction of **8** with ethyl 2-methylacetoacetate gives the phenylhydrazone of ethyl pyruvate.[8] The scope and applications of this procedure have been reviewed.[9]

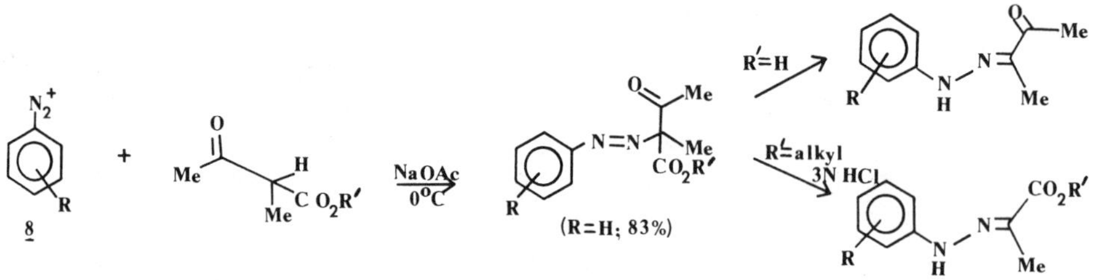

N_2^+ R <u>8</u> + Me O H Me CO_2R' $\xrightarrow[0^oC]{NaOAc}$ R N=N O Me Me CO_2R' (R=H; 83%)

R′=H: R N H N O Me Me

R′=alkyl, 3N HCl: R N H N CO_2R' Me

The mechanism of the Fischer indole synthesis has been studied in great detail and is summarized in Scheme 5.2. Judicious selection of the catalyst can control product formation, as demonstrated by the transformation of **9** to

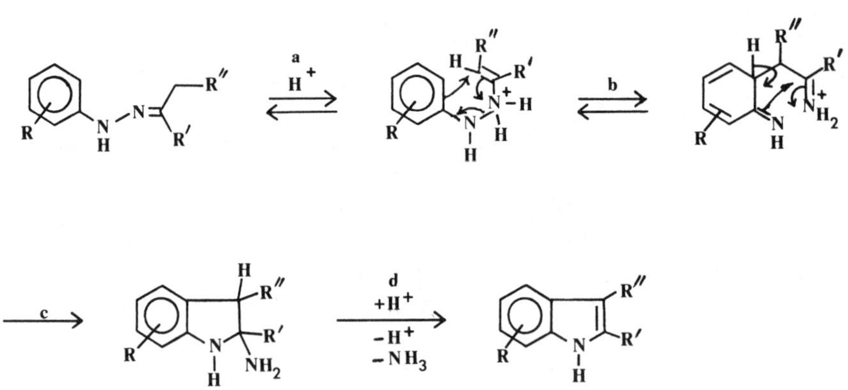

R N H N R″ R′ $\underset{a}{\overset{H^+}{\rightleftharpoons}}$ R N H H H R″ R′ N^+–H H $\underset{b}{\rightleftharpoons}$ R N H H R″ R′ NH_2^+

$\xrightarrow{c}$ R N H H R″ R′ NH_2 $\xrightarrow[-H^+, -NH_3]{d, +H^+}$ R N H R″ R′

Scheme 5.2 (*a*) Hydrazone-enehydrazine tautomerism. (*b*) C_3—C_{3a} cyclization (*o*-benzidine rearrangment). (*c*) N_1—C_2 cyclization. (*d*) Elimination of ammonia; aromatization.

either **10** or **11.**[10] Hydrazone **12,** on treatment with zinc chloride, gives indole **13;** treatment of **12** with polyphosphoric acid (PPA) gives a 1 : 5 mixture of **13** and **14,** respectively.[11] Indolenines **(15)** are in some instances isolable intermediates, but generally rearrange in acidic media to the more stable indole system.[12]

These acid-catalyzed rearrangements, to convert **15** to **16,** probably proceed via a Wagner-Meerwein rearrangement as shown in Scheme 5.3. Control of temperature, reaction time, and the acid catalyst used in these cyclization reactions is critical to the formation of the desired product(s). Structure proof of the resultant products is often necessary to eliminate any ambiguity, especially when different substituents are present on the starting hydrazone.

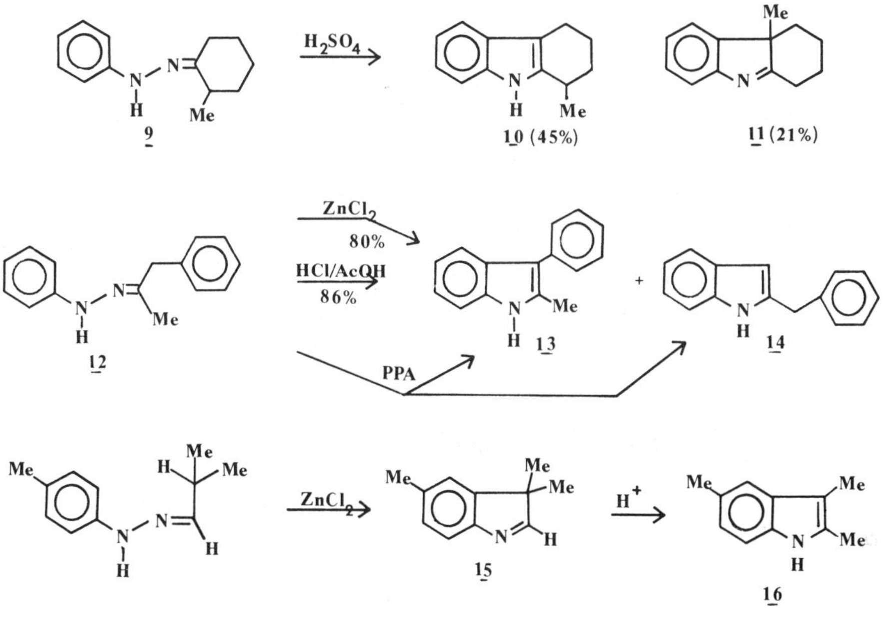

Treatment of arylamines with α-halo-, α-hydroxy-, or α-anilinoketones gives rise to the indole nucleus; this is known as the *Bischler indole synthesis.*[13] Numerous modifications of this reaction have been reported in which glycidic acid esters, 1-phenacylpyridinium salts, diazoketones, and α-alkyl-α-arylaminoacetaldehyde acetals replace the previously mentioned ketones. Thus treat-

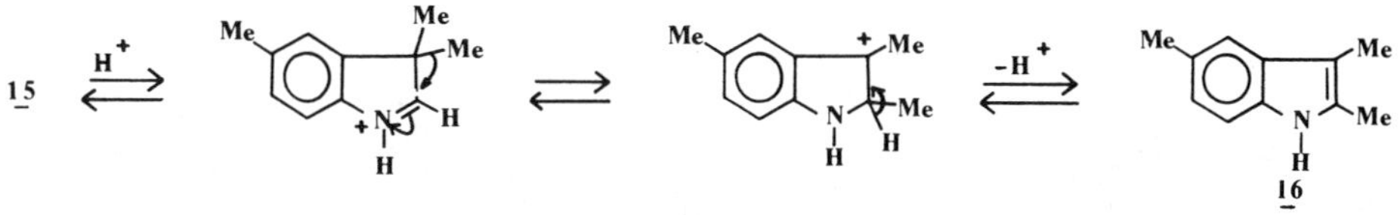

Scheme 5.3

ment of amine **17** with **18** gives indole **19**[14] via several discrete steps: (*a*) *N*-alkylation, (*b*) enamine cyclization to generate the 3—3*a* bond, and (*c*) aromatization by loss of HX. If R and R′ are different, general acid catalysis causes ready scrambling of the substituents. Alternative mechanisms that have been proposed are the "ortho shift" hypothesis (Scheme 5.4) and the hypothesis that involves direct cyclization of the α-arylaminoketone (Scheme 5.5). The specifics for the various mechanisms are described.[13c]

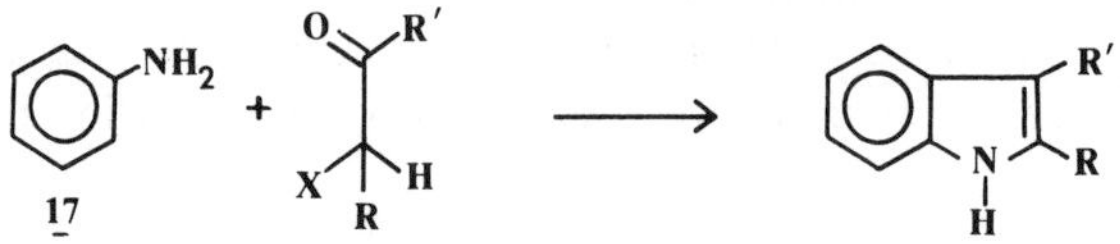

17 + 18 (X = OH; Cl; Br) → 19

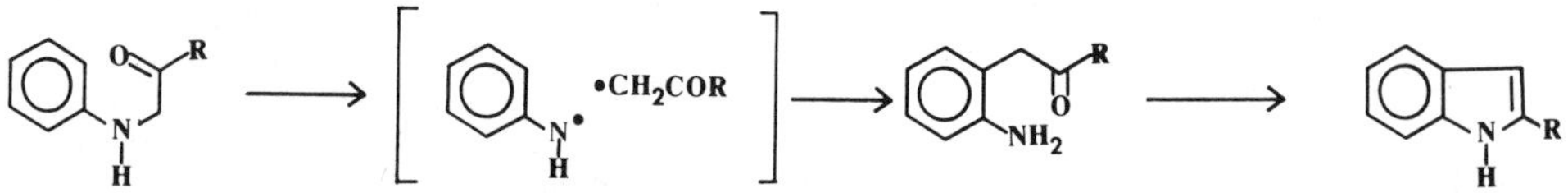

Scheme 5.4

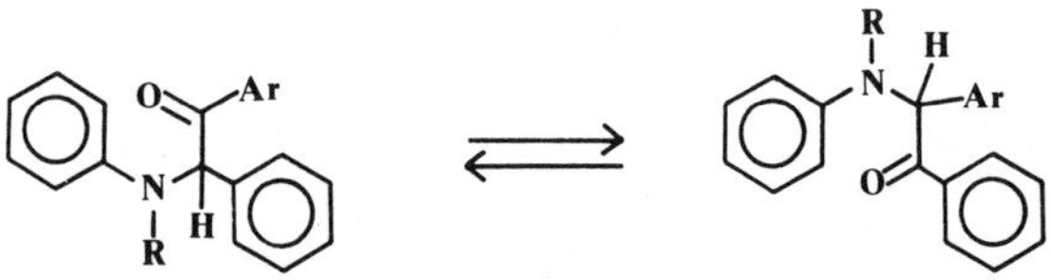

Scheme 5.5

Intramolecular cyclization of *N*-acylated *o*-alkylanilines is known as the *Madelung indole synthesis.* Amide **20** with strong base at elevated temperatures and in the absence of air gives 2-methylindole (**21**).[15] In the *Verley modification,*[16] the use of sodamide instead of alkoxide bases often improves the yields of **21.** In order to avoid these drastic reaction conditions, alkylidenephosphoranes (**22**) are prepared and subsequently cyclized in excellent overall yields.[17]

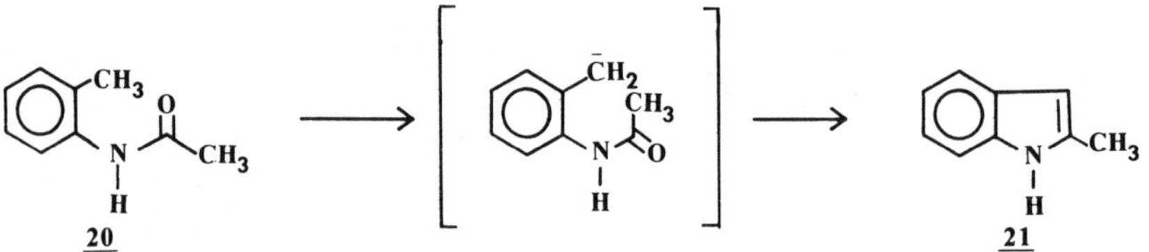

20 → 21

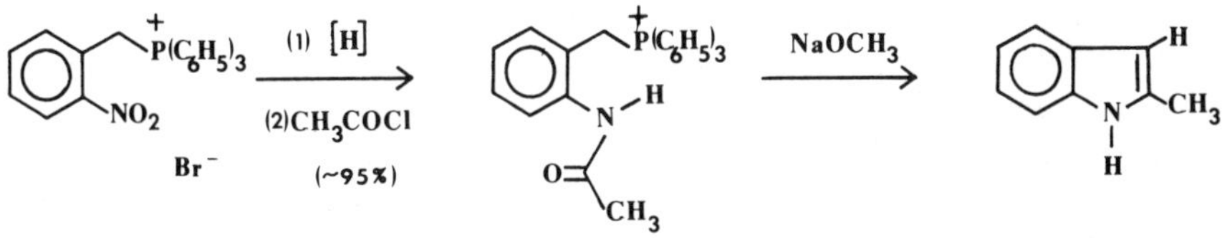

22 21 (96%)

The Reissert indole synthesis is the base-catalyzed intramolecular cyclization of phenylpyruvic acid derivatives, which are generated by the action of strong base on *o*-nitrotoluenes (**23**) in the presence of oxalate esters. Subsequent chemical reduction of the nitro substituent affords an *o*-amino intermediate **24,** which cyclizes and dehydrates to give the corresponding indole-2-carboxylic acid (**25**).[18]

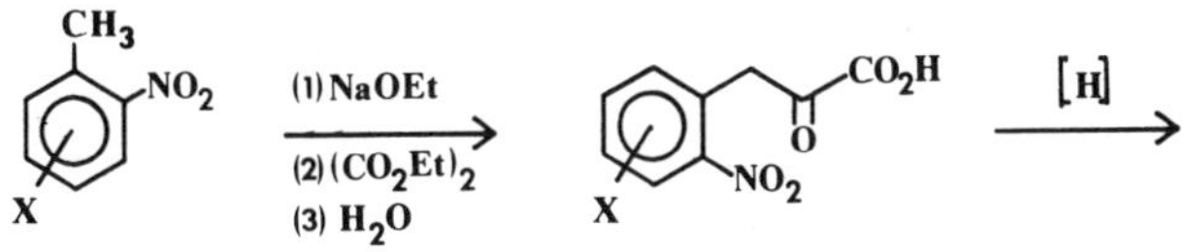

23 (41%)

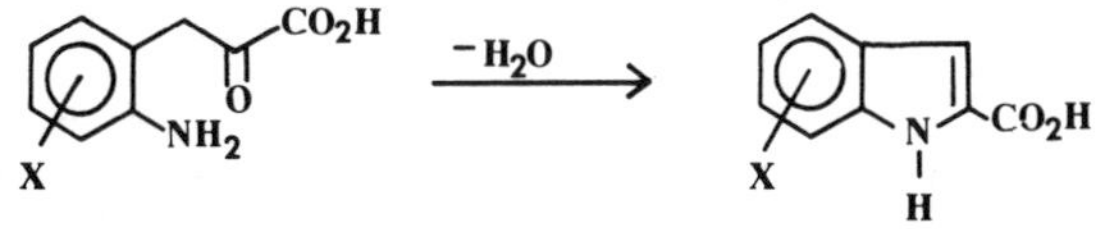

24 25 (90%)

Direct aromatic substitution of halide in *o*-haloanilines by strong alkyl anions allows access to aminoketones similar to those generated in the Reissert procedure. The photochemical reaction of the acetone enolate on *o*-iodoaniline in liquid ammonia affords intermediate **26,** which readily cyclizes to indole (**21**) in quantitative yields.[19] Wolfe et al.[20] have further extended this procedure to the synthesis of oxindoles (**27**) and azaoxindoles.

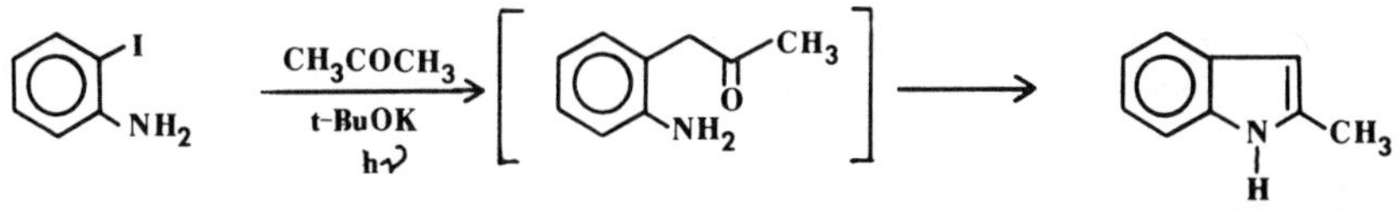

26 21 (100%)

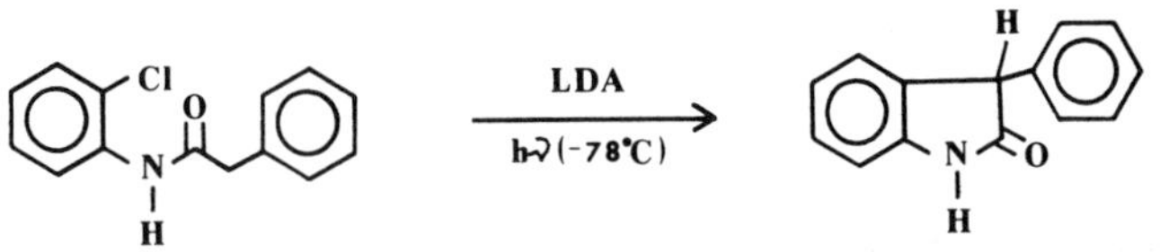

27 (60–80%)

The reaction of *p*-benzoquinone (**28**) with alkyl β-aminocrotonates (**29**) affords the substituted indole-3-carboxylic acid (**30**).[21] This procedure is known as the *Nenitzescu indole synthesis* and produces 5-hydroxyindoles after the facile decarboxylation of the 3-carboxylic acid functionality. Scheme 5.6 shows the proposed mechanism, in which the first step is an enamine addition to the α,β-unsaturated ketone. The details and supportive data for this mechanism have been summarized.[13c]

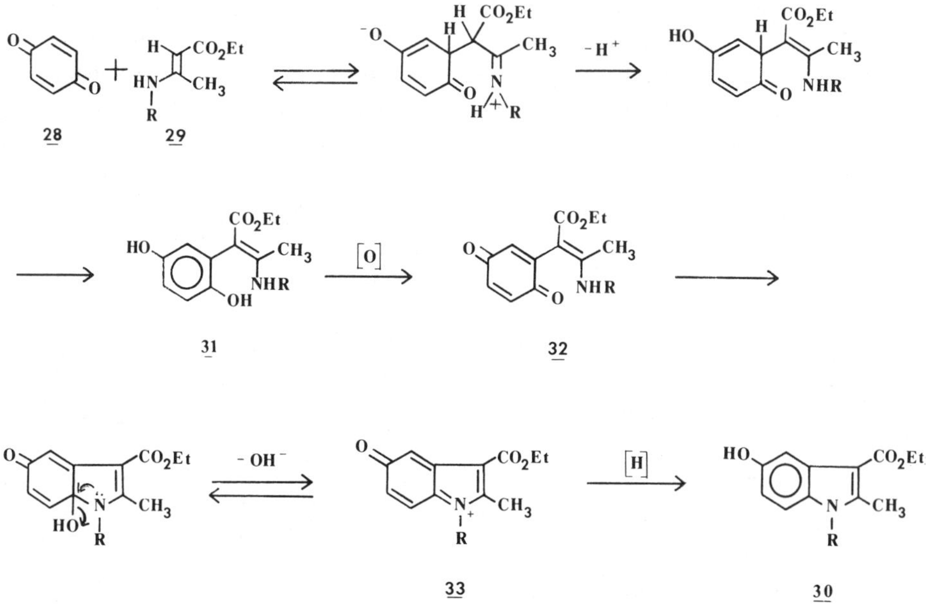

Scheme 5.6

Because of the facile oxidation of the hydroquinone intermediate **31,** *in situ* generation of quinone intermediates, such as **32** and **33,** provides a simple route to indoles of biological importance.[22] Epinochrome (**35**), derived from the oxidation of epine **34,** is rapidly transformed to 5,6-dihydroxy-1-methylindole (**36**) in the presence of strong base.[23]

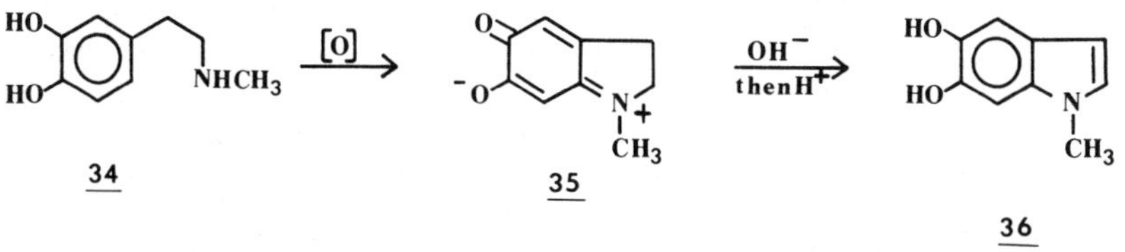

As demonstrated by the synthesis of 2-methylindole (**21**) *palladium-assisted heterocyclization reactions*[24] have been applied to the generation of the indole nucleus. When *o*-allylaniline is treated with $PdCl_2(CH_3CN)_2$, the σ-alkylpalladium(II) intermediate (**37**) is formed. Subsequent reaction with a tertiary amine generates **38**, which isomerizes to **21** in 84% overall yield.[24]

NH_2 → $PdCl_2(CH_3CN)_2$ → [N–Pd–Cl, H_2, Cl] → Et_3N →

Cl, Pd–Cl, NEt_3, +N, H, H, H → −HCl, −"PdH" → N, H, CH_2 → N, H, CH_3

37 38 21 (84%)

Application of nickel complexes to the cyclization of 2-chloro-*N*-methyl-*N*-allylaniline (**40**) gives 3-methylindole.[25] The $Ni[P(C_6H_5)_3]_4$ complex is readily obtained by reduction of $Ni(acac)_2$ in the presence of four equivalents of triphenylphosphine.[26] Generation of the indole nucleus proceeds via initial formation of a nickel-carbon sigma complex (**41**), which undergoes an intramolecular carbometalation, loss of a nickel hydride complex, and bond reorganization. Similar reactions occur under the influence of Grignard reagents in the presence of catalytic amounts of $NiCl_2[P(C_6H_5)_3]_2$.[25]

Cl, N, CH_3 → $NiCl_4$ → Cl, NiL_2, N, CH_3 → NiL_2Cl, N, CH_3

40 41

−NiL_2, −HCl → CH_2, N, CH_3 → CH_3, N, CH_3

$Co_2(CO)_8$,[27] tetracarbonyldichlorodirhodium, $[Rh(CO)_2Cl]_2$, or chlorocarbonylbis(triphenylphosphine)rhodium, $RhCl(CO)(PPh_3)_2$, reacts with 2-arylazirines in benzene at room temperature to give 2-styrylindole (**42**).[28] Palladium(II) complexes have also been utilized to transform 2*H*-azirines (for example, **43**) to indoles **44.** The reaction is proposed to proceed through a 2*H*-azirine palladium(II) complex (**45**).[29]

$[Rh(CO)_2Cl]_2$; rt; C_6H_6 → **42** (39–80%)

$PdCl_2(C_6H_5CN)_2$, 30° → **45**; ca 80% → **44**; **43** → 170° w/o Catalyst → **44**

Pyrolysis of azirine **46** leads to a substituted vinylnitrene by carbon-nitrogen bond cleavage. β-Styrylnitrene (**47**) gives an equal mixture of indole (**1**) and phenylacetonitrile (**48**).[30] If no hydrogen atom is adjacent to the nitrene function, substituted indoles **49** are formed in quantitative yields.[31]

46 → Δ → [$C_6H_5CH{=}CH$–N:] **47** → **1** + $C_6H_5CH_2CN$ **48**

480°, He → **49**

Oxidation-reduction sequences have been employed in the production of indoles. However, the yields are generally poor when compared to those attainable with the procedures described above. Indole can be synthesized from *o*-nitrophenylethanol by initial catalytic reduction, followed by an oxidative dehydroxylation-dehydrogenation.[32] Similar aldehydic intermediates (for example, **50**) are probably generated during the chemical reduction of the dinitrostyrene **51**[33] and methyl *o*-nitrostyrylcarbamate (**52**).[34] Other unsaturated side-chain functions that can undergo facile nucleophilic cyclization include ketones, nitriles, imines,[35] and α,β-unsaturated electron-poor olefins.

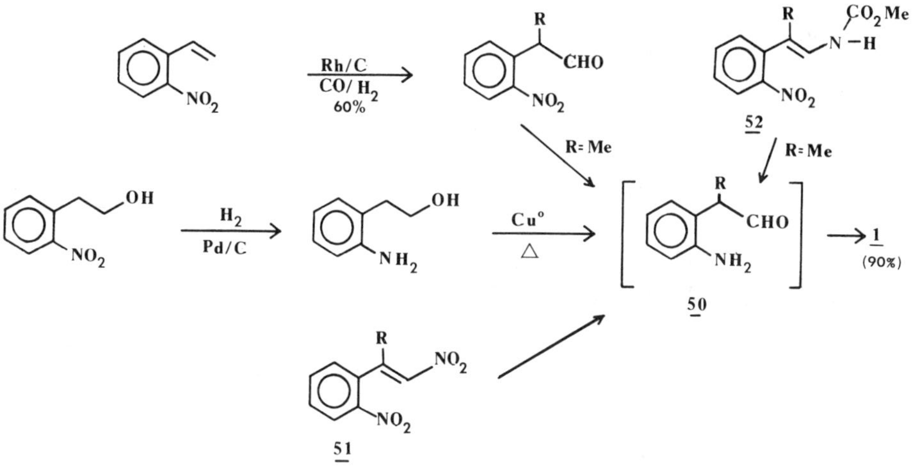

More specialized cyclization procedures have recently been proposed. For example, when *o*-tolylisocyanide (**53**) is treated with lithium diethylamide (LDA) at −78°C, anion **54** is formed and then trapped with an alkylating agent to afford either **55** or **56** depending on the cationic counterion.[36] Upon treatment with thiomethyl ketones or aldehydes, *N*-chloroaniline (**57**) generates the azasulfonium salt **58,** which undergoes a spontaneous *Sommelet–Hauser* type rearrangement to give a dienone imine. Hydride transfer and rearomatization give an α-aminoketone that cyclizes and dehydrates to indolenine **59.**[37] A related sulfur reagent, dimethylsulfonium methylide, reacts with aromatic *o*-aminocarbonyl compounds (**60**) to provide indoles[38] (**61** and **62**) in variable yields by the mechanism shown in Scheme 5.7. The *N*-methylindole **62** can result as in Scheme 5.7 by the direct methylation of **61** through reaction with dimethylsulfonium methylide.

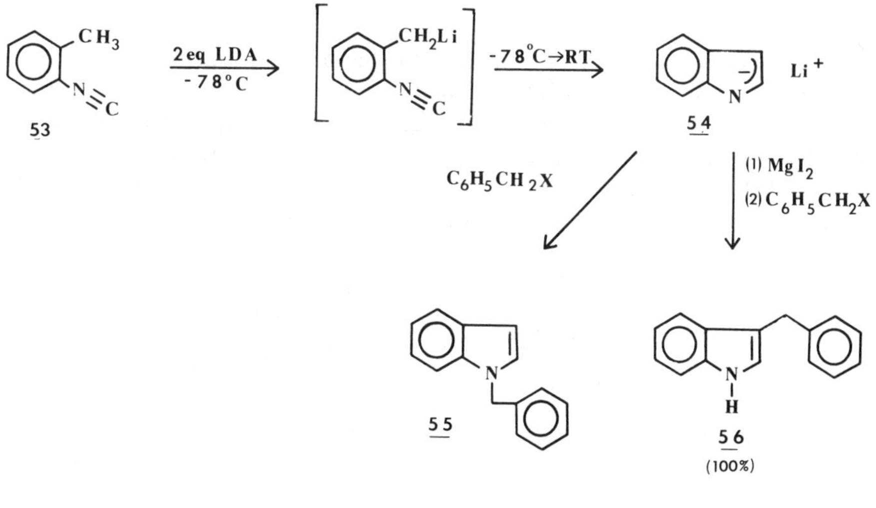

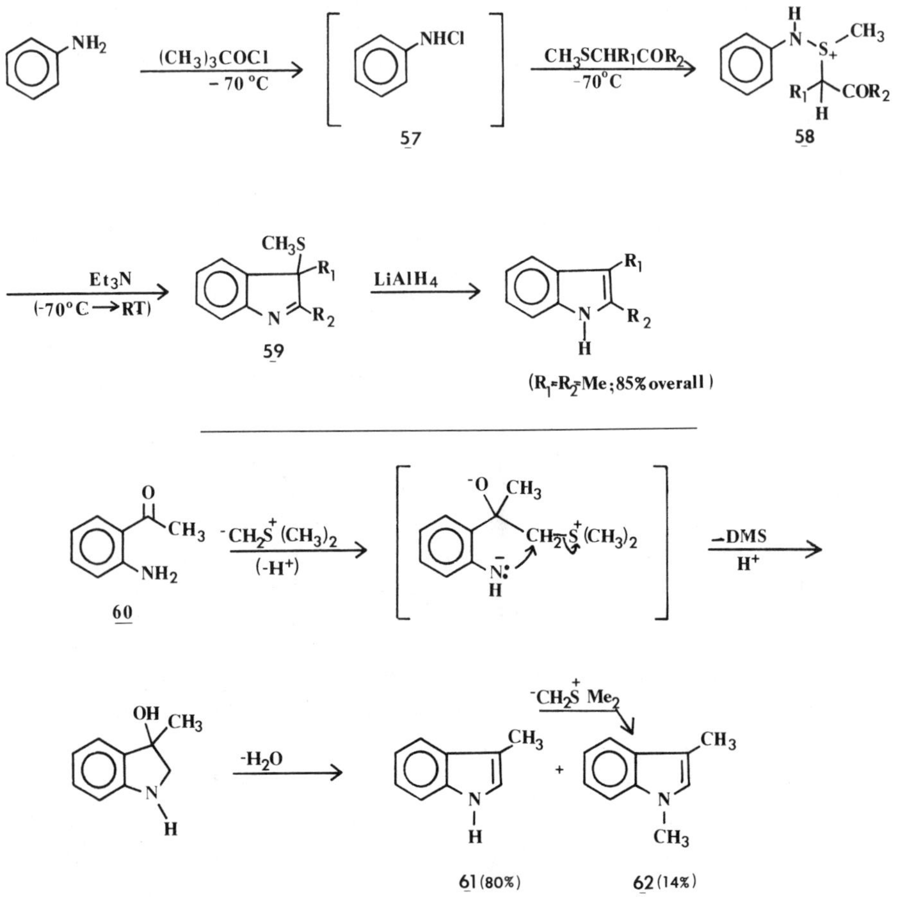

$(CH_3)_3COCl$, −70 °C; **57**; $CH_3SCHR_1COR_2$, −70°C; **58**; Et_3N (−70°C → RT); **59**; $LiAlH_4$; ($R_1=R_2=Me$; 85% overall)

60; $^-CH_2\overset{+}{S}(CH_3)_2$ (−H⁺); −DMS, H⁺; −H₂O; $^-CH_2\overset{+}{S}Me_2$; **61** (80%); **62** (14%)

Scheme 5.7

Cycloaddition reactions of *N*-methylpyrrole (**63**) and dimethyl acetylenedicarboxylate (**64**) give an unstable bicyclic intermediate **65**, which readily interacts with a second equivalent of **64** to generate **66.**[39] This adduct, when oxidized with bromine in methanol, affords trimethyl *N*-methylindole-2,3,4-tricarboxylate (**67**). Indoles can also react in a stepwise manner with **64** to afford carbazoles **68** possessing acidic functional groups.[40]

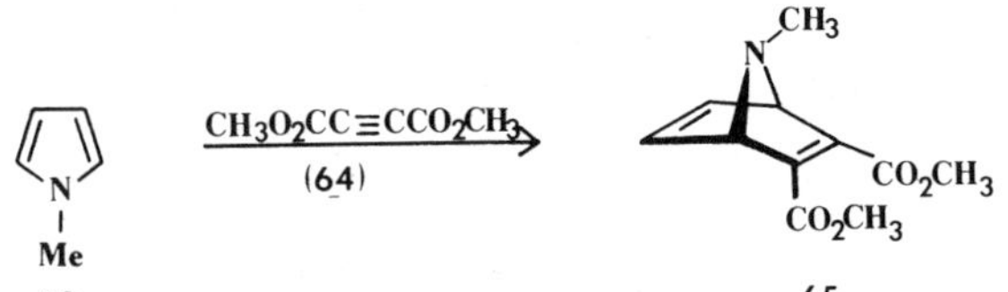

63; $CH_3O_2CC{\equiv}CCO_2CH_3$ (**64**); **65**

65 →(+64) 66 →(Br2, CH3OH) 67

Indole →(+64) 68 (63%)

5.1.2. ISOINDOLES

The parent isoindole, as well as the substituted 2*H* derivatives, readily tautomerize to the 1*H* isomer **69** (isoindolenine, 1*H*-isoindole). In general, **3** is very unstable and rapidly oxidizes in air to give complex mixtures and polymers. Although **3** had been prepared in dilute solutions,[41] it was not until 1972 that its characterization and isolation were accomplished.[41,42] Pure, unsubstituted isoindole (**3**) can be prepared free of isomers by the flash pyrolysis of **70** at 600°C[41,43] via a [4 + 2] retrocycloaddition. However, rapid isomerization takes place and gives rise to the isolation of a mixture of **3** and **69**.

70 →(−(CH_2=CH_2), 600°C) 3 + 69

N-Methylisoindole was first prepared by Wittig et al.[44] through an *elimination reaction* of 2,2-dimethylisoindolinium bromide (**71**) with phenyllithium; methane gas is liberated. With **72**, isoindole formation (**72** → **73**) is only one of several competitive reactions which occur. An increase of the reaction temperature from 0 to 100°C results in a 1,2-migration of a benzyl group via a Stevens rearrangement[45] (**72** → **74**). At 200°C, an alternative migration occurs to give an ortho-substituted phenyl ring via a Sommelet rearrangement[45] (**72** → **75**).

71 →(C_6H_5Li) 73 (74%)

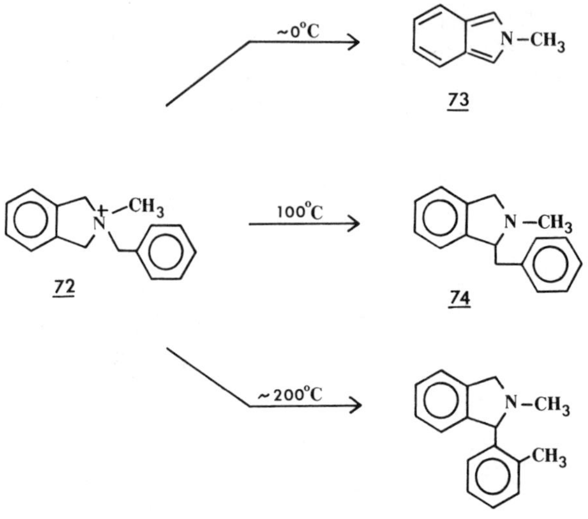

~0°C
73
72
100°C
74
~200°C
75

The reaction of substituted isoindoline *N*-oxides (**76**) with acetic anhydride and triethylamine at 0 to −10°C gives rise to *N*-alkyl- and *N*-arylisoindoles.[46] This procedure is particularly convenient in view of the synthetic ease with which the starting materials can be made. An alternative elimination reaction has been employed, in which the *N*-substituents are good leaving groups,[43] such as tosylate.[47] In 1928, Fenton and Ingold first attempted the synthesis of **3** by treatment of **77** with potassium hydroxide[47]; however, isoindoline **78** was the only product isolated. 1,3-Diphenylisoindole (**79**) was subsequently prepared via this procedure.[48]

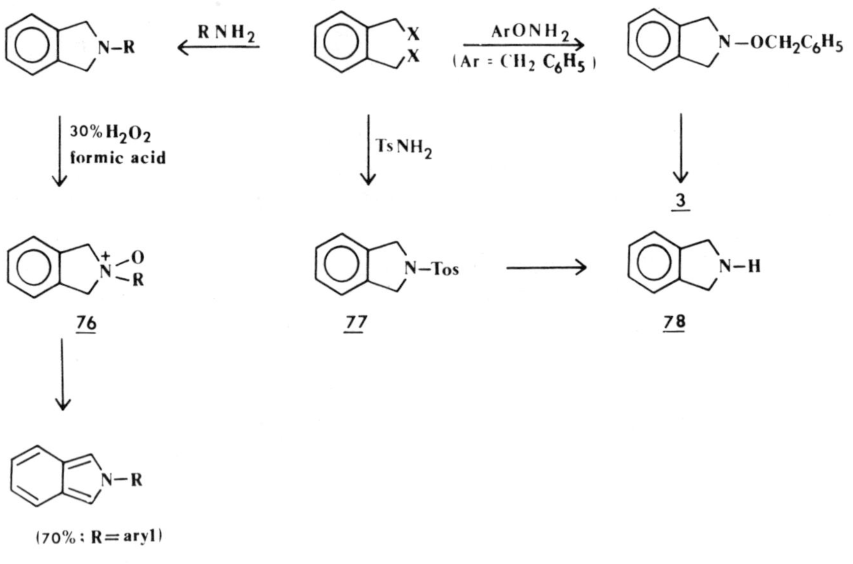

RNH_2
$ArONH_2$ ($Ar = CH_2C_6H_5$)
$N-OCH_2C_6H_5$
30% H_2O_2 formic acid
$TsNH_2$
3
76
77
78
(70%: R = aryl)

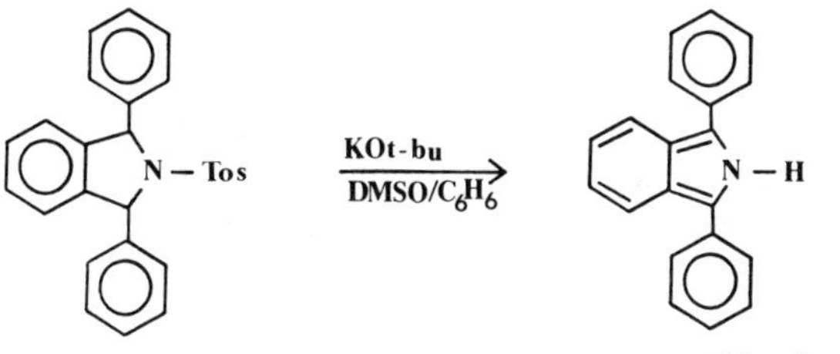

N−Tos ; KOt-bu, DMSO/C_6H_6 ; N−H

79 (38%)

Reduction[49] *or reductive alkylation*[50] *of phthalimidines* (**80**) affords isoindoles. This transformation involves initial nucleophilic addition to the carbonyl group, then facile elimination of water. Similarly, *N*-methylphthalimide (**81**) is reduced by lithium in liquid ammonia at −78°C to give **82.** Trapping experiments of this deep purple dianion with various alkylating agents result in the formation of **83** and **84** rather than in the formation of the expected 1,3-dialkoxyisoindole derivatives.[51]

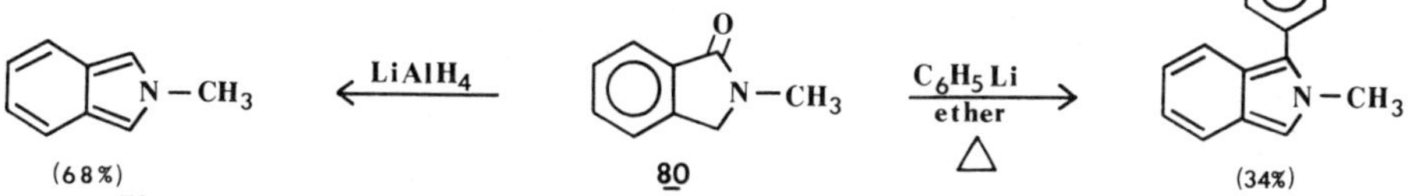

N−CH_3 ; $LiAlH_4$; O ; N−CH_3 ; C_6H_5Li, ether, Δ ; N−CH_3

(68%) 73 ; 80 ; (34%)

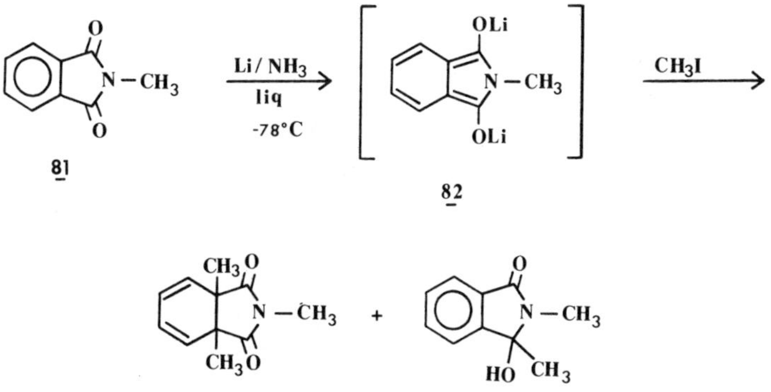

O ; N−CH_3 ; O ; Li/NH_3 liq −78°C ; OLi ; N−CH_3 ; OLi ; CH_3I

81 ; 82

CH_3 O ; N−CH_3 ; CH_3 O ; + ; O ; N−CH_3 ; HO CH_3

83 (36%) ; 84 (37%)

In a procedure similar to that used in the synthesis of pyrroles from 1,4-diketones, 1,3-diphenylisoindole (**79**)[48,52] is prepared from γ-dibenzoylbenzene on treatment with ammonium formate or with *N,N*-dimethylhydrazine. *N*-Methyl-1,3-diphenylisoindole is prepared from the same 1,4-diketone when methylammonium formate is used.[52] An improved procedure utilizes a more direct analogue of the 1,4-diketone, in which **85,** available in the bis-enol orientation **86,** is employed. In this compound the unique relationship to vinyl halides, generated by the thermal conrotatory ring opening of benzocyclobutane precursors,[44] is apparent and convenient. Treatment of **85** or **87** (X = Br) with ammonia or aniline readily affords **79** or *N*-phenylisoindole, respectively. Other related structures such as substituted isobenzofurans[53] are converted to trisubstituted isoindoles in a similar manner (see below).

79 ← $NH_4(HCO_2)$ (44%) or $(CH_3)_2NNH_2$ — $\xrightarrow{CH_3\overset{+}{N}H_3\ HCO_2^-}$ N—CH_3 (28%)

85 ⇌ 86 ← Δ 87 (Br, Br); X, X; RNH_2 → N–R 79 (R=H)

5.1.3. INDOLIZINES

Substituted indolizines are most easily prepared by the reaction of pyridine (**88**) with dimethyl acetylenedicarboxylate (**64**).[54] The mechanism is envisioned as proceeding by the stepwise sequence shown in Scheme 5.8. Initial nucleophilic addition gives zwitterion **89,** which is protonated, followed by a Michael-type addition of methoxide ion to generate the zwitterion **90**. Addition of **90** to a second equivalent of **64** is followed by cyclization to afford **91.** Subsequent aromatization yields one of two products, **92** or **93**, depending on the oxidative procedure used.

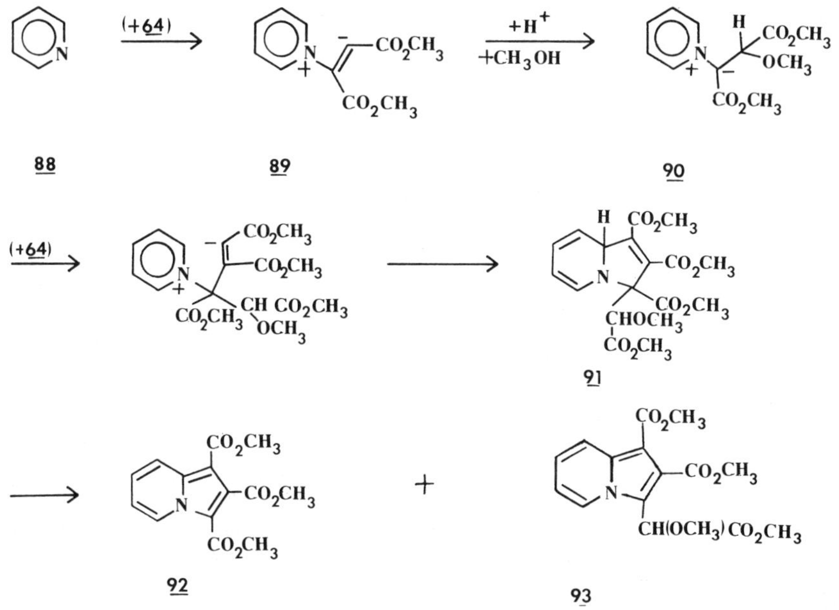

Scheme 5.8

Condensation reactions have commonly been used to generate the indolizine nucleus. Ethyl 2-pyridylacetate (**94**) reacts with ethyl bromopyruvate (**95**) to give a quaternary salt which readily cyclizes in the presence of base.[55] Subsequent hydrolysis and thermolysis afford the parent compound **4.** *N*-Oxides, such as **96,** upon treatment with acetic anhydride are converted to the α-acetoxy derivatives **97,** which upon heating cyclize and aromatize by loss of acetic acid to give rise to **4.**[56]

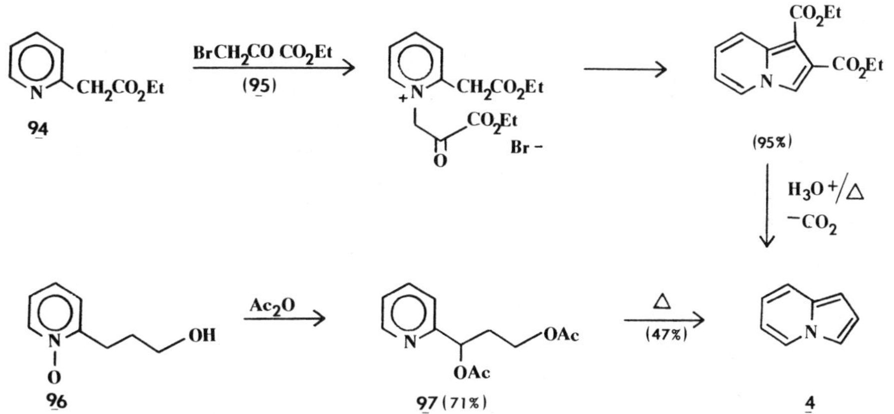

One of the most useful syntheses of indolizine (**4**) involves the reaction of 2-pyridyllithium with 2-chloromethyloxirane (**98**) to produce 2-hydroxy-2,3-

dihydro-1*H*-indolizinium chloride (**99**), which is converted to **4** by treatment with sodium hydroxide.[57] Dipolar cyclizations and condensation procedures leading to diverse indolizine have been reviewed.[58]

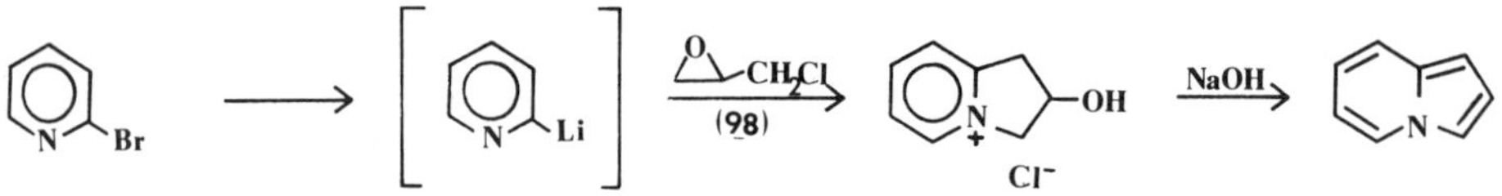

99(47%) 4(88%)

5.2. BENZOFURANS

Two isomeric benzofurans exist. Benzo[*b*]furan (**100**), which is the most common isomer, was synthesized for the first time by Fittig and Ebert in 1883,[59] and is the subject of several reviews,[60] Benzo[*c*]furan (**101**), or isobenzofuran, is an essentially stable molecule that is, nevertheless, quite reactive toward dienophiles and oxygen. One dibenzofuran is known and has structure **102**.

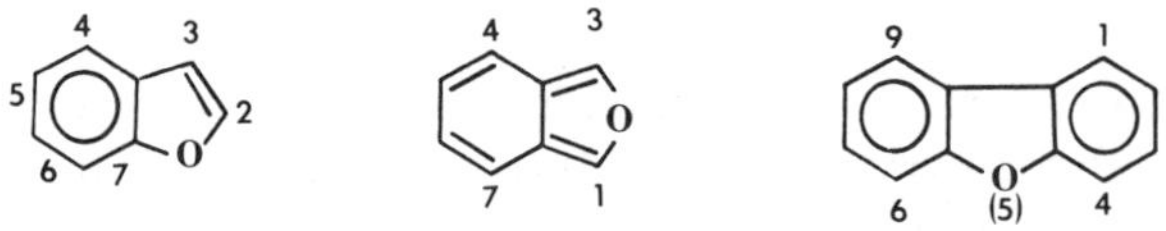

100 101 102

5.2.1. BENZO[*b*]FURANS

Cyclodehydration reactions (**103** → **104**) provide a common route to benzo[*b*]-furans. The cyclodehydration occurs either thermally or chemically and can result in rearrangement products. When aryloxyacetone **105** is treated with standard dehydrating reagents (H_2SO_4, $ZnCl_2$, $POCl_3$, KOH, or PPA), the corresponding 3-alkylbenzofuran **106** is isolated.[61] With the *p*-chloroethers **107** or **108**, rearrangements and cyclization have been observed depending on the conditions of thermolysis or dehydration; thus **108** gives **109** upon treatment with PPA and **110** upon pyrolysis.[62]

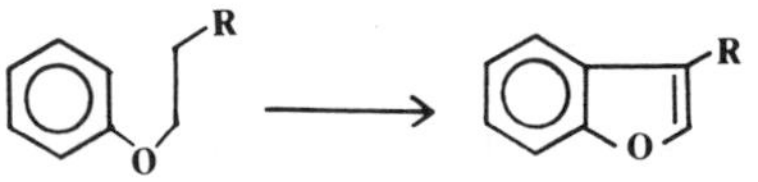

103 104

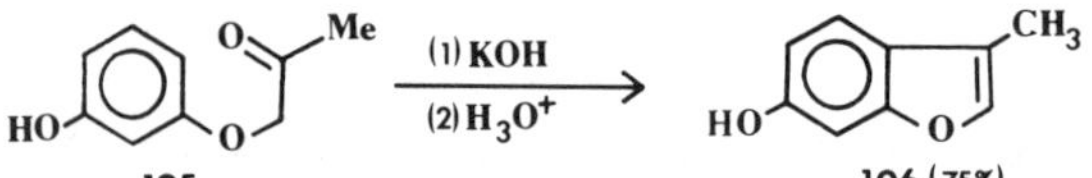

105 106 (75%)

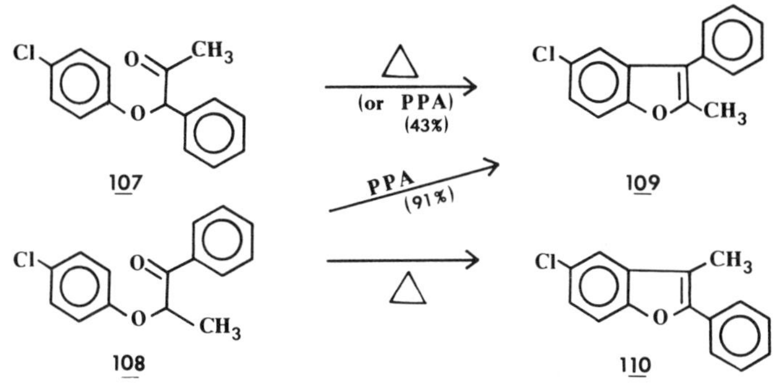

The key step in the total synthesis of secofuranoeremophilane (**111**), isolated from the aerial parts of the South African composite *Euryops hebecarpus* (DC) B. Nord, utilizes titanium trichloride in ethanol to cyclize ether **112** to benzofuran **113.** The remaining steps in the transformation are shown in Scheme 5.9.[63]

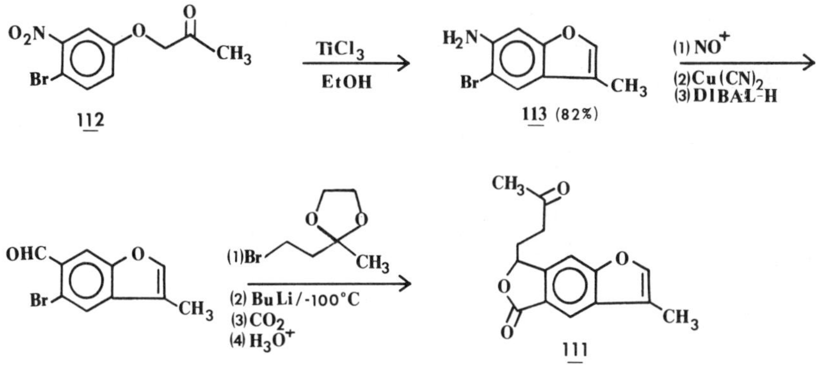

Scheme 5.9

Thermal ring closure with subsequent cyclodehydrogenation of ortho-substituted phenols is known as the *Hansch reaction* and can best be demonstrated by the acid-catalyzed cyclization of **114.** The key aldehyde **114** is conveniently obtained by ozonolysis of allylphenol (via a Claisen rearrangement of the allyloxy compounds).[64] Numerous modifications of this general procedure, which start with *o*-hydroxybenzalkyl or aryl ketones, are known. One such facile route to these aldehydes and ketones is by the regiospecific $S_{RN}1$ photo-stimulated reaction of 2-iodoanisole with enolates in liquid ammonia.[65] The ketonic ether **115** is generated in quantitative yield when R = *t*-butyl (ca. 60% for other alkyl groups); demethylation[66a] with iodotrimethylsilane[66b] affords phenol intermediate **116,** which subsequently cyclizes (100°C) to give the desired benzofuran **117.**

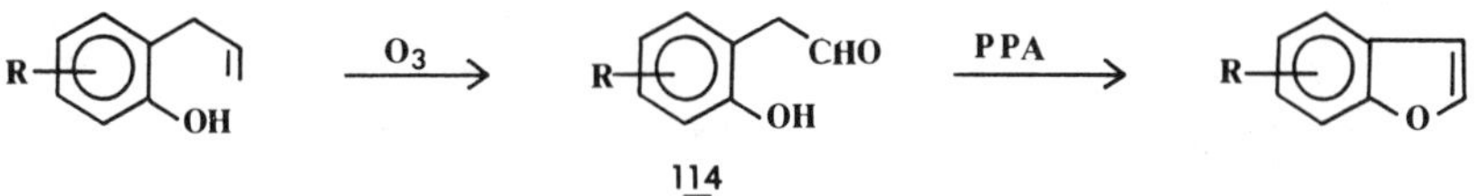

R OH O_3 → R CHO OH PPA → R O

114

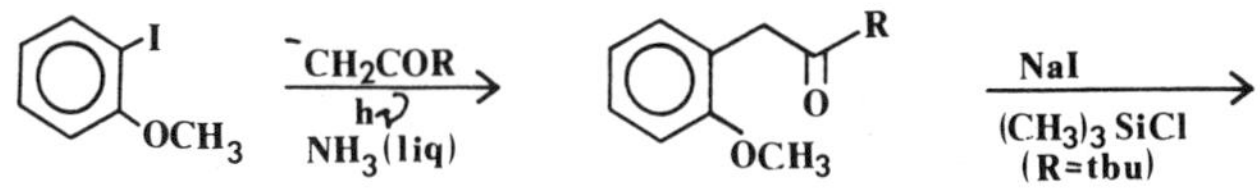

I OCH_3 $^{-}CH_2COR$ hν NH_3(liq) → R O OCH_3 NaI $(CH_3)_3SiCl$ (R=tbu) →

115

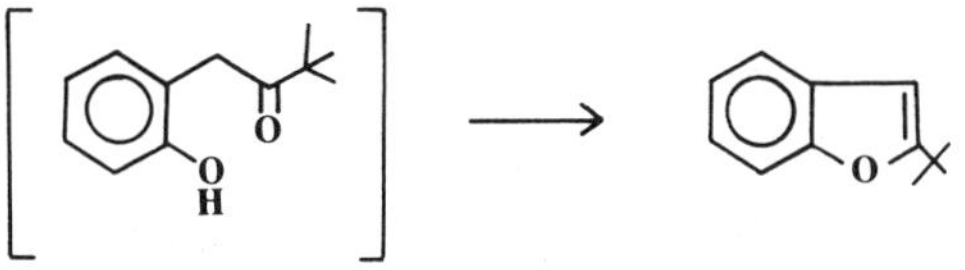

O OH H → O

116 117(100%)

Cyclization of *o*-iodophenols with unsaturated side chains is demonstrated by a reaction of **118** with base which results in a nucleophilic addition to the acetylenic moiety. Since copper reagents offer easy access to arylacetylenes, this provides a simple, one-step source of benzofurans (for example, **119**).[67]

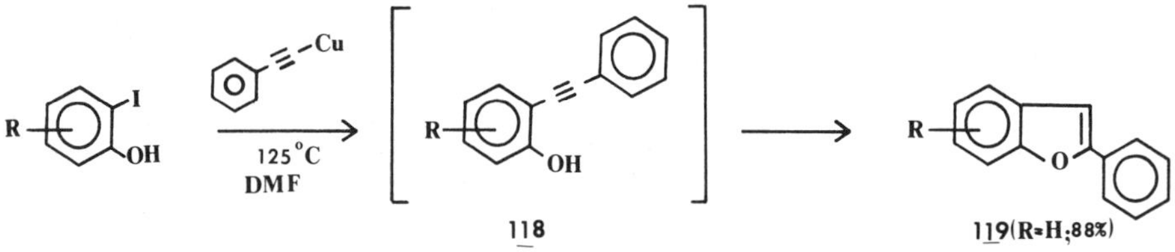

R I OH Cu 125°C DMF → R OH → R O

118 119(R=H;88%)

The cyclization of 2-(but-2′-enyl)phenol in the presence of (+)-[(3,2,10-η-pinine)PdOAc)]$_2$, cupric acetate, and oxygen gives an optically active dihydrobenzofuran (**120**) and 2-ethylbenzofuran (**121**).[68] It has been proposed that the cyclization reaction occurs by an intramolecular oxypalladation and loss of acetic acid, followed by Pd-H elimination. Benzo- and naphthofuran derivatives are also acquired (20–50%) by use of the $PdCl_2 \cdot 2C_6H_5CN$ complex.[69]

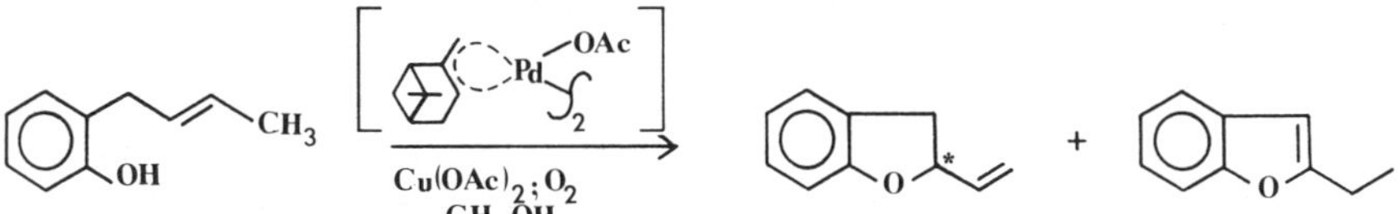

OH CH_3 Pd OAc 2 $Cu(OAc)_2$; O_2 CH_3OH → O * (S) - (+) + O

120 121

5.2.2. BENZO[*c*]FURANS

Because of the reactivity of benzo[*c*]furan (**101**) toward dienophiles, the most common route to this compound is via a *retro Diels–Alder* reaction. Treatment of naphthalene oxide (**122**) with **123** at 120°C *in vacuo* affords adduct **124**, which thermally fragments to give **101** and 3,6-di(2′-pyridyl)pyrazine.[70]

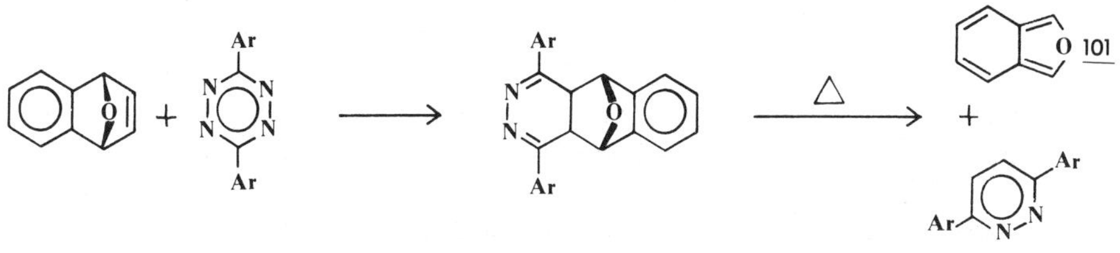

122 + 123 (Ar=2-pyr) → 124 (65%) —Δ→ 101 + Ar, Ar-pyridazine

Based on the synthesis of isoindole, the 1,2-dibenzoylcyclohexa-1,4-diene intermediate (**85**) can be dehydrated by treatment with *p*-toluenesulfonic acid to give the more stable 1,3-diarylisobenzofuran (**125**).[53]

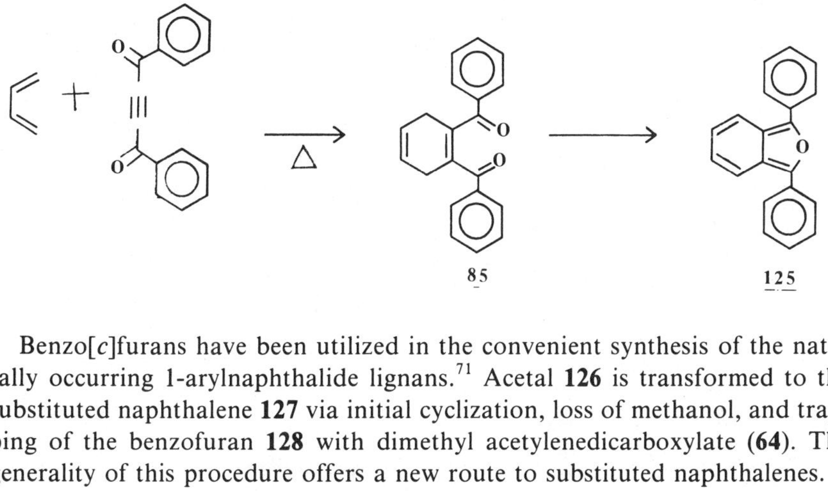

—Δ→ 85 → 125

Benzo[*c*]furans have been utilized in the convenient synthesis of the naturally occurring 1-arylnaphthalide lignans.[71] Acetal **126** is transformed to the substituted naphthalene **127** via initial cyclization, loss of methanol, and trapping of the benzofuran **128** with dimethyl acetylenedicarboxylate (**64**). The generality of this procedure offers a new route to substituted naphthalenes.

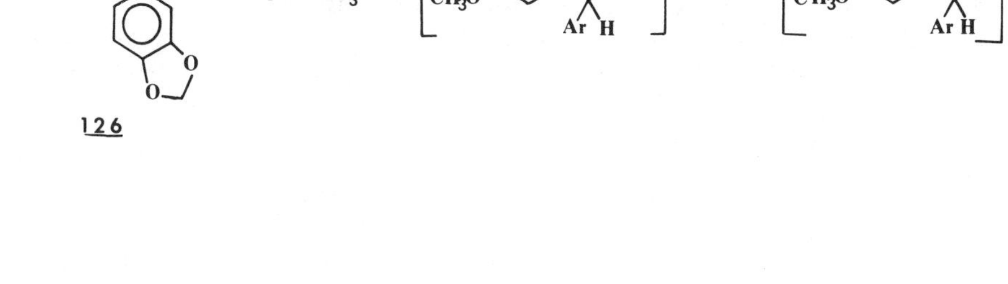

CH_3O, CH_3O, $CH(OCH_3)_2$, H, OH — C_6H_6 / Δ, *p*-tol SO_3H → [CH_3O, CH_3O, OCH_3, O, Ar H] → [CH_3O, CH_3O, +O, Ar H]

126

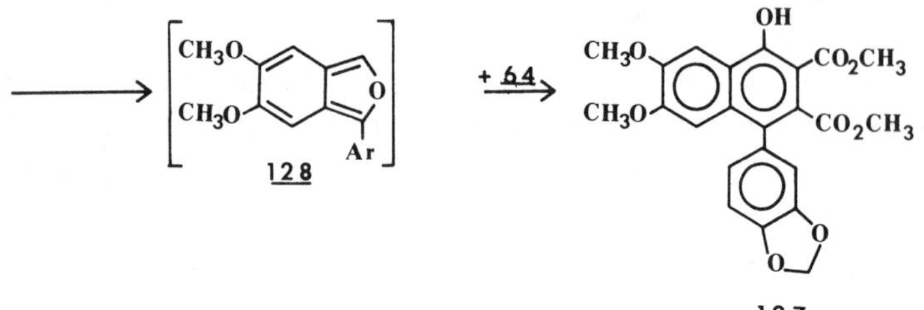

Isobenzo[*c*]furan is converted into isobenzo[*c*]indole by a novel, one-step replacement of the oxygen atom with nitrogen.[53] This interesting reaction occurs when **125** is heated with *N*-sulfinylaniline (**129**) in the presence of boron trifluoride-etherate. The initial Diels–Alder adduct rearranges to the sultam **130,** followed by extrusion of sulfur dioxide (a common reaction of unsaturated δ-sultams) to give **131.**

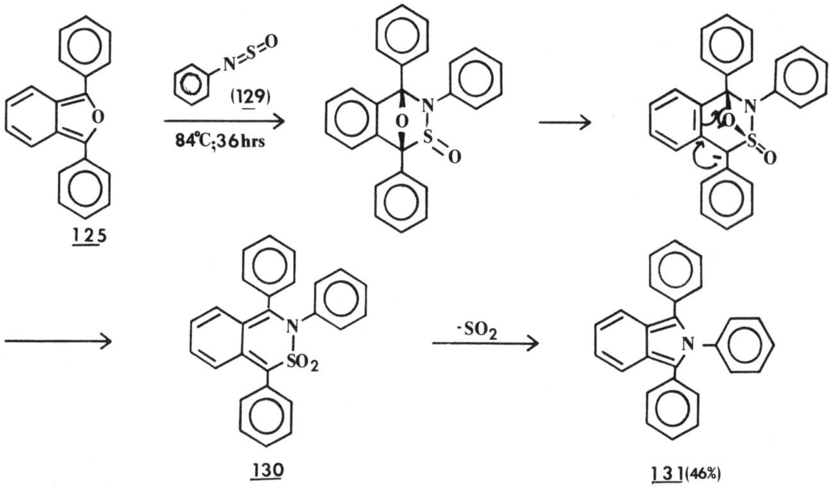

5.2.3. DIBENZOFURANS

Although very little research has been directed toward the preparation of dibenzofuran (**102**), the recent discovery that chlorinated dibenzofurans are toxic atmospheric contaminants[72] has generated new interest in their syntheses. Their availability is, however, still limited. The photocyclization of substituted diphenylethers **132** gives direct access to substituted dibenzofurans.[73] Similarly, stoichiometric coupling of **132** with $Pd(OAc)_2$ in acetic acid gives the benzofurans in greater than 65% yield.[74] Cycloaddition reactions of vinyl benzo[*b*]furans (**133**) with dienophiles **64,**[75] and conversely of 2,3-dehydrobenzofuran (**134**) with dienes,[76] offer access to the unsymmetrically substituted dibenzofurans.

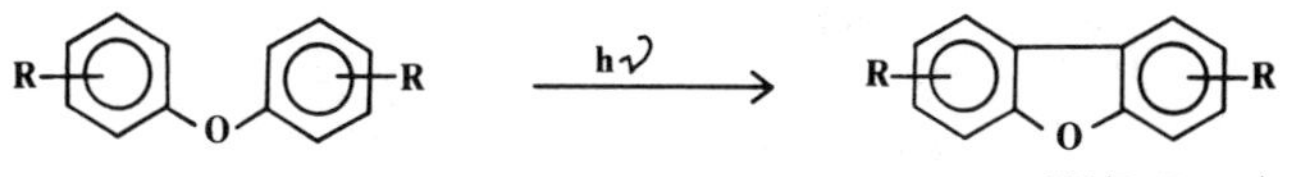

132 → (hν) 102 (R=H; 60%)

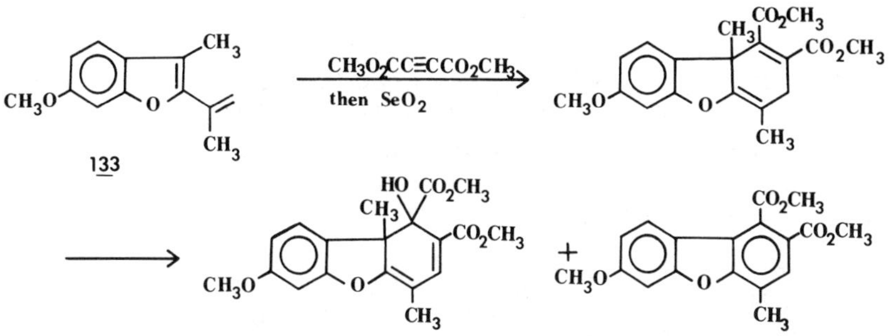

133 → ($CH_3O_2CC{\equiv}CCO_2CH_3$, then SeO_2)

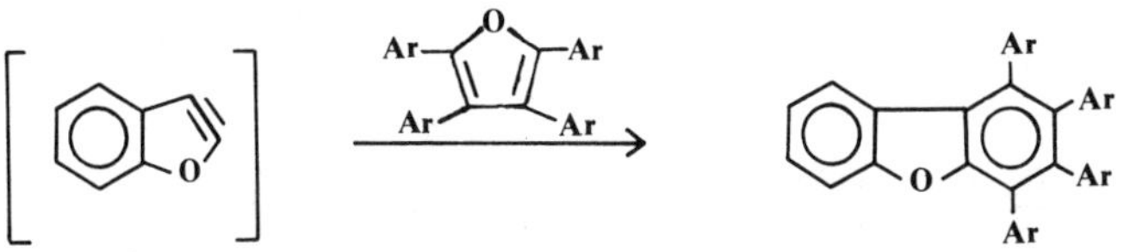

134

5.3. BENZOTHIOPHENES

The most common benzo-fused thiophene, benzo[*b*]thiophene (**135**), has recently been reviewed.[77] Though both benzo[*c*]thiophene (**136**) and dibenzothiophene (**137**) are known, they have been studied only to a limited extent.

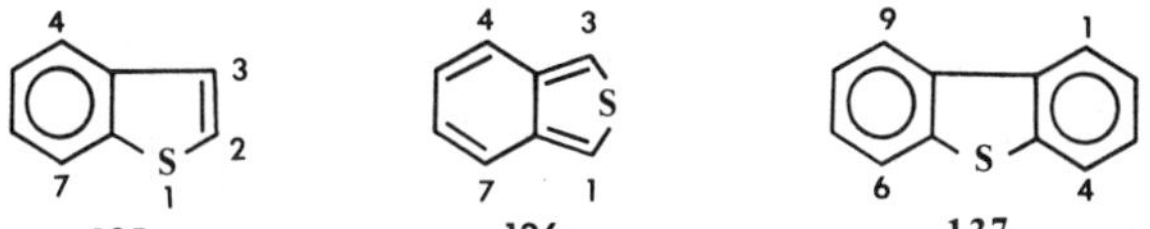

135 136 137

5.3.1. BENZO[*b*]THIOPHENES

Although benzo[*b*]thiophene can be synthesized by treatment of thiophenol or thioanisole with alkynyl or alkenylbenzenes, in the presence of hydrogen sulfide and a heated catalyst, these methods are generally not applicable to laboratory syntheses. Most procedures available for the preparation of benzo[*b*]furans can be applied to the syntheses of benzo[*b*]thiophenes. Thus (arylthio)acetaldehyde acetals **138** are readily cyclized in the presence of PPA to give substituted benzo[*b*]thiophenes.[78] (Arylthio)acetones, arylphenacylsulfides, and *S*-arylthioglycolic acids react similarly.

138 (80.5%)

The Krollpfeiffer synthesis of benzo[*b*]thiophenes refers to the cyclization and dealkylation of (*o*-acylphenyl)dialkylsulfonium salts. Treatment of methyl *p*-tolylsulfide (**139**) with acetyl chloride under Friedel–Crafts conditions gives **140.** This compound is *S*-methylated, cyclized in the presence of sodium hydroxide, and finally aromatized by the use of hydrogen bromide to afford **141.**[79]

139 140

141 (ca 90%)

The parent **135** can be synthesized by the direct pyrolysis of thiophene via the thiophyne intermediate **142.**[80] Although several mechanisms have been proposed, the most reasonable one seems to involve thiophyne (**142**), as an intermediate, which undergoes a [4 + 2] cycloaddition, followed by cheleotropic loss of sulfur.

142 135

5.3.2. BENZO[*c*]THIOPHENES

Numerous one-step reactions have been used to prepare **136** from an appropriate saturated derivative. For example, vapor-phase catalytic dehydrogenation of 1,3-dihydrobenzo[*c*]thiophene (**143**),[81] decarboxylation of **144** with copper in quinoline,[82] and dehydration of 2-*S*-oxide **145**[83] have been used to prepare the parent compound. The dehydration of *S*-oxide **145** in acetic anhydride is proposed[84] to proceed via a *Pummerer rearrangement* to give an α-acetoxysulfide **146,** which subsequently loses acetic acid to form benzo[*c*]thiophene. Thermolysis of 1,3-dihydronaphtho[2,3-*c*]thiophene 2-*S*-oxide at 800°C in a quartz tube gives the related naphtho[2,3-*c*]thiophene (**147**) as a bright yellow solid, which is more reactive than **136.** This increased reactivity is probably due to the extended quinoid structure.[85]

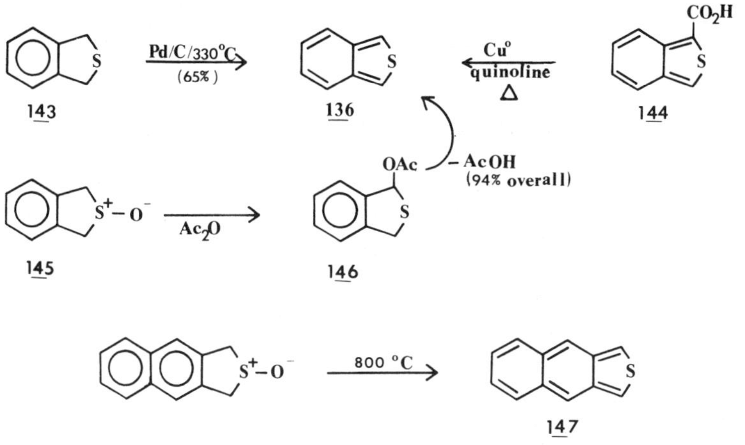

In a procedure similar to that used in the preparation of *N*-methylisoindole (**70** → **71**), 1,3-dihydrobenzo[*c*]thiophene methylsulfonium iodide (**148**) is treated with excess phenyllithium; however, **136** is *not* formed, but rather, **149** and **150** are isolated. These products probably arise via ring cleavage of the five-membered ring to give the transient *o*-quinodimethane **151,** which dimerizes to the isolated products.[86]

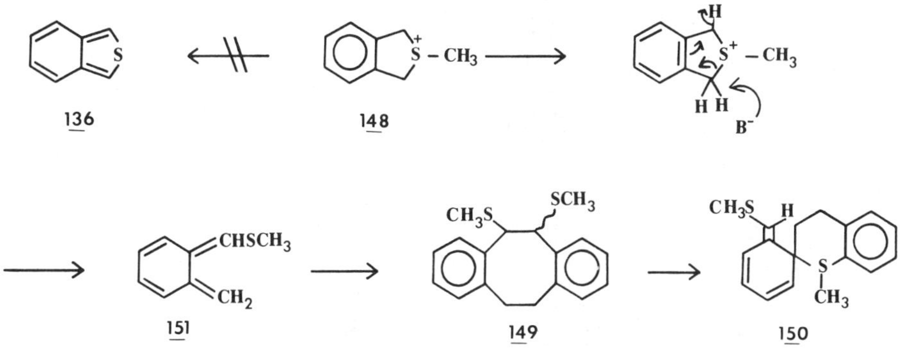

The general methods previously described for the syntheses of substituted benzo[*c*]indoles and benzo[*c*]furans are also applicable to the preparation of substituted benzo[*c*]thiophenes. Thus **85,** when treated with P_2S_5 in refluxing toluene, gives **152.**[87]

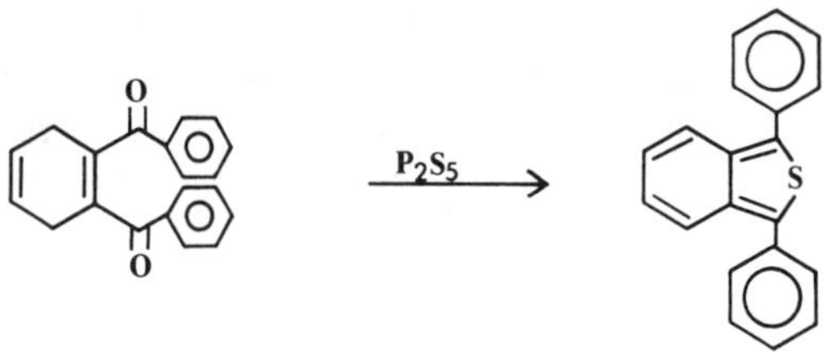

P_2S_5

85 152 (90%)

5.3.3. DIBENZOTHIOPHENES

Stenhouse first prepared **137** in 1870 by the dehydrogenation-cyclization of diphenyl sulfide (**153**).[88] The structure of **137** was elucidated by Graebe,[89] an incredible 10 years before the discovery of thiophene! In a related vein, biphenyl (**154**) in the presence of sulfur and aluminum chloride gives **137.** Numerous modifications have been devised to afford similar cyclized products.[90]

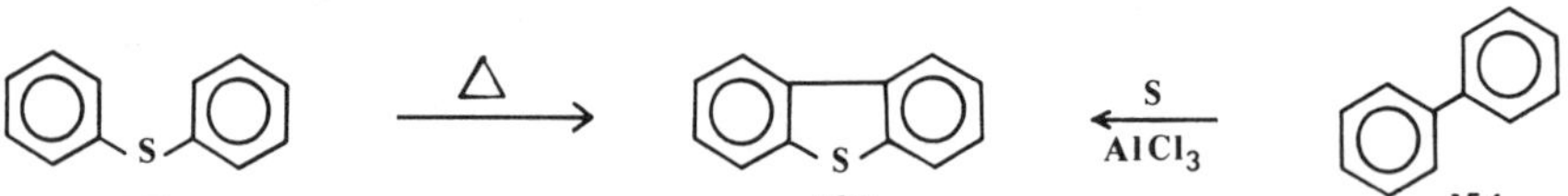

Δ S, $AlCl_3$

153 137 154

5.4. BENZOSELENOPHENES

The general procedures used to prepare the benzofurans and benzothiophenes have been applied to the synthesis of the benzo[*b*]- and dibenzoselenophenes. For example, **155** upon treatment with P_2O_5 gives **156,**[91] **157** gives **158** when subjected to potassium acetate in acetic anhydride followed by hydrolysis,[92] and 2-acetylphenylselenacetic acid (**159**) cyclizes under basic conditions to give **160.**[98] Decarboxylation of **160** with copper bronze in refluxing quinoline affords **161.**

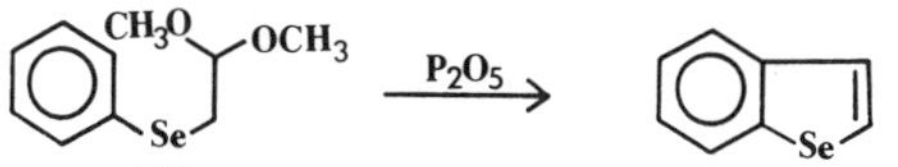

P_2O_5

155 156

$-CO_2$ Na/Hg, EtOH

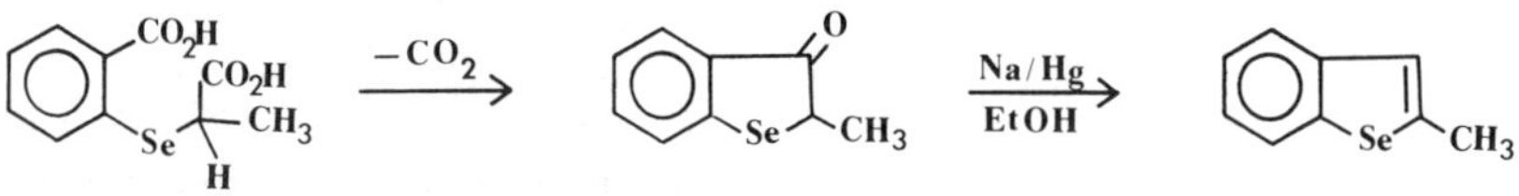

157 158

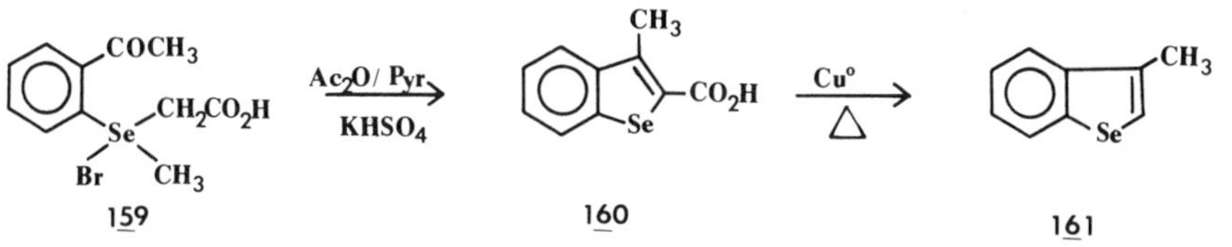

Thermolysis of selenanthrene (**162**) over copper bronze gives dibenzoselenophene (**163**).[94] The diazonium salt of **164,** on decomposition, undergoes a coupling reaction to afford **163**[94] in low yield. Other novel cyclizations have also been shown to give **163,**[95] as for example, the ring expansion of **165** with selenium.[96]

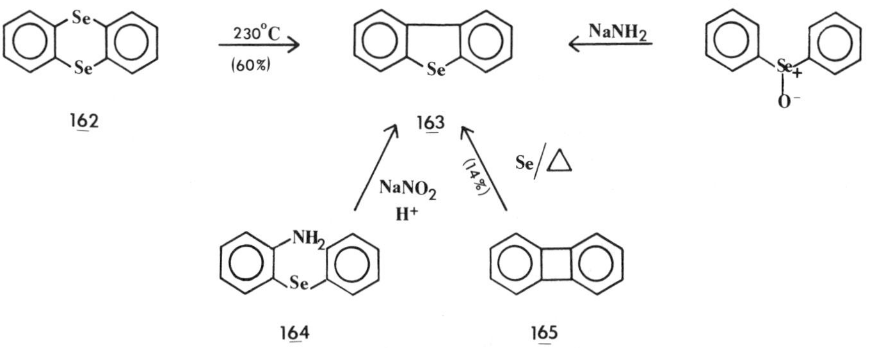

Ortho metalation of phenylacetylene (**166**), initially with butyllithium followed by potassium *tert*-butoxide, gives the bis-organometallic **167,** which with powdered selenium at −20 to −5°C in the presence of hexamethylphosphoric triamide affords benzo[*b*]selenophene.[97]

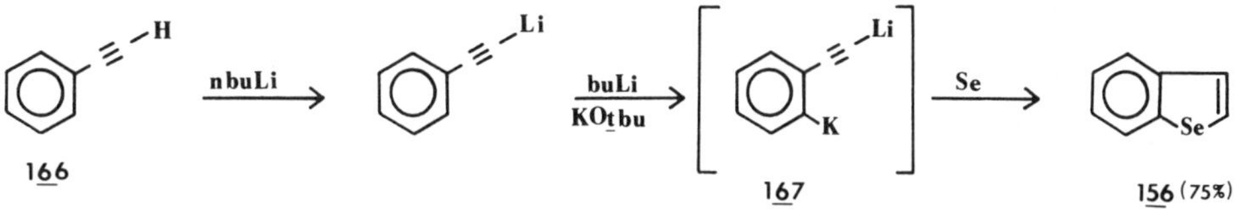

5.5. BENZOTELLUROPHENES

The substituted acetophenone **168** is tellurium-alkylated with bromoacetic acid, thermally dealkylated, and cyclized via a base-catalyzed condensation. Benzo[*b*]tellurophenes are easily obtained by application of either the Wittig or Wittig–Horner reaction on telluroindoxyl (**169**).[98] Numerous substituted benzotellurophenes have been reported.[98]

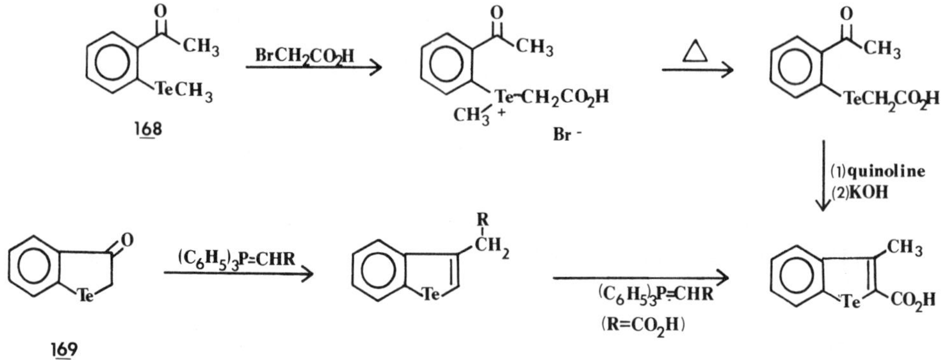

5.6. REACTIONS OF INDOLES

5.6.1. GENERAL COMMENTS

The reactivity of pyrrole would certainly be expected to be modified when fused to the "π-neutral" benzene ring (to form indole) or, more drastically so, when fused to a π-deficient ring, such as pyridine. Clearly, the presence of the benzo group in indole (**1**) deactivates the C-2 position toward electrophilic attack in comparison with the C-2 in pyrrole [in order to consider a resonance structure placing a negative charge on C-2, the benzenoid aromaticity would be interrupted (**1c**).[99]

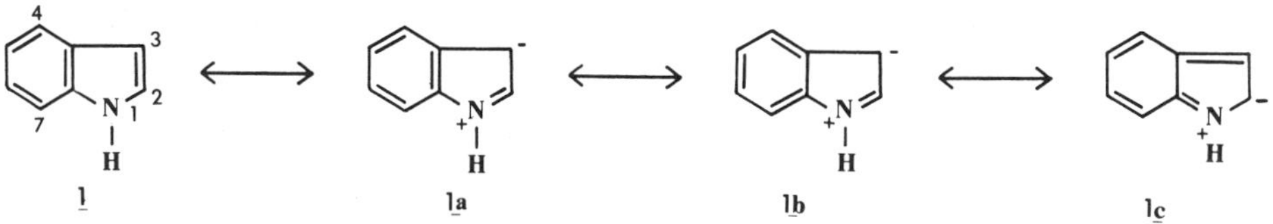

5.6.2. ELECTROPHILIC SUBSTITUTIONS

The decreased reactivity of indole (**1**) with respect to pyrrole is nicely exemplified by H ⇌ D exchange reactions. As may be recalled, pyrrole decomposes in acidic media; although normal nitration media cause decomposition of indole, it can be nitrated with benzoyl nitrate to yield the 3-nitro derivative **170.**[100,101]

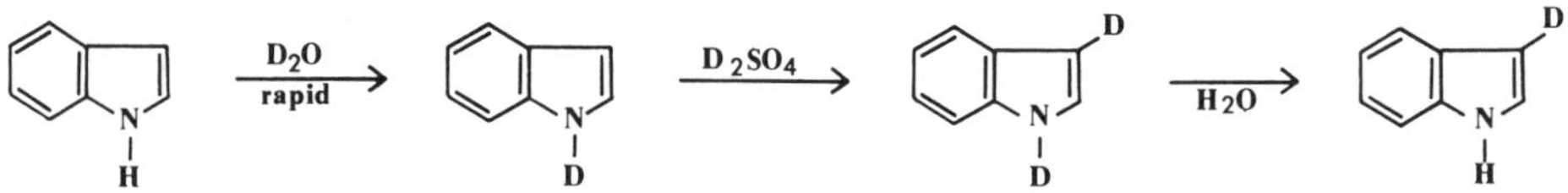

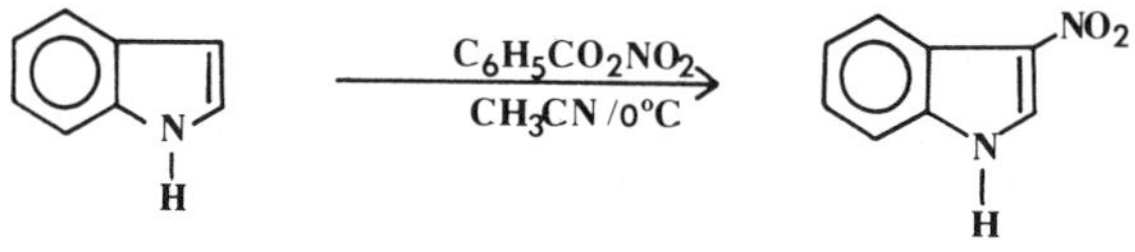

$C_6H_5CO_2NO_2$ / CH_3CN /0°C

NO_2

170 (35%)

When the acid stability of the indole nucleus is increased by introduction of a 2-methyl group (**21**), along with **171**, several minor nitro derivatives with the substituent on the benzo ring (**172–174**) can be isolated.[102]

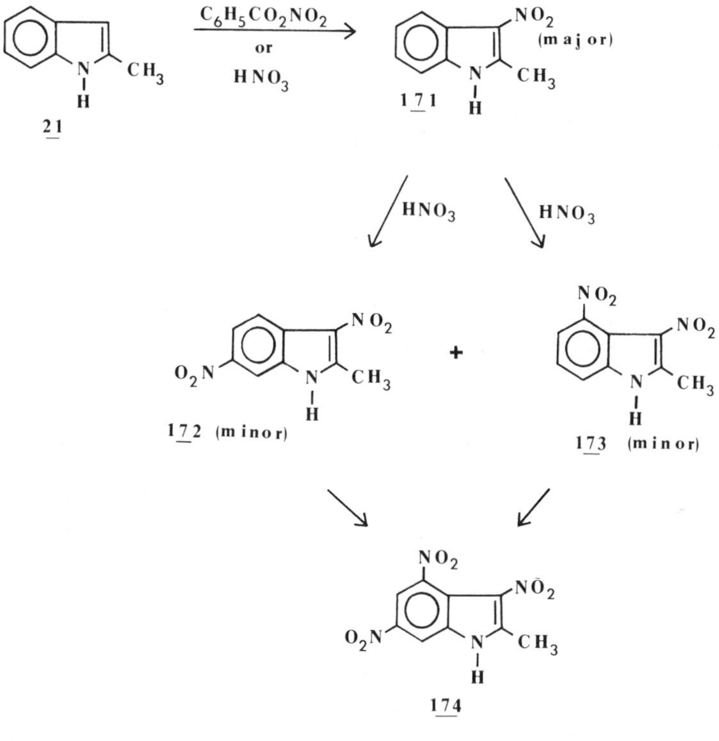

21

$C_6H_5CO_2NO_2$ or HNO_3

171 (major)

HNO_3 HNO_3

172 (minor) + **173** (minor)

174

The sites of nitration, once C-3 has been nitrated, reflect the involvement of intermediates **175** and **176.** Although nitrosation of indole forms rather complex dimeric mixtures, the less sensitive (toward acids) 2-methylindole (**21**) can be nitrosated under acidic conditions. The resulting nitroso derivative **177** amazingly exists largely in the oximino form **178.** (This compound, as well as the parent 3-nitroso derivative, can also be obtained under basic conditions.[102,103]) The fact that 3-methylindole (**61**) when nitrosated affords an *N*-nitroso derivative (**179**) also lends some credence to the idea that **178** is formed via its *N*-nitroso derivative as well.[102,104]

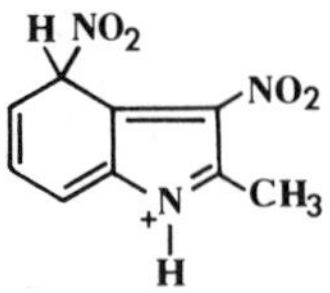

175 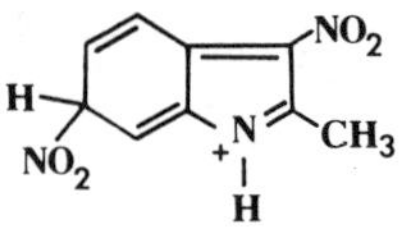 176

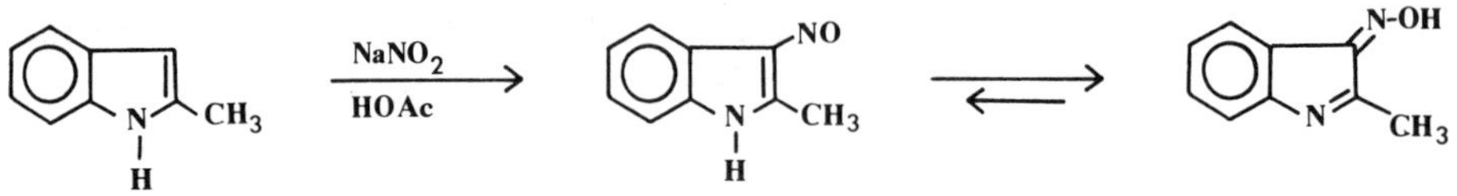

$NaNO_2$ / HOAc

21 177 178

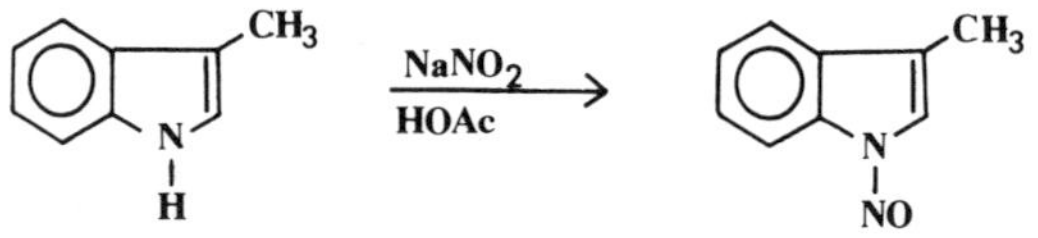

$NaNO_2$ / HOAc

61 179

When the 3-position is *not* blocked, other electrophilic substitution reactions, such as coupling with diazonium salts, halogenation, and sulfonation, occur at that site.[104–108]

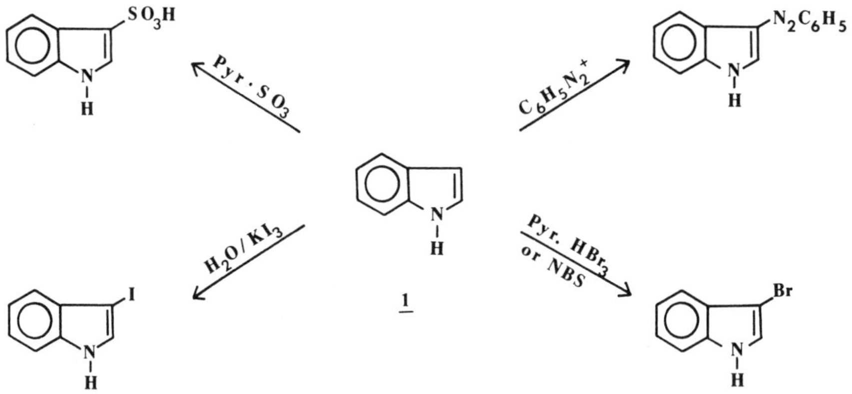

Pyr · SO_3 $C_6H_5N_2^+$ H_2O/KI_3 Pyr. HBr_3 or NBS

1

An intriguing set of reactions is exemplified by the hydrolysis of 2- and 3-haloindoles (**180** and **181,** respectively), both of which afford α-oxoindoles (**182**).[109] The proposed transformations are as follows:

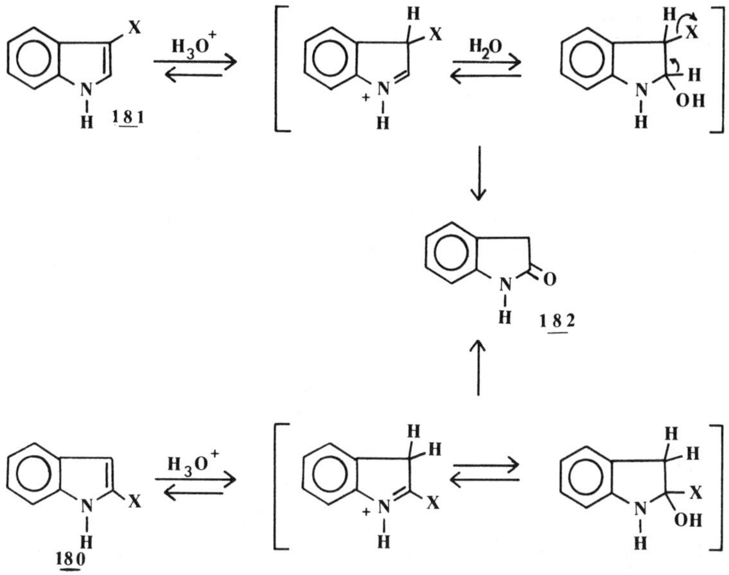

Forced alkylation of indoles with alkyl halides (above 100°C) follows the same pattern as has already been described for pyrrole itself (for example, formation of **183**).[110]

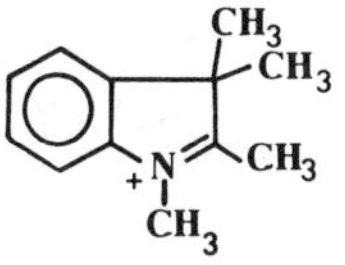

The Mannich reaction is a frequently employed route in indole chemistry, resulting in the formation of *β*-alkylaminoindoles (**184**), although *N*-alkylamino intermediates (**185**) can occasionally be isolated.[111] The formation of 3-formylindole (**186**) by the Vilsmeier–Haack procedure is yet another example of an efficient electrophilic substitution in indole chemistry.[112]

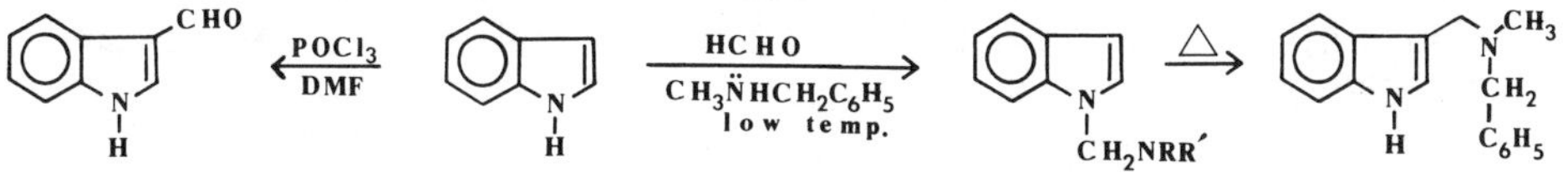

186 (95.5%) ← $POCl_3$, DMF — 1 — HCHO, $CH_3\ddot{N}HCH_2C_6H_5$, low temp. → 185 (CH_2NRR') —△→ 184 (90%)

5.6.3. NUCLEOPHILIC SUBSTITUTIONS AND DEPROTONATIONS

As is true for pyrrole, indoles form salts by replacement of the hydrogen bonded to nitrogen. Reagents such as sodium hydride, alkyl Grignards, and potassium *tert*-butoxide readily accomplish this task. The resulting anion is a resonance-stabilized structure such as **187.** Depending on the nature of the metal ion, these anions afford products from attack either at nitrogen (e.g., **188**) or carbon (e.g., **189**).[113,114]

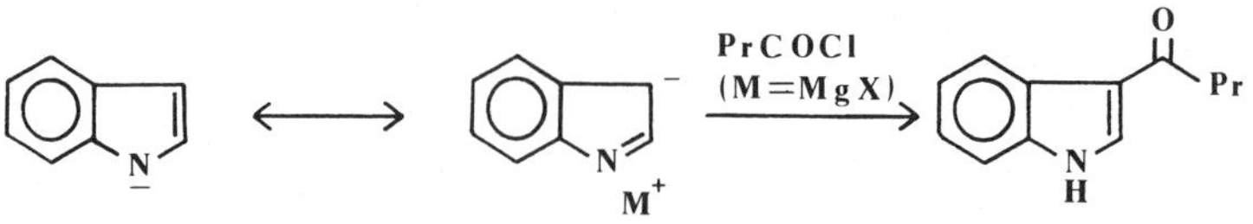
PrCOCl
(M=MgX)
O
Pr
N
M+
N
H

187a 187b 189

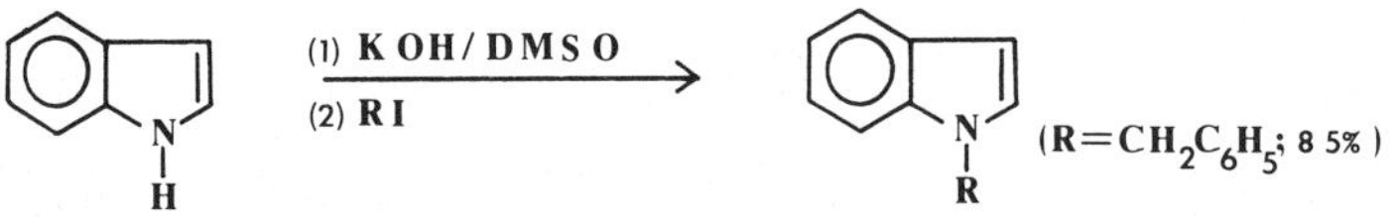
(1) KOH/DMSO
(2) RI
N
H
N
R
(R=CH2C6H5; 85%)

188

N-Alkylindoles, when deprotonation is no longer competitive, react with butyllithium to yield the 2-lithio derivative **190.**[115]

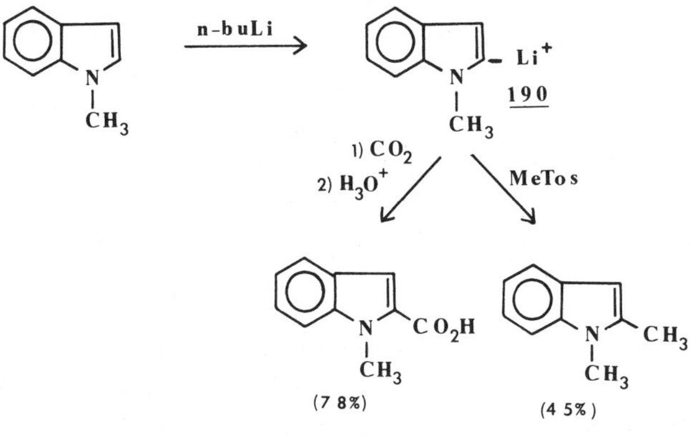
n-buLi
N
CH3
Li+
190
1) CO2
2) H3O+
MeTos
CO2H
(78%)
CH3
(45%)

5.6.4. FREE RADICAL REACTIONS

As is true for pyrrole, very little is known about the free radical reactions on indole. Attack of benzyl radicals on **1** generates a mixture of 1-, 3-, 1,3-, and 2,3-substituted indoles.[116]

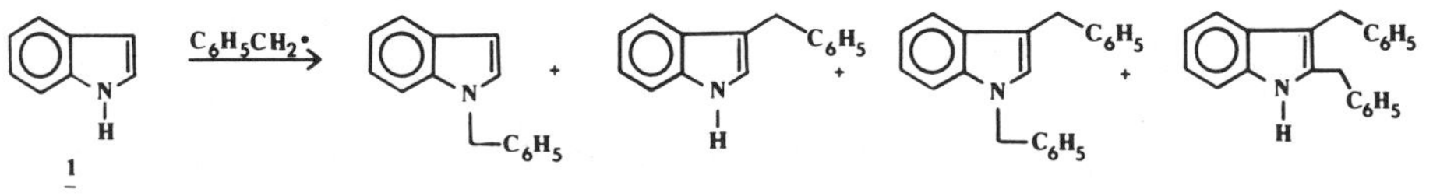
C6H5CH2•
N
H
1
C6H5
+

Interestingly, homolytic attack by ·OH on the indole nucleus affords mainly 4-hydroxyindole (**191**).[117] When *N*-methylindole reacts with benzoyl peroxide, the 3-substituted product **192** is obtained. However, when the 3-position is blocked, as is the case in 1,3-dimethylindole (**193**), the 2-substituted derivative is formed.[118]

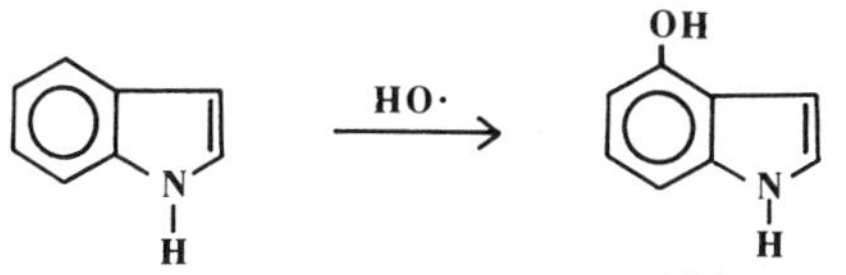

REFERENCES

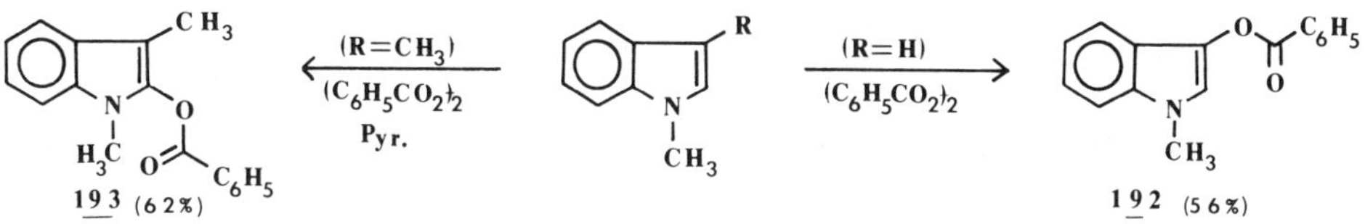

1. B. Robinson, *Chem. Rev.* **63,** 373 (1963); **69,** 227 (1969).
2. N. Campbell and B. M. Barclay, *Chem. Rev.* **40,** 359 (1947).
3. (a) A. E. Arbusow and W. M. Tichwinsky, *Chem. Ber.* **43,** 2301 (1910); (b) G. Baccolini and P. E. Todeseo, *J. Chem. Soc. Chem. Commun.* **1981,** 563.
4. W. W. Hartman and L. J. Roll, *Org. Syntheses, Coll. Vol.* **2,** 418 (1943).
5. G. E. Ficken and J. D. Kendall, *J. Chem. Soc.* **1959,** 3202; **1961,** 584.
6. D. P. Ainsworth and H. Suschitzky, *J. Chem. Soc. (C)* **1967,** 315.
7. F. R. Japp and F. Klingemann, *Chem. Ber.* **21,** 549 (1888); *Ann. Chem.* **247,** 190 (1888).
8. B. Heath-Brown and P. G. Philpott, *J. Chem. Soc.* **1965,** 7185.
9. R. R. Phillips, *Org. Reactions* **10,** 143 (1959).
10. K. H. Pausacker, *J. Chem. Soc.* **1950,** 621.
11. N. P. Buu-Hoi, P. Jacquignon, and O. Perin-Roussel, *Bull. Soc. Chim. Fr.* **32,** 2849 (1965).
12. E. Fischer and T. Schmitt, *Chem. Ber.* **21,** 1071, 1811 (1888).
13. (a) P. L. Julian, E. W. Meyer, and H. C. Printy, in *Heterocyclic Compounds,* Vol. 3, R. C. Elderfield, Ed., Wiley, New York, 1952, Chap. 1. (b) W. C. Sumpter and F. M. Miller, *Heterocyclic Compounds with Indole and Carbazole Systems,* A. Weissberger, Ed., Interscience, New York, 1954. (c) R. K. Brown, in *Heterocyclic Compounds,* Vol. 25, W. J. Houlinhan, Ed., Wiley-Interscience, New York, 1972, pp. 317–385.
14. H.-J. Opgenorth and H. Scheuermann, *Ann. Chem.* **1979,** 1503; also see J. E. Norlander, D. B. Catalane, K. D. Kotian, R. M. Stevens, and J. E. Haky, *J. Org. Chem.* **46,** 778 (1981).
15. W. Madelung, *Chem. Ber.* **45,** 1128 (1912).
16. M. A. Verlay, *Bull. Soc. Chim. Fr.* **35,** 1039 (1924).
17. M. Le Corre, A. Hercouet, and H. Le Baron, *J. Chem. Soc. Chem. Commun.* **1981,** 14.
18. H. N. Rydon and J. C. Tweddle, *J. Chem. Soc.* **1955,** 3499.
19. R. Beugelmans and G. Roussi, *J. Chem. Soc. Chem. Commun.* **1979,** 950; R. R. Bard and J. F. Bunnett, *J. Org. Chem.* **45,** 1547 (1980).
20. J. F. Wolfe, M. C. Sleevi, and R. R. Goehring, *J. Am. Chem. Soc.* **102,** 3646 (1980).

21. C. D. Nenitzescu, *Bull. Soc. Chim. Romania* **11,** 37 (1929); also see V. Aggarwal, A. Kumar, H. Ila, and H. Junjappa, *Synthesis* **1981,** 157.
22. R. A. Heacock, *Chem. Rev.* **59,** 181 (1959).
23. H. Burton, *J. Chem. Soc.* **1932,** 546.
24. L. S. Hegedus, G. F. Allen, and D. J. Olsen, *J. Am. Chem. Soc.* **102,** 3583 (1980); L. S. Hegedus, G. F. Allen, and E. L. Waterman, *J. Am. Chem. Soc.* **98,** 2674 (1976).
25. M. Mori and Y. Ban, *Tetrahedron Lett.* **1976,** 1803.
26. B. Bogdanovic, P. Heimback, M. Kröner, and G. Wilke, *Ann. Chem.* **727,** 143 (1969); A. S. Kende, L. S. Liebeskind, and D. M. Braitsch, *Tetrahedron Lett.* **1975,** 3375.
27. H. Alper and J. E. Prickett, *Tetrahedron Lett.* **1976,** 2589.
28. H. Alper and J. E. Prickett, *J. Chem. Soc. Chem. Commun.* **1976,** 483.
29. K. Isomura, K. Uto, and H. Taniguchi, *J. Chem. Soc. Chem. Commun.* **1977,** 664.
30. K. Isomura, S. Kobayashi, and H. Taniguchi, *Tetrahedron Lett.* **1968,** 349.
31. L. A. Wendling and R. G. Bergman, *J. Org. Chem.* **41,** 831 (1976).
32. J. Bakke, H. Heikman, and E. B. Hellgren, *Acta Chem. Scand.* **B28,** 393 (1974); J. Bakke, *Acta Chem. Scand.* **B28,** 134 (1974).
33. C. Nenitzescu, *Chem. Ber.* **58B,** 1063 (1925); also see E. Ucciana and A. Bonfand, *J. Chem. Soc. Chem. Commun.* **1981,** 82.
34. R. A. Weerman, *Rec. Trav. Chim.* **29,** 18 (1910).
35. H. R. Snyder, E. P. Merica, C. G. Force, and E. G. White, *J. Am. Chem. Soc.* **80,** 4622 (1958).
36. Y. Ito, K. Kobayashi, N. Seko, and T. Saegusa, *Chem. Lett.* **1979,** 1273.
37. P. G. Gassman, D. P. Gilbert, and T. J. van Bergen, *J. Chem. Soc. Chem. Commun.* **1974,** 201.
38. P. Bravo, G. Gaudiano, and A. Umani-Ronchi, *Tetrahedron Lett.* **1969,** 679.
39. R. M. Acheson, A. R. Hands, and J. M. Vernon, *Proc. Chem. Soc.* **1961,** 164; R. M. Acheson and J. M. Vernon, *J. Chem. Soc.* **1962,** 1148; also see W. E. Noland and C. K. Lee, *J. Org. Chem.* **45,** 4573 (1980).
40. W. E. Noland, W. C. Kuryla, and R. F. Lange, *J. Am. Chem. Soc.* **81,** 6010 (1959).
41. J. E. Bornstein, D. E. Remy, and J. E. Shields, *J. Chem. Soc. Chem. Commun.* **1972,** 1149.
42. R. Bonnett and R. F. C. Brown, *J. Chem. Soc. Chem. Commun.* **1972,** 393; R. Bonnett, R. F. C. Brown, and R. G. Smith, *J. Chem. Soc. Perkin I* **1973,** 1432.
43. R. Kreher and J. Seubert, *Z. Naturforsch.* **20b,** 75 (1966).
44. G. Wittig, H. Tenhaeff, W. Schoch, and G. Koenig, *Ann. Chem.* **572,** 1 (1951).
45. S. H. Pine, *Org. Reactions* **18,** 403 (1970).
46. R. Kreher and J. Seubert, *Angew. Chem. Int. Ed.* **3,** 639 (1964); **5,** 967 (1966).
47. G. W. Fenton and C. K. Ingold, *J. Chem. Soc.* **1928,** 3295.
48. J. C. Emmett, D. F. Veber, and W. Lwowski, *Chem. Commun.* **1965,** 272.
49. G. Wittig, G. Closs, and F. Mindermann, *Ann. Chem.* **594,** 89 (1955).
50. W. Theilacker and H. Kalenda, *Ann. Chem.* **584,** 87 (1953).
51. G. A. Flynn, *J. Chem. Soc. Chem. Commun.* **1980,** 862.
52. J. C. Emmett and W. Lwowski, *Tetrahedron* **22,** 1011 (1966).
53. M. P. Cava and R. H. Schlessinger, *J. Org. Chem.* **28,** 2464 (1963).
54. A. Crabtree, A. W. Johnson, and J. C. Tebby, *J. Chem. Soc.* **1961,** 3497; O. Diels and R. Meyer, *Ann. Chem.* **513,** 129 (1934).
55. D. R. Bragg and D. G. Wibberley, *J. Chem. Soc.* **1963,** 3277.
56. V. Boekelheide and W. Feely, *J. Org. Chem.* **22,** 589 (1957).

57. W. Flitsch and E. Gerstmann, *Chem. Ber.* **102,** 1309 (1969).

58. T. Uchida and K. Matsumoto, *Synthesis* **1976,** 209.

59. R. Fittig and G. Ebert, *Ann. Chem.* **216,** 162 (1883).

60. (a) A. Mustafa, *Benzofurans,* Vol. 29, Wiley, New York, 1974; (b) P. Cagniant and D. Cagniant, in *Advances in Heterocyclic Chemistry,* Vol. 18, A. R. Katritzky and A. J. Boulton, Eds., Academic Press, New York, 1975, Chap. 6. (c) W. Friedrichsen, in *Advances in Heterocyclic Chemistry,* Vol. 26, A. R. Katritzky and A. J. Boulton, Eds., Academic Press, New York, 1980, pp. 135–241.

61. J. K. MacLeod and B. R. Worth, *Tetrahedron Lett.* **1972,** 237; J. K. MacLeod, B. R. Worth, and R. J. Wells, *Tetrahedron Lett.* **1972,** 241.

62. E. Bisagni and C. Rivalle, *Bull. Soc. Chim. Fr.* **1969,** 2463.

63. F. Bohlmann and G. Fritz, *Tetrahedron Lett.* **1981,** 95.

64. R. Aneja, S. K. Mukerjee, and T. R. Seshadri, *Tetrahedron* **2,** 203 (1958).

65. R. Beugelmans and H. Ginsburg, *J. Chem. Soc. Chem. Commun.* **1980,** 508.

66. (a) G. A. Olah, S. C. Narang, B. G. B. Gupta, and R. Malhotra, *J. Org. Chem.* **44,** 1247 (1979); (b) A. H. Schmidt, *Aldrichemia Acta* **14,** 31, (1981).

67. C. E. Castro and R. D. Stephens, *J. Org. Chem.* **28,** 2163 (1963); R. D. Stephens and C. E. Castro, *J. Org. Chem.* **28,** 3313 (1963); C. E. Castro, F. J. Gaughan, and D. C. Owsley, *J. Org. Chem.* **31,** 4071 (1966).

68. T. Hosokawa, T. Uno, and S.-i. Murahashi, *J. Chem. Soc. Chem. Commun.* **1979,** 475.

69. T. Hosokawa, K. Maeda, K. Koga, and I. Moritani, *Tetrahedron Lett.* **1973,** 739; T. Hosokawa, H. Ohkata, and J. Moritani, *Bull. Chem. Soc. Japan* **48,** 1533 (1975).

70. R. N. Warrener, *J. Am. Chem. Soc.* **93,** 2346 (1971).

71. H. P. Plaumann, J. G. Smith, and R. Rodrigo, *J. Chem. Soc. Commun.* **1980,** 354.

72. G. W. Bowes, M. J. Mulvihill, M. R. DeCamp, and A. S. Kende, *J. Agric. Food Chem.* **23,** 1222 (1957); K. D. Bartle, M. L. Lee, and S. A. Wise, *Chem. Soc. Rev.* **10,** 113 (1981).

73. K.-P. Zeller and H. Peterson, *Synthesis* **1975,** 532.

74. B. Akermark, L. Eberson, E. Jonsson, and E. Pattersson, *J. Org. Chem.* **40,** 1365 (1975).

75. J. A. Elix and D. Tronson, *Aust. J. Chem.* **26,** 1093 (1973).

76. G. Wittig, *Angew. Chem. Int. Ed.* **1,** 415 (1962).

77. B. Iddon and R. M. Scrowston, *Advances in Heterocyclic Chemistry,* Vol. 11, A. R. Katritzky and A. J. Boulton, Eds., Academic Press, 1970, pp. 177–381; B. Iddon, in *New Trends in Heterocyclic Chemistry,* R. B. Mitra, N. R. Ayyanger, V. N. Gogte, R. M. Acheson, and N. Cromwell, Eds., Elsevier, New York, 1979, pp. 250–289.

78. M. Pailer and E. Romberger, *Monatsh. Chem.* **91,** 1070 (1960).

79. R. A. Guerra, *Acta Salmanticensia, Ser. Cienc.* (N. S.) **6,** 7 (1963) [*Chem. Abstr.* **63,** 5581 (1965)].

80. E. K. Fields and S. Meyerson, *Chem. Commun.* **1966,** 708.

81. R. Mayer, H. Kleinert, S. Richter, and K. Gewald, *Angew. Chem. Int. Ed.* **1,** 115 (1962); *J. Prakt. Chem.* **20,** 244 (1963).

82. B. D. Tilak and S. S. Gupte, *Indian J. Chem.* **7,** 9 (1969).

83. M. P. Cava and N. M. Pollack, *J. Am. Chem. Soc.* **88,** 4112 (1966).

84. J. M. Holland and D. W. Jones, *J. Chem. Soc. (C)* **1970,** 536.

85. J. Bornstein, R. P. Hardy, and D. E. Remy, *J. Chem. Soc. Chem. Commun.* **1980,** 612.

86. J. Bornstein and J. H. Supple, *Chem. Ind. (London)* **1960,** 1333; J. Bornstein, J. E. Shields, and J. J. Supple, *J. Org. Chem.* **32,** 1499 (1967).

87. J. D. White, M. E. Mann, H. D. Kirshenbaum, and A. Mitra, *J. Org. Chem.* **36,** 1048 (1971); M. E. Mann and J. D. White, *Chem. Commun.* **1969,** 420.

88. J. Stenhouse, *Ann. Chem.* **156,** 332 (1870).
89. C. Graebe, *Ann. Chem.* **174,** 177 (1874); *Chem. Ber.* **7,** 50 (1874).
90. H. D. Hartough and S. L. Meisel, *Compounds with Condensed Thiophene Rings,* Interscience, New York, 1954, Chap. IV.
91. R. B. Mitra, K. Rabindran, and B. D. Tilak, *Current Sci.* **23,** 263 (1954).
92. R. Lesser and R. Weiss, *Chem. Ber.* **45,** 1835 (1912).
93. L. Christiaens and M. Renson, *Bull. Soc. Chim. Belg.* **77,** 153 (1968).
94. N. M. Cullinane, A. G. Rees, and C. A. J. Plummer, *J. Chem. Soc.* **1939,** 151.
95. C. Courtot and A. Monytamedi, *Compt. Rend.* **199,** 531 (1934).
96. J. M. Gaidis, *J. Org. Chem.* **35,** 2811 (1970).
97. H. Hommes, H. D. Verkruijsse, and L. Brandsma, *J. Chem. Soc. Chem. Commun.* **1981,** 366.
98. J. M. Talbot, J. L. Piette, and M. Renson, *Bull. Soc. Chim. Belg.* **89,** 763 (1980).
99. R. J. Sundberg, *The Chemistry of Indoles,* Academic Press, New York, 1970.
100. R. L. Hinman and C. P. Bauman, *J. Org. Chem.* **29,** 2437 (1964).
101. G. Berti, A. Da Settimo, and E. Nannipieri, *J. Chem. Soc.* (*C*) **1968,** 2145.
102. W. E. Noland, L. R. Smith, and K. R. Rush, *J. Org. Chem.* **30,** 3457 (1965).
103. B. C. Challis and A. J. Lawson, *J. Chem. Soc. Perkin II* **1973,** 918.
104. H. F. Hodson and G. F. Smith, *J. Chem. Soc.* **1957,** 3546.
105. W. Madelung and O. Wilhelmi, *Chem. Ber.* **57,** 234 (1924).
106. J. C. Powers, *J. Org. Chem.* **31,** 2627 (1966).
107. R. D. Arnold, W. M. Nutter, and W. L. Stepp, *J. Org. Chem.* **24,** 117 (1959).
108. M. Mousseron-Canet and J.-P. Boca, *Bull. Soc. Chim. Fr.* **1967,** 1294.
109. R. L. Hinman and C. P. Bauman, *J. Org. Chem.* **29,** 1206 (1964).
110. B. Witkop and J. B. Patrick, *J. Am. Chem. Soc.* **75,** 2572 (1953).
111. W. J. Brehm and H. Lindwall, *J. Org. Chem.* **15,** 685 (1950).
112. G. F. Smith, *J. Chem. Soc.* **1954,** 3842.
113. J. I. DeGraw, J. G. Kennedy, and W. A. Skinner, *J. Heterocycl. Chem.* **3,** 67 (1966).
114. H. Heaney and S. V. Ley, *Org. Syn.* **54,** 58 (1974).
115. D. A. Shirley and P. A. Roussel, *J. Am. Chem. Soc.* **75,** 375 (1953).
116. J. Hutton and W. A. Waters, *J. Chem. Soc.* **1965,** 4253.
117. M. Julia and F. Ricalens, *Compt. Rend.* (*C*) **275,** 613 (1972).
118. Y. Kanaoka, M. Aiura, and S. Hariya, *J. Org. Chem.* **36,** 458 (1971).

Chapter 6

π-EXCESSIVE AMINO AND OXO COMPOUNDS

6.1. AMINO DERIVATIVES

Consideration must be given to the possible existence (Scheme 6.1) of tautomeric equilibria in all of the 2- and 3-amino-substituted, π-excessive, five-membered heteroaromatic compounds (**1** and **4,** respectively).

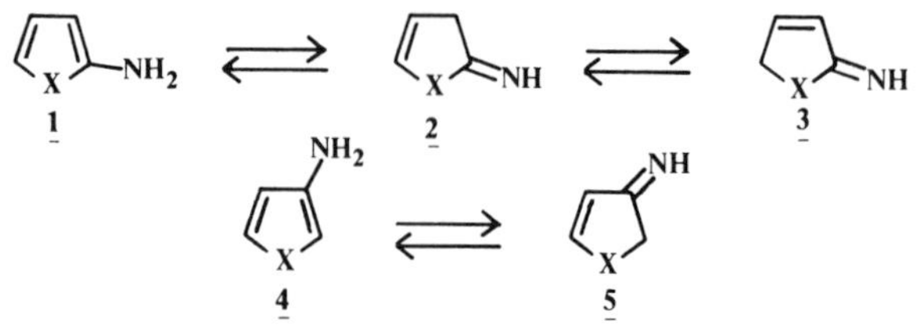

Scheme 6.1

6.1.1. AMINOPYRROLES

Only a limited number of aminopyrroles have been synthesized and little spectroscopic evidence is available to answer questions about tautomerism. In an infrared study of the ring-nitrogen-alkylated 2-amino-4-cyanopyrrole (**6**), it has been established that, at least for this derivative, no *imine* (**2** or **3**) form is present.[1] Furthermore, when **6** is heated with 2*N* NaOH, *no ammonia is evolved*! This observation has been construed as additional evidence that the imino forms are not present.

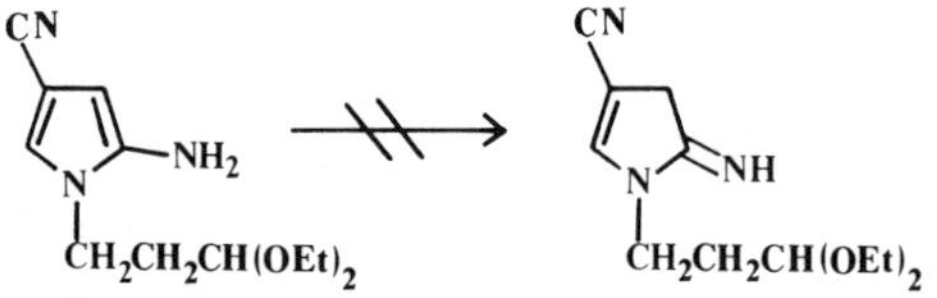

Even though 2,5-diaminopyrrole (**7**) is known, its precise structure has not been conclusively established. The existence of an amino-imino tautomerism appears, however, to be excluded by its ultraviolet spectrum.[2]

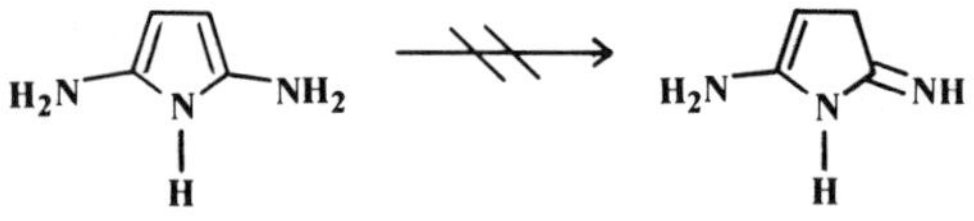

6.1.2. AMINOFURANS

The majority of aminofuran derivatives are 3-amino isomers. All attempts to obtain 2-aminofuran in its pure form have met with failure.[3] Despite this, **8** does appear to have been obtained in solution by treatment of furanimide with hydrazine (Gabriel synthesis). NMR evidence has been offered in support of the conclusion that no imino form is present when **8** is in solution.[4] When electron-withdrawing groups are present, 2-aminofurans are stable compounds and exist predominantly in the amino-tautomeric forms **9** or **10.**[5,6] Mild hydrolysis of the cyano derivative **10** with HCl-ethanol affords the unsaturated γ-lactone **11.** This hydrolytic behavior is contrary to that expected for an "aromatic" amino group; thus under these reaction conditions, the tautomeric imino isomer must play an active role.

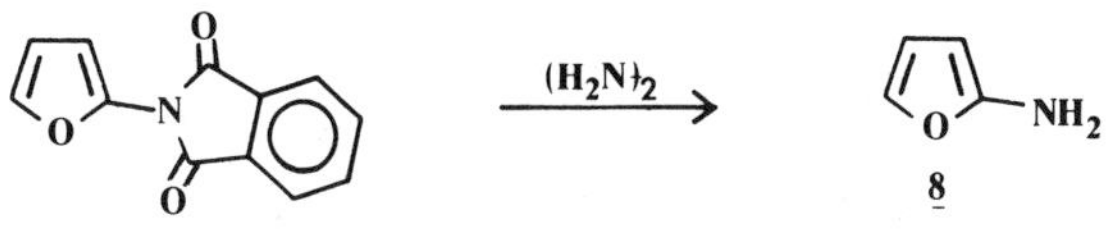

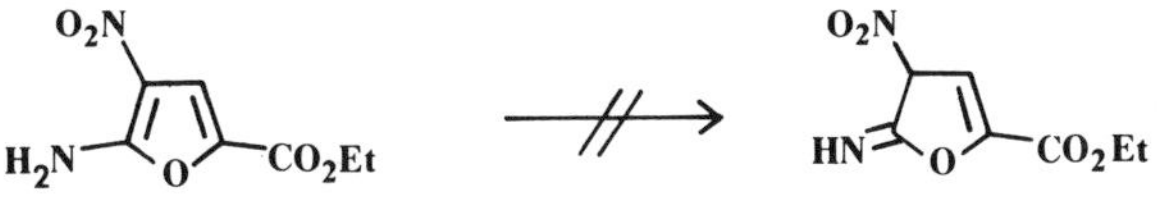

Chemically, 3-aminofurans appear to behave as "normal" arylamines in that they can be diazotized and coupled with electron-rich aromatic compounds.[2,3] Again, limited information is available on the 3-amino-imino tautomerism question, and the fact that hydrolysis of **12** affords ammonia has been interpreted to mean that a tautomeric mixture (**12** ⇌ **13**) exists.

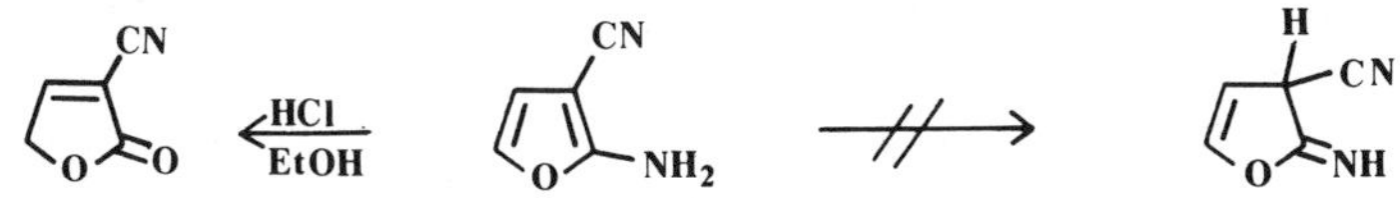

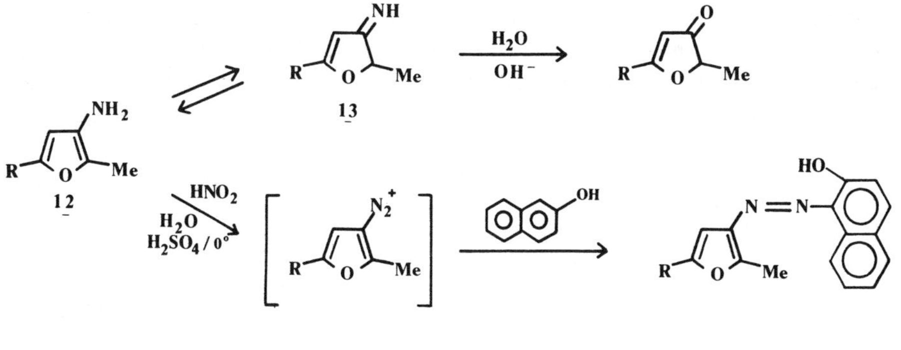

6.1.3. AMINOTHIOPHENES

Aminothiophenes have been more thoroughly studied. Both the 2- and 3-aminothiophenes are known[7]; however, neither is particularly stable. The 3-aminothiophene (**14**) appears to be somewhat less stable than the isomeric **15,** and both isomers are best stored as their stannic chloride double salts. With respect to the tautomerism question, the limited data indicate that both **14** and **15** exist predominantly, if not exclusively, in the amino form.

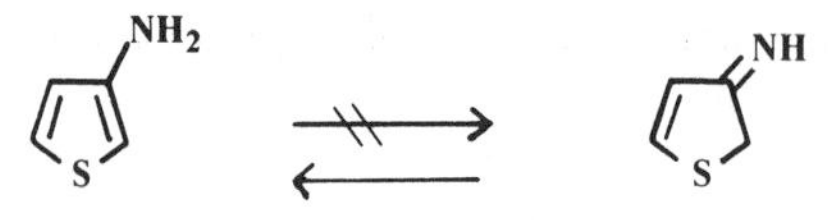

14 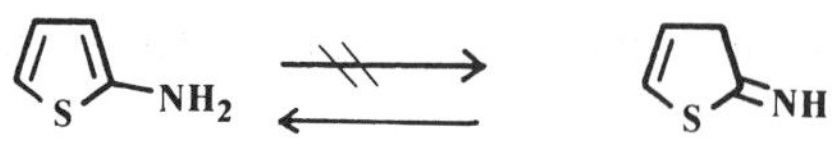 15

6.1.4. AMINOSELENOPHENES AND AMINOTELLUROPHENES

No aminoselenophenes or aminotellurophenes are currently known.

6.1.5. SUMMARY

The general conclusions that can be drawn from the available data are as follows: (a) 2-aminopyrroles and 2-aminofurans probably exist as the amino tautomer (**1**); (b) 3-aminofurans may well exist as a tautomeric mixture; and (c) 2- and 3-aminothiophenes exist largely (or exclusively) in the amino forms (**1**). Substituents may play a dominant role in this tautomeric equilibrium; thus care must be exercised in structural conclusions about the polysubstituted electron-rich systems.

6.2. OXO DERIVATIVES

In the oxo series of π-excessive heteroaromatics, the possible presence of tautomeric isomers shown in Scheme 6.2 must be considered. Significant differences between the π-deficient and π-excessive hydroxy compounds are immediately obvious from an examination of the possible tautomeric structures. With the latter, a prototropic shift involves the transformation of an sp^2 to an sp^3 hybridized *carbon*; in the former, no rehybridization of a carbon occurs. In this series, as in the π-deficient systems, ultraviolet spectroscopy has been used to differentiate between the "hydroxy" and the "oxo" tautomeric forms.

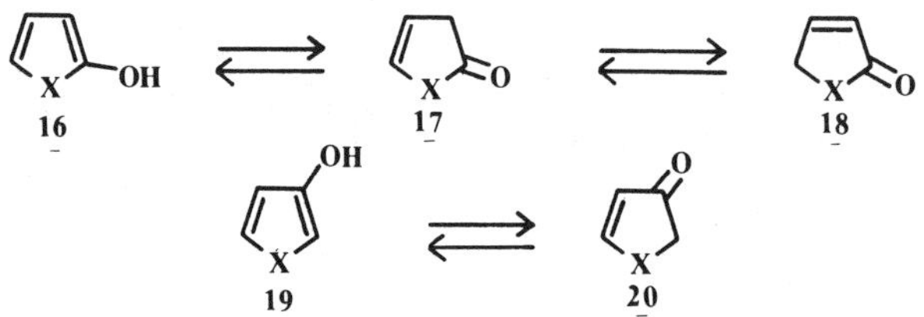

16 17 18

19 20

Scheme 6.2

6.2.1. OXOPYRROLES

The 2-hydroxypyrroles exist predominantly in the tautomeric pyrrol-2-one form (**21–23,** an α,β-unsaturated lactam), in which the double bond is situated between C-3 and C-4, that is, the Δ^3-position.[8–10] When there is an acyl group or an ester function at C-4, the Δ^4-isomer **24** predominates. All the 5-unsubstituted Δ^3-pyrrolin-2-ones readily react with aldehydes and ketones to form products expected from the condensation of these reagents with an active methylene group (**25** → **26**).[10] Treatment of the oxopyrrole **25** with acetic anhydride in pyridine affords the *O*-acylated product **27**.[11]

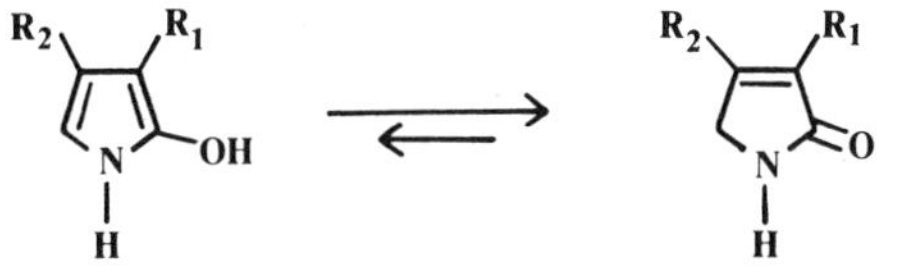

21 (R_1=R_2=H)
22 (R_1=H; R_2=CH_3)
23 (R_1=Acyl; R_2=H)

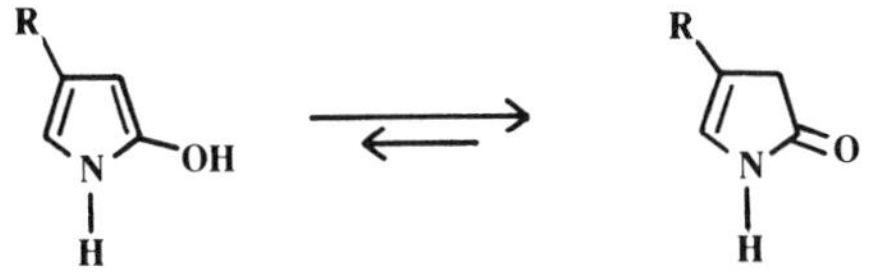

24 (R=CH_3CO; CO_2R')

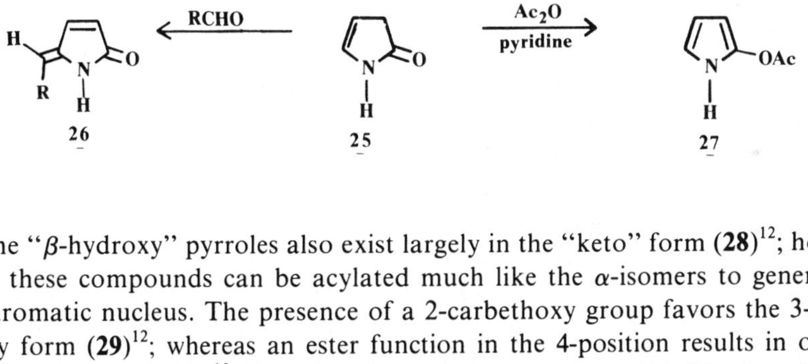

RCHO

Ac_2O
pyridine

26 25 27

The "β-hydroxy" pyrroles also exist largely in the "keto" form (**28**)[12]; however, these compounds can be acylated much like the α-isomers to generate the aromatic nucleus. The presence of a 2-carbethoxy group favors the 3-hydroxy form (**29**)[12]; whereas an ester function in the 4-position results in oxo-form stabilization (**30**).[13] There is, of course, one more possible "hydroxy" isomer of pyrrole: *N*-hydroxypyrrole (**31**). These compounds, which cannot tautomerize, have been prepared[14] and are readily reduced to pyrroles by zinc and acetic acid (See also 6.2.5.1).

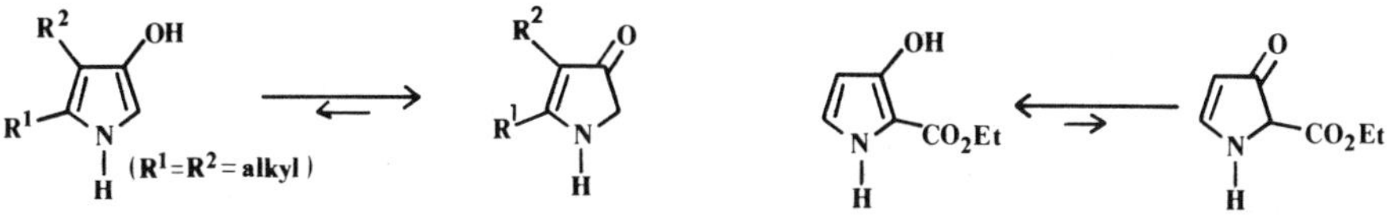

(R^1=R^2=alkyl)

28 29

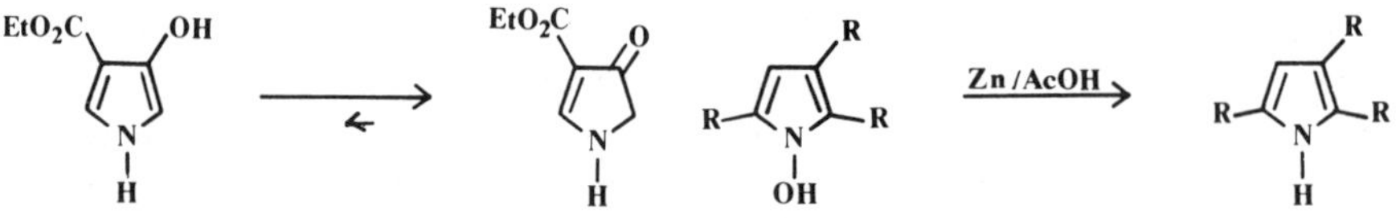

30 31

6.2.2. OXOFURANS

Considerable confusion has existed in the literature with regard to both the 2- and 3-"hydroxy" furans (**32** and **33**); however, it has finally been established that the oxo forms (**34** and **35**, respectively) are the lone tautomers in this series.[15]

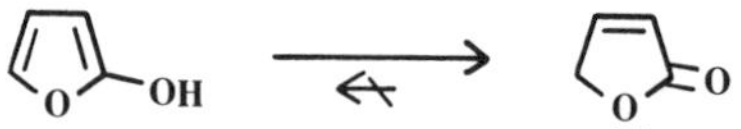

32 34 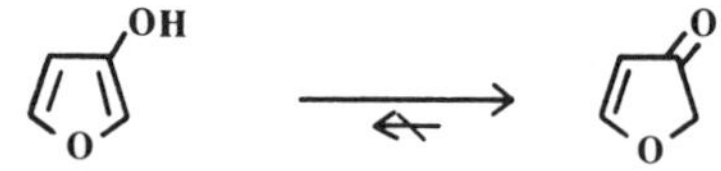 33 35

The 2(5*H*)-furanone (**34**; $\Delta^{\alpha,\beta}$-butenolide) as well as the 3(2*H*)-furanone (**35**) show little or no evidence of enolic behavior. The 4-hydroxy-2(5*H*)-furanones [the parent is known as tetronic acid (**36**)] exist in the form indicated. These furanones in contrast to the thiophene and pyrrole analogues, generally fail to undergo *O*-alkylation. For example, α-angelica lactone (**37**) when treated with methyl iodide affords the *C*-alkylated products **38** and **39**.[16] 3-"Hydroxy" furan (**40**) is a minor contributor to the tautomeric equilibrium and thus 3(2*H*)-furanone (**41**) is the major isomer. These compounds exist predominantly in the keto form *unless* there is a carbonyl substituent at the 2-position; then the hydroxy form (**42**) is favored. Such a dramatic shift in the tautomeric behavior has been attributed to the presence of strong hydrogen bonding between the hydroxyl and carbonyl functions (for example, **43**).[17] The 3(2*H*)-furanones (**41**) do not give typical carbonyl derivatives such as semicarbazones **44**. Chemically, reaction conditions greatly influence the product outcome, as exemplified by the treatment of **41** with acetic anhydride and sodium acetate to give *O*-acylation and with methyl iodide and strong base to give *C*-alkylation products.[18]

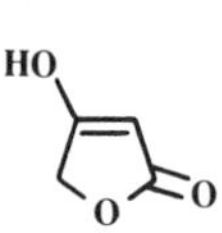

36

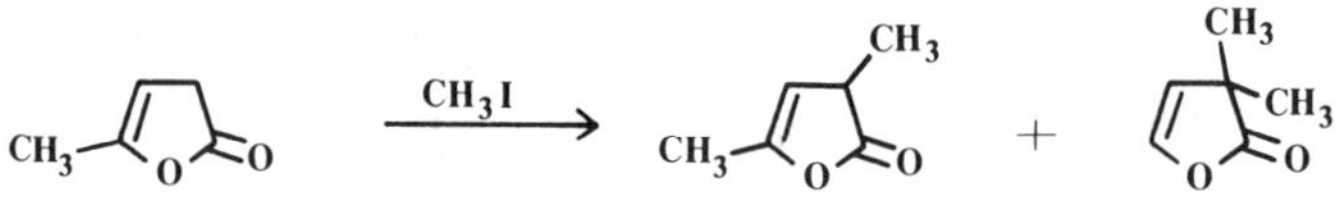

CH_3I

37 38 39

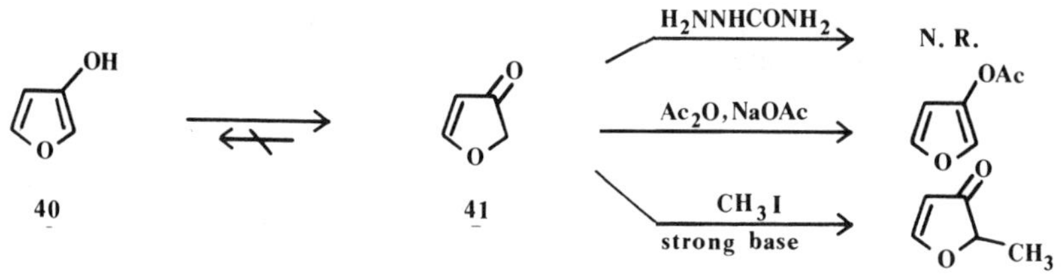
OH
O
H2NNHCONH2
N. R.
Ac2O, NaOAc
OAc
CH3I
strong base
CH3

40 41

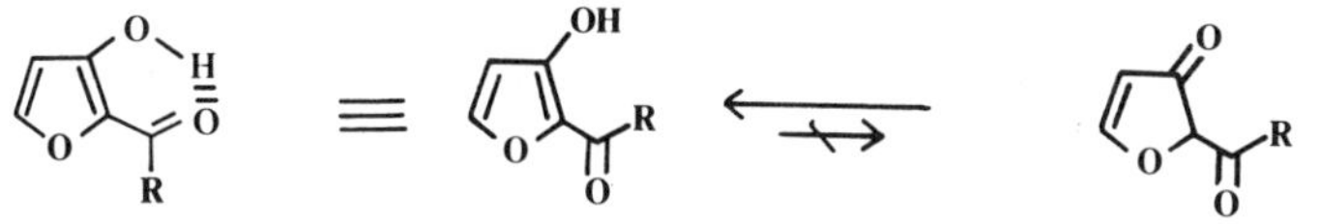
H
O
OH
R
O

43 42

An interesting transformation of a 3-oxofuran **45** to the γ-lactone **46** can be envisioned to occur by the following mechanistic sequence[18]:

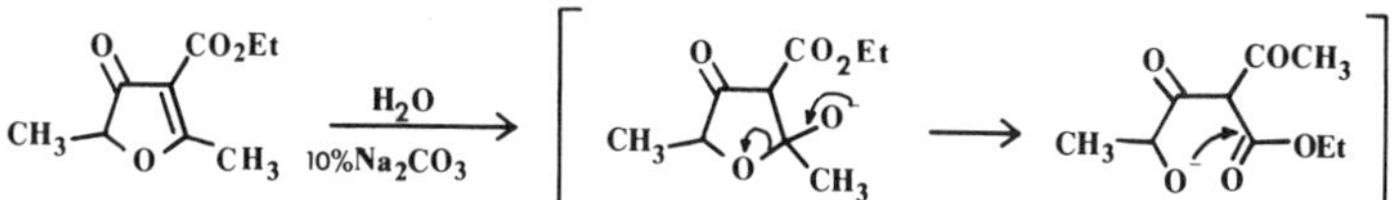
CO2Et
CH3
H2O
10%Na2CO3
COCH3
OEt

45

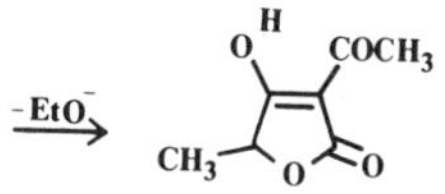
-EtO-
H
COCH3
CH3

46

6.2.3. OXOTHIOPHENES

The tautomeric behavior of thienols has been studied in detail. The existence of the three possible tautomers in the 2-hydroxy series has been found to depend on the structural features. Analogous to the simple 2-oxofurans, the nonaromatic γ-thiolactones (**47**) are generally preferred.[19-21] Formation of the $\Delta^{3,4}$-thiophenone is observed when a 5-methyl (**48**) or, more especially, a 5-phenyl (**49**) group is present. When an electron-withdrawing group is present at the 3-position (**50**), the hydroxy form predominates, as was observed in the furan analogues. A 5-carbethoxy group (for example, **51**) is not as efficient in bringing about complete hydroxy tautomer stabilization in carbon tetrachloride, since only 85% of this form is present and the remaining 15% is present as 2-thiophenone (**52**).[21] In the 3-hydroxythiophenes (**53**) the hydroxy form is generally preferred over the 3-oxo isomer (**54**).[22-25] In the parent compound (**53**, R = H) both isomers are present in the neat sample; however, when R is 2-Me, 80% of the compound exists as the hydroxy tautomer (**53**) in carbon disulfide solution, whereas a 2-*tert*-butyl group decreases **53** to 55% in the same solvent. In the 2,5-di-*tert*-butylthiophene analogue (**55**), either neat or in carbon disulfide, the oxo compound is the sole isomer. Electron-withdrawing substituents, on the other hand, force the equilibrium totally to the hydroxy side (**56**).[24,25]

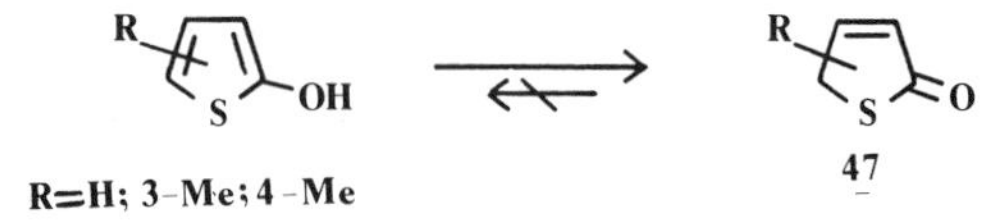

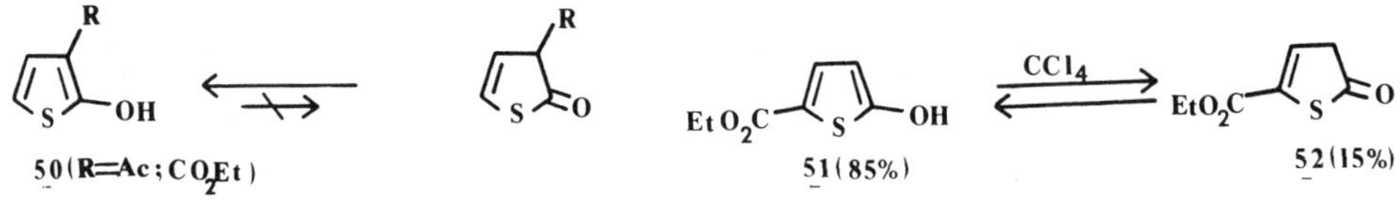

R=H; 3-Me; 4-Me

47

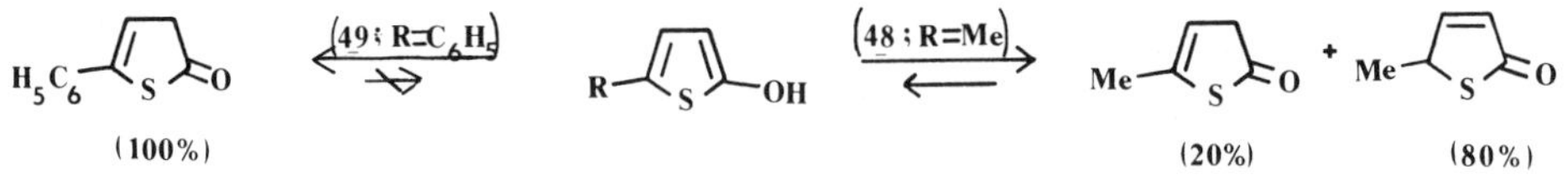

(49; R=C_6H_5) (48; R=Me)

H_5C_6 (100%) (20%) (80%)

50 (R=Ac; CO_2Et)

CCl4

51 (85%) 52 (15%)

53 54 55 (100%)

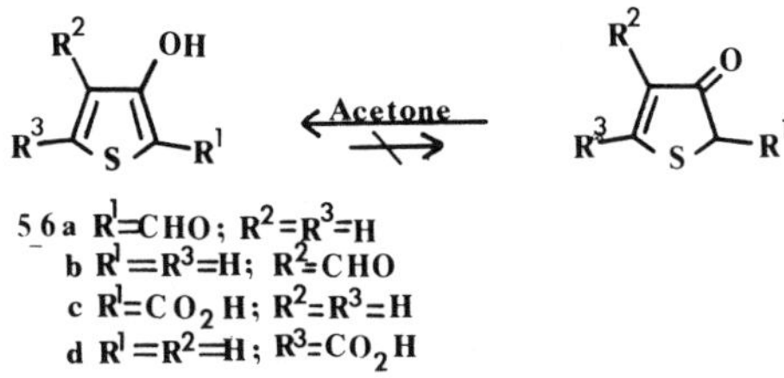

Acetone

56 a R^1=CHO; R^2=R^3=H
b R^1=R^3=H; R^2=CHO
c R^1=CO_2H; R^2=R^3=H
d R^1=R^2=H; R^3=CO_2H

With the exception of the 2,5-"dihydroxy" thiophene, which exists totally in the oxo form (**57**), all the other dihydroxy thiophenes exist in the monohydroxy, mono-oxo form (**58–60**).[22–25]

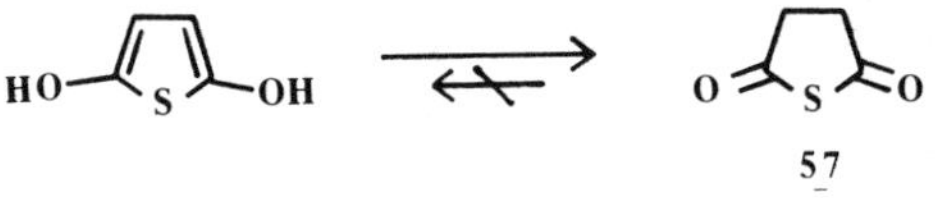

57

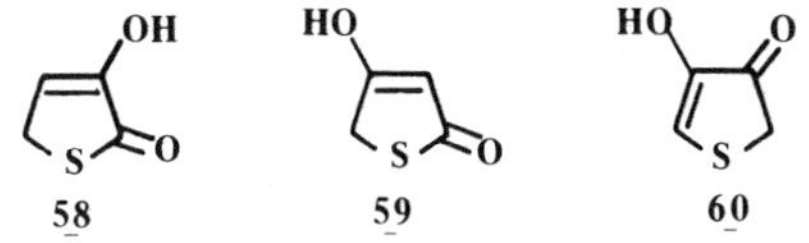

58 59 60

6.2.4. OXOSELENOPHENES

Although the question of the amino-imino tautomerism in aminoselenophenes has not yet been examined, it has been shown that 2-"hydroxy" selenophenes (**61**) exist as a tautomeric mixture of the two possible oxo forms (**62–63**).[26]

R–(selenophene)–$OC(CH_3)_3$ $\xrightarrow{H^+}$ **62** ⇌ (crossed) **61**; **62** ⇌ **63**

6.2.5. HETEROATOM-BONDED OXO COMPOUNDS

Among the monohetero π-excessive systems under discussion, formation of a heteroatom-oxygen bond is only possible in pyrrole (**64**), thiophene (**65**), and selenophene (**66**). In the thiophene and selenophene instances, the formation of dihetero-oxygenated species (**67, 68**) is also feasible.

6.2.5.1. *N*-Hydroxypyrroles. *N*-Hydroxypyrroles (**69**) are known[14,27] and are prepared by the following reaction. These compounds are all readily reduced to the parent pyrroles by zinc and acetic acid.[14]

$BrCH_2C(=NOH)R$ + $ROCCH_2COR$ ⟶ **69**

6.2.5.2. Thiophene Sulfoxides and Sulfones. Thus far, all attempts at preparation of thiophene sulfoxide appear to have failed. The thiophene sulfones (**70**) can be prepared by perbenzoic acid oxidation of thiophenes.[28] In the case of thiophene itself, the sulfone is not stable but undergoes an autocondensation to give ultimately **71.** The corresponding 2,5-dimethyl analogue [**70** R = Me)] is somewhat more stable and reacts with maleic anhydride to give a stable Diels–Alder adduct **72.** The ultraviolet spectra of these sulfones are very similar to that of cyclopentadiene; thus they are not aromatic compounds but very reactive dienes.

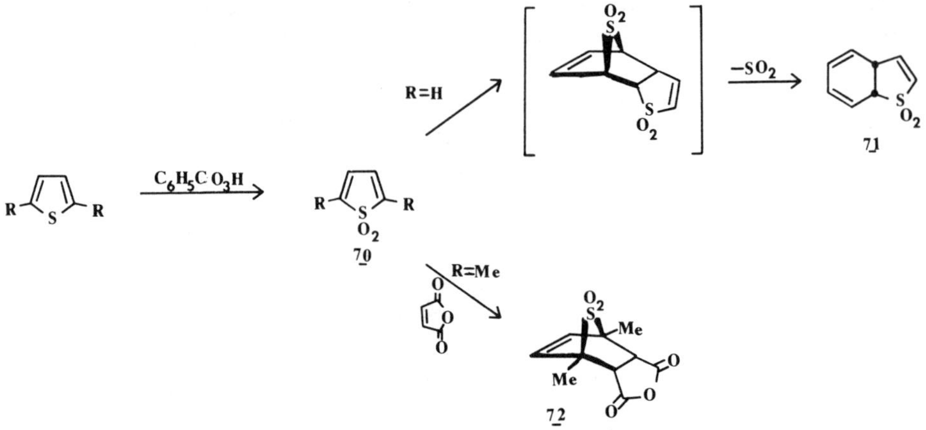

REFERENCES

1. C. A. Grob and H. Utzinger, *Helv. Chim. Acta* **37,** 1256 (1954).
2. R. M. Acheson, *An Introduction to the Chemistry of Heterocyclic Compounds,* Interscience, New York, 1967, p. 79.
3. A. P. Dunlop and F. N. Peters, *The Furans,* Reinhold, New York, 1953, p. 185.
4. J. H. Reisch and H. Labitzke, *Arch. Pharm.* **308,** 713 (1975).
5. H. Schafer, H. Hartmann, and K. Gewald, *J. Prakt. Chem.* **315,** 497 (1973).
6. K. P. C. Vollhardt and R. G. Bergman, *J. Am. Chem. Soc.* **95,** 7538 (1973).
7. Ya. L. Gol'dfarb, M. M. Polonskaya, B. P. Fabrichnyi, and I. G. Shalavina, *Proc. Acad. Sci. (U.S.S.R.)* **126,** 331 (1959).
8. J. H. Atkinson, R. S. Atkinson, and A. W. Johnson, *J. Chem. Soc.* **1964,** 5999.
9. C. A. Grob and A. P. Ankli, *Helv. Chim. Acta* **32,** 2023 (1949).
10. H. Plieninger and M. Decker, *Ann. Chem.* **598,** 198 (1956).
11. H. Plieninger, H. Boiver, W. Buhler, J. Kurze, and U. Lerch, *Ann. Chem.* **680,** 69 (1964).
12. R. Chong and P. S. Clezy, *Aust. J. Chem.* **20,** 935 (1967).
13. A. Treibs and A. Ohorodnik, *Ann. Chem.* **611,** 149 (1958).
14. R. Ramasseul and A. Rassat, *Bull. Soc. Chim. Fr.* **1970,** 4330.
15. A. Hofmann, W. V. Philipsborn, and C. H. Eugster, *Helv. Chim. Acta* **48,** 1322 (1965).
16. B. Cederlund, A. Jespersen, and A. -B. Hornfeldt, *Acta Chem. Scand.* **25,** 3656 (1971).
17. T. P. C. Mulholland, R. Foster, and D. B. Haydock, *J. Chem. Soc. Perkin I* **1972,** 1225.
18. J. Elguero, C. Marzin, A. R. Katritzky, and P. Linda, *Adv. Heterocycl. Chem. Suppl.* **1,** 214 (1976).
19. S. Gronowitz and R. A. Hoffman, *Ark. Kem.* **15,** 499 (1960).
20. A. -B. Hornfeldt and S. Gronowitz, *Acta Chem. Scand.* **16,** 789 (1962).
21. H. J. Jakobsen, E. H. Larsen, and S. O. Lawesson, *Tetrahedron* **19,** 1867 (1963).
22. A. -B. Hornfeldt, *Acta Chem. Scand.* **19,** 1249 (1965).

23. R. Lantz and A. -B. Hornfeldt, *Chem. Scripta* **2,** 9 (1972).
24. S. Gronowitz and A. Bugge, *Acta Chem. Scand.* **20,** 261 (1966).
25. C. Paulmier, J. Morel, D. Semard, and P. Pastour, *Bull. Soc. Chim. Fr.* **1973,** 2434.
26. J. Morel, C. Paulmier, D. Semard, and P. Pastour, *Compt. Rend.* (*C*) **270,** 825 (1970).
27. V. Sprio and I. Fabra, *Ann. Chim.* (*Rome*) **50,** 1635 (1960).
28. J. A. Joule and G. F. Smith, *Heterocyclic Chemistry,* Van Nostrand-Reinhold, New York, 1972, p. 226.

Chapter 7

ELECTROPHILIC SUBSTITUTION. FIVE-MEMBERED π-EXCESSIVE HETEROCYCLIC RINGS

7.1. GENERAL COMMENTS

The most characteristic reactions of the π-excessive heteroaromatic systems are those of electrophilic substitutions. These reactions are comparable, in a general sense, to those of aniline and phenol. A consideration of the various possible structures of the intermediates obtained from electrophilic attack at the 2- and 3-positions of these compounds (**1a–1c,** and **2a, 2b,** respectively) suggests that the 2-position (α) should be more reactive than the 3-position (β) toward electrophilic substitution.

1a ⟷ 1b ⟷ 1c

2a ⟷ 2b

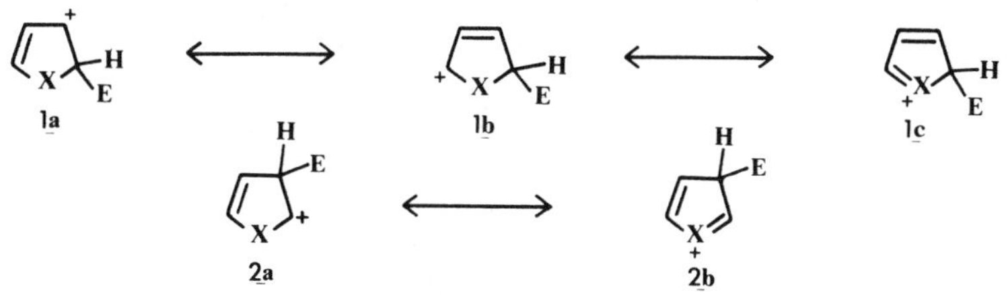

The energy profile for reactions of this type (cf. Figure 7.1) also demonstrates that the σ-complex which is involved in α-substitution is stabilized to a greater extent than the one involved in β-substitution. In fact, most monoheteroatomic, π-excessive, five-membered aromatic rings undergo preferential electrophilic substitution on the carbon α to the heteroatom.

A comparative study of some electrophilic substitution reactions on pyrrole, furan, tellurophene, selenophene, and thiophene has established that these ring systems are decreasingly reactive in the order shown:

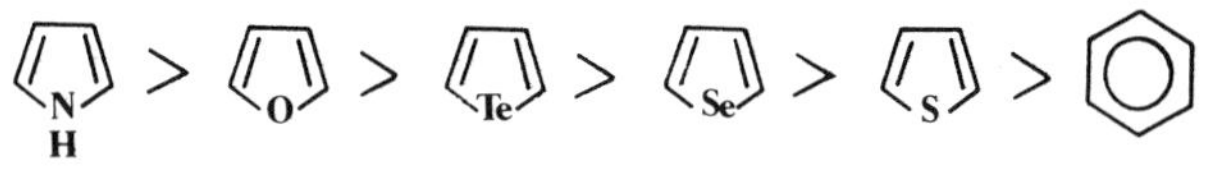

pyrrole (NH) > furan (O) > tellurophene (Te) > selenophene (Se) > thiophene (S) > benzene

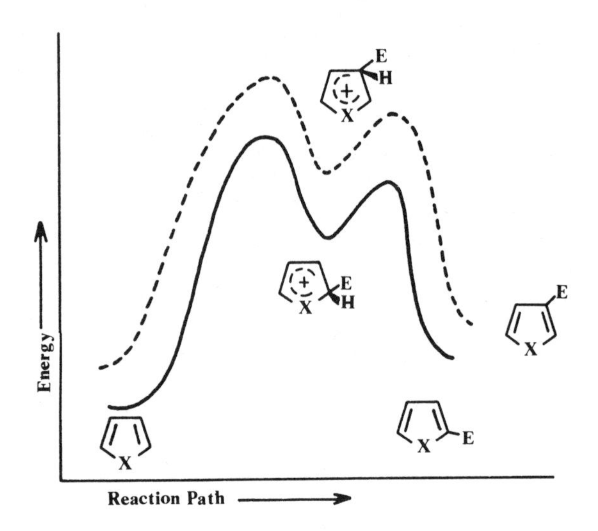

Figure 7.1. Energy profile for electrophilic substitutions in five-membered heteroatomic systems.

TABLE 7.1. Relative Rates of Electrophilic Substitution at the 2-Position of (X) (3) (2)

Parent Compound	2-Substituent			
	Br_2[a]	CF_3CO	H_3CO	HCO
N–H	4.9×10^{6}	3.8×10^{5}	—	—
O	1	1	1	1
Te	—	3.3×10^{-1}	6.3×10^{-1}	3.5
Se	4.9×10^{-1}	5.2×10^{-3}	1.8×10^{-1}	3.6×10^{-1}
S	8.3×10^{-3}	7.1×10^{-3}	8.4×10^{-2}	9.7×10^{-2}

[a] As the 2-carbomethoxy derivatives.

Table 7.1 lists the relative rate constants for bromination, trifluoroacetylation, acetylation, and formylation of these ring systems.[1]

The relative rates of these reactions reflect more than just the aromaticity and ground-state electron distributions of these systems (see Appendix A). Consideration must be given not only to the stabilities of the starting materials and products but also to that of the transition state, the "position" of this complex on the reaction coordinate, solvent effects, and other factors, many of which are not yet clearly understood. The one statement that can be made is that the more electronegative heteroatoms (nitrogen and oxygen), for which *d*-orbital participation *is not* possible, facilitate these reactions more significantly than do those heteroatoms for which *d*-orbital participation *is* possible (tellurium, selenium, sulfur).

7.2. REACTIVITIES TOWARD ACIDS

7.2.1. PYRROLES

The proton attached to the nitrogen atom of pyrrole undergoes very rapid exchange in either acidic or basic media. Exchange of the other ring hydrogens, however, occurs only under drastic conditions.[2] In aprotic solvents, complete protonation of the α-position can occur. For example, when gaseous hydrogen chloride is passed into an ethereal solution of pyrrole **3**, a crystalline hydrochloride (**4**) is obtained. Although the β-protonated salts (**5**) have not been isolated, they are most probably intermediates in the H ⇌ D exchange processes that occur, albeit slowly, at C-3.

+ D_2O —neutral, acidic or basic media→

CH3 Et … CH3 —HCl, ether→ … Cl⁻

3 4

—DCl→ […] →

5

When pyrrole itself is allowed to remain in contact with aqueous mineral acids, a number of polymeric materials are formed. It is possible to isolate from this mixture a trimer (**6**),[3] formation of which can be envisioned by the following sequence:

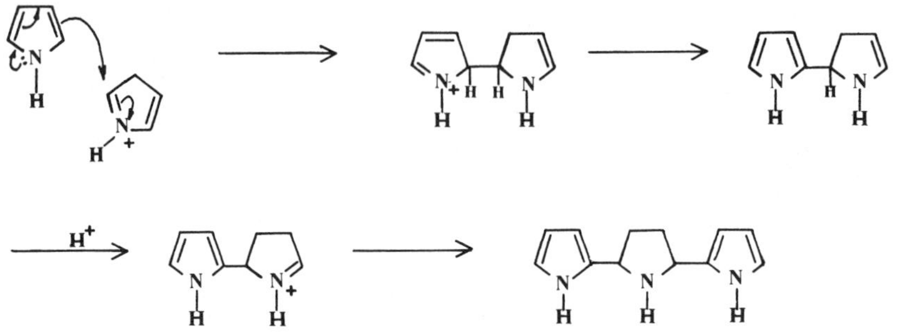

6

7.2.2. FURANS

Furans are easily attacked by acids and are somewhat analogous to enol ethers; thus hydrolysis of furan affords succindialdehyde (**7**). In all the instances studied, even under the mildest possible acidic conditions, hydrolysis occurs and polymers are inevitably obtained from the intermediate cation. Examination of hydrogen-deuterium exchange reactions of 2-methylfuran (**8**)[4] indicates, that, as in pyrrole, the reaction involves an *A-SE2* (addition-substitution electrophilic-second order) mechanism.

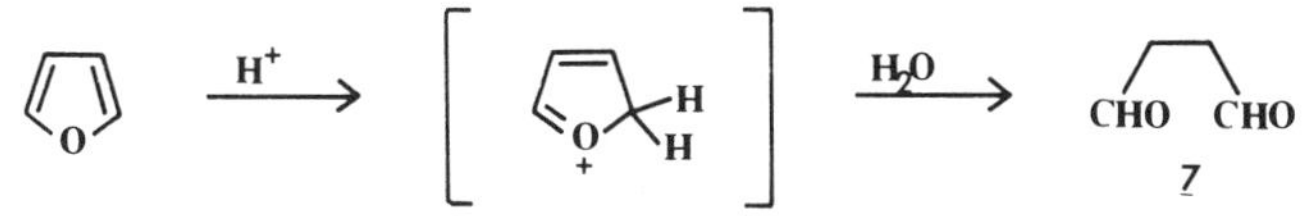

7

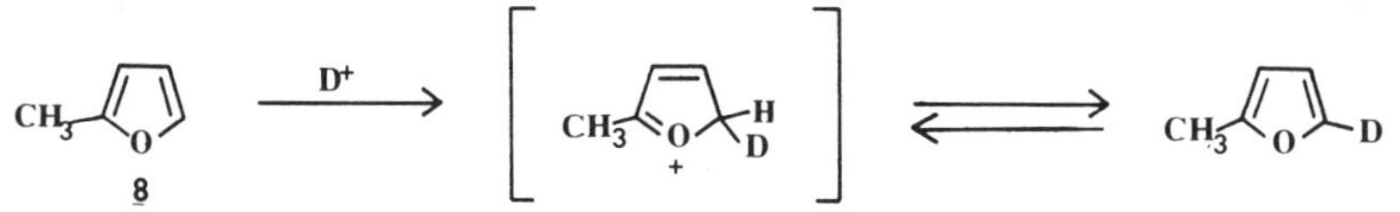

8

7.2.3. THIOPHENES

Hydrogen-deuterium exchange on thiophenes in trifluoroacetic acid or aqueous sulfuric acid has been examined.[5,6] The α-hydrogen exchanges about 3000 times faster than does the hydrogen at the β-position. Thiophene itself is stable to aqueous but not to anhydrous mineral acids; however, it is cleaved to some extent in boiling 100% phosphoric acid, and is polymerized to dark amorphous solids with sulfuric acid and hydrogen fluoride. Under milder conditions, thiophene with orthophosphoric acid gives a trimer (**9**), as well as a pentamer. Interestingly, the structure of this thiophene trimer is different (2,4-disubstituted central ring) than the one obtained when pyrrole is treated with acid (2,5-disubstituted central ring). The structure of the pentamer is yet unknown.

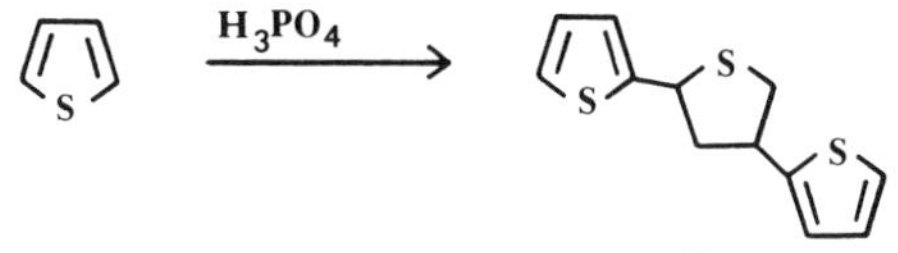

H_3PO_4

9

7.2.4. SELENOPHENES

Selenophene is relatively stable under moderately acidic conditions and forms a greenish-blue indophenine product on treatment with isatin and concentrated sulfuric acid.

7.2.5. TELLUROPHENES

Tellurophene is very sensitive to mineral acids and is readily hydrolyzed. The hydrolysis has not been studied in any detail.

7.3. HALOGENATIONS

7.3.1. PYRROLES

Pyrrole itself reacts with the halogens to afford tetrahalo derivatives (**10**). Monohalopyrroles are formed only under very special conditions and decompose rapidly in air and in light.[7] In the presence of electron-withdrawing substituents, nuclear monosubstitution occurs more cleanly. For example, 2-carbomethoxypyrrole upon treatment with bromine affords the monobrominated derivatives **11** and **12,** as well as some 4,5-dibrominated compound.[8]

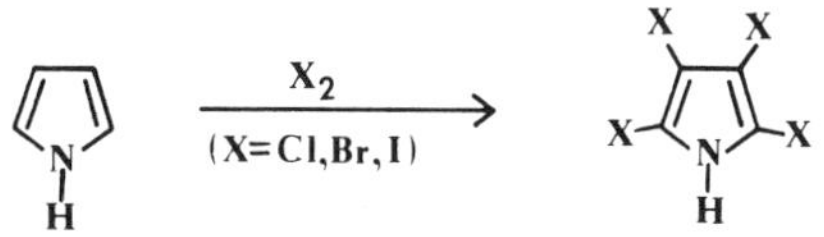

X_2 (X=Cl,Br,I)

10

Br_2/HOAc r.t./dark

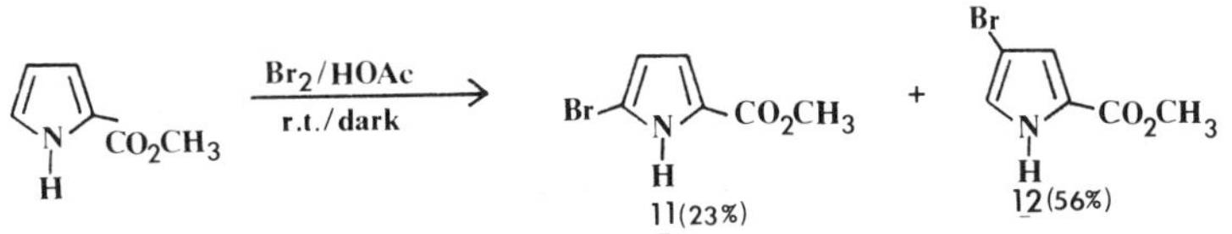

11(23%) 12(56%)

Carboxylic acid and acyl groups on substituted pyrroles (for example, **15**) are often replaced by bromine (**13**) or iodine (**14**).[9] It is noteworthy that a methyl group α to the pyrrole nitrogen is also monobrominated, even under mild conditions.

14 ← (KI, K_2CO_3) 15 → (Br_2, $NaHCO_3$) 13 (11%)

7.3.2. FURANS

Early studies on the direct bromination of furan have led to considerable confusion because of the poorly characterized products that had been isolated. Use of the bromine-dioxan complex at −5°C does, however, afford 2-bromofuran (**16**)[10] in good yield. When bromination is effected in acetic acid with added potassium acetate, a mixture of 2,5-dihydro-2,5-diacetoxyfurans (**17**) is obtained.[11]

17 ← (AcOH, AcOK) furan → (dioxane · Br_2, -5°C) 16

An NMR study of the bromination of furan at −50°C in carbon disulfide[12] has established that the *cis*- and *trans*-2,5-dibromo-2,5-dihydrofurans (**18, 19**) are formed under these conditions. On warming, these dihalides are transformed to 2-bromofuran (**16**). This strongly suggests, but does not prove, that 2-bromofuran is formed from the dibromo derivatives by loss of HBr.

furan → (Br_2 · CS_2, -50°C) [18 + 19] → (warming) 16

Substituted furans (for example, **20, 21**) that have been halogenated are those with electron-withdrawing substituents at the 2- or 3-position.[13] These reactions apparently *do* occur by an electrophilic substitution process.

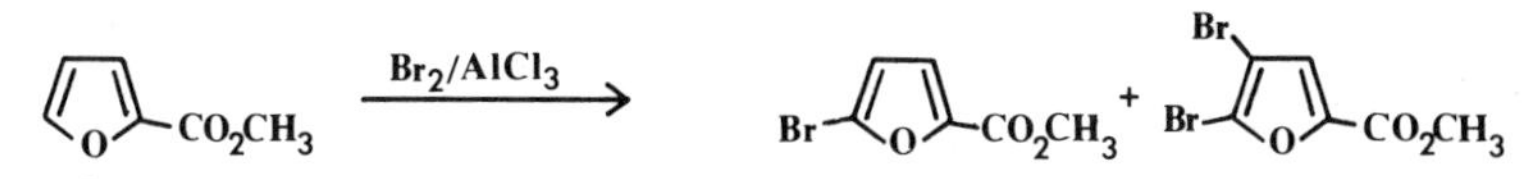

20

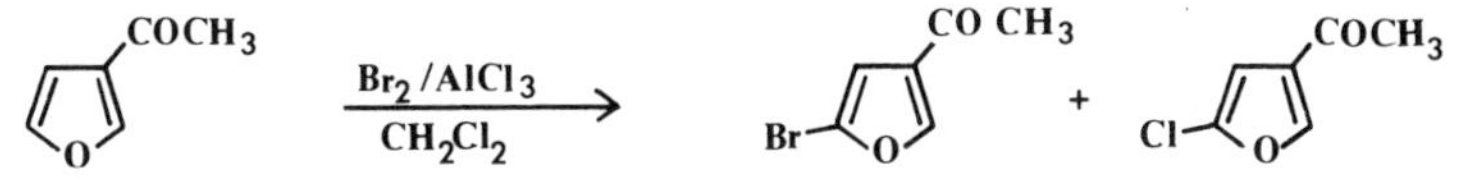

21

Treatment of furan with 1.6 moles of chlorine in methylene chloride affords mainly 2-chlorofuran (**22**).[14] Whether this product is formed via the intermediate, 2,5-dichloro-2,5-dihydrofuran, is not yet established.

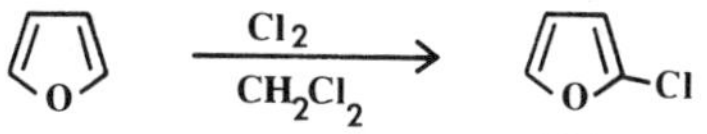

22 (64%)

7.3.3. THIOPHENES

As in furan, halogenation of thiophene occurs preferentially at the α-position. Chlorination, even at $-30°C$, generally gives a mixture of both substitution and addition products. The reaction mixture from the chlorination of thiophene, after dehydrochlorination of the addition products, gives eight different halogenated thiophenes (**23–30**) in the yields indicated.[15]

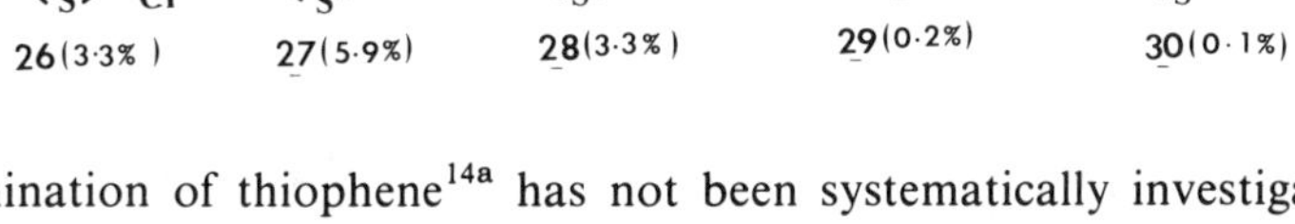

23 (36.8%) 24 (27.1%) 25 (7.3%)

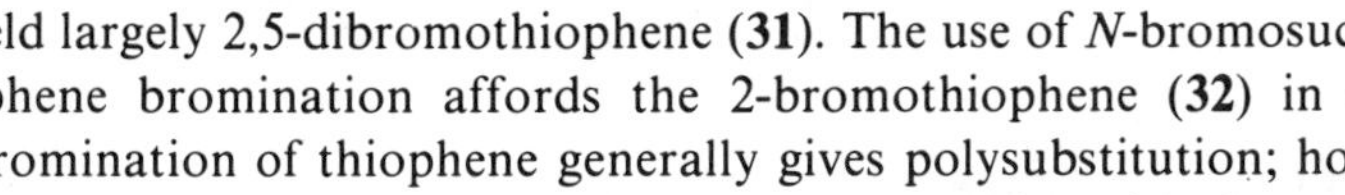

26 (3.3%) 27 (5.9%) 28 (3.3%) 29 (0.2%) 30 (0.1%)

Bromination of thiophene[14a] has not been systematically investigated, although the addition products obtained from this reaction, after treatment with base, yield largely 2,5-dibromothiophene (**31**). The use of *N*-bromosuccinimide in thiophene bromination affords the 2-bromothiophene (**32**) in excellent yield. Bromination of thiophene generally gives polysubstitution; however, if sodium acetate is added to neutralize the hydrogen bromide formed, subsequent reduction with zinc dust affords (>50%) 3-bromothiophene.[14b] This procedure is also applicable to selenophenes. Direct iodination of thiophene affords 2-iodothiophene (**33**) in good yield. As these halogenation results demonstrate, the α-position in thiophene is much more reactive to substitution than is the β-position. In fact, the relative rate difference[16] between α- and β-bromination in thiophene has been shown to be 490:1.

(1) Br_2 (2) Base → 31

N-bromosuccinimide → 32 (77%)

I_2 → 33

7.3.4. SELENOPHENES

As is the case with both furan and thiophene, halogenation of selenophene (**34**) occurs preferentially at the α-position.[17]

34: Br_2 → 35; SO_2Cl_2 → 36; I_2, HgO (yellow) → 37

36: X_2 → 38

Sulfuryl chloride also effects 2-chlorination of selenophene.[18] Depending upon the reaction conditions, further halogenation of the monohalogenated selenophenes can take place to give 2,5-disubstituted products (**38**).[17] The 3-halogenated selenophenes (**40**) are not available by direct halogenation; however, they can be obtained from zinc dust reduction of 2,3,5-trihaloselenophene (**39**).[19]

39: Zn, AcOH → 40

7.3.5. TELLUROPHENES

Although many electrophilic substitution reactions on tellurophene (**41**) occur at the "normal" 2-position, the reaction of **41** with halogens affords 1,1-dihalogenated derivatives (**41**).[20]

41 $\xrightarrow{X_2}$ 42

7.4. NITRATIONS

7.4.1. PYRROLES

Because of the great acid lability of pyrrole, use of the reagents commonly employed in the nitration of benzenes results in extensive decomposition. At lower temperatures, acetyl nitrate can be used to afford 2-nitropyrrole (**43**) in relatively good yield, along with smaller amounts of the 3-nitro isomer (**44**).[21] A comprehensive study of the nitration of pyrrole has been conducted.[21]

CH_3CONO_3, −10°C → 43 (50%) + 44 (13%)

In contrast, *N*-alkylated pyrroles afford larger amounts of 3-nitro derivatives,[22] whereas 2-methylpyrrole (**45**) gives the 5-nitro derivative (**46**) in larger proportion than the 3-nitro isomer (**47**).[23]

45 $\xrightarrow{MeCONO_3}$ 46 + 47

7.4.2. FURANS

As is the case in the halogenation of furan, nitration with a mixture of fuming nitric acid and acetic anhydride occurs largely by an addition-elimination process.[14,24,25] An addition product has been isolated which has been assigned structure **48.** In the presence of electron-withdrawing group(s), for example in 2-furancarboxaldehyde (**49**), two different addition products (**50, 51**) have been isolated.[25] The expected nitro compound **52** is also obtained.

HNO_3 fuming, Ac_2O; −5°C → 48 → warm →

49 $\xrightarrow{HNO_3\ \text{fuming},\ Ac_2O}$ 50 + 51 + 52

7.4.3. THIOPHENES

Direct nitration of thiophene[26] occurs 850 times faster than does that of benzene and affords an α versus β ratio of 6.2:1. The α-selectivity in nitration decreases with increasing reaction vigor, affording up to 15% of the 3-nitrothiophene (**53**). The nitration of 2-methylthiophene (**54**) affords a mixture of the 3- (**55**) and 5-nitro (**56**) derivatives[27]; in contrast, a *tert*-butyl group at the 2-position (**57**) presents sufficient steric deterrent to prevent the formation of the 3-nitro derivative and the 5-nitro compound (**58**) is the sole nitro product obtained.

HNO_3 / H_2SO_4 → 2-nitrothiophene + **53** (6.2:1)

54 → HNO_3 / H_2SO_4 → **55** + **56** (3:7)

57 → HNO_3 / H_2SO_4 → **58**

2-Halothiophenes (**59**; X = Cl, Br, I) suffer an equal deterrence with respect to nitration at the 3-position, and the sole product obtained is the 5-nitro derivative (**60**). On the other hand, the electron-donating methoxy group (**61**) activates the 3-position towards nitration and both the 3- and the 5-nitrothiophenes (**62** and **63**) are isolated.[14]

59 (X Cl, Br, or I) → HNO_3/H_2SO_4 → **60**

61 → HNO_3 / H_2SO_4 → **62** + **63** (4:1)

7.4.4. SELENOPHENES

Nitration of selenophene with nitric acid in acetic anhydride affords mainly the 2-nitro derivative (**64**), along with only minor amounts of the 3-nitro isomer (**65**).[28] If the 2-position in selenophene is blocked by an electron-withdrawing group, nitration takes place preferentially at the 4-position (**66**).[29] This behavior is of synthetic significance since, if the electron-withdrawing group can be readily removed (as in **66** → **67**), easy access to the 3-nitroselenophenes becomes available.

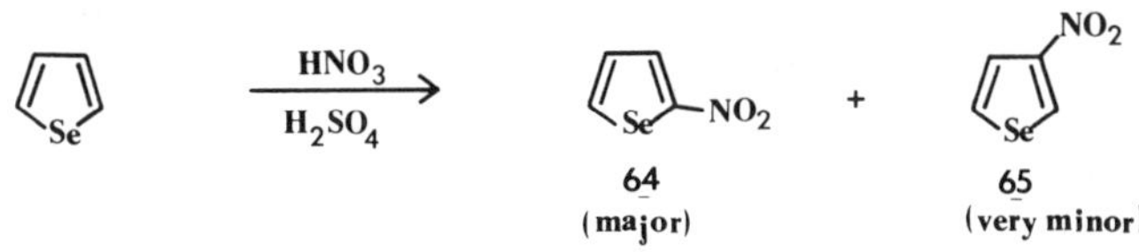

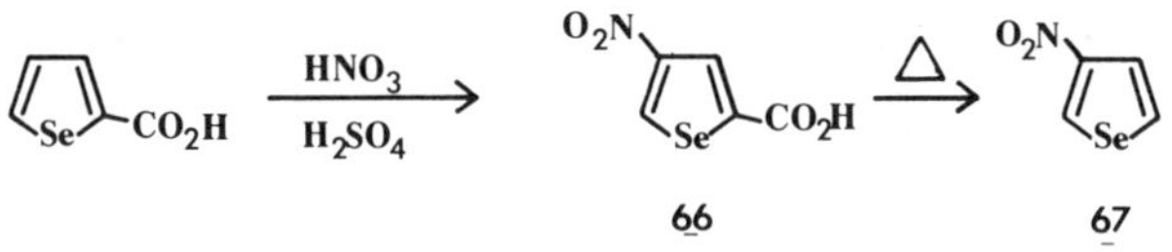

64 (major)

65 (very minor)

66

67

7.4.5. TELLUROPHENES

As might be anticipated from the great acid lability of tellurophene, 2-nitrotellurophene has not been obtained.[30]

7.5 SULFONATIONS

7.5.1. PYRROLES

Despite the fact that the pyrrole nucleus can be destroyed by sulfuric acid, pyrrole (**68**) is sulfonated by heating with a pyridine-sulfur trioxide complex. When both α-positions are blocked, sulfonation occurs at the remaining β-position (**68** → **69**).

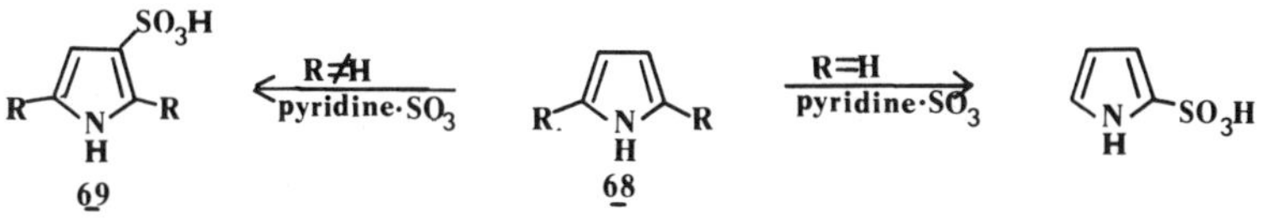

69

68

7.5.2. FURANS

The great acid lability of furan precludes its direct sulfonation by sulfuric acid.[1] Furan can, however, be converted to the 2-sulfonic acid derivative (**70**) by treatment with sulfur dioxide in dioxane or in pyridine.[31] When the furan ring is deactivated, as in the case of furan-2-carboxaldehyde (furfural, **71**), the ring is more stable toward hydrolysis by oleum and the 5-sulfonic acid derivative **72** is obtained.

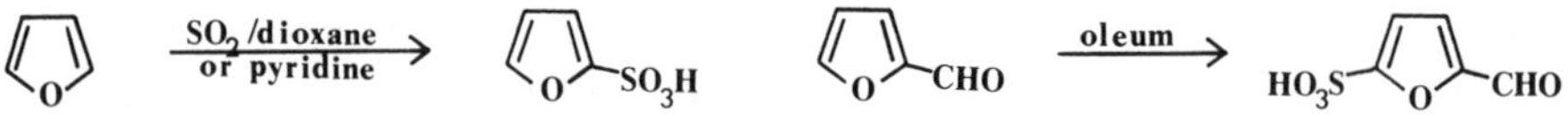

SO_2/dioxane or pyridine → 70; 71 oleum → 72

7.5.3. THIOPHENES

As already noted, thiophene is considerably more stable toward acids than is furan. Thus it is not surprising that thiophene is readily sulfonated at room temperature with 95% sulfuric acid; thiophene-2-sulfonic acid (**73**) is obtained in good yield.[32] 2-Halothiophenes are sulfonated almost exclusively at the open α-position (**74**).[2]

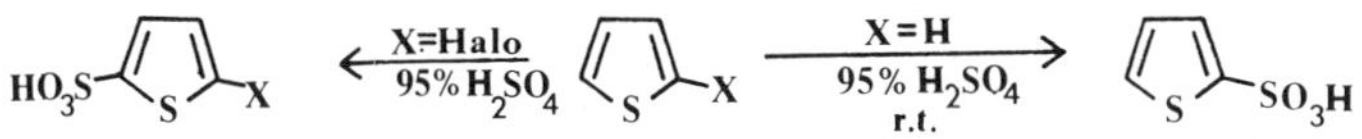

74;(X=Cl; Br; or I) ← X=Halo, 95% H_2SO_4 — X=H, 95% H_2SO_4, r.t. → 73 (69-76%)

7.5.4. SELENOPHENES

Direct sulfonation of selenophene with sulfuric acid[33] or with a pyridine-sulfur trioxide complex[34] affords the selenophene-2-sulfonic acid (**75**) in good yield. Blocking of the 2-position with a carboxylic acid group, as in **76,** directs the sulfonic acid function almost exclusively to the remaining α-position (**77**).[34] The sulfonic acid group is readily replaced by a nitro function when **75** is treated with fuming nitric acid.

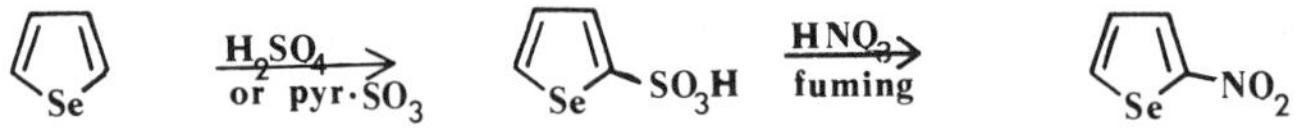

H_2SO_4 or pyr·SO_3 → 75; HNO_3 fuming → 78

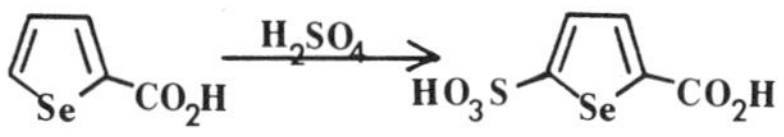

76 H_2SO_4 → 77

7.5.5. TELLUROPHENES

As in the case with attempted nitrations, sulfonation of tellurophene appears to destroy the ring system.[30]

7.6. ALKYLATIONS AND ACYLATIONS

7.6.1. PYRROLES

In contrast to the other ring systems in this class, pyrroles react readily with benzylic as well as allylic halides. These reactions are generally uncontrollable and afford polyalkylated and polymeric products which result from both *N*-

and *C*-alkylation (**79**).[8] When less reactive halides are used, they also react directly with pyrrole at higher temperatures (100–150°C) to yield, ultimately, polyalkylated products. Acylation of pyrrole itself affords either *N*- or *C*-acylated products, depending upon the conditions. Thus when pyrrole (**80**) is treated with acetic anhydride in the presence of sodium acetate, *N*-acetylpyrrole (**81**) is isolated, albeit in low yield. When the alkali metal salts of pyrrole (**82**) are treated with acid chlorides, the *N*-acylpyrroles (**83**) are obtained in high yields.

CH_3I, 100-150°C → → → → **79**

80 —(Ac_2O, NaOAc)→ **81**; **82** (M^+ = alkali metal ion) —(RCOCl)→ **83**

N-Alkylations and *N*-acylations become the minor reaction course, however, when pyrrole Grignard reagents **84** are the starting material; acylation takes place at C-2 (**85**). These reactions become much more controllable when the pyrrole nucleus has been deactivated by the presence of an electron-withdrawing group. For example, Friedel–Crafts propylation of ethyl pyrrole-2-carboxylate (**86**) gives the 4-isopropyl derivative (**87**), along with the 4,5-diisopropyl compound (**88**), in good yields.[35] Ester **86** can also be readily acylated with alkyl dichloromethyl ether under Friedel–Crafts conditions to afford the 5-formyl derivative **89.**[36] Direct *C*-formylation of pyrrole itself can be accomplished by the application of the Gattermann reaction.[37] The same

84 —((1) Ac_2O, (2) H_2O)→ **85**

89 ← **86** —($CH_2=CHCH_3$, $AlCl_3$)→ **87** + **88**

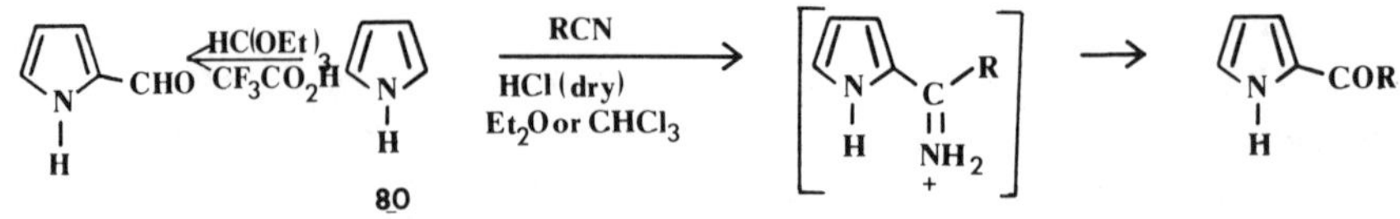

results, a worthwhile alternative to the use of RCN compounds, can also be accomplished by the use of triethyl orthoformate in the presence of trifluoroacetic acid.[38]

Among all of these acylation procedures, the Vilsmeier–Haack reaction is the most useful one for formylation[39] and acylation[40] of pyrroles. In this reaction, a complex of any *N,N*-dialkylamide with phosphoryl chloride is treated with pyrrole. The intermediate imine salt (**90**) is hydrolyzed under mildly alkaline conditions to form the acylated products in very high yields.

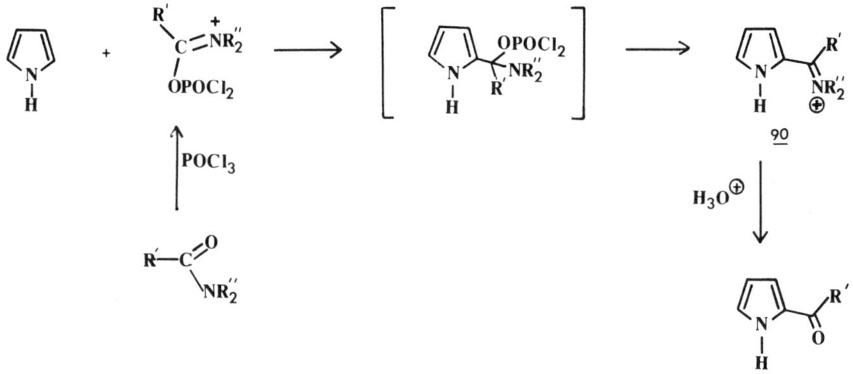

7.6.2. FURANS

Generally, reaction of furan with either an alkylcarbonium or acylonium ion affords the 2-substituted products **91** and **92,** respectively. The carbonium ion (R^+) can be generated simply by treatment of an olefin with phosphoric acid or boron trifluoride. The use of anhydrides in the presence of either boron trifluoride etherate or phosphoric acid affords the desired acylonium ions.[31] An interesting recent variant of these electrophilic substitution reactions involves the use of phosphoric acid on kieselguhr.[41] When this is used as the catalyst for the reaction of furan with isobutene, a mixture of the 2- and 3-*tert*-butylfurans (**93** and **94**) is obtained. The relative proportion of the two isomers obtained in this reaction is temperature dependent. In fact, it is highly likely that the 3-isomer (**94**) is formed from the corresponding 2-isomer (**93**).

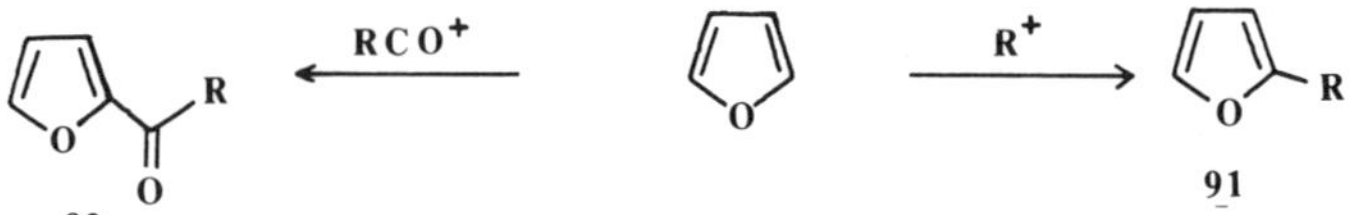

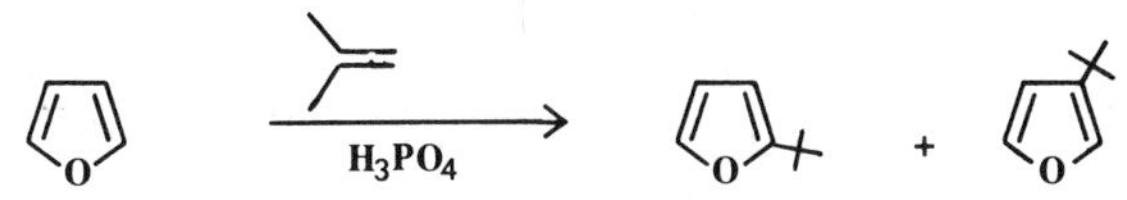

H_3PO_4

93 94

When the furan ring is deactivated by the presence of an electron-withdrawing substituent, such as a carboxaldehyde function (**49**)[42] or an acetyl group (**95**),[43] the isopropylation of these compounds affords a mixture of isopropylfurans. The apparent absence of the 5-isopropyl derivative from **49** is noteworthy.

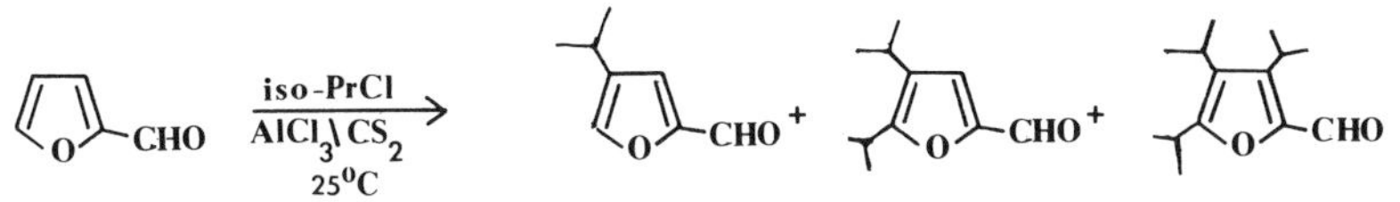

CHO iso-PrCl $AlCl_3$\ CS_2 25°C CHO + CHO + CHO

49

(9:7:5)

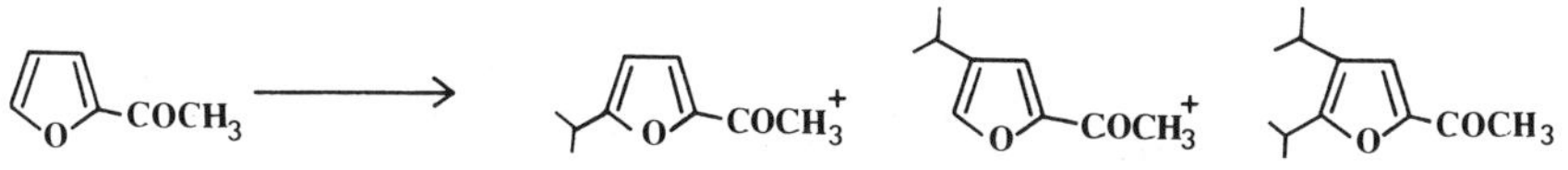

$COCH_3$ $COCH_3$ + $COCH_3$ + $COCH_3$

95

(4:10:7)

Both isopropylation and *tert*-butylation of methyl furan-2-carboxylate (**96**) affords the 5- as well as 4-isomers (**97** and **98,** respectively). The 5-isomers **97** have been shown to rearrange to the 4-isomers in these instances.[43]

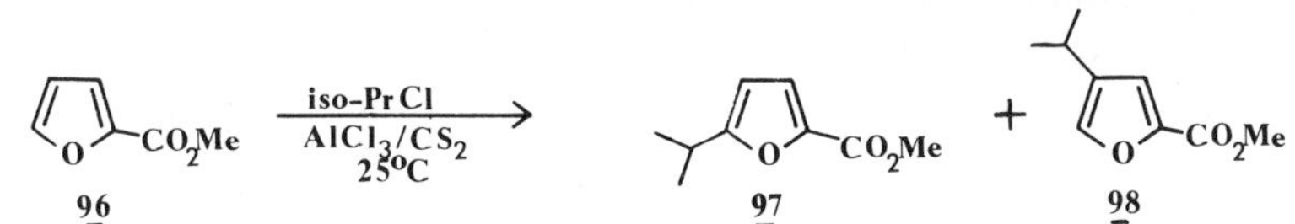

CO_2Me iso-PrCl $AlCl_3/CS_2$ 25°C CO_2Me + CO_2Me

96 97 98

7.6.3. THIOPHENES

As is true for the formylation and acylation of pyrrole, the Vilsmeier–Haack procedure is superior to other procedures of this type for substituting the thiophene ring.[39,40] Acetylation of thiophene can also be accomplished by use of acetic anhydride in the presence of stannic chloride.[16] The 2-acetyl derivative is formed at a rate 200 times greater than is the 3-isomer, and the 2-selectivity in alkylation and acylation reactions is not affected by the presence of either a 2-methyl (**99**), 2-*tert*-butyl (**100**), or 2-methylthio (**101**) substituent.[44] When a methoxy group is substituted at the 2-position, formylation still affords predominantly the α-substituted product (**102**), with only traces of the 3-substituted isomer (**103**) being detected.

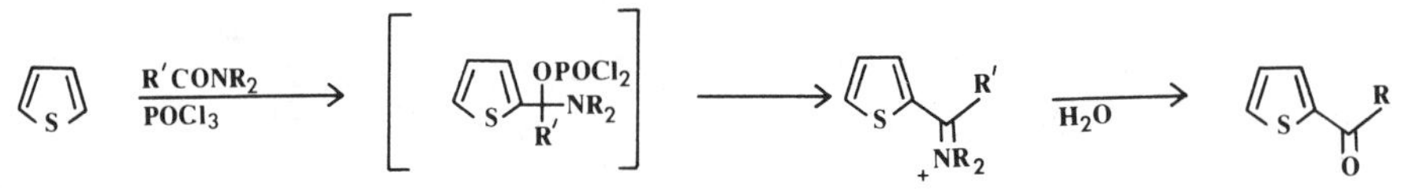

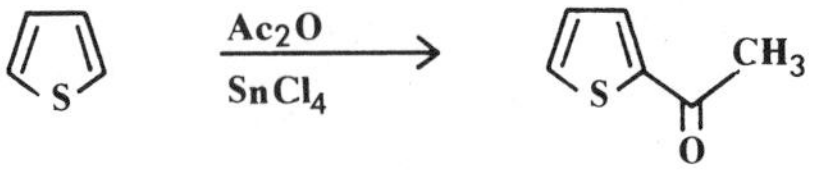

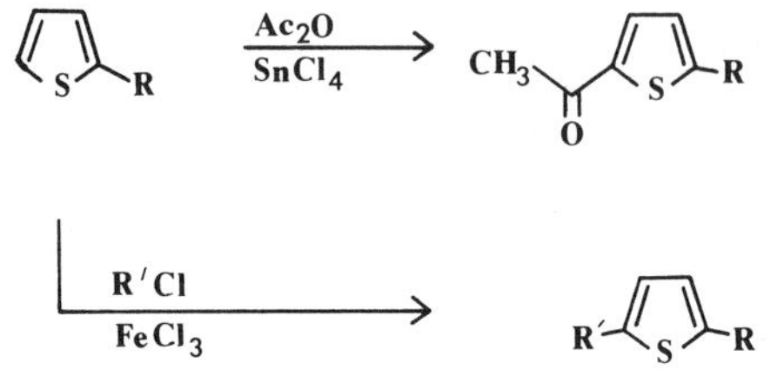

$R'CONR_2$ / $POCl_3$ → $OPOCl_2$, NR_2, R′ → R′, $^+NR_2$ → H_2O → R, O

Ac_2O / $SnCl_4$ → CH_3, O

R → Ac_2O / $SnCl_4$ → CH_3, O, R

R′Cl / $FeCl_3$ → R′, R

99 (R=CH_3)
100 (R= *tert*-bu)
101 (R=SCH_3)

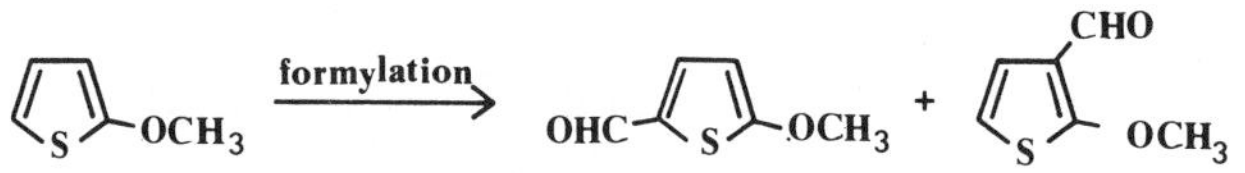

OCH_3 → formylation → OHC, OCH_3 + CHO, OCH_3

102 103

The presence of a β-phenyl group appears to activate the α-position and adjacent β-position to a much smaller extent: **104** gives **105** and **106** as the major products.[45] Electron-withdrawing substituents at the 2-position of thiophene (for example, **107**) have a pronounced effect on the substitution patterns.[46,47] The use of an excess of aluminum chloride, the so-called "swamping catalyst effect",[48] decreases the formation of derivatives with substituents at the 3- and 5-positions (**108**) and thus increases the formation of the 4-substituted products.

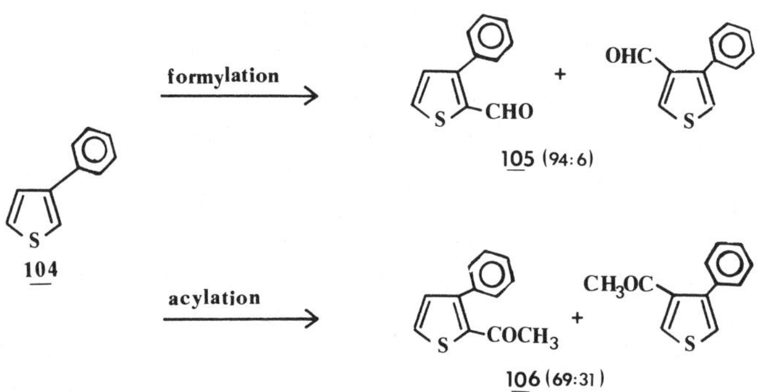

104

formylation → CHO + OHC

105 (94:6)

acylation → $COCH_3$ + CH_3OC

106 (69:31)

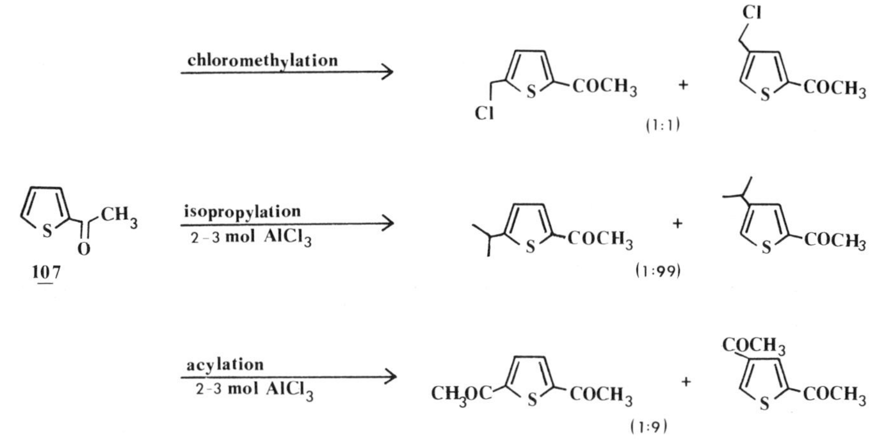

7.6.4. SELENOPHENES

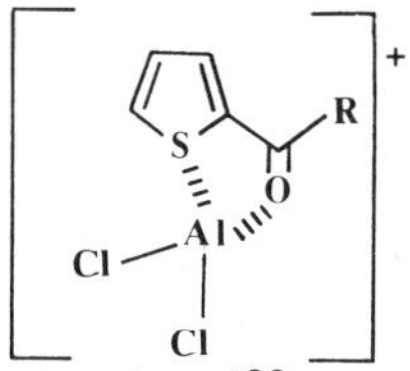

Formylation of selenophene by the Vilsmeier procedure affords the 2-formyl derivative **109** in high yield.[49a] The 2- and 3-methylselenophenes (**110** and **111**) behave similarly.[49b]

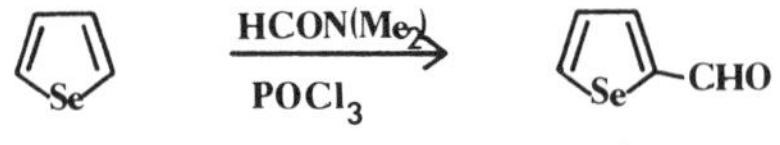

Friedel–Crafts conditions result in the formation of 2-acetylselenophene (**112**); other, modified procedures afford the same product.[50–52] The chloromethylated selenophene **113** is readily obtained when hydrogen chloride is passed into a mixture of selenophene and formalin in ethylene dichloride.[53]

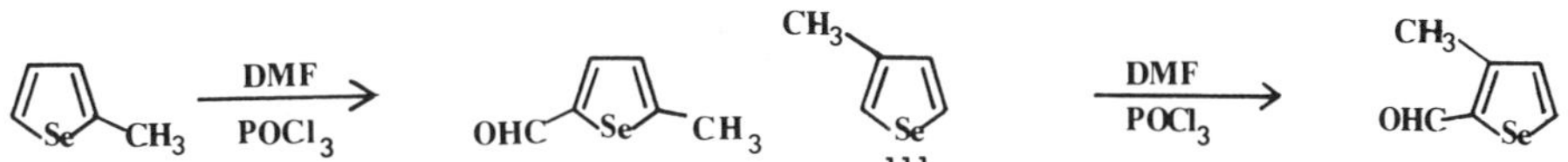

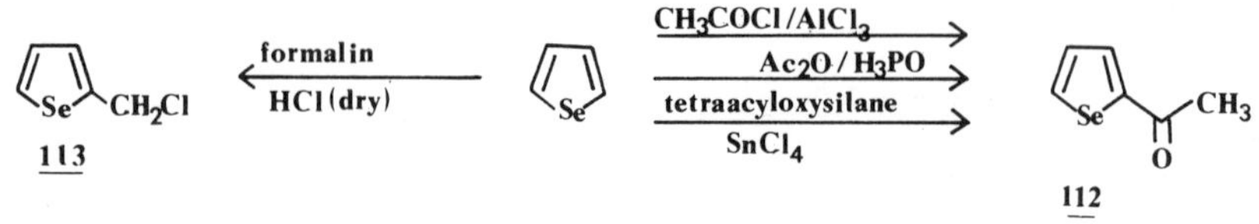

113 (Se, CH_2Cl) ← formalin, HCl (dry) — selenophene — $CH_3COCl/AlCl_3$; Ac_2O/H_3PO_4; tetraacyloxysilane, $SnCl_4$ → 112 (Se, $COCH_3$)

7.6.5. TELLUROPHENES

Under Friedel-Crafts conditions tellurophene reacts with acetic anhydride in the presence of stannic chloride to produce the expected 2-acetyl derivative (**114**).[20,30] When an electron-withdrawing group is present (as in **115**) acetylation still occurs at the remaining α-position.

Tellurophene —Ac_2O / $SnCl_4$→ 114 (Te, Ac); 115 (Te, CO_2Me) —Ac_2O / $SnCl_4$→ Ac (Te) CO_2Me

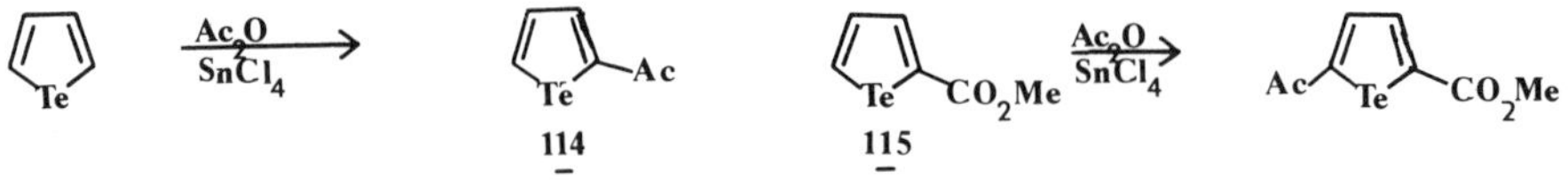

7.7. SUMMARY OF SUBSTITUENT EFFECTS

1. Electron-withdrawing substituents (Y) at the position α to the heteroatom generally cause substitution at the 4- and/or 5-position.

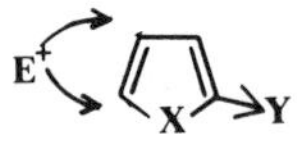

E+ X Y

2. Electron-donating substituents (Z) at the position α to the heteroatom tend to cause substitution at both the remaining α-position (C-5) and at the 3-position.

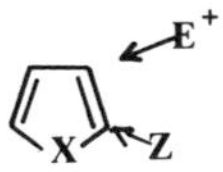

E+ X Z

3. Electron-withdrawing substituents (Y) at the position β to the heteroatom facilitate substitution at position 5.

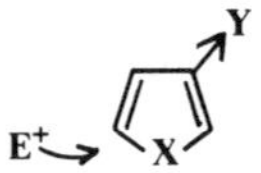

Y E+ X

4. Electron-donating substituents (Z) at the position β to the heteroatom generally cause substitution to take place at the 2-position.

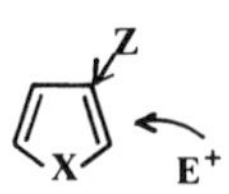

REFERENCES

1. G. Marino, *Adv. Heterocycl. Chem.* **13,** 235 (1971).
2. R. J. Abraham and H. J. Bernstein, *Can. J. Chem.* **37,** 1056 (1959).
3. A. Pieroni and A. Moggi, *Gazzetta* **53,** 120 (1923).
4. P. Salomaa, A. Kankaaupera, E. Nikander, K. Kaipaineu, and R. Aaltonen, *Acta Chem. Scand.* **27,** 153 (1973).
5. E. N. Zvyagintesva, T. A. Yukushina, and A. I. Shatenstein, *J. Gen. Chem.* (*USSR*) **38,** 1933 (1968).
6. B. Ostman and S. Olsson, *Ark. Kem.* **15,** 275 (1960).
7. H. Fischer and H. Orth, *Die Chemie des Pyrrols,* Akademischer Verlap Leipzig, Vols. II and III, 1937 and 1940; H. W. Gilow and D. E. Barton, *J. Org. Chem.* **46,** 2221 (1981).
8. R. A. Jones and G. P. Bean, *The Chemistry of Pyrroles,* Academic Press, New York, 1977, p. 134.
9. A. H. Corwin and G. G. Kleinspehn, *J. Am. Chem. Soc.* **75,** 2089, 5295 (1953).
10. A. P. Tenet'ev, L. I. Belen'ki, and L. A. Yanovskaya, *Zh. Obshch. Khim.* **24,** 1265 (1955).
11. N. Clauson-Kaas, *Acta Chem. Scand.* **1,** 379 (1947).
12. E. Baciocchi, S. Clementi, and G. V. Sebastiani, *J. Chem. Soc. Chem. Commun.* **1975,** 875.
13. B. P. Roques, M. C. Fournie-Zaluski, and R. Oberlin, *Bull. Soc. Chim. Fr.* **1975,** 2334.
14. A. P. Dunlap and F. N. Peters, *The Furans,* Reinhold, New York, 1953.
15. (a) H. L. Coonradt, H. D. Hartough, and G. C. Johnson, *J. Am. Chem. Soc.* **70,** 2564 (1948); (b) A. Hallberg, S. Liljefors, and P. Pedaja, *Syn. Commun.* **11,** 25 (1981).
16. P. Linda and G. Marino, *Ric. Sci.* **37,** 424 (1967); S. Clementi, P. Linda, and G. Marino, *J. Chem. Soc.* (*B*) **1968,** 397.
17. H. Suginome and S. Umezawa, *Bull. Chem. Soc. Japan* **11,** 157 (1936).
18. Yu. K. Yur'ev, N. N. Magdesieva, and A. T. Monakhova, *Zh. Obshch. Khim.* **30,** 2726 (1960).
19. H. Gerding, G. Milazzo, and H. H. K. Rossmark, *Rec. Trav. Chim.* **72,** 957 (1953).
20. F. Fringuelli and A. Taticchi, *J. Chem. Soc. Perkin I* **1972,** 199.
21. A. R. Cooksey, K. J. Morgan, and D. P. Morrey, *Tetrahedron* **26,** 5101 (1970).
22. H. J. Anderson and S. J. Griffiths, *Can. J. Chem.* **45,** 2227 (1967).
23. P. E. Sonnet, *J. Heterocycl. Chem.* **7,** 399 (1970).
24. D. Lola, K. Venters, E. Liepius, M. Trusule, and S. Hillers, *Khim. Geterotsikl. Soedin.* **1976,** 601 [*Chem. Abstr.* **85,** 77957 (1976)].
25. D. Lola, K. Venters, E. Liepius, and S. Hillers, *Khim. Geterotsikl. Soedin,* **1975,** 883 [*Chem. Abstr.* **84,** 4749 (1976)].
26. B. Ostman, *Ark. Kem.* **19,** 499 (1962).

27. J. Fabian, S. Scheithauer, and R. Mayer, *J. Prakt. Chem.* **311,** 45 (1969).
28. Yu. K. Yur'ev, E. L. Zaitseua, and G. G. Rosantsev, *Zh. Obshch. Khim.* **30,** 2207 (1960).
29. Yu. K. Yur'ev and E. L. Zaitseva, *Zh. Obshch. Khim.* **30,** 859 (1960).
30. F. Fringuelli, G. Marino, and A. Taticci, *Adv. Heterocycl. Chem.* **21,** 119 (1977).
31. P. Bosshard and C. H. Eugte, *Adv. Heterocycl. Chem.* **1,** 377 (1966).
32. W. Stienkopt and W. Ohse, *Ann. Chem.* **437,** 14 (1924).
33. S. Vinezawa, *Bull. Soc. Chem. Japan* **11,** 775 (1936).
34. F. G. Kataev and A. E. Zimkin, *Uch. Zap. Kasansk. Gos. Univ.* **117,** 174 (1957).
35. J. K. Groves, H. J. Anderson, and H. Nagy, *Can. J. Chem.* **49,** 2427 (1971).
36. P. E. Sonnet, *J. Med. Chem.* **15,** 97 (1971).
37. H. Fischer and W. Zerweck, *Chem. Ber.* **55,** 1942 (1922).
38. P. S. Clezy, C. J. R. Fookes, and A. J. Liepa, *Aust. J. Chem.* **25,** 1979 (1972).
39. P. A. Burbridge, G. L. Collier, A. H. Jackson, and G. W. Kenner, *J. Chem. Soc.* (*B*) **1967,** 930.
40. P. Rothemunt, *J. Am. Chem. Soc.* **58,** 625 (1936).
41. L. I. Belen'kii, *Isv. Akad. Nauk. S.S.S.R., Ser. Khim* **1975,** 344.
42. M. Valenta and I. Koubek, *Coll. Czech. Chem. Commun.* **41,** 78 (1976).
43. H. J. Anderson and C. W. Huang, *Can. J. Chem.* **48,** 1550 (1970).
44. S. Clementi, P. Linda, and G. Marino, *J. Chem. Soc.* (*B*) **1970,** 1153, and refs. cited therein.
45. N. Gjøs and S. Gronowitz, *Acta Chem. Scand.* **24,** 99 (1970).
46. D. J. Chadwick, J. Chambers, H. E. Hargraves, G. D. Meakins, and R. L. Snowden, *J. Chem. Soc. Perkin I* **1973,** 2327.
47. L. I. Belen'kii, I. B. Karmanova, Yu. B. Volkenshtein, and Ya. L. Gol'dfarb, *Bull. Acad. Sci. USSR* **1971,** 878.
48. Ya. L. Gol'dfarb, Yu. B. Volkenstein, and L. I. Belen'kii, *Angew. Chem. Int. Ed.* **7,** 519 (1968).
49. (a) Yu. K. Yur'ev and N. N. Mezentsova, *Zh. Obshch. Khim.* **27,** 179 (1957); (b) *ibid.,* **29,** 1917 (1959).
50. N. P. Buu-Hoi, P. Demerseman, and R. Royer, *Compt. Rend.* **237,** 397 (1953).
51. E. G. Kataeva, M. V. Palkina, E. G. Kataev and M. V. Palkina, *Uch. Zap. Kazan. Gosudarst. Univ. im. V. I. Ulyanova-Lenina, Khim.* **113,** 115 (1953) [*Chem. Abstr.* **52,** 3762 (1958)].
52. Yu. K. Yur'ev, N. K. Sadovaya, and V. V. Titov, *Zh. Obshch. Khim* **28,** 3036 (1958).
53. Yu. K. Yur'ev, N. K. Sadovaya, and V. V. Titov, *Zh. Obshch. Khim* **32,** 259 (1962).

Chapter 8

NUCLEOPHILIC SUBSTITUTION. FIVE-MEMBERED π-EXCESSIVE HETEROAROMATIC SYSTEMS

8.1. GENERAL COMMENTS

Direct nucleophilic substitutions of hydrogen in pyrrole are unknown, but have been observed in tellurophenes and thiophenes. Nucleophilic substitution of halogen in some of these ring systems has been described, although relatively harsh conditions must be employed. *A priori,* consideration of Wheland intermediates in these reactions suggests that in the case of thiophene, and probably selenophene and tellurophene, the heteroatom may cause additional stabilization of the intermediates by involvement of its *d*-orbitals (**1**)

X Y S

1

8.2. PYRROLES

Although electrophilic substitutions of pyrrole have been studied in great detail, there is a paucity of studies dealing with nucleophilic substitutions. It is known, however, that both 2-halo- and 3-halopyrroles behave in nucleophilic displacement reactions in a manner similar to aryl halides. Thus 2-chloropyrrole (**2**) reacts with neither potassium *tert*-butoxide nor sodium in liquid ammonia, nor is the halogen reduced by lithium aluminum hydride.[1]

KOtbu; $Na/NH_3(l)$; $LiAlH_4$ → N. R.

2

8.3. FURANS

As is true for pyrrole, nucleophilic substitutions of halofurans have not been extensively studied. It is known that 2-bromo- and 2-chlorofuran (**3** and **4**) react with piperidine (**6**) about 10 times faster than do the corresponding benzene derivatives.[2] On the other hand, neither 2-bromo- nor 2-iodofuran (**5**) reacts with sodium methoxide at 100°C.[3,4] The reaction of 2-iodofuran (**5**) with **6** occurs 2.2 times faster than does the same reaction on 2-iodo-5-methylfuran (**7**).[4] 3-Iodofuran (**8**) is transformed to 3-methoxyfuran upon treatment with sodium methoxide with cuprous oxide in methanol and to 3-cyanofuran with cuprous cyanide in quinoline.[5]

NR (X= Br, I) ← $NaOCH_3$, 100°C — 3 (X= Br), 4 (X= Cl), 5 (X= I) — (6), 200°C →

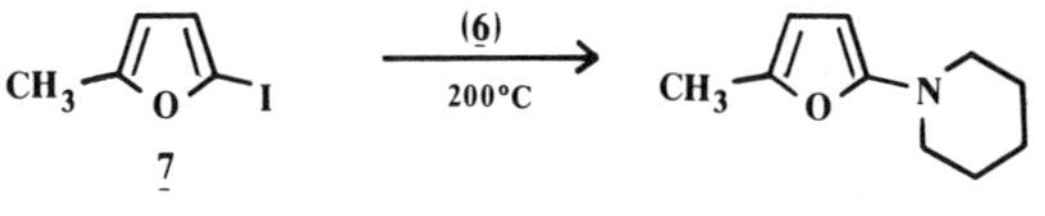

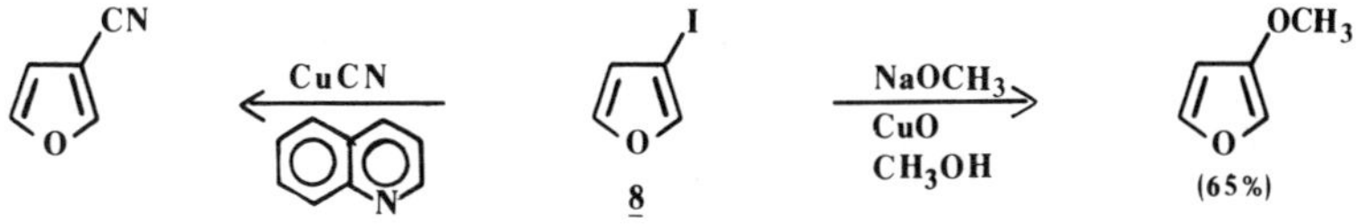

Consistent with the behavior of haloaromatics, the presence of electron-withdrawing groups on the furan nucleus facilitates nucleophilic displacement. Thus 5-halo-2-nitrofurans (**9**) and methyl 5-halofuran-2-carboxylates (**10**) react readily with nucleophiles.[2,6] It is of considerable interest that in these reactions, the furans react approximately 10 times faster than the corresponding benzene analogues! This apparent incongruity has also been noted in the thiophene series.

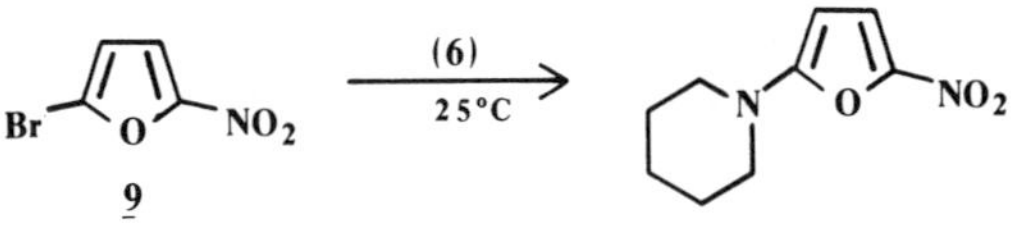

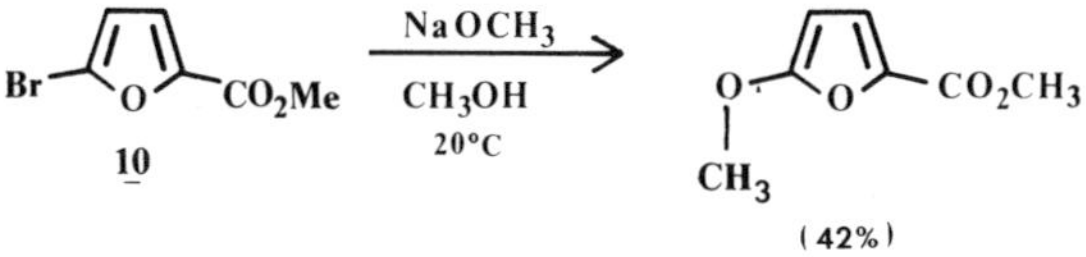

8.4. THIOPHENES

Most of the nucleophilic substitutions that have been examined in the thiophene ring system have involved halonitrothiophenes. These studies have shown that nucleophilic substitution occurs much more readily within the thiophene series than it does for the corresponding benzene compounds. Table 8.1 lists the relative reaction rates for piperidine substitution of a number of bromothiophenes and related benzenoid derivatives. A glance at these data shows that, regardless of the relative position of the bromo substitutent with respect to the nitro group or heteroatom, the thiophenes are at least 1000 times more reactive than the corresponding benzene analogues. This increased reactivity has also been noted in the electrophilic substitutions of thiophene, with respect to the corresponding benzenoid compounds. A comparison of the Wheland intermediates involved in the nucleophilic displacement of the bromo substituent in 2-bromo-5-nitrothiophene (**11**) and 4-bromonitrobenzene (**12**) brings to light the reason that the thiophenes are more reactive. Clearly, a nitro group can delocalize the negative charge in the σ-complex with greater

TABLE 8.1. Relative Nucleophilic Substitution Rates of a Series of Bromothiophenes and Bromobenzenes with Piperidine at 25°C

Compound	Relative Rate	Compound	Relative Rate
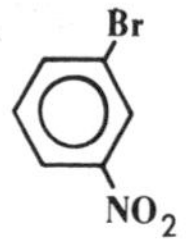 Br, NO_2 (benzene)	1	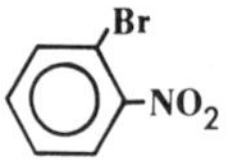 Br, NO_2 (benzene)	1620
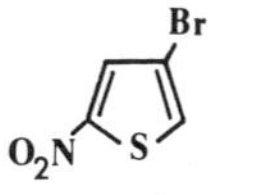 Br, O_2N, S	1360	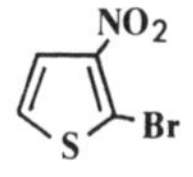 NO_2, S, Br	632, 000
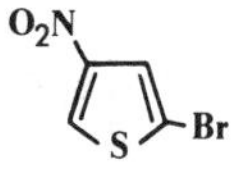 O_2N, S, Br	"very fast"	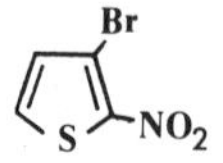 Br, S, NO_2	2,500,000
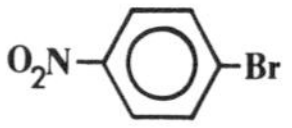 O_2N–(benzene)–Br	185	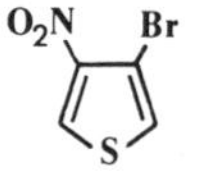 O_2N, Br, S	"very fast"
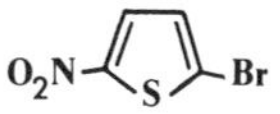 O_2N–(S)–Br	28,400		

efficiency in thiophene than in benzene. The involvement of charged species, such as **13** and **14**, also greatly stabilizes the Wheland intermediate in the thiophene instances.[8]

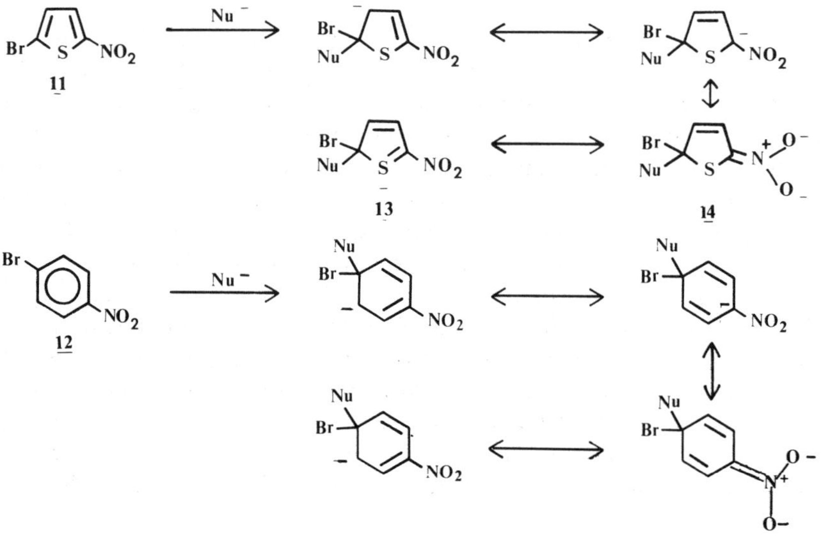

11 $\xrightarrow{Nu^-}$ ⟷ 13 ⟷ 14

12 $\xrightarrow{Nu^-}$ ⟷ ⟷

Halothiophenes undergo several intriguing nucleophilic transformations. For example, when 2-bromothiophene (**15**) is treated briefly with sodium amide in liquid ammonia, it is converted to 3-bromothiophene (**16**); when potassium amide is used, 3-aminothiophene (**17**) is formed instead.[9–11] 2-Iodothiophene (**18**) upon treatment with the potassium salt of *N*-methylaniline (**19**) affords 3-iodothiophene (**20**) in excellent yield. In fact, these reactions are of great synthetic utility for the formation of 3-substituted thiophenes.

Br NaNH$_2$ (6 equiv.) NH$_3$ (l) 16 Br 15 KNH$_2$ (6 equiv.) NH$_3$(l) NH$_2$ 17

I 18 KNMe(C$_6$H$_5$) (19) (6 moles) I 20

There is considerable evidence that these transformations involve a very complex series of reactions initiated by formation of a carbanion **21,** rather than an aryne-type intermediate **22.**

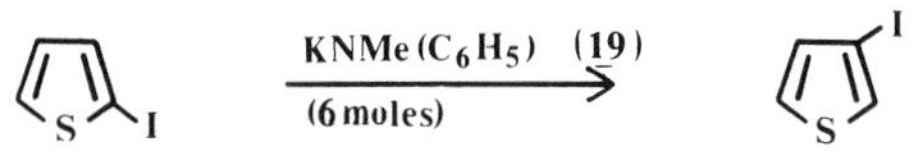

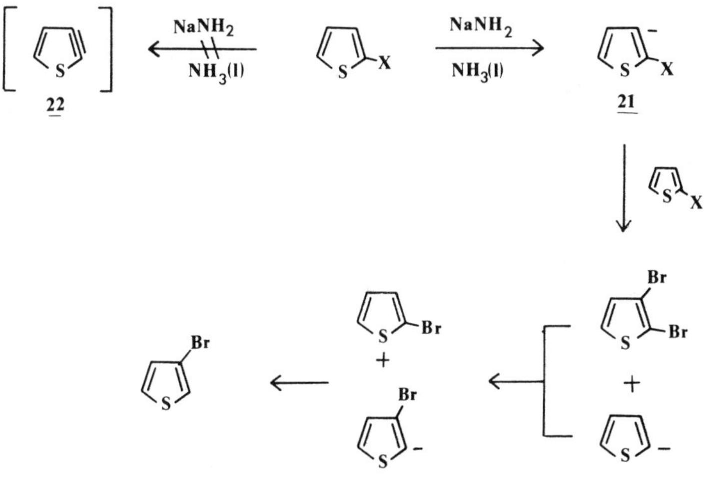

22 NaNH$_2$ NH$_3$(l) X NaNH$_2$ NH$_3$(l) X 21

X

Br Br + Br Br Br + Br

Copper-mediated nucleophilic substitutions of halothiophenes are also of great synthetic utility. Examples of some typical transformations follow[12–14]:

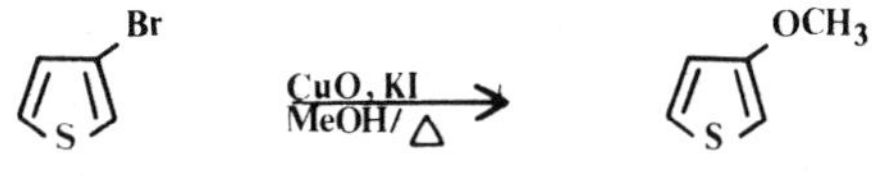

Br 16 CuO, KI MeOH/Δ OCH$_3$ (81%)

Metal-halogen exchange reactions have found great synthetic utility in thiophene chemistry. Bromo- as well as iodothiophenes react rapidly with butyllithium at −100°C in ether[15-19]; because fluoro- and chlorothiophenes are generally inert, selective interchanges can be observed (for example, **23**). That the 2-bromo substitutent is more labile to metal-halogen exchange than the corresponding 3-isomer is demonstrated by **24.** These lithio derivatives (also obtainable by nucleophilic attack on a ring H when no suitable halogen is present) can be converted to the corresponding carboxylic acids by this procedure.

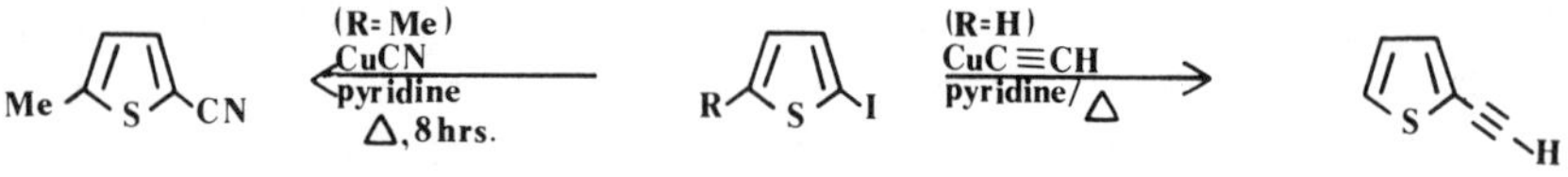

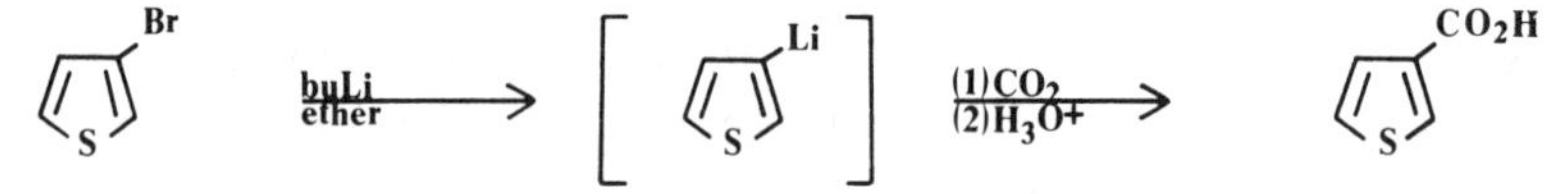

8.5. SELENOPHENES

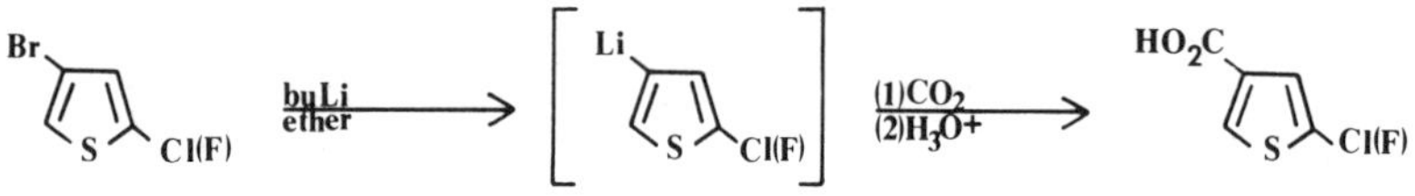

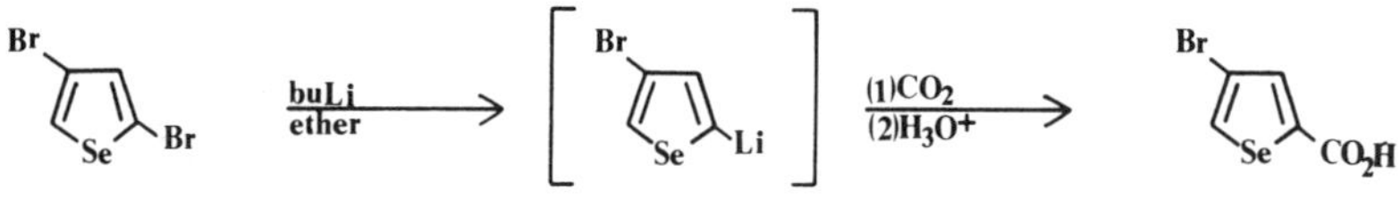

Treatment of bromoselenophene (**25**) with cuprous cyanide in pyridine affords the corresponding nucleophilic displacement product **26.**[20] As is the case in thiophene chemistry, the haloselenophenes react with alkyl- and aryllithium reagents to afford the corresponding lithioselenophenes (**27**) which can be transformed, by appropriate electrophiles, into a variety of substituted selenophenes.[21,22]

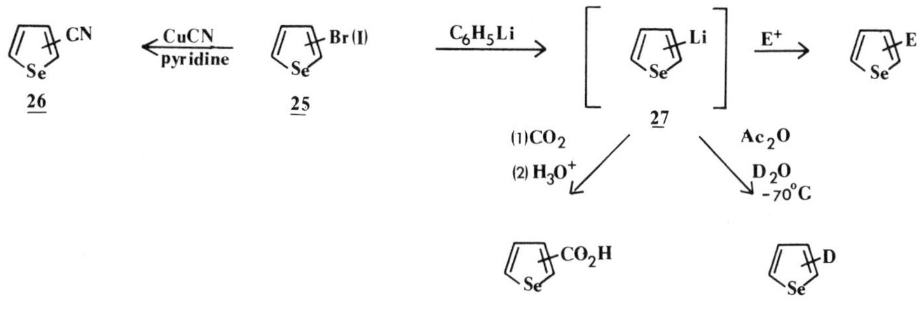

8.6. TELLUROPHENES

Halotellurophenes (**28–30**)[22] have been synthesized by a very circuitous route, since direct halogenation[23] of tellurophenes affords the 1,1-dihalo compounds **31**. The key 2-lithiotellurophene (**32**) is most simply prepared by H—Li exchange (nucleophilic attack of an α-ring *hydrogen*). Such lithio intermediates, as is the case for all of the other π-excessive, five-membered, heteroaromatic rings described so far, lend themselves to useful synthetic transformations.[22]

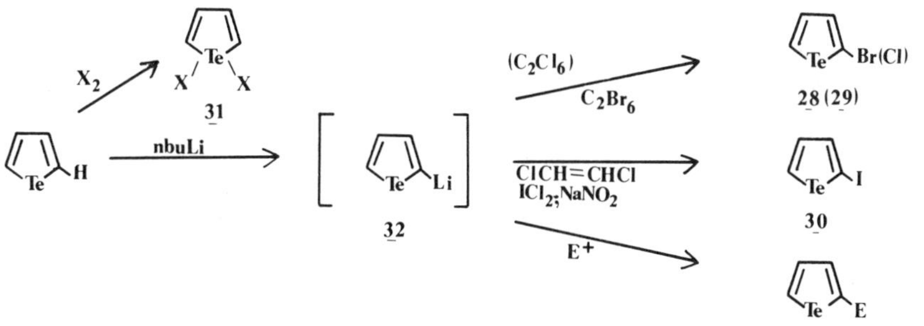

[E^+: $(CF_3CO)_2O$; Me_2S_2; CO_2; $HCONHC_6H_5$; Me_2SO_4; MeCHO; Ac_2O]

REFERENCES

1. J. A. Joule and G. F. Smith, *Heterocyclic Chemistry,* Van Nostrand-Reinhold, London, 1972, p. 211.
2. G. Illuminati, *Adv. Heterocycl. Chem.* **3,** 285 (1964).
3. D. G. Manly and E. D. Amstutz, *J. Org. Chem.* **21,** 516 (1956).
4. D. G. Manly and E. D. Amstutz, *J. Org. Chem.* **22,** 133 (1957).
5. S. Gronowitz and G. Sörlin, *Ark. Kem.* **19,** 515 (1962).
6. D. Spinelli, G. Guanti, and C. Dell'Erba, *Boll. Sci. Fac. Chim. Ind. Bologna* **25,** 71 (1967) [*Chem. Abstr.* **68,** 2389 (1968)].

7. G. Doddi, A. Poretti, and F. Stegel, *J. Heterocycl. Chem.* **11,** 97 (1974).
8. G. Illuminati, *Adv. Heterocycl. Chem.* **3,** 69 (1964).
9. M. G. Reinecke and H. W. Adickes, *J. Am. Chem. Soc.* **90,** 511 (1968).
10. M. G. Reinecke, H. W. Adickes, and C. Pyun, *J. Org. Chem.* **36,** 3820 (1971).
11. H. C. van der Plas, D. A. de Bie, G. Geurtsen, M. G. Reinecke, and H. W. Adickes, *Rec. Trav. Chim.* **93,** 33 (1974).
12. A. Vecchi and G. Melone, *J. Org. Chem.* **22,** 1636 (1957).
13. S. Gronowitz, *Ark. Khim.* **12,** 239 (1958).
14. R. F. Curtis and J. A. Taylor, *J. Chem. Soc.* (*C*) **1969,** 1813.
15. S. Gronowitz and A. Bugge, *Acta Chem. Scand.* **19,** 1271 (1965).
16. P. Moses and S. Gronowitz, *Ark. Khim.* **18,** 119 (1962).
17. S. Gronowitz, A. B. Hornfelt, and R. Hakensson, *Ark. Khim.* **17,** 1652 (1961).
18. S. Gronowitz and B. Holm, *Acta Chem. Scand.* (*B*) **36,** 505 (1976).
19. S. Gronowitz, *Adv. Heterocycl. Chem.* **1,** 1 (1963).
20. Yu. K. Yur'ev, N. K. Sadovaya, and E. A. Grekova, *Zh. Obshch. Khim.* **34,** 847 (1964).
21. Yu. K. Yur'ev and N. K. Sadovaya, *Zh. Obshch. Khim.* **26,** 3154 (1956).
22. F. Fringuelli, G. Marino, and A. Tatticchi, *Adv. Heterocycl. Chem.* **21,** 119 (1977).
23. E. H. Braye, W. Hübel, and I. Caplier, *J. Am. Chem. Soc.* **83,** 4406 (1961); W. Mack, *Angew. Chem. Int. Ed.* **4,** 245 (1965); **5,** 896 (1966).

Chapter 9

FREE RADICAL AND PHOTOCHEMICAL REACTIONS. FIVE-MEMBERED π-EXCESSIVE HETEROAROMATIC SYSTEMS

9.1. GENERAL COMMENTS

Since it has been amply demonstrated that all the π-excessive unsaturated heterocycles discussed so far are aromatic, although to differing degrees, it is certainly anticipated that these compounds will also undergo free radical substitutions and photochemical transformations typical of benzenoid systems. These reactions are, of course, expected to be altered, often significantly, by the presence of the heteroatoms.

It is anticipated *a priori* that, as in the nucleophilic and electrophilic reactions, some selectivity with respect to the position of free radical attack should exist. Thus, whereas free radical substitution of benzene itself can give only one monosubstitution product, pyrrole, for example, might be expected to give any one or all three possible monosubstitution products (**1–3**). On the other hand, only two different monosubstitution products (**4, 5**) are possible for furan, thiophene, selenophone, and tellurophene.

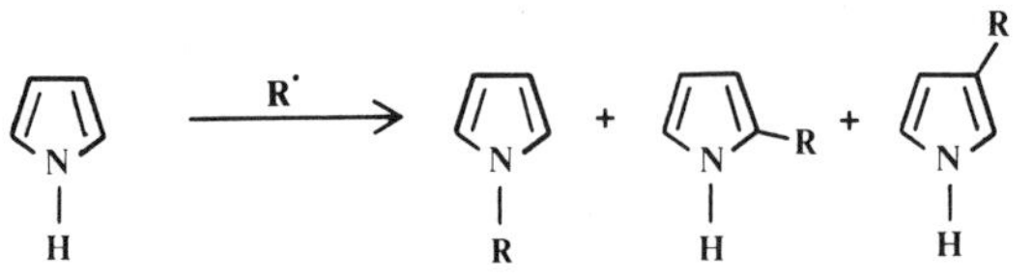

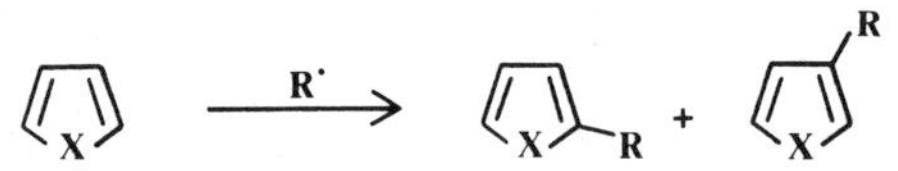

Unfortunately, no comprehensive study has yet been undertaken to answer the questions concerning regioselectivity in this series. Free radical arylation of furan and thiophene has been examined with reference to benzene reactivity under the same reaction conditions.[1] This study shows that furan undergoes arylation exclusively at the 2-position (**6**), whereas thiophene gives mainly 2-arylthiophene (**7**) along with some of the 3-isomer (**8**). Furan is 12 times more reactive than benzene in this reaction and 4.5 times more reactive than thiophene.[1]

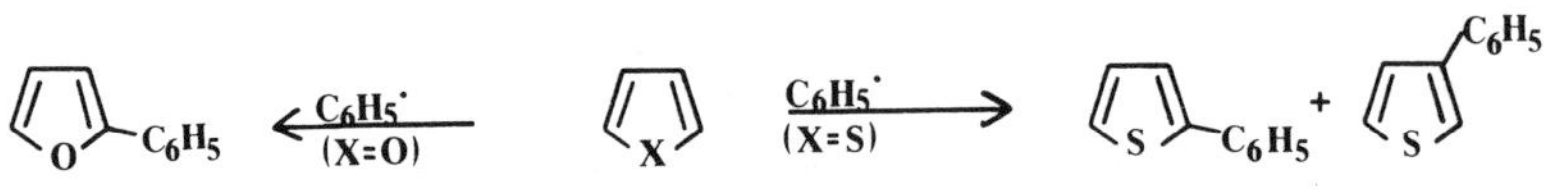

9.2. PYRROLES

Pyrrole itself, when treated with triphenylmethyl radicals, affords the dihydropyrrole **9**[2] rather than a simple substitution product. Pyrrole, when allowed to react with *tert*-butylperoxyl radical, yields the quasi dimer **10.**

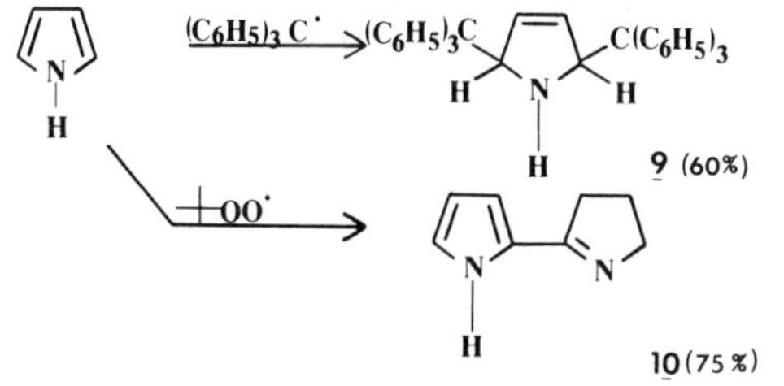

N-Methylpyrrole (**11**), on the other hand, behaves in a more "normal" manner when treated with benzoyl peroxide in that it forms the 2-benzoyloxy derivative **12.**[3] This "normalcy" is not carried over into the reaction of *N*-methylpyrrole (**11**) with *tert*-butyl hydroperoxide, where only traces of **13** and **14** are isolated.[4] These compounds are clearly products resulting from initial hydrogen abstraction from the methyl group, whereupon the resulting radical dimerizes to form **13** or attacks another molecule of starting material (**11**) to form **14.**

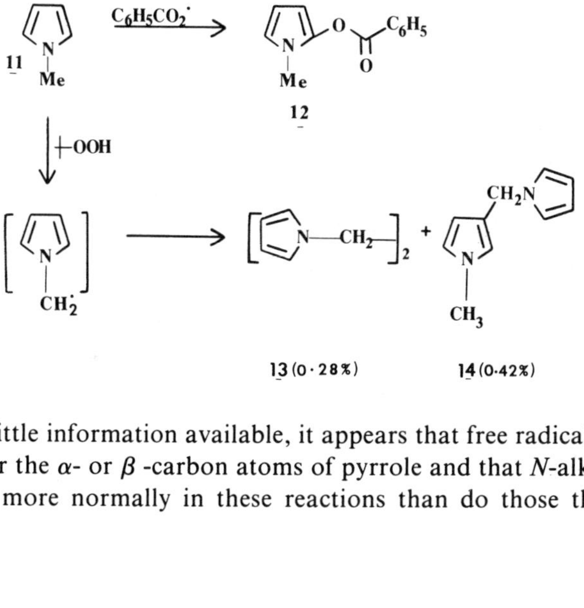

From the little information available, it appears that free radical attack can occur at either the α- or β -carbon atoms of pyrrole and that *N*-alkylated pyrroles behave more normally in these reactions than do those that are not *N*-alkylated.

9.3. FURANS

Furan undergoes both addition and substitution reactions with radicals. Thus furan reacts with aryl radicals, generated either by the decomposition of *N*-nitrosoacetanilide (Gomberg reaction) or by diazotization of aniline under aprotic conditions,[5–7] to afford exclusively 2-phenylfuran (**15**).

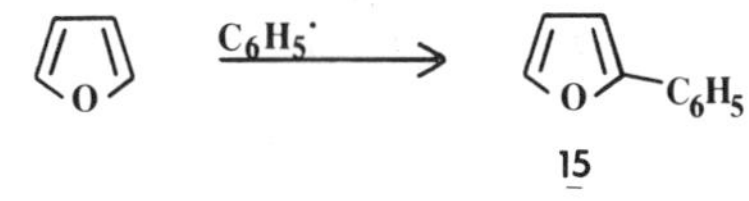

In contrast, when treated with either dibenzoyl peroxide[6] or phenylazotriphenylmethane,[7] furan affords the *cis*- and *trans*-2,5-disubstitution products (**16, 17** and **18, 19**, respectively). The suggestion has been made that these 2,5-addition products may be formed because of the relatively greater stability of the intermediate free radicals (**20** and **21**, respectively). However, this does not account for the formation of 2-phenylfuran when the phenyl radical is generated by other means.

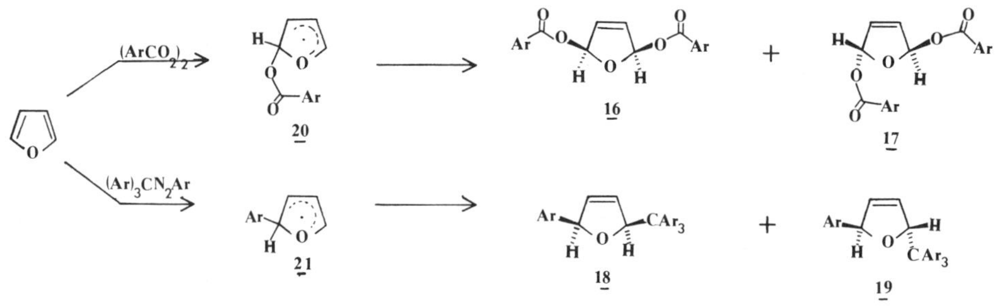

The 2-substituted furans **22**, when allowed to react with aryl radicals, yield 2,5-disubstituted products **23**.[5] When 2,5-dimethylfuran reacts with *N*-nitrosoacetanilide, no 3-substituted-2,5-dimethylfuran (**24**) is obtained; rather, attack at the methyl group takes place to form **25**.[5]

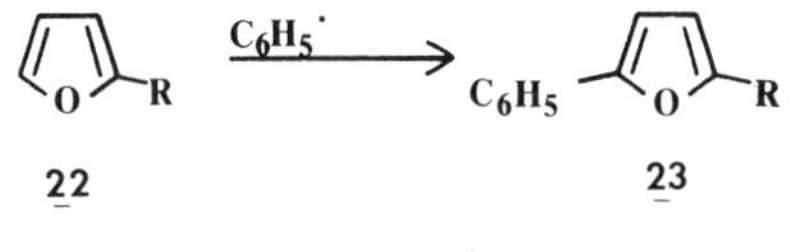

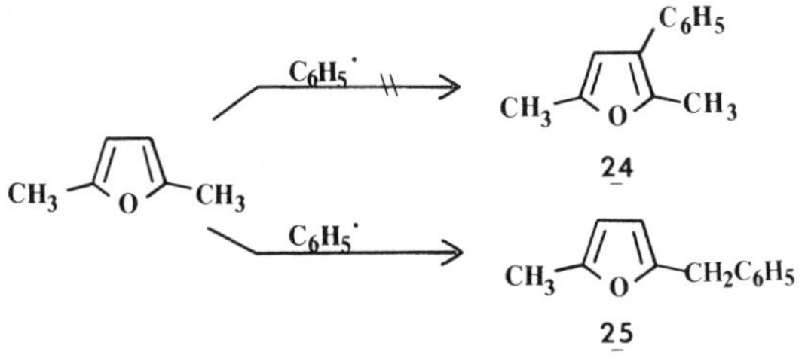

Furan can undergo photochemical [4 + 4] and [2 + 2] cycloaddition reactions. Thus when 2-acetylfuran is irradiated in the presence of 2,3-dimethyl-2-butene, two major products (**26** and **27**) arise via the [2 + 2] pathway.[8] In one case the addition of the olefin is to the furan nucleus and in the other the allowed [2 + 2] addition is to the carbonyl group. When furan is irradiated in the presence of benzene, the [4 + 4] adduct **28** is formed as the initial major product along with **29** and **30.**[9] Adduct **28** undergoes either a thermal Cope rearrangement to generate **31** or a photochemically allowed [2 + 2] intramolecular cyclization to give **32.**

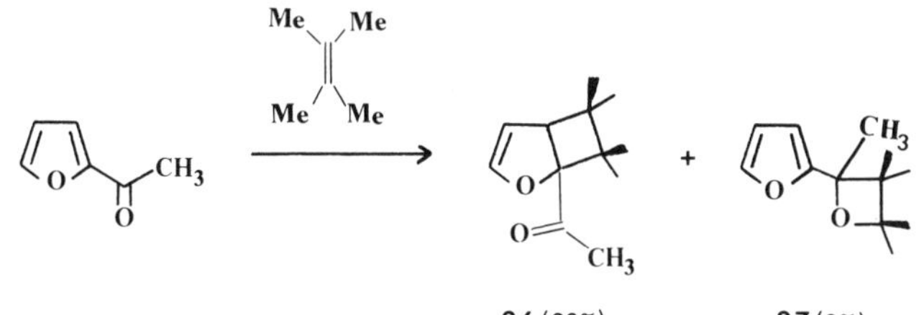

Me Me Me Me CH_3 O O O CH_3 O CH_3 O O

26 (33%) 27 (8%)

hν

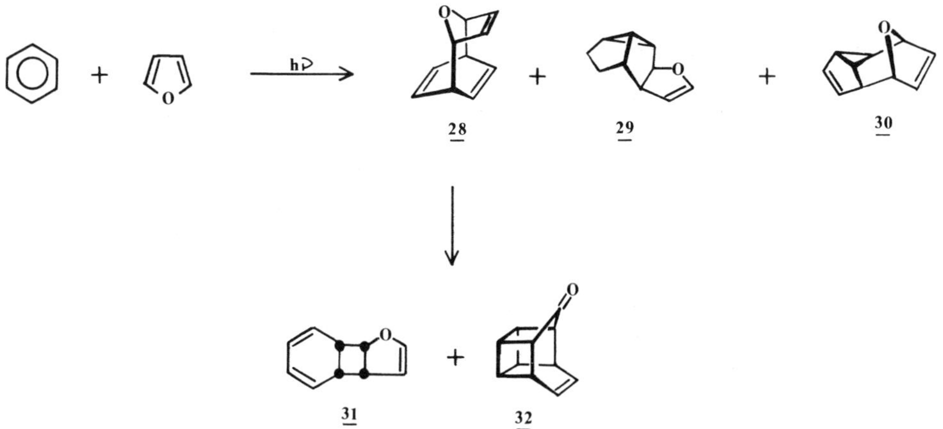

28 29 30

31 32

Vapor-phase sensitized photoirradiation of furan and methylfurans[10,11] involves a C—O bond homolysis, followed by an overall ring contraction to give cyclopropenes of the general structure **33.** In the case of 2,5-dimethylfuran, *13* different compounds were isolated.[10] The photochemical transformation of 2,5-di-*tert*-butylfuran **(34)** affords as the major products **35, 36,** and **37,**[12] presumably by competitive recombination processes.

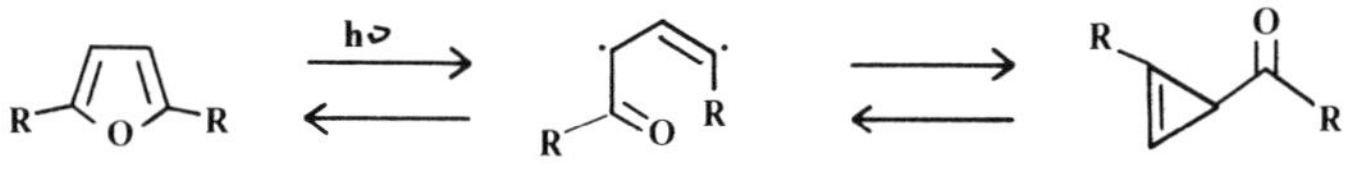

hν R O R R O R R O R

33

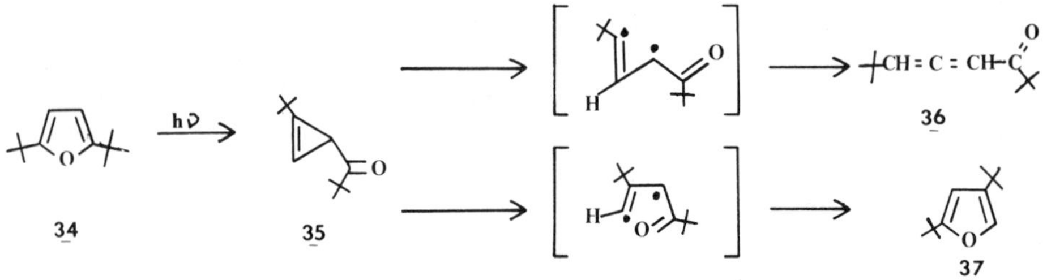

34 35 36 37

An alternative mode of ring opening has been observed, when 2,3,5-tri-*tert*-butylfuran (**38**) is irradiated[12] to afford **39** and **40** via the following proposed mechanism:

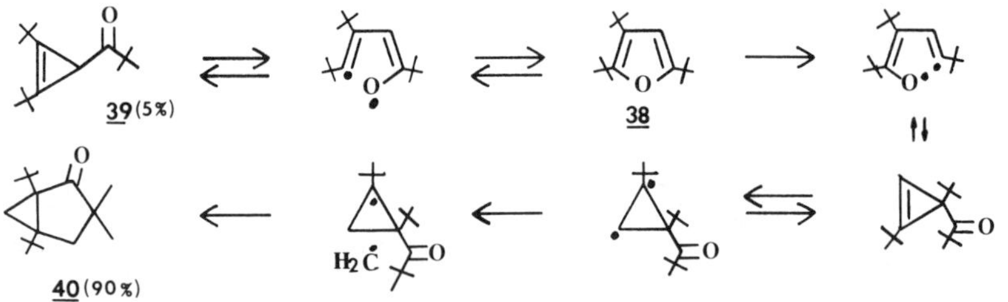

39 (5%) 38 40 (90%)

An interesting photochemical transformation involves the conversion of alkylfurans **41** to *N*-alkylpyrroles by irradiation of the former in an excess of propylamine[13]; bond reorganization is also possible as demonstrated by the photolysis of **42** to give a 3-methylpyrrole derivative.

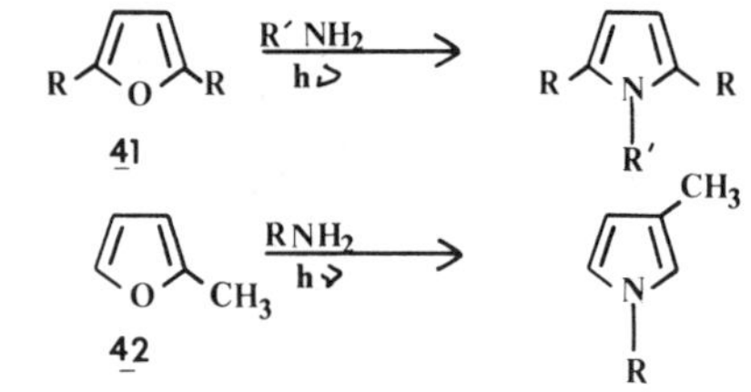

41 42

9.4. THIOPHENES

As in the case of furan and pyrrole, relatively little is known about the reactivity of thiophenes toward free radicals. The Gomberg–Bachman reaction of thiophene affords arylthiophenes (**43, 44**) in moderate yields (11%),[14] the α-substitution product **43** invariably being the major isomer formed.

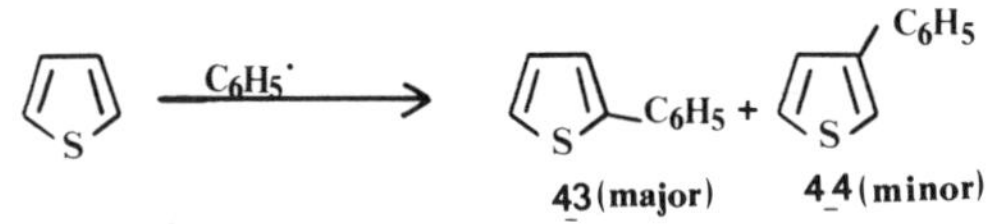

43 (major) 44 (minor)

A study of the reactivity of thiophene toward heteroaromatic radicals follows the same pattern as for the carbocyclic aromatic radicals.[15] Phenylazotriphenylmethane is somewhat less selective, affording the α- and β-phenylthiophenes in a 4 : 1 ratio.[16] When strongly electrophilic diazonium salts,[17] such as **45**, are employed, thiophene as well as the 2- and 3-methylthiophenes (**46** and **47**) are phenylated "normally," whereas 2-phenyl- (**48**), 2-*tert*-butyl- (**49**), and 2,4-dimethyl- (**50**) thiophenes anomalously afford the azo compounds **51–53**, respectively.[18]

NO_2 O_2N N_2^+ (45) R S NO_2 NO_2

46(R = H, CH_3) (R=H, 20%) (R= CH_3, 27%)

CH_3 S 45 CH_3 NO_2 S (30%) NO_2

47

R_2 S R_1 45 R_2 NO_2 R^1 S N_2 NO_2

48 (R_1= C_6H_5; R_2= H) 51 [R_1= C_6H_5; R_2=H (13%)]

49 (R_1=C$(CH_3)_3$; R_2= H) 52 [R_1= C$(CH_3)_3$; R_2=H (20%)]

50 (R_1= R_2= CH_3) 53 [R_1= R_2= CH_3 (60%)]

When both of the α-positions are blocked, as in 2,5-dimethylthiophene (**54**), hydrogen abstraction from one of the methyl groups occurs and the corresponding azo-coupled product **55** is formed, in addition to the 3-azo isomer **56**.

CH_3 S CH_3 45 N_2Ar CH_3 S CH_3 + CH_3 S CH=N NHAr

54 56 (13.5%) 55 (13.5%)

As is observed for furan, treatment of thiophene with benzoyl peroxide produces 2,2-bithienyl (**57**) rather than any addition products.[19] Mechanistically, an addition-dimerization-elimination sequence can be invoked to account for the formation of **57**.

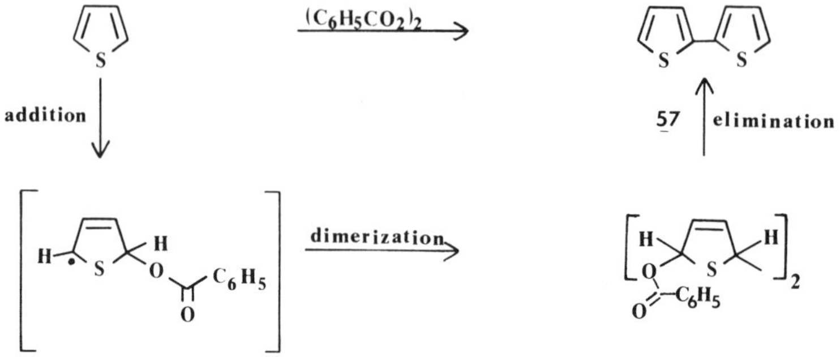

$(C_6H_5CO_2)_2$

addition

57 elimination

dimerization

C_6H_5

2

Photochemical studies of thiophenes have given some intriguing results.[20] For example, 2-substituted thiophenes **58** upon irradiation afford the corresponding 3-substituted isomers **59.** Although the mechanism for this fascinating transformation has not yet been conclusively established, several intermediates appear to be possible.[12,21] Arylthiophenes readily undergo this rearrangement. In the case of 2-phenylthiophene, it has been shown that although the C-aryl group remains intact during the reorganization, hydrogen scrambling (shown by deuterium labeling experiments) also occurs. Thus scrambling is observed in both the starting 2-phenylthiophene as well as the 3-phenyl photoproduct. Consequently, valence isomerization via any or all of the postulated intermediates must precede the rearrangement step. Although photogenerated singlet oxygen does not attack thiophene, in contrast to furan and pyrrole, it does react with alkylthiophenes[22,23] to form the ring-cleaved products **60** and **61.** The formation of these products can be explained either by initial *S*-peroxide **62** or zwitterion **63** formation or via the 1,4-addition product **64.**[7,22–24]

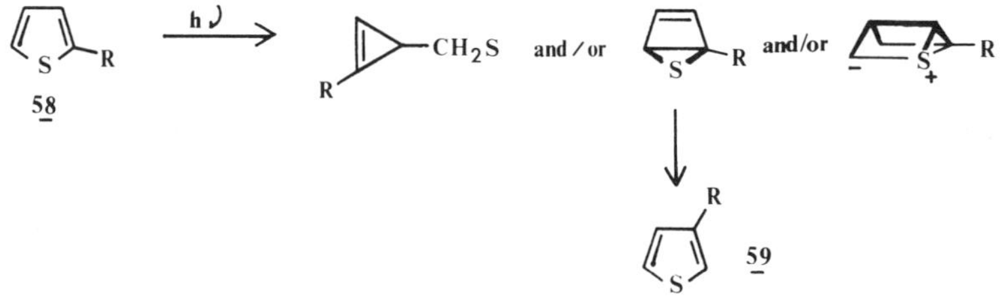

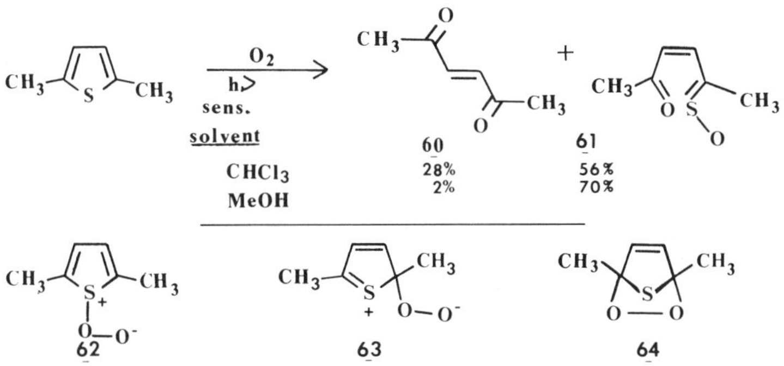

58 —hν→ CH_2S (R) and/or (R) and/or (R)

59

O_2, hν, sens.

solvent	60	61
$CHCl_3$	28%	56%
MeOH	2%	70%

62 63 64

The photochemical addition of alkenes, alkynes, and ketones[7,23–25] to some thiophenes (for example, **65**) affords interesting bicyclic compounds, analogous to the ones previously described in the furan series. Thus **66** and **68** result from [2 + 2] cycloadditions and **67** is formed by a (thermal) [2 + 4] addition of the olefin to the thiophene ring.

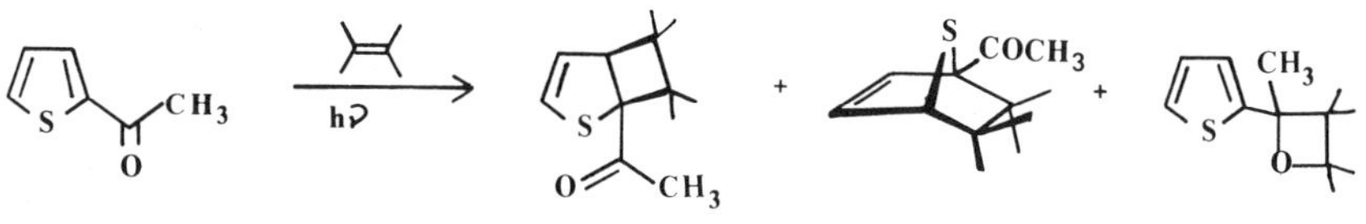

65 66 (10%) 67 (38%) 68 (11%)

REFERENCES

1. L. Benati, C. M. Camaggi, M. Tiecco, and A. Tundo, *J. Heterocycl. Chem.* **9,** 919 (1972).
2. R. A. Jones and G. P. Bean, *The Chemistry of Pyrroles,* Academic Press, London, 1977.
3. M. Aiura and Y. Kanaoka, *Heterocycles* **1,** 237 (1973); *Chem. Pharm. Bull.* **23,** 2835 (1975).
4. R. J. Gritter and R. J. Chriss, *J. Org. Chem.* **29,** 1163 (1964).
5. K. C. Bass and P. Nebasing, *Adv. Free Radical Chem.* **4,** 1 (1971).
6. K. E. Kolb and W. A. Black, *Chem. Commun.* **1969,** 1119.
7. L. Benati, M. Tiecco, A. Tundo, and F. Taddei, *J. Chem. Soc.* (*B*) **1970,** 1443.
8. T. S. Cantrell, *J. Org. Chem.* **39,** 2242 (1974).
9. J. C. Berridge, D. Bryce-Smith, A. Gilbert, and T. S. Cantrell, *J. Chem. Soc. Chem. Commun.* **1975,** 611.
10. S. Boué and R. Srinivasan, *J. Am. Chem. Soc.* **92,** 1824 (1970).
11. H. Hiraoka, *J. Phys. Chem.* **74,** 574 (1970).
12. E. E. van Tamalen and T. H. Whitesides, *J. Am. Chem. Soc.* **90,** 3894 (1968).
13. A. Couture, A. Delevallee, A. Leblache-Combier, and C. Párkányi, *Tetrahedron* **31,** 785 (1975).
14. M. Gomberg and W. E. Bachmann, *J. Am. Chem. Soc.* **46,** 2339 (1924).
15. G. Vernin, J. Metzger, and C. Párkányi, *J. Org. Chem.* **40,** 3183 (1975).
16. C. M. Camaggi, R. Leardini, M. Tiecco, and A. Tundo, *J. Chem. Soc.* (*B*) **1969,** 1251.
17. M. Bartle, S. T. Gore, R. K. Mackie, and J. M. Tedder, *J. Chem. Soc. Perkin I* **1976,** 1636.
18. S. T. Gore, R. K. Mackie, and J. M. Tedder, *J. Chem. Soc. Perkin I* **1976,** 1639.
19. F. Minisci and O. Porta, *Adv. Heterocycl. Chem.* **16,** 123 (1974).
20. H. Wynberg, *Acc. Chem. Res.* **4,** 65 (1971).
21. H. A. Wiebe, S. Braslavsky, and J. Heicklen, *Can. J. Chem.* **50,** 2721 (1972).
22. C. N. Skold and R. H. Schlessinger, *Tetrahedron Lett.* **1970,** 791.
23. H. H. Wasserman and W. Strehlow, *Tetrahedron Lett.* **1970,** 795.
24. C. Rivas, M. Vélez, and O. Crescente, *Chem. Commun.* **1970,** 1474.
25. H. J. Kuhn and K. Gollnick, *Tetrahedron Lett.* **1972,** 1909.

Chapter 10

CYCLOADDITION REACTIONS. FIVE-MEMBERED π-EXCESSIVE HETEROAROMATIC RINGS

10.1. GENERAL COMMENTS

A consideration of the general structure of monohetero-π-excessive five-membered rings with respect to their possible utilization in various cycloaddition reactions suggests, among many possibilities, the following reactions:

1. X + H R / R H ⟶ X R R

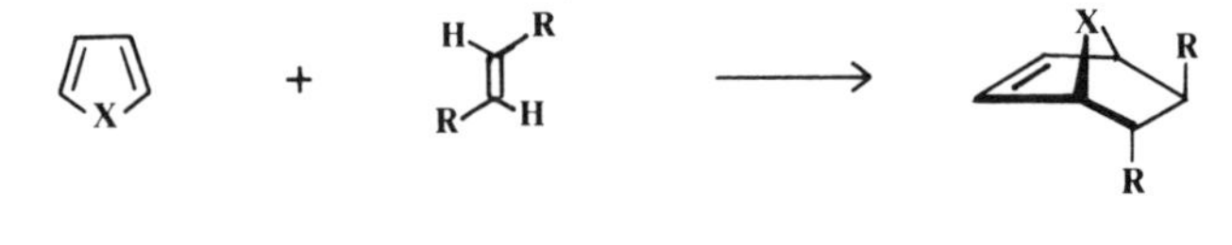

2. 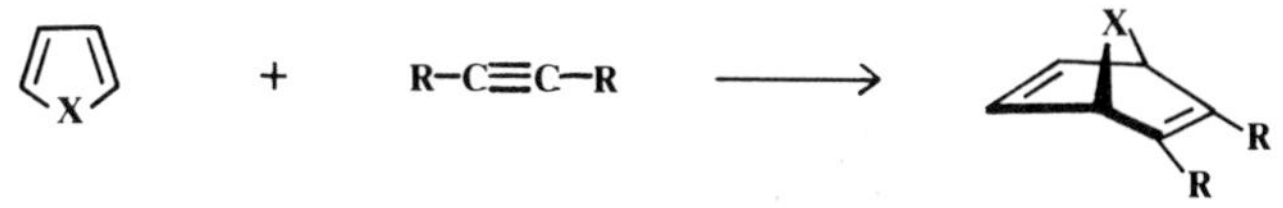 [furan-type ring X] + R–C≡C–R ⟶ [bicyclic adduct]

3.  [furan-type ring X] + O=O ⟶ [bicyclic peroxide]

4.  [furan-type ring X] + :CR_2 ⟶ [cyclopropane adduct]

The processes labeled 1 through 3 are examples of Diels–Alder reactions, whereas reaction 4 is a typical carbene insertion into a double bond. Thus [4 + 2] and [2 + 1] cycloadditions can be expected for these ring systems.

It is the purpose of this chapter to examine, in a general way, some of the pertinent studies that have been performed on these ring systems in connection with cycloaddition reactions.

10.2. PYRROLES

Treatment of pyrrole with chloroform and a strong base affords a mixture of 2-formylpyrrole (**1**) and 3-chloropyridine (**2**).[1] These transformations are envisioned to occur via the initial formation of dichlorocarbene and its subsequent insertion in the 2,3-bond of pyrrole (**3**).

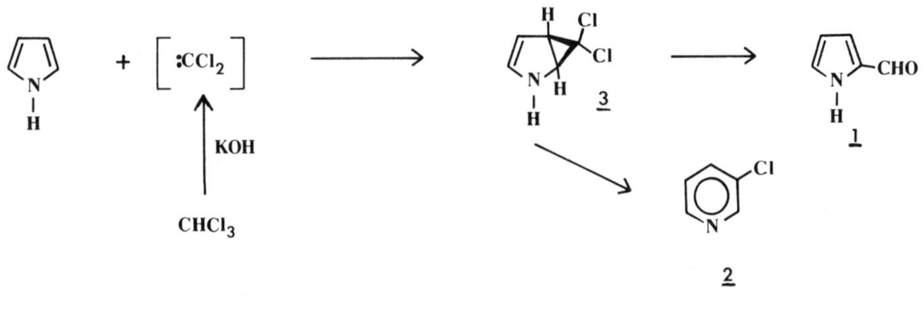

Transformation of the bicyclic intermediate **3** to **1** and **2** is interpretable by the following processes:

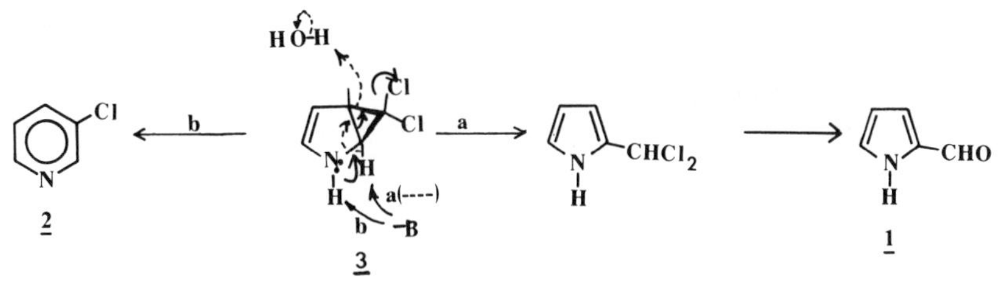

Higher yields of pyridine derivatives can be obtained in this reaction if the carbenes are generated from dichloromethane and butyllithium or from the pyrolysis of sodium trichloroacetate. When the carbene used in the insertion reaction is obtained from the light- or copper-catalyzed decomposition of diazoacetate ester,[2] only the α-substituted pyrrolylacetate esters (**4**) are obtained. When the α-positions of pyrrole are blocked [for example, **5** ($R \neq H$)], the corresponding β-substituted acetate esters (**6**) are obtained.[2] *N*-Alkoxycarbonylpyrroles, when treated with either diazomethane or diazoacetate ester and copper(I) salts, afford the bicyclic products **7.** The tricyclic compound **8** is also obtained when *N*-alkoxycarbonylpyrrole itself is used.[3]

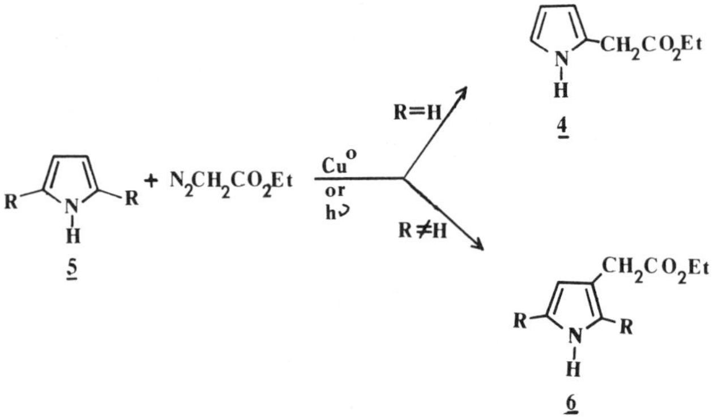

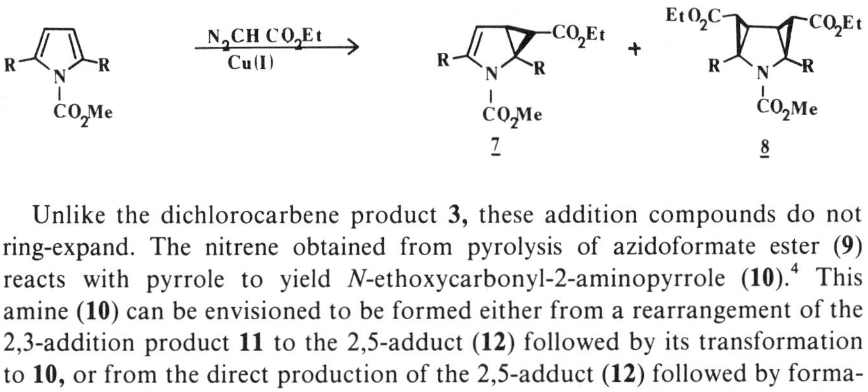

Unlike the dichlorocarbene product **3,** these addition compounds do not ring-expand. The nitrene obtained from pyrolysis of azidoformate ester (**9**) reacts with pyrrole to yield *N*-ethoxycarbonyl-2-aminopyrrole (**10**).[4] This amine (**10**) can be envisioned to be formed either from a rearrangement of the 2,3-addition product **11** to the 2,5-adduct (**12**) followed by its transformation to **10,** or from the direct production of the 2,5-adduct (**12**) followed by formation of **10.** The reaction of pyrrole itself with typical dienophiles such as maleic acid, alkyl acetylenedicarboxylates, and azodicarboxylate esters affords Michael-type addition products.[5]

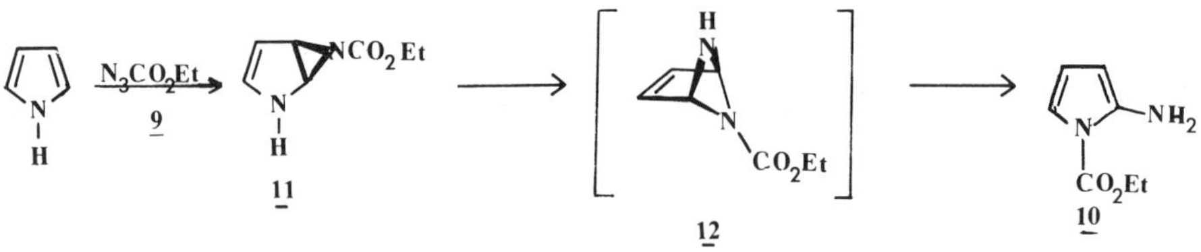

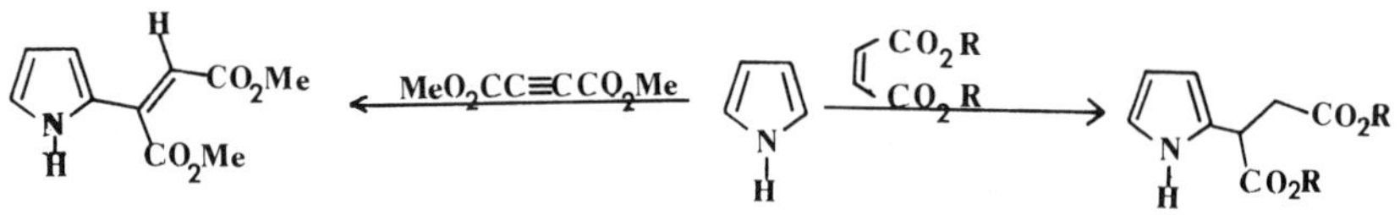

Reaction of the *N*-substituted pyrrole **13** with acetylenedicarboxylic acid generates the 2,5-addition product **14** in minor yield, along with **15** and **16**, which result from electrophilic α-substitution on the pyrrole ring.[6]

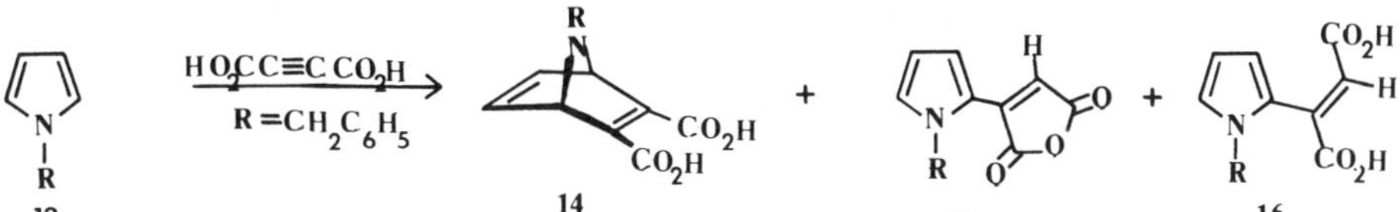

The presence of a carbomethoxy function (an electron-withdrawing group, which tends to decrease the "aromaticity" of pyrrole) on the pyrrole nitrogen (**17**) greatly facilitates the [2 + 4] addition to dimethyl acetylenedicarboxylate.[7] The product **18** of this reaction is readily explained in terms of a Diels–Alder-retro Diels–Alder sequence.

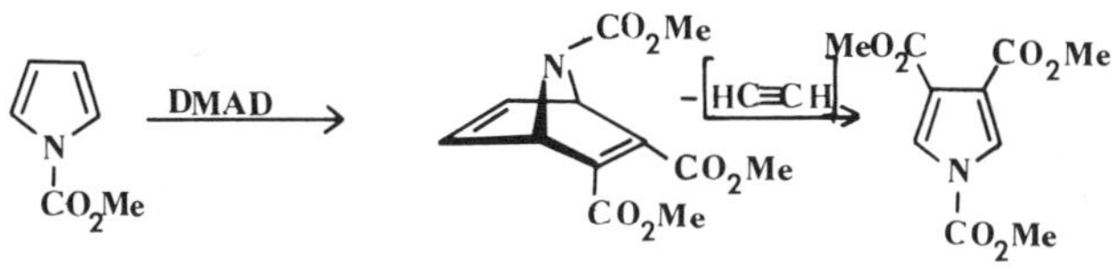

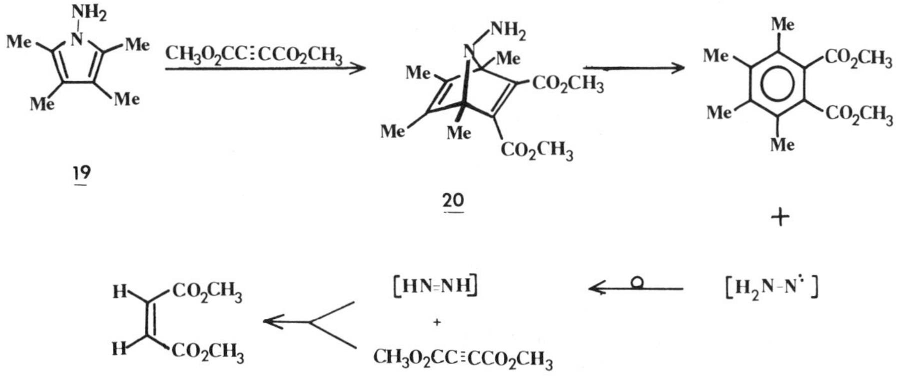

When *N*-aminopyrrole **19** is utilized in the reaction with dimethyl acetylenedicarboxylate, the anticipated substituted benzene derivative is isolated along with dimethyl maleate (**21**).[8] The presence of **21** suggests that the decomposition of intermediate **20** generates the aminonitrene, which after rearrangement to diimide (HN=NH) reduces the activated acetylene.[8] Use of different dialkylamino pyrroles under similar conditions gives **22** and **23.**

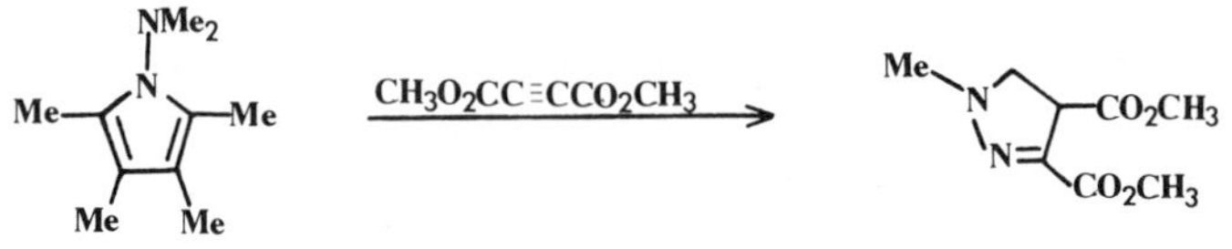

$CH_3O_2CC{\equiv}CCO_2CH_3$

22

$CH_3O_2CC{\equiv}CCO_2CH_3$

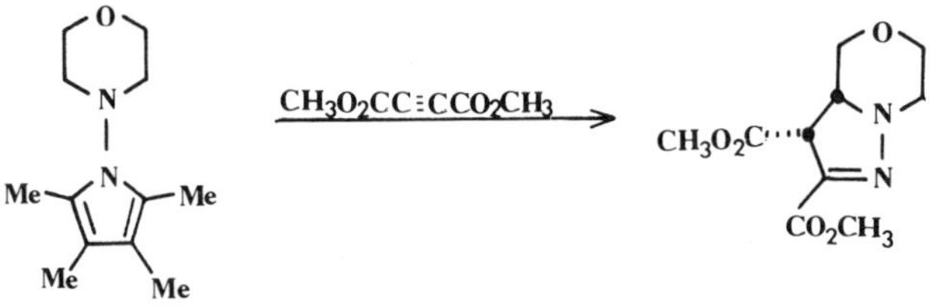

23

N-Carbalkoxypyrroles as well as *N*-alkylpyrroles react with benzynes to afford azanorbornadiene derivatives **24.**[9] The norbornadiene derivative **24** (R = CH_3) can be converted to naphthalene **25** by successive treatment with methyl iodide and silver oxide. When pyrrole reacts with benzyne, the only product isolated is 2-phenylpyrrole (**26**); *no* bicyclic intermediates are detected in this reaction.[9]

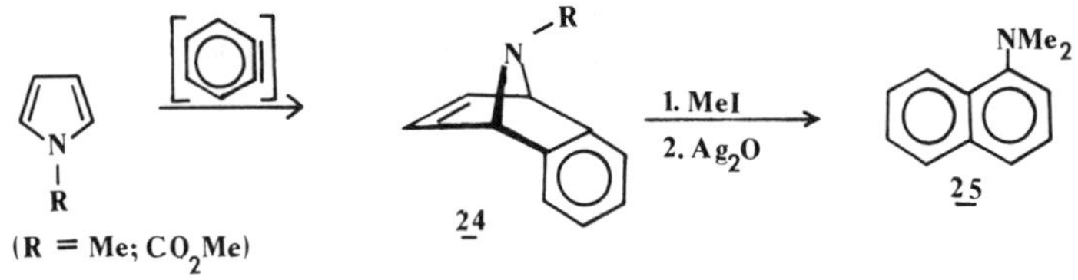

(R = Me; CO_2Me)

1. MeI
2. Ag_2O

24

25

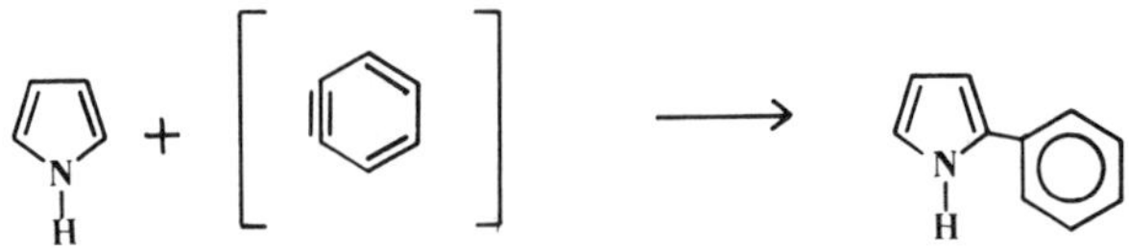

26

The very powerful dipolarophile hexafluorobicyclo[2.2.0]hexa-2,5-diene (hexafluoro-Dewar benzene, **27**) reacts in the Diels–Alder fashion with pyrrole to yield both a mono- (**28**) and a diadduct (**29**).[10]

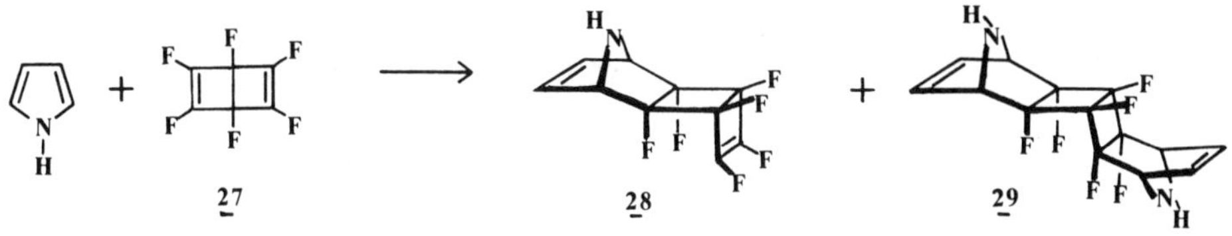

27 28 29

2,2-Dimethylcyclopropanone, a dienophile which acts as the oxallyl zwitterion, reacts with *N*-methylpyrrole in a fashion similar to that of hexafluoro-Dewar benzene to generate the azabicycloheptanone **30.**[11] During attempts to isolate this compound by preparative vapor phase chromatography, it rearranges to the 2-substituted pyrroles **31** and **32.**

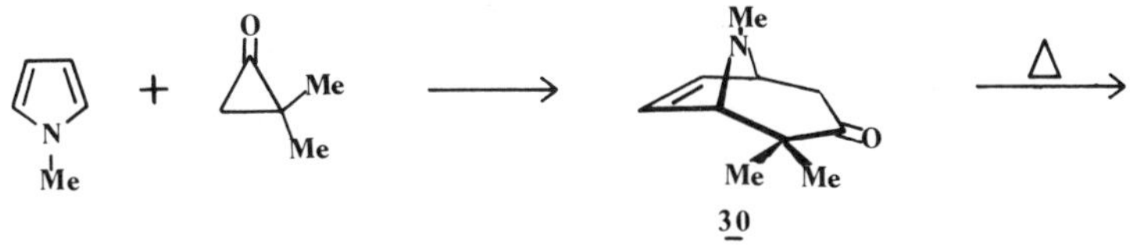

30

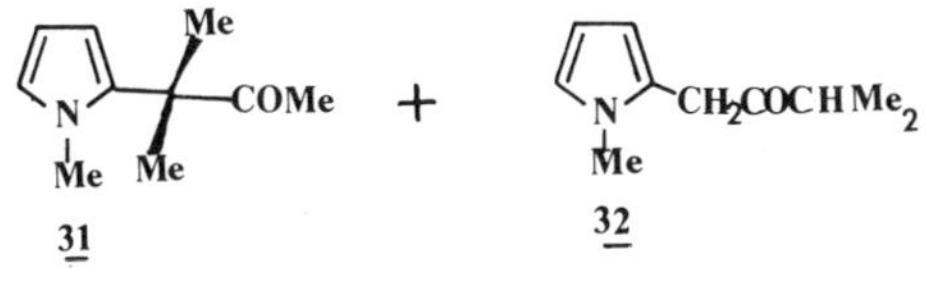

31 32

Sensitized photooxidation of pyrrole affords 5-hydroxy-2-pyrrolone (**33**),[12] which may well be formed via the Diels–Alder intermediate **34,** obtained by addition of singlet oxygen to pyrrole or its derivatives.[13,14]

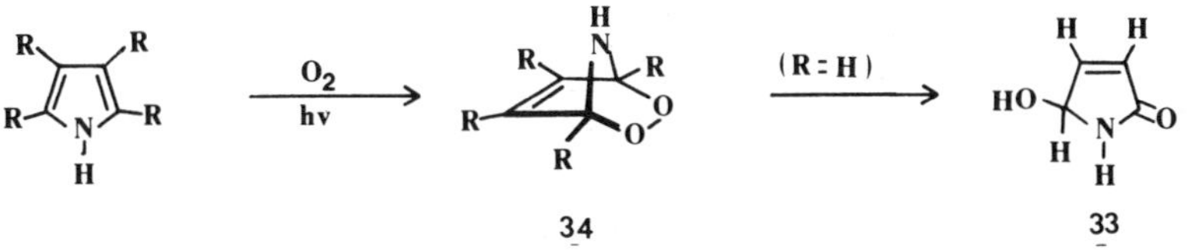

34 33

10.3. FURANS

Unlike pyrrole, furans undergo Diels–Alder reactions very readily. Where applicable, the exo adducts (**35**) are, in general, thermodynamically more stable than are the kinetic endo adducts (**36**).[15,16]

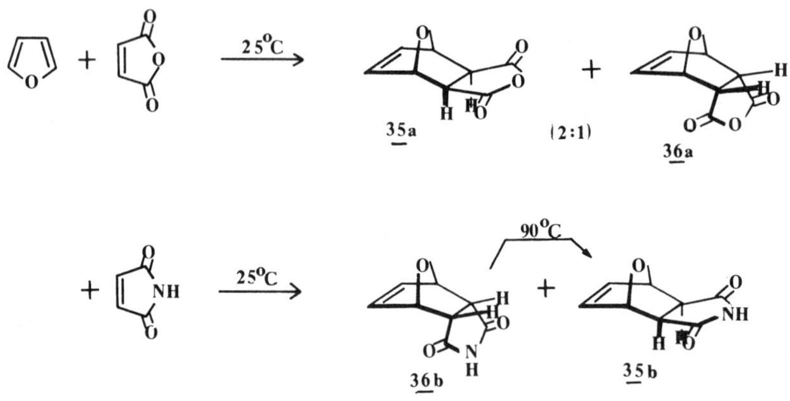

25°C

35a (2:1) 36a

25°C 90°C

36b 35b

Cyclopropene reacts with furan with great ease to form a 1 : 1 mixture of the exo (**37**) and endo (**38**) products.[17] Halocyclopropenes, such as **39,** afford the endo products (**40**) exclusively. This type of fused cyclopropane has been shown[18] to undergo an electrocyclic ring-opening which involves a stereospecific 1,2-halogen migration to form **41.**

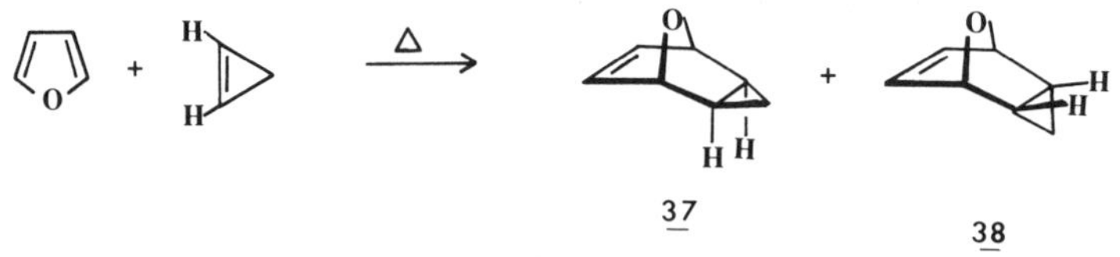

Δ

37 38

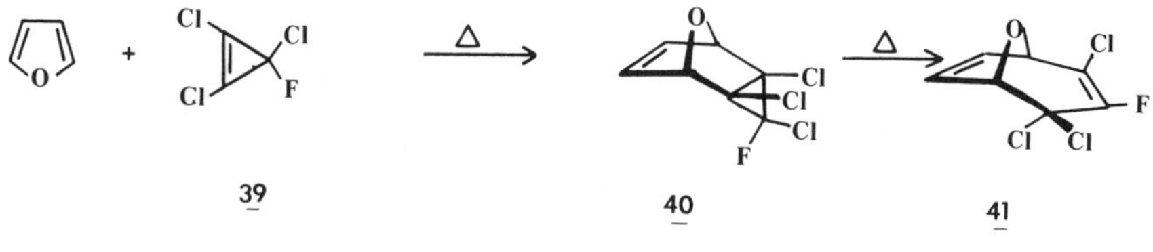

Δ Δ

39 40 41

If the dienophile has only one electron-withdrawing group, the Diels–Alder reaction on furans, at room temperature, is considerably slowed down. However, the use of high pressures in these reactions "restores" the reactivity.[19]

As might well be expected, furans react readily with acetylenic dienophiles. The oxanorbornadienes (**42, 43**) obtained in these reactions have found use in many synthetic sequences[20]:

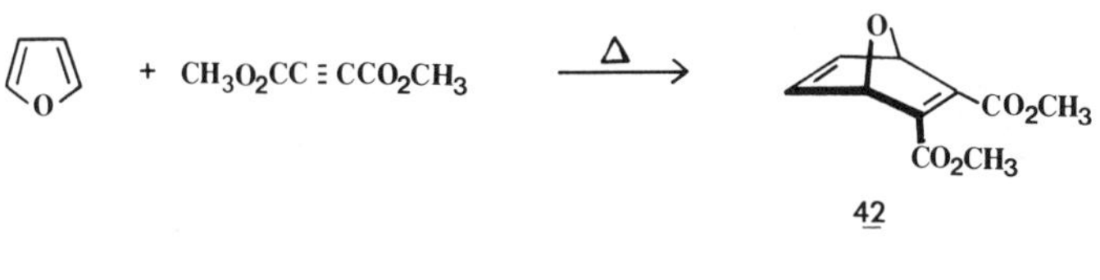

+ $CH_3O_2CC \equiv CCO_2CH_3$ Δ → 42 (CO_2CH_3, CO_2CH_3)

+ $CF_3C{\equiv}CCF_3$ ⟶ 43 $\xrightarrow[\Delta]{H_2/Pt}$

An unusual-looking product (**44**) is obtained from straightforward reactions when the dienophile ethyl propiolate is treated with furan at 130°C.[21] Compound **44** probably arises from the [4 + 2] reaction of ethyl propiolate with intermediate **45,** the primary Diels–Alder adduct of this reaction. Arynes also react with furan in the expected manner to yield compounds of general structure **46.**[21]

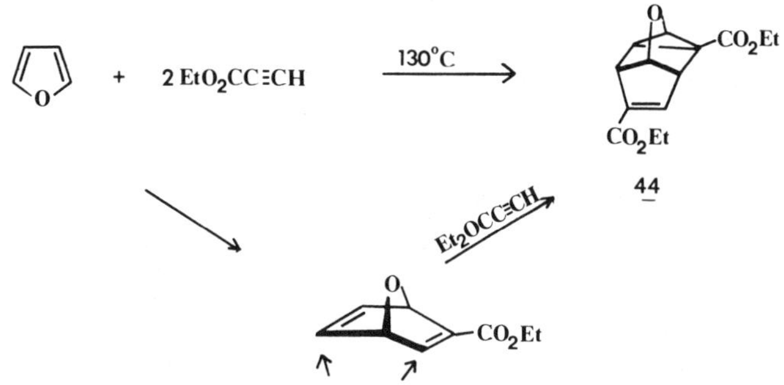

+ $2\ EtO_2CC{\equiv}CH$ $\xrightarrow{130°C}$ 44

$Et_2OCC{\equiv}CH$

45

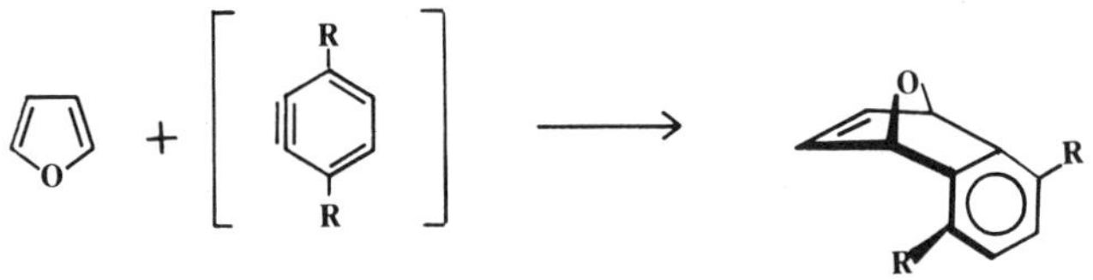

46

As is the case with *N*-substituted pyrroles, furans react with cyclopropanones as well. These reactions are envisioned to occur via the oxallyl zwitterion (**47**) and afford the appropriate addition compounds (**48**).[11] *trans*-2,4-Dimethylthietan-3-one (**49**) is ring-opened by triphenylphosphine in methanol-furan (1 : 1) to give **50**, which is produced from the expected [4 + 3] cycloaddition of furan to 1,3-dimethylallylium-2-olate (**51**) or its *O*-protonated equivalent (**52**).[22]

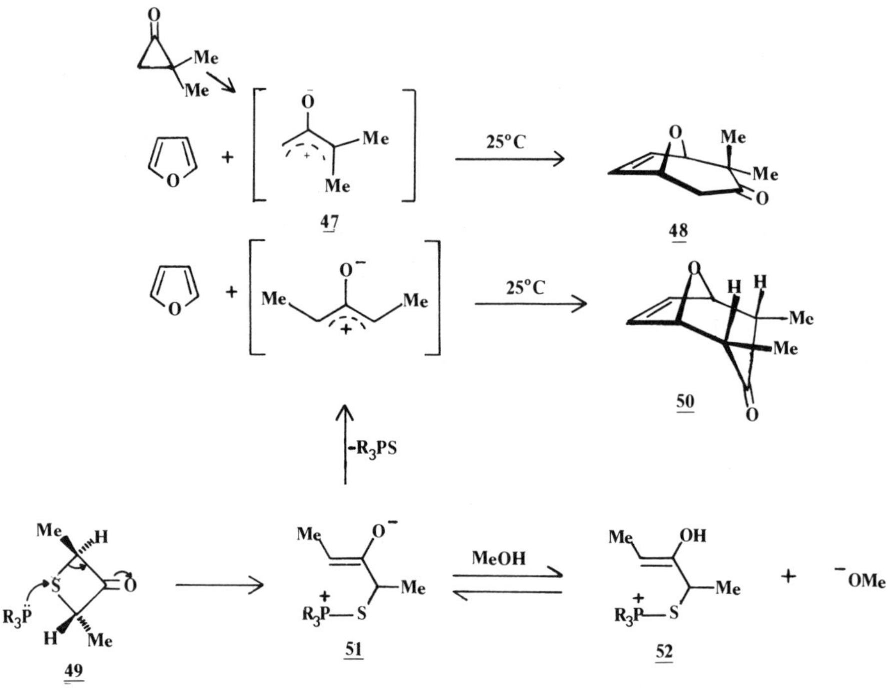

47 48 50 49 51 52

Furan reacts as a diene with carbenes.[23] Methyl diazoacetate dissolved in furan and subjected to ultraviolet radiation yields adduct **53,** which when heated briefly undergoes a valence-tautomeric transformation to the ring-opened aldehyde **54.**

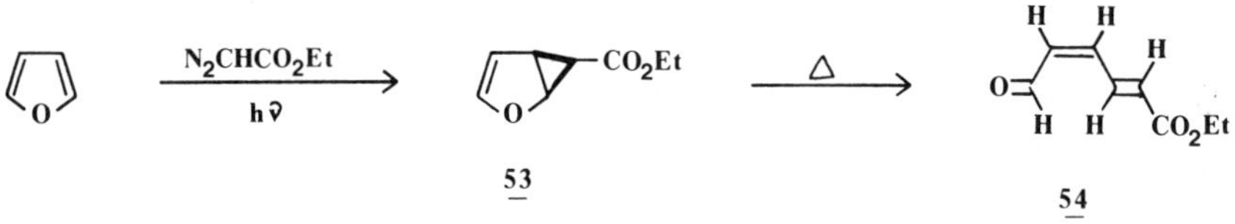

53 54

Thermolysis of ethyl azidoformate in the presence of furan yields either lactam **55** or **56.** It is assumed that the aziridine **57** initially formed in this reaction ring-cleaves to **58** and recyclizes to afford the isolated lactam.[24]

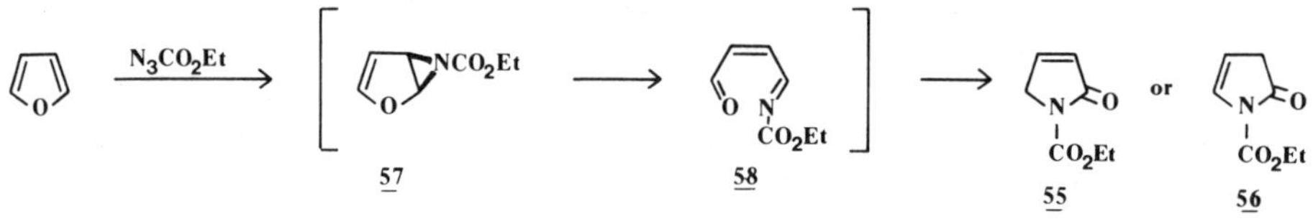

57 58 55 56

10.4. THIOPHENES

Thiophenes, contrary to earlier reports, have now been shown to undergo the Diels–Alder reaction with acetylenes and other dienophiles.[25] Thiophene reacts with maleic anhydride in a variety of solvents at 100°C and 15 kbar to form the exo adduct (**59**).[26]

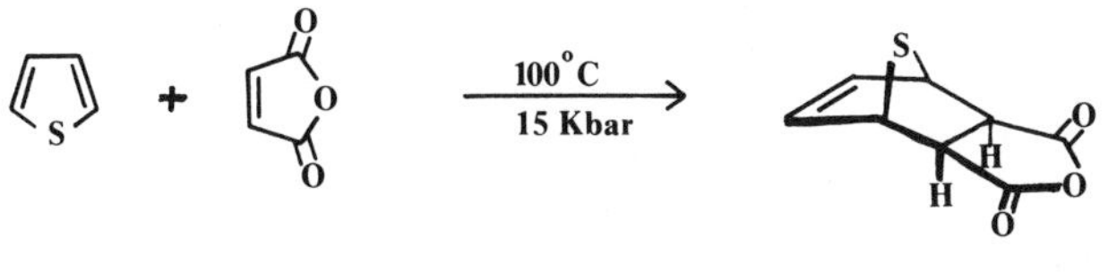

59

Thiophenes do, however, behave somewhat differently in these reactions than, for example, the furans. In many instances the intermediate addition products (**60**) are unstable and undergo cheleotropic extrusion of sulfur.

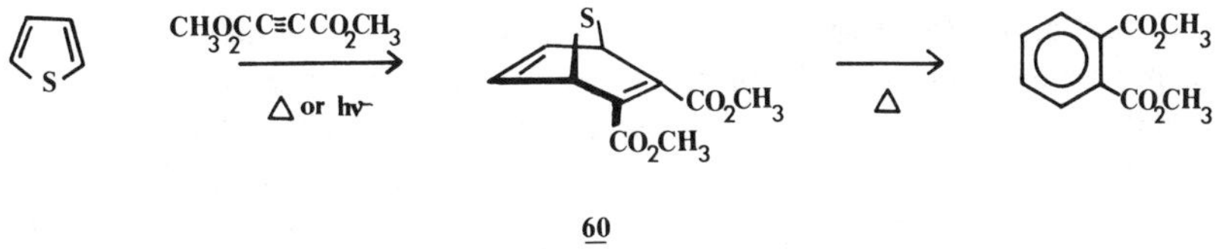

60

When tetramethylthiophene (**61**) reacts with dicyanoacetylene in the presence of aluminum trichloride, the reaction proceeds via a [2 + 2] addition to afford **62.**

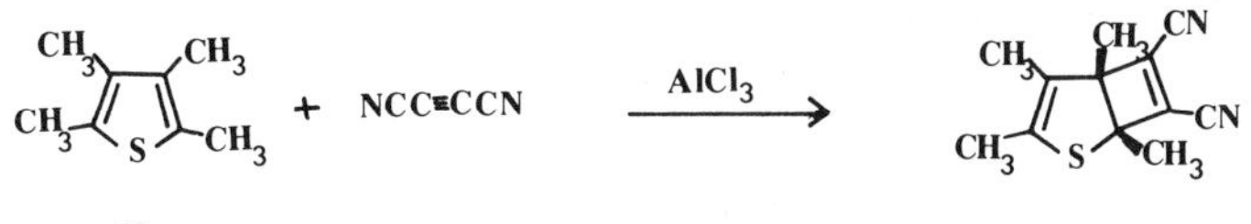

61 62

The tendency of enamine-like 3-aminothiophene (**63**) to behave as a nucleophile in addition reactions is most intriguing. In polar solvents, **63** undergoes electrophilic substitution at C-2, whereas in nonpolar solvents, a [2 + 2] addition reaction takes place.[27] Tetracyanoethylene oxide (**64**) reacts with thiophene itself in a [2 + 3] manner.[28]

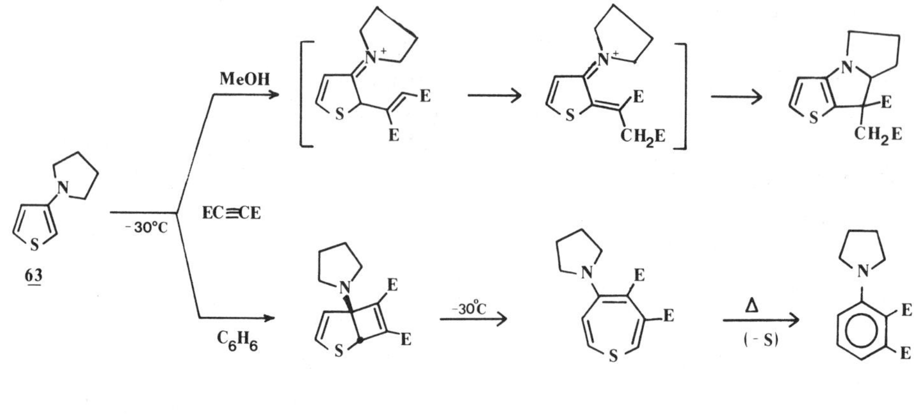

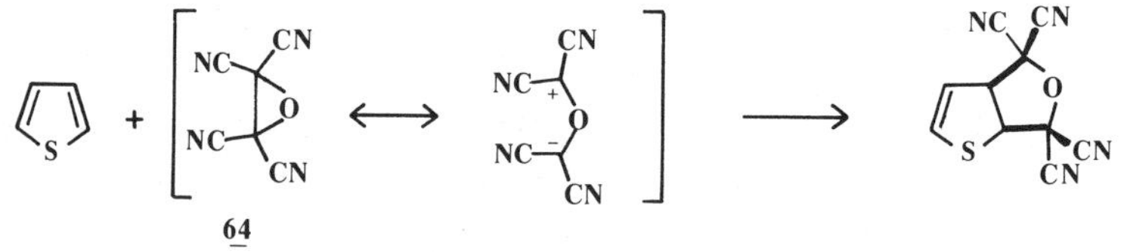

Addition and substitution reactions of thiophenes with carbenes and nitrenes have been described.[29] The formation of the bicyclic ester **65** follows the same pattern as observed for pyrrole: cycloaddition to the 2,3-bond of the heterocycle. The formation of **66,** however, can be envisioned to occur either via Diels–Alder adduct **67** or through the rearrangement of intermediate **68.**

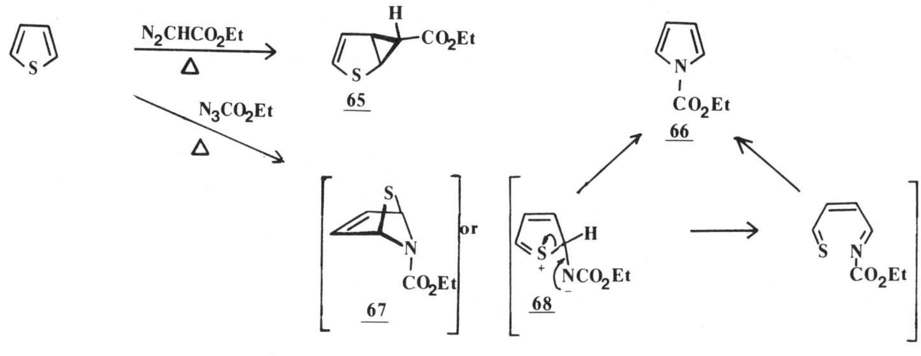

Unlike furan and pyrrole, thiophene does not react with photogenerated singlet oxygen.[30] Alkylthiophenes (**69**), on the other hand, do react. The [4 + 2] cycloadduct **70** is a likely candidate for the intermediate in this reaction.

Me O Me Me S Me 69 O_2 hν S Me O Me O 70 + Me Me O S O

10.5. OXAZOLES

Among the dienes most commonly used in the Diels–Alder reaction are the substituted oxazoles.[31] The synthesis of a novel furan prostanoid **71** employed oxazole **72** as the pivotal starting material; thus treatment of **72** with acetylene **73** gives furan **74**, which via standard manipulation affords **71**.

N O + $MeO_2C(CH_2)_6C{\equiv}CCHO$ 72 73 N O $(CH_2)_6CO_2CH_3$ CHO − CN Δ

O $(CH_2)_6CO_2CH_3$ CHO 74(73%) O P Me_2 O O $(CH_2)_6CO_2CH_3$ $(CH_2)_3CH_3$ O (1)[H] (2) LiOH/THF - H_2O

O $(CH_2)_6CO_2H$ $(CH_2)_3CH_3$ H OH 71

The synthesis of a new class of saluretic agents has recently been reported[32] in which oxazole **75** (the diene) reacts with olefin **76** to give pyridine **77** rather than the furan heteroaromatic.

75 $\xrightarrow[(70\text{-}75°C)\ (76)]{CH_2{:}CHCN}$ [ ] → 77 $\xrightarrow[(2)\ ICl]{(1)\ H_2/Pd}$

REFERENCES

1. R. L. Jones and C. W. Rees, *J. Chem. Soc.* (*C*) **1969,** 2249, 2255.
2. G. M. Badger, J. A. Elix, and G. E. Lewis, *Aust. J. Chem.* **20,** 1777 (1967).
3. F. W. Fowler, *Angew. Chem. Int. Ed.* **10,** 135 (1971).
4. K. Hafner and W. Kaiser, *Tetrahedron Lett.* **1964,** 2185.
5. O. Diels and K. Alder, *Ann. Chem.* **490,** 267 (1931).
6. L. Mandell and W. A. Blanchard, *J. Am. Chem. Soc.* **79,** 6198 (1957).
7. R. M. Acheson and J. M. Vernon, *J. Chem. Soc.* **1961,** 457; **1963,** 1008.
8. A. G. Schultz, M. Shen, and R. Ravichandran, *Tetrahedron Lett.* **22,** 1767 (1981).
9. E. Wolthuis, D. V. Jagt, S. Mels, and A. DeBoer, *J. Org. Chem.* **30,** 190 (1965).
10. M. G. Barlow, R. N. Haszeldine, and R. Hubbard, *J. Chem. Soc.* (*C*) **1971,** 90.
11. N. J. Turro, S. S. Edelson, J. R. Williams, T. R. Darling, and W. B. Hammond, *J. Am. Chem. Soc.* **91,** 2283 (1969).
12. P. DeMayo and S. T. Reid, *Chem. Ind.* (*London*) **1962,** 1576.
13. C. Duffraisse, G. Rio, and A. Ranjon, *Compt. Rend.* (*C*) **265,** 310 (1967).
14. H. H. Wasserman and A. H. Miller, *Chem. Commun.* **1969,** 199.
15. A. S. Onishchenko, *Diene Synthesis,* Israel Program for Scientific Translations, Jerusalem, 1964.
16. P. Bosshard and C. H. Eugster, *Adv. Heterocycl. Chem.* **7,** 377 (1966).
17. R. W. LaRochelle and B. M. Trost, *Chem. Commun.* **1970,** 1353.
18. D. C. F. Law and S. W. Tobey, *J. Am. Chem. Soc.* **90,** 2376 (1968).
19. W. G. Dauben and H. O. Krabbenhoft, *J. Am. Chem. Soc.* **98,** 1992 (1976); K. Matsumoto, T. Uchida, and R. M. Acheson, *Heterocycles.* **16,** 1367 (1981).
20. W. S. Wilson and R. N. Warrener, *J. Chem. Soc. Chem. Commun.* **1972,** 211.
21. A. W. McCulloch and A. G. McInnes, *Can. J. Chem.* **53,** 1496 (1975).
22. B. Föhlisch and W. Gottstein, *J. Chem. Res.* (*S*) **1981,** 94; (*M*) **1981,** 1132, and refs. cited therein.
23. E. Müller, H. Kessler, H. Fricke, and H. Suhr, *Tetrahedron Lett.* **1963,** 1047.
24. D. W. Jones, *J. Chem. Soc. Perkin I* **1972,** 2728.
25. D. N. Reinhoudt and C. G. Kouwenhoven, *Tetrahedron* **30,** 2093 (1974).
26. H. Kotsuki, S. Kitagawa, H. Nishizawa, and T. Tokoroyama, *J. Org. Chem.* **43,** 1471 (1978).
27. D. N. Reinhoudt, W. P. Trompenaars, and J. Geevers, *Tetrahedron Lett.* **1976,** 4777, and refs. cited therein.
28. S. Gronowitz and B. Uppstrom, *Acta Chem. Scand.* **29,** 441 (1975).
29. K. Hafner and W. Kaiser, *Tetrahedron Lett.* **1964,** 2185.
30. H. H. Wasserman and W. Strehlow, *Tetrahedron Lett.* **1970,** 795.
31. M. F. Ansell, M. P. L. Caton, and P. C. North, *Tetrahedron Lett.* **22,** 1727 (1981).
32. G. E. Stokker, R. L. Smith, E. J. Cragoe, C. T. Ludden, H. F. Russo, C. S. Sweet, and L. S. Watson, *J. Med. Chem.* **24,** 115 (1981).

Chapter 11

REDUCED π-EXCESSIVE HETEROAROMATIC COMPOUNDS

11.1. GENERAL COMMENTS

Reduction of any of the π-excessive, heteroaromatic, five-membered ring compounds can, in principle, yield three dihydro (**2, 3**) and tetrahydro (**4**, perhydro) derivatives.

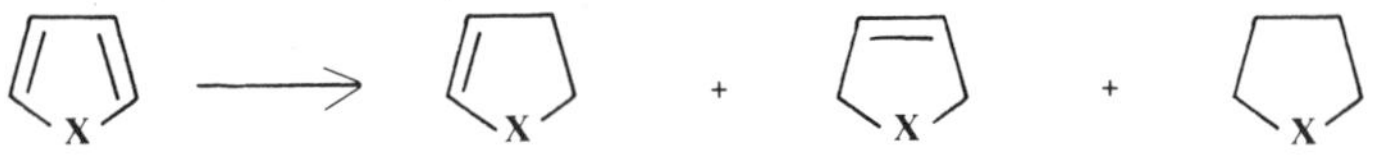

1 2 3 4

When these compounds are, for example, 2,5-symmetrically disubstituted (**5**), the number of different isomers is increased by two (**5** → **6–10**). In this instance, the *cis*- and *trans*-dihydro (**7** and **8**) and the *cis*- and *trans*-tetrahydro (**9** and **10**) isomers are possible. Clearly, the number of possible isomers increases with decreasing symmetry and increasing degree of substitution of these heterocyclic compounds.

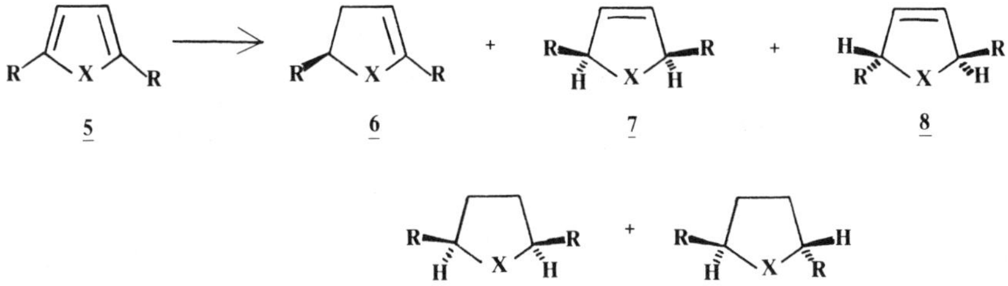

5 6 7 8

9 10

11.2. PYRROLES

The chemical reduction of pyrrole (**11**) and of *N*-alkylpyrroles (**12**) with zinc and hydrochloric acid affords the Δ^3-pyrrolines (**13** and **14**) and the pyrrolidines (**15** and **16**).[1] These reductions are envisioned to occur via the α- and β-pyrrolenine salts (**17** and **18**).

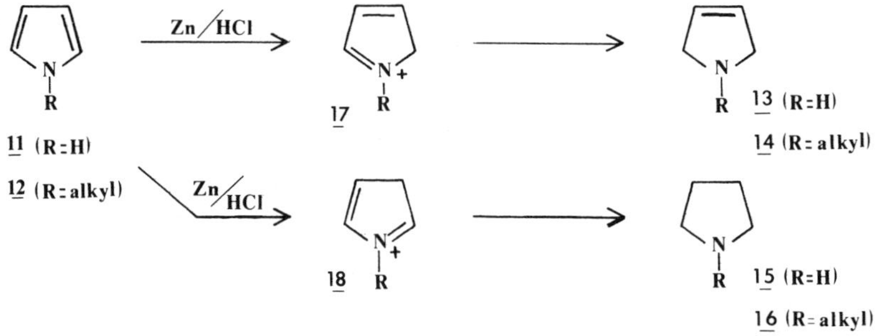

11 (R=H)
12 (R=alkyl)
17
18
13 (R=H)
14 (R=alkyl)
15 (R=H)
16 (R=alkyl)

A similar reduction of 2,5-dimethylpyrrole (**19**) affords the *cis*-**20** and *trans*-**21** (Δ^3-isomers) in a 1:3.5 ratio.[2]

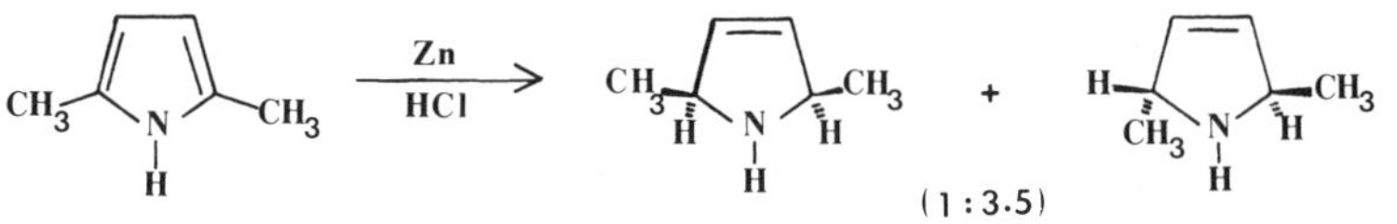

Catalytic reduction of pyrroles to substituted pyrrolidines can be accomplished quantitatively at moderate temperatures with a variety of heterogeneous catalysts (Pd, Pt, Raney Ni, other noble metals).[3] The presence of a carbalkoxy or an allyl group on pyrrole results in preferential catalytic reduction (**22** and **23,** respectively) of the substituents. Under acidic conditions, **23** is further catalytically reduced to the corresponding pyrrolidine derivative **24.**[4]

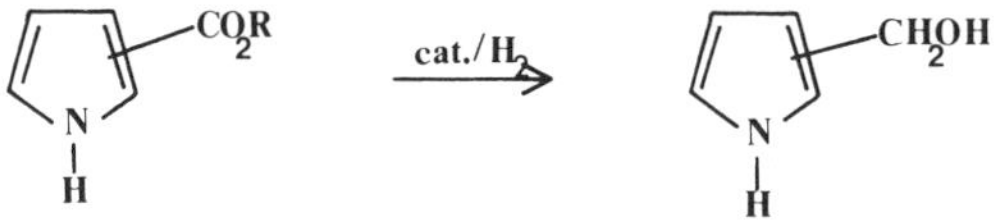

On the other hand, an *N*-carbalkoxy group facilitates ring reduction to **25.** The electron-withdrawing nature of the *N*-substituent greatly decreases the contribution of the *N*-electrons to the aromatic system, thus diminishing the overall aromatic character of the pyrrole nucleus.[5] Reduction of the "butadiene" moiety is thus quite facile.

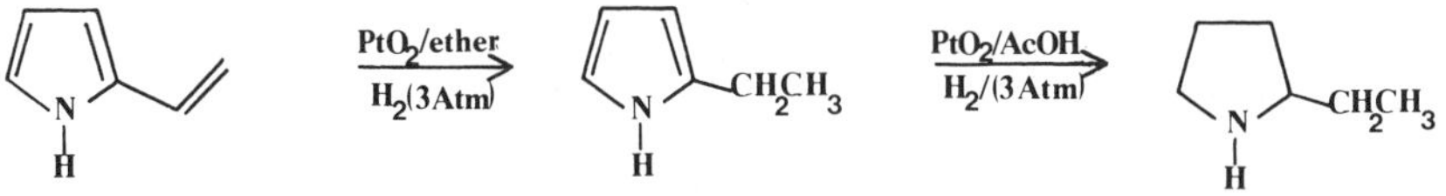

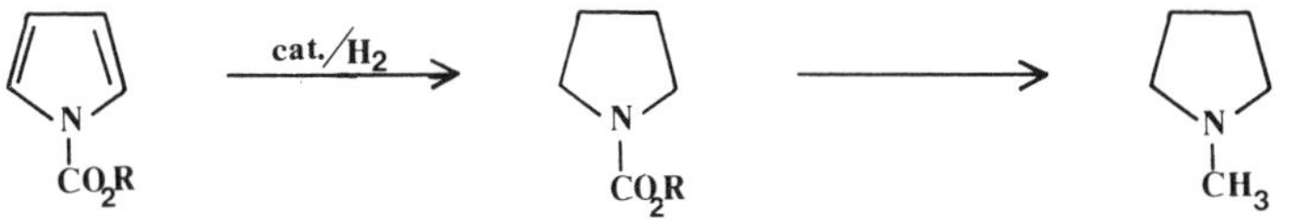

11.3. FURANS

The furan ring is more resistant to catalytic reduction than is pyrrole.[6,7] Furan itself can be reduced over Raney nickel at 80–160°C at up to 160 atmospheres to afford tetrahydrofuran (THF, **26**), an excellent solvent. Conversion of tetrahydrofuran to the commercially important adipic acid **(27)** by means of carbon monoxide and nickel carbonyl catalyst is of significance. Side-chain unsaturation can generally be reduced without loss of the aromatic character. If the reaction conditions are properly chosen,[7] the ring can be reduced, leaving functional groups intact.

R/Ni ($H_2 \leqslant 160$ Atm) 80–160°C → 26 → (1) CO, $Ni(CO)_4$; (2) H_2O → $CO_2H(CH_2)_4CO_2H$ 27

When cyclic acetal **28** is treated with sodium, it affords 2,3-dihydrofuran (**29**), which ring-contracts upon heating to give cyclopropanecarboxaldehyde (**30**).[7] The double bond in the cyclic vinyl ether **29** is quite labile and readily reacts with water to form hemiacetal **31**.

Cl, OEt 28 → Na° → 29 → Δ → CHO 30; 29 → H_2O → OH 31

Alkali metal/liquid ammonia reduction of furans (for example, **32**) affords the 2,5-dihydro derivatives.[8]

R, CO_2H (cis,trans) ← (R≠H) Li/NH_3(liq) MeOH ← R, CO_2H 32 → (R = H) (1) Na^0/NH_3(−78°C) (2) NH_4Cl/H_2O → CO_2H (80%)

Furfural (**33**) is catalytically reduced to give 2-hydroxymethyltetrahydrofuran (**34**) which, upon treatment with alumina gel at 400°C, undergoes an interesting transformation via a ring-expansion and subsequent dehydration to afford the vinyl ether **35**. Because of the importance of this reaction, numerous labeling and mechanistic studies have been conducted and summarized.[9]

CHO 33 → H_2/Ra(Ni) → CH_2OH 34 → Δ Al_2O_3 → 35

The isomeric 2,5-dihydrofuran (**37**) is readily synthesized by the selective reduction of **36**, followed by thermal dehydration to give **37.**[10]

$HOCH_2C{\equiv}CCH_2OH$ (**36**) —H_2/Ni→ $HOCH_2CH{=}CHCH_2OH$ (cis & trans) —Al_2O_3, 240°C→ **37**

2,5-Dialkoxy- and 2,5-diacetoxydihydrofurans (**38** and **39**) are obtained by addition reactions previously described. These compounds are readily reduced to their tetrahydro derivatives (**40**), which are easily converted to pyrroles (**12**) by treatment with alkyl or aryl amines, or to unsubstituted pyrrole (**11**) by reaction with ammonia in acetic acid.[11]

38 (R=alkyl), **39** (R=acetyl) —H_2/Ra(Ni), CH_3OH→ **40** —RNH_2→ **12**

11.4. THIOPHENES

The presence of the sulfur atom in thiophene essentially precludes catalytic reductions because of the catalyst-poisoning nature of sulfur. This type of reaction proceeds only under conditions that employ a great excess of catalyst. Thus perhydrothiophene (**41**) is obtained (71%) as long as the ratio of catalyst to thiophene is at least 2 : 1.[12] Interestingly, Raney nickel catalysts are *not* poisoned and can be used to reduce the nitro group in nitrothiophenes (**42**).[13,14]

Thiophene —Pd/C→ **41** **42** —R(Ni),H_2, Ac_2O→ 2-NHAc-thiophene

Other substituted thiophenes **43** can be hydrogenatively desulfurized[14,15] by Raney nickel. An elegant example of this desulfurization process is the conversion of biotin (**44**) to desthiobiotin (**45**), which allowed the subsequent structure elucidation of this vitamin.[16]

R(Ni), H_2 / toluene; **43**; H_2 / Ni; HN, NH, O, S, $(CH_2)_4$, HO_2C **44**; CH_3, $(CH_2)_5$, CO_2H **45**

Thiophene is reduced by sodium in methanol/liquid ammonia at −40°C to afford a mixture of compounds.[17] The 2,3- and 2,5-dihydrothiophenes (**46** and **47**) are the major products and can be separated by steam distillation. 2,3-Dihydrothiophene (**46**) polymerizes on standing, whereas the 2,5-dihydro isomer (**47**) is stable.

Na°/MeOH / NH_3 (liq); **46** + **47** + butanethiols + n-butane + H_2S

Tetrahydrothiophene (thiolane, **41**) can be synthesized directly by treatment of 1,4-dichlorobutane (**48**) with sodium sulfide.[18] 1,4-Butanediol (**49**)[19] and tetrahydrofuran (**26**)[19,20] are equally amenable to conversion to thiolane.

OH OH **49**; H_2S/Al_2O_3 400°C; Cl Cl **48**; Na_2S/80°C (85%); H_2S/Al_2O_3 350 - 400°C; **26**

11.5. SELENOPHENES

Tetrahydroselenophene (selenolane, **50**) can be prepared in several different ways.[21-23]

Br Br; Na_2Se/H_2O; Se(powder); Se **50**; H_2Se/Al_2O_3 400°C; O

Reaction of 2,5-dibromoadipic acid (**51**) with potassium selenide in acetone affords the dicarboxylic acids **52** and **53**. Use of *meso*-α,α'-dibromoadipic acid affords the *cis*-dicarboxylic acid **52,** whereas racemic α,α'-dibromoadipic acid yields the trans isomer **53.**[24]

K_2Se, acetone

51 → 52 (from meso) + 53 (from dl)

11.6. TELLUROPHENES

Tetrahydrotellurophene, (tellurane, **54**) is similarly obtained from 1,4-dibromobutane and sodium telluride, which is prepared *in situ* by the reaction of tellurium with a hot aqueous solution of sodium formaldehyde sulfoxylate (Rongalite).[25] Tellurane (**54**) readily reacts with bromine to afford 1,1-dibromotelluran (**55**).

Na_2Te → 54 ; Br_2 → 55

11.7. MISCELLANEOUS LESS COMMON SATURATED FIVE-MEMBERED HETEROCYCLES

Phospholan (**56**) is formed by the diborane reduction of *p*-dimethylaminocyclophosphine.[26] 1-Phenylarsolan (**57**),[27] germacyclopentane (**58**),[28] and silolan (**59**)[29] are synthesized by the reaction of 1,4-butanedimagnesium bromide with phenyldichloroarsine, germanium tetrachloride, and silicon tetrachloride, respectively.

$C_6H_5AsCl_2$ → 57 ; (1) XCl_4 (2) $LiAlH_4$ → 58 (X = Ge), 59 (X = Si)

REFERENCES

1. C. B. Hudson and A. V. Robertson, *Tetrahedron Lett.* **1967,** 4015.
2. D. M. Lemal and S. D. McGregor, *J. Am. Chem. Soc.* **88,** 1335 (1966), and refs. cited therein.
3. H. A. Bates and H. Rapoport, *J. Am. Chem. Soc.* **101,** 1259 (1979).
4. P. A. Cantor and C. A. Vanderwerf, *J. Am. Chem. Soc.* **80,** 970 (1958).
5. N. W. Gabel, *J. Org. Chem.* **27,** 301 (1962).
6. A. P. Dunlop and F. N. Peters, *The Furans,* Reinhold, New York, 1953.
7. P. Bosshard and C. H. Eugster, *Adv. Heterocycl. Chem.* **7,** 377 (1966).
8. A. J. Birch and J. Slobbe, *Heterocycles* **5,** 905 (1976).
9. H. C. van der Plas, in *Ring Transformations of Heterocycles,* Vol. 1, Academic Press, New York, pp. 167–170.
10. S. Holand, F. Mercier, N. Le Goff, and R. Epsztein, *Bull. Soc. Chim. Fr.* **1972,** 4357.
11. N. Elming and N. Clauson-Kaas, *Acta Chem. Scand.* **6,** 867 (1952).
12. R. Mozingo, S. A. Harris, D. E. Wolf, C. E. H. Hoffhine, Jr., N. R. Easton, and K. Folkers, *J. Am. Chem. Soc.* **67,** 2092 (1945).
13. A. I. Meyers, *Heterocycles in Organic Synthesis,* Wiley-Interscience, New York, **1974,** pp. 15, 228, 266.
14. Ya. L. Gol'dfarb, B. P. Fabrichnyi, and I. G. Shalavina, *Tetrahedron* **18,** 21 (1962).
15. G. R. Pettit and E. E. van Tamelen, *Org. Reactions* **12,** 356 (1962).
16. V. du Vigneaud, D. B. Melville, K. Folkers, D. E. Wolf, R. Mozingo, J. C. Keresztesy, and S. A. Harris, *J. Biol. Chem.* **146,** 475 (1942).
17. S. F. Birch and D. T. McAllan, *J. Chem. Soc.* **1951,** 2556.
18. B. K. Menon and P. C. Guha, *Chem. Ber.* **64,** 544 (1931); C. S. Marvel and W. W. Williams, *J. Am. Chem. Soc.* **61,** 2714 (1939).
19. H. D. Hartough, in *Thiophene and its Derivatives,* A. Weissberger, Ed., Interscience, New York, 1952, pp. 76–77.
20. Yu. K. Yur'ev *et al., Zh. Obshch. Khim.* **19,** 724 (1950) [*Chem. Abstr.* **44,** 1092 (1950)]; *Zh. Obshch. Khim.* **22,** 339 (1952) [*Chem. Abstr.* **46,** 11177 (1952)]; *Zh. Obshch. Khim.* **23,** 1944 (1953) [*Chem. Abstr.* **49,** 281 (1955)].
21. G. T. Morgan and F. H. Burstall, *J. Chem. Soc.* **1929,** 1096.
22. J. D. McCullough and A. Lefohn, *Inorg. Chem.* **5,** 150 (1966).
23. Yu. K. Yue'ev, *J. Gen. Chem.* (*U.S.S.R.*) **16,** 851 (1946).
24. A. Fredga, *J. Prakt. Chem.* **130,** 180 (1931).
25. W. V. Farrar and J. M. Gulland, *J. Chem. Soc.* **1945,** 11.
26. A. B. Berg and P. J. Slota, Jr., *J. Am. Chem. Soc.* **82,** 2148 (1960).
27. J. J. Monagle, *J. Org. Chem.* **27,** 3851 (1962).
28. P. Mazerolles, *Bull. Soc. Chim. Fr.* **1962,** 1907.
29. R. West, *J. Am. Chem. Soc.* **76,** 6012 (1954).

Chapter 12

ALKYL DERIVATIVES OF π-EXCESSIVE HETEROAROMATIC SYSTEMS

12.1. GENERAL COMMENTS

In stark contrast to the π-deficient systems, in which anions are generated by proton removal from an α- or β-substituted methyl group and are stabilized by involvement of the heteroatom, such is not possible for the π-excessive systems. These alkyl groups thus bear some resemblance to the methyl group in toluene. Since these compounds are already "electron rich," the addition of C-alkyl substituents (**1**) should facilitate electrophilic substitution.

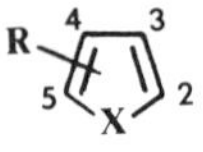

12.2. ALKYLPYRROLES

A most amazing reaction occurs when certain pyrroles (for example, **2**) are treated with sodium alkoxide[1] or phenoxide[2] at 220°C: alkylated or arylated pyrroles are obtained directly. Another interesting transformation is the thermal rearrangement of *N*-methyl- and *N*-phenylpyrrole (**3** and **4**, respectively) to give the corresponding 2- and 3-substituted pyrroles.[3] 1-Acetylpyrrole and 1-benzoylpyrrole rearrange similarly at 250–280°C.

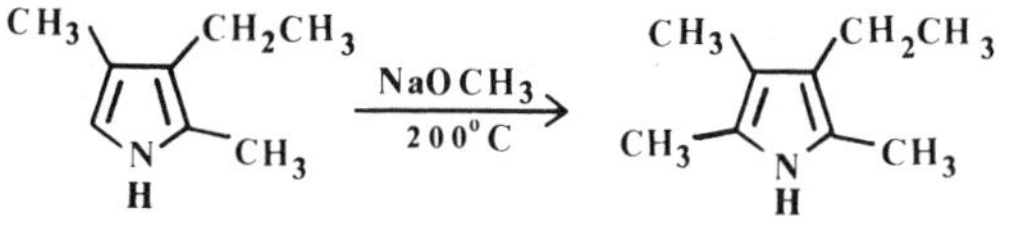

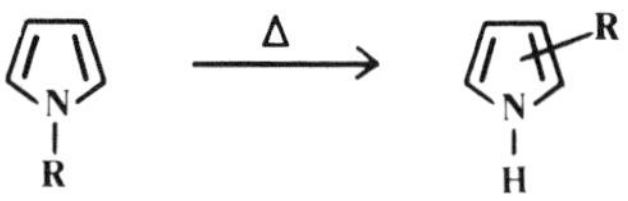

In general, alkylpyrroles can be prepared by the standard pyrrole synthetic routes, previously considered in Chapter 4. Alkylation by the reaction of alkyl halides with pyrroles generally gives a mixture of products that depends upon numerous factors, such as solvent, temperature, alkyl source, and metal cation. It has been suggested, in related alkylation studies, that dissociation of salts favors *N*-substitution whereas ionic association favors *C*-substitution in a homogeneous medium.[4]

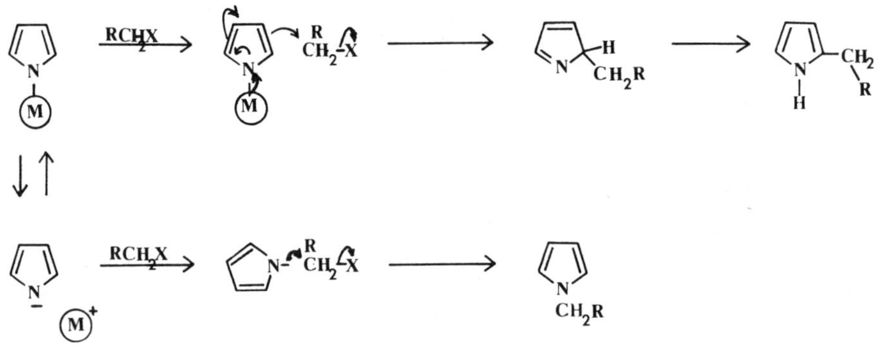

It is abundantly clear that, as long as any ring carbon is free, reactions will occur preferentially on the ring. When all the ring carbons are substituted (for example, **5**), a free methyl group can be readily halogenated and acetoxylated at reduced temperature.[5,6] α-Alkylpyrroles tend to undergo radical substitution α to the ring as well as ring substitution.

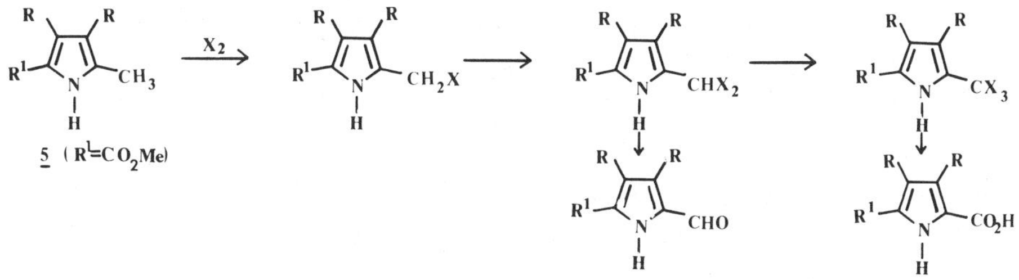

As a demonstration of the enhanced reactivity of *C*- and *N*-alkylpyrroles (for example, **6** and **7**), acylation by acetic anhydride without catalysts readily occurs[7]; alkylpyrroles possessing electron-withdrawing substituents normally require a Friedel–Crafts catalyst to induce electrophilic substitution.

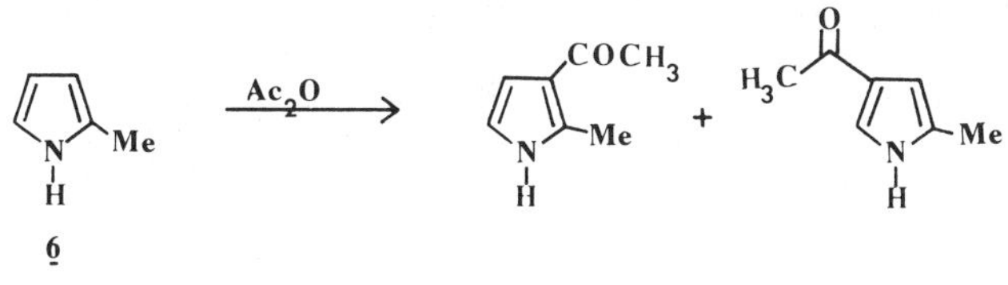

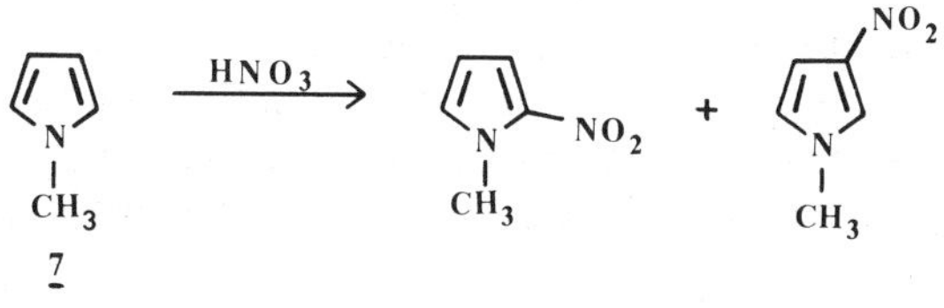

12.3. ALKYLFURANS

Syntheses of alkylfurans via ring-construction are described in Chapter 4; direct substitution of furans by means of Friedel–Crafts alkylation or Grignard reactions is also a standard procedure. When 2-furan-carboxaldehyde (**8,** furfural), a versatile, readily available starting material, is subjected to the Wolff–Kishner reduction, the 2,5-dihydro compound **9** is obtained along with the anticipated 2-methylfuran (**10**).[8]

3-Methylfuran (**11**) undergoes facile electrophilic substitution at the 2-position upon treatment with acetic anhydride and boron trifluoride etherate.[9] 2-Alkylfurans are formylated using Vilsmeier reaction conditions (*N,N*-dimethylformamide, $POCl_3$, 0–10°C) in the 5-position,[10] and if both α-positions are alkylated then 3-substitution occurs.

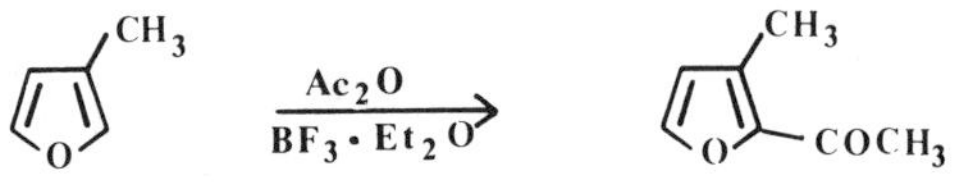

The increased reactivity of the alkylfurans is further demonstrated by the facile hydrolysis of 2-methylfuran (**12**) to levulinic aldehyde (**13**),[11] or under hydrolytic-reductive conditions to **14.**[12]

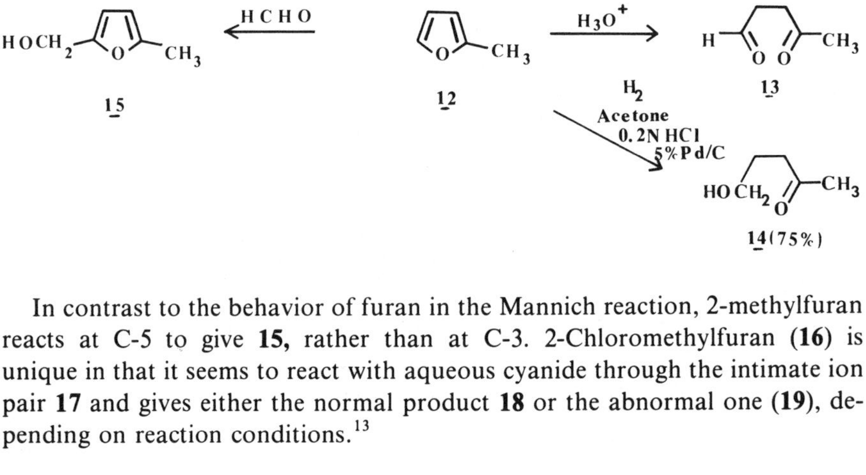

In contrast to the behavior of furan in the Mannich reaction, 2-methylfuran reacts at C-5 to give **15,** rather than at C-3. 2-Chloromethylfuran (**16**) is unique in that it seems to react with aqueous cyanide through the intimate ion pair **17** and gives either the normal product **18** or the abnormal one (**19**), depending on reaction conditions.[13]

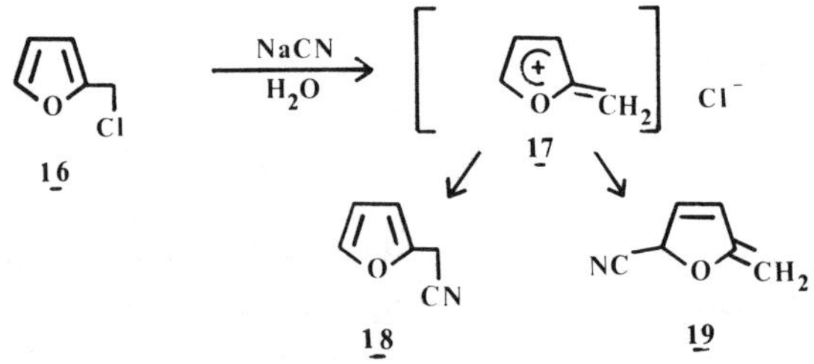

2,3-Dimethyl-5-*tert*-butylfuran (**20**), when treated with *tert*-butyl chloride and aluminum chloride, affords the methylene derivative **21,** an isomer of the expected furan **22.**[14] It appears that the steric strain, expected to exist between the two *tert*-butyl groups in **22,** is decreased by its tautomerization to the exocyclic isomer. 2,5-Dimethylfuran (**23**) upon treatment with bromine in the presence of the free radical initiator azobisisobutyronitrile (AIBN) affords the side-chain brominated compounds **24** and **25.**[15]

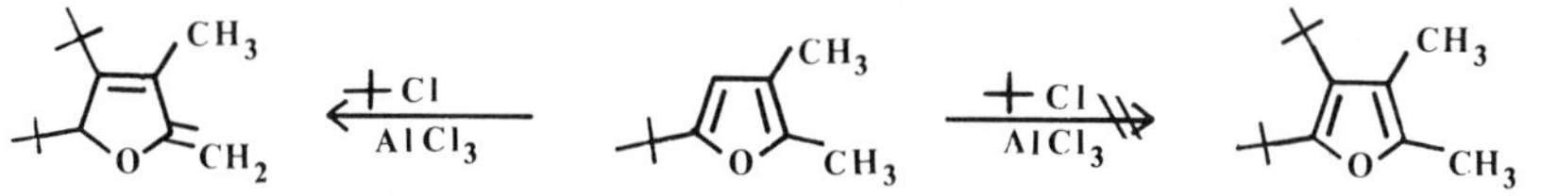

12.4. ALKYLTHIOPHENES

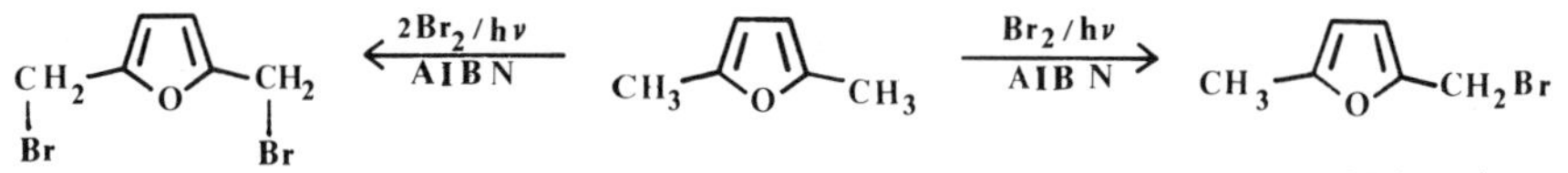

Alkylthiophenes are synthetically available from the appropriate cyclization reactions previously considered. In contrast to alkylpyrroles and alkylfurans, the methylthiophenes react more like toluenes.

Generally, all common electrophilic substitution reactions on 2-alkylthiophenes give rise to substitution in the 5-position exclusively, except for nitration, which affords considerable amounts of 3-substituted products. 3-Alkylthiophenes (**27**) undergo electrophilic substitution predominantly at the 2-position, except where steric conditions force 5-substitution.

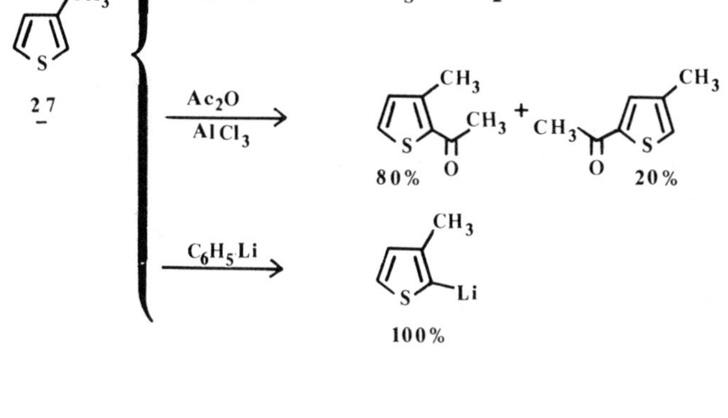

Although ring halogenation is still a problem, side-chain halogenation of the alkylthiophenes **26–28** can be accomplished under carefully controlled homolytic conditions[15,16] which require the addition of AIBN as the radical catalyst. Under these conditions, **26** affords a low yield of 2-bromomethylthiophene (**29**), whereas the major product is the dibromo derivative **30.** Unfortunately, addition of only 1 mole of bromine by this procedure is not yet reported; however, it does seem reasonable that higher yields of **29** should be obtainable by variation of the reaction parameters. For bromination of **27** in the presence of AIBN, the major product is **31,** but the yield is lower and some 3-dibromomethylthiophene (**32**) is also produced. With *N*-bromosuccinimide (NBS), **27** affords **31** in high yields. In this reaction, when one equivalent of bromine is used, minimal ring-substituted products (for example, **33**) are observed. When both the α-positions of thiophene (**28**) are blocked, **34** is the sole product under the reaction conditions indicated.[15,16]

26 —($2Br_2$/hν, AIBN)→ **29** (16%) + **30** (83%)

27 —(Br_2/hν, AIBN)→ **31** (47%) + **32** (19%)

27 —(NBS/CCl_4, $(C_6H_5CO_2)_2$)→ **31** (77%) + **33** (10%)

28 —(Br_2/hν, AIBN)→ **34** (70%) + **35** (0%)

28 —($2Br_2$/hν, AIBN)→ **34** (8%) + **35** (80%)

The condensation reaction of 2-methylthiophene (**26**) with benzaldehyde in the presence of zinc chloride does not afford the benzal derivative (**36**) but rather the ring-substituted product **37.** Thus the alkylthiophene ring is even more reactive to substitution than the side chain, but less so than the corresponding rings of the alkylfurans and alkylpyrroles. It is surprising that 2- and 3-methylthiophenes are oxidized under mild conditions and in high yields to the corresponding carboxylic acids (**38**).[17]

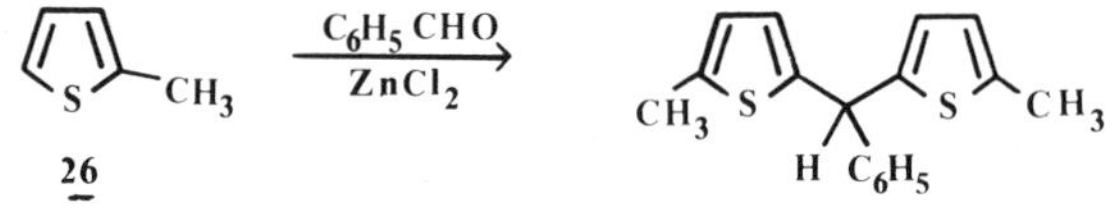

26 37 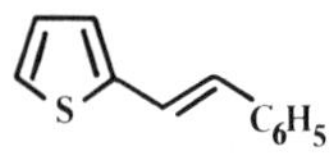 36 (0%)

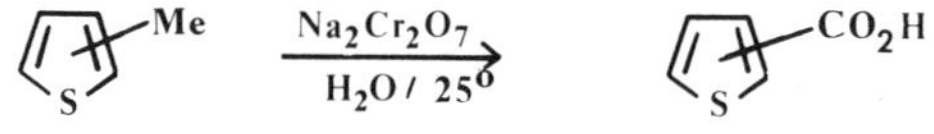

38

12.5. ALKYLSELENOPHENES

2-Methyl-3,4,5-trisubstituted selenophenes (**39**) can be obtained by a facile cyclization procedure of an alkynylketone with hydrogen selenide[18]; the 3-methylselenophenes (**40**) are similarly available from the reaction of the acetylenic epoxides **41** with the same reagent.[19]

39

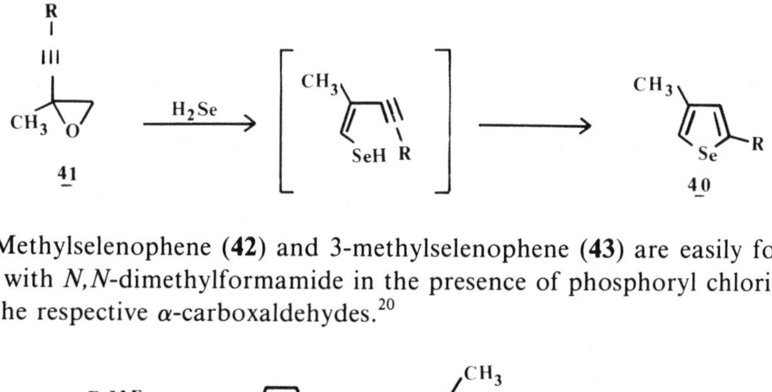

41 40

2-Methylselenophene (**42**) and 3-methylselenophene (**43**) are easily formylated with *N,N*-dimethylformamide in the presence of phosphoryl chloride to give the respective α-carboxaldehydes.[20]

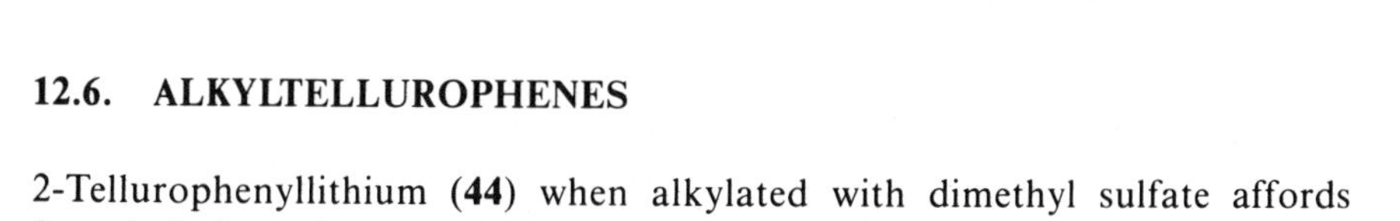

42 43

12.6. ALKYLTELLUROPHENES

2-Tellurophenyllithium (**44**) when alkylated with dimethyl sulfate affords 2-methyltellurophene (**45**), along with what appears to be a trace of the

3-methyl isomer (**46**).[21] Subsequent treatment of **45** with butyllithium generates the 5-lithio intermediate (**47**), which may be quenched with carbon dioxide to give **48** as the major product.[21]

44 $\xrightarrow{(CH_3)_2SO_4}$ **45** + **46** (Trace)

45 $\xrightarrow{\text{buLi}}$ **47** $\xrightarrow{CO_2,\text{ then } H_3O^+}$ **48**

REFERENCES

1. H. Fischer and E. Bartholomäus, *Chem. Ber.* **45,** 466 (1912); H. Fischer and K. Eismayer, *Chem. Ber.* **47,** 1820 (1914).
2. J. W. Cornforth, R. H. Cornforth, and R. J. Robinson, *J. Chem. Soc.* **1942,** 682.
3. J. P. Wibaut and J. Dhont, *Rec. Trav. Chim.* **62,** 272 (1943), and refs. cited therein; E. Späth and P. Kainrath, *Chem. Ber.* **71,** 1276 (1938).
4. P. A. Cantor and C. A. Vanderwerf, *J. Am. Chem. Soc.* **80,** 970 (1958); C. F. Hobbs, C. K. McMillin, E. P. Papadopoulos, and C. A. Vanderwerf, *J. Am. Chem. Soc.* **84,** 43 (1962).
5. A. Hayes, G. W. Kenners, and N. R. Williams, *J. Chem. Soc.* **1958,** 3779.
6. A. H. Jackson, G. W. Kenner, and D. Warburton, *J. Chem. Soc.* **1965,** 1328.
7. S. Clementi and G. Marino, *Tetrahedron* **25,** 4599 (1969); *J. Chem. Soc. Perkin II* **1972,** 71.
8. H. L. Rice, *J. Am. Chem. Soc.* **74,** 3193 (1952).
9. P. A. Finan and G. A. Fothergill, *J. Chem. Soc.* **1963,** 2723.
10. V. J. Traynelis, J. J. Miskel, and J. R. Sowa, *J. Org. Chem.* **22,** 1269 (1957).
11. D. S. P. Eftax and A. P. Dunlop, *J. Org. Chem.* **30,** 1317 (1965).
12. T. E. Londergan, N. L. Hause, and W. R. Schmitz, *J. Am. Chem. Soc.* **75,** 4456 (1953).
13. S. Divald, M. C. Chun, and M. M. Jouillie, *J. Org. Chem.* **41,** 2835 (1976).
14. H. Wynberg and U. E. Wiersum, *Tetrahedron Lett.* **1975,** 3619.
15. E. Campaigne and B. F. Tullar, *Org. Synthesis* Coll. Vol. **4,** 921 (1963).
16. J. A. Clarke and O. Meth-Cohn, *Tetrahedron Lett.* **1975,** 4705.
17. L. Friedman, D. L. Fishel, and H. Shechter, *J. Org. Chem.* **30,** 1453 (1965).
18. K. E. Schulte, J. Reisch, and D. Bergenthal, *Chem. Ber.* **101,** 1540 (1968).
19. F. Ya. Perveev, N. I. Kudryashova, and D. N. Glebovskii, *J. Gen. Chem. U.S.S.R.* **26,** 3707 (1956).
20. Yu. K. Yusyer and N. N. Mezentsova, *J. Gen. Chem. U.S.S.R.* **27,** 201 (1957).
21. F. Fringuelli and A. Taticchi, *J. Chem. Soc. Perkin I* **1972,** 199.

Chapter 13

SYNTHESES OF SOME NATURALLY OCCURRING AND PHARMACEUTICALLY IMPORTANT FIVE-MEMBERED π-EXCESSIVE HETEROAROMATIC COMPOUNDS

13.1. GENERAL COMMENTS

One of the major forces that sustains interest in heterocyclic chemistry is the fact that nature elaborates many of these ring systems. A large number of the compounds are of biological interest because they offer insights into the life processes of plants and animals; others are of medicinal value. Many of the preparations and reactions so far discussed have been applied to the syntheses of natural products and synthetic drugs. Thus it is only reasonable that a discussion of some of these syntheses should be included in a book of this type.

13.2. NATURALLY OCCURRING PYRROLES

Perhaps the most important naturally occurring monopyrrole is *porphobilinogen* (**1**). It is a biological precursor to the chlorophylls, vitamin B_{12}, and the porphyrins. Among the various reported syntheses, the following is an excellent example of the construction and manipulation of complex heterocyclic compounds.[2,3]

$EtO_2CCH_2CH_2COCH(COCH_3)CH_2CO_2Et$ + EtO_2C–C(H)=$\overset{+}{N}$–OH $\xrightarrow[AcOH]{Zn}$ **2**

2: $EtO_2C(CH_2)_2$, CH_2CO_2Et, EtO_2C, CH_3, N–H

[O] ↓

3: $EtO_2C(CH_2)_2$, CH_2CO_2Et, EtO_2C, CHO, N–H

(1) OH^- (2) H_3O^+ →

4: [$HO_2C(CH_2)_2$, CH_2CO_2H, HO_2C, CHO, N–H]

Warm ↓

5: $HO_2C(CH_2)_2$, CH_2CO_2H, H, CHO, N–H

$H_2NOH·HCl$ →

6: $HO_2C(CH_2)_2$, CH_2CO_2H, CH=NOH, N–H

[H] ↓

1: $HO_2C(CH_2)_2$, CH_2CO_2H, CH_2NH_2, N–H

The Knorr pyrrole synthesis affords the tetrasubstituted pyrrole **2**, which is di-α-chlorinated and hydrolyzed to give aldehyde **3**. Saponification of the aldehydic triester **3** by aqueous base is successful since pyrrolecarboxaldehydes are relatively stable to Cannizzaro-type (acidic) reactions. Decarboxylation of **4** to **5** is facile since π-excessive, heteroaromatic, α-carboxylic acids are notoriously unstable. The conversion of **5** to **1,** via the intermediate oxime **6,** is a routine reaction sequence for the synthesis of primary amines.

Pyoluteorin (**7**) is produced by certain strains of the bacterium *Pseudomonas aeruginosa,* and because of its antibiotic activity it has been the object of several syntheses,[4–6] one of which is described:

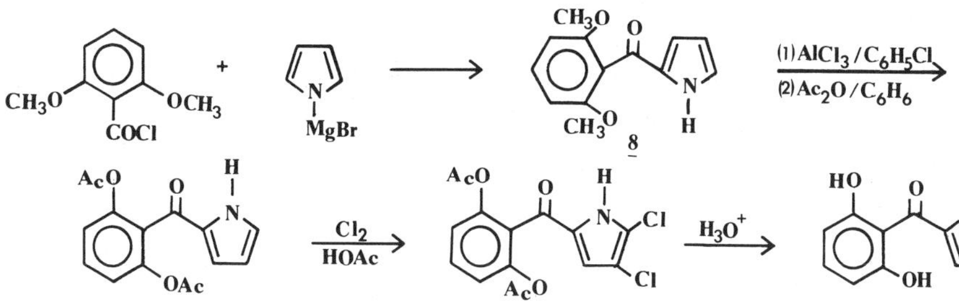

CH_3O OCH$_3$ COCl + MgBr → CH_3O O N H CH_3O 8 $\xrightarrow[(2) Ac_2O/C_6H_6]{(1) AlCl_3/C_6H_5Cl}$

AcO O H N OAc 9 $\xrightarrow[HOAc]{Cl_2}$ AcO O H N Cl Cl AcO 10 $\xrightarrow{H_3O^+}$ HO O H N Cl Cl OH 7

The initial condensation reaction to afford **8** is based on the well-known reactivity of *N*-metal pyrrole derivatives, and the dealkylation of ethers in the presence of Lewis acids and subsequent acetylation (**8** → **9**) finds ample precedent in carbocyclic chemistry. Formation of the dichloro derivative **10** is equally reasonable, based on the electrophilic reactivity of pyrrole.

The prodigiosins, based on the pyrryldipyrrylmethene skeleton (**11**), are a series of red antifungal and antibacterial compounds distributed widely in nonpathogenic bacteria which reside in both soil and water.

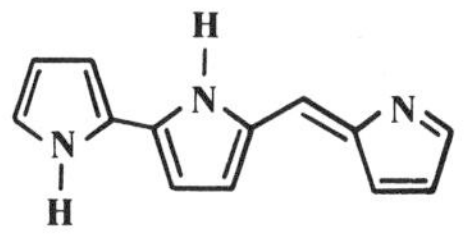

H N N H N

11

The synthesis of *prodiogosin* (**12**) itself is accomplished in a very clever sequence of reactions founded on the pyrrole chemistry described in earlier chapters,[7] and begins with the ethereal ester **13.** Condensation of **13** with Δ^1-pyrroline (**14**) generates the tetrahydro dipyrridyl derivative **15,** which is thermally dehydrogenated (aromatized) with palladium/carbon to **16.** Subsequent conversion of the ester to the aldehyde **17** follows the standard transformations indicated. The acid-catalyzed condensation of **18** with **17** completes the overall synthesis of **12.**

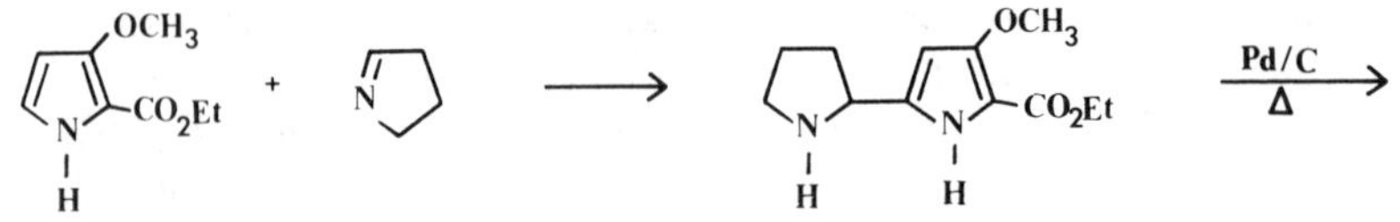

13 14 15

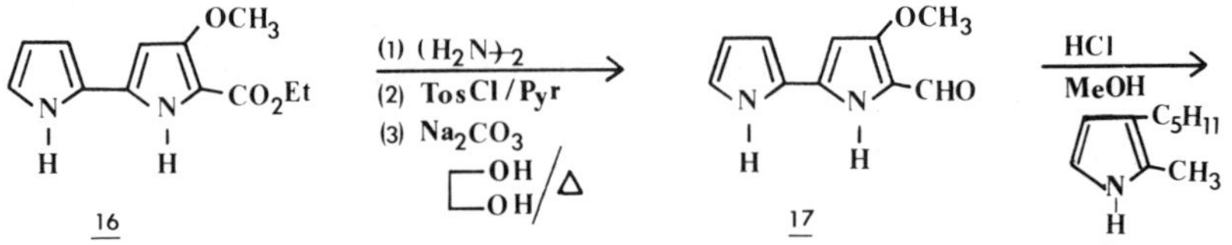

16 17 18

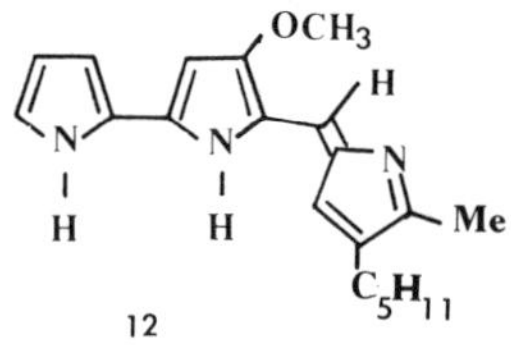

12

Porphyrins (and chlorins) are perhaps the most widely distributed pigments found in nature, with a general skeletal structure such as shown in **19** (M = metal such as Fe, Mg).

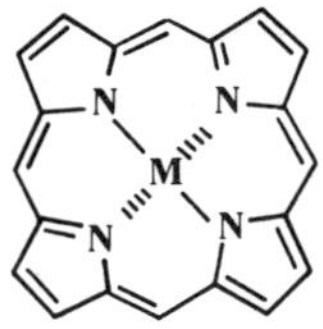

19

One of the many brilliant syntheses of these compounds is that of *deoxophylloerythroetioporphyrin* (**20**) by Flaugh and Rapoport[8] (see Scheme 13.1), from which a number of very interesting transformations can be discerned on careful scrutiny. The formation of **23** by the decarboxylative condensation of **21** with **22** utilizes the facile decarboxylation of α-carboxylic acids in the pyrrole series. The same reactivity facilitates conversion of **23** to **24.** Use of malononitrile, as a protecting group in the synthesis of **25,** is noteworthy. Formation of **27** from **26** is an example of an electrophilic substitution on the pyrrole nucleus, and the conversion of **27** to **28** again is a facile decarboxylation.

Since the aldehyde function in **25** is protected, its condensation with **28** occurs readily to afford **29.** Subsequent conversion of **29** to the imino derivative

30 is followed by condensation of the carbonyl compounds **30** with **24** to afford the quasi model condensation product **31.** The sequence of hydrolytic debenzylation–decarboxylation of **31** (removal of R′) and cyclocondensation finally affords the desired product **20.**

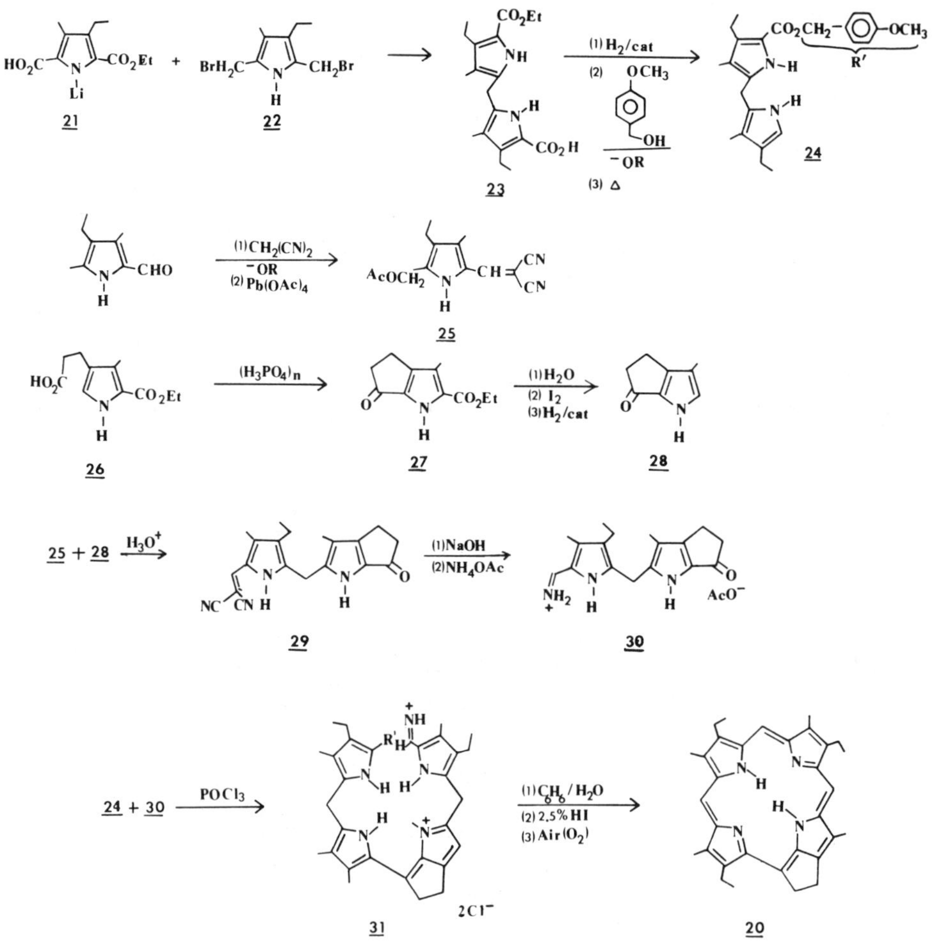

Scheme 13.1

The tetramerization of pyrroles is a simple but effective approach to the syntheses of porphyrins and is appliable to pyrroles in which the 3- and 4-substituents are identical; otherwise different substituents may produce a number of different isomers. The simplest example of this reaction involves the condensation of pyrrole itself with benzaldehyde in propionic acid.[9] The resulting *meso*-tetraphenylporphin (**32**) is obtained in amazingly good yield.

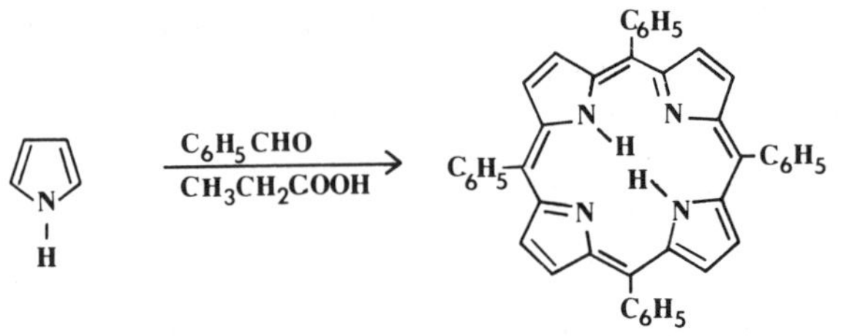

C_6H_5CHO / CH_3CH_2COOH

32

Among the large number of indole derivatives found in nature,[10] *vincamine* (**33**), from *Vinca minor L.*,[10a] is one whose synthesis reflects some of the indole nucleus reactivity (Scheme 13.2). Tryptamine (**36**) is formed from indole via an initial Mannich reaction, followed by several other well-known reactions. Conversion of **36** to **38** can be accomplished by the acid-catalyzed condensation of **36** with **37** (prepared by the enamine alkylation of butyrylaldehyde with two equivalents of methyl acrylate in methanol). The initially formed Schiff's base is cyclized upon heating to generate epimeric tetracyclic lactam esters.[11] An oxidation-reduction-esterification sequence completes the synthesis.

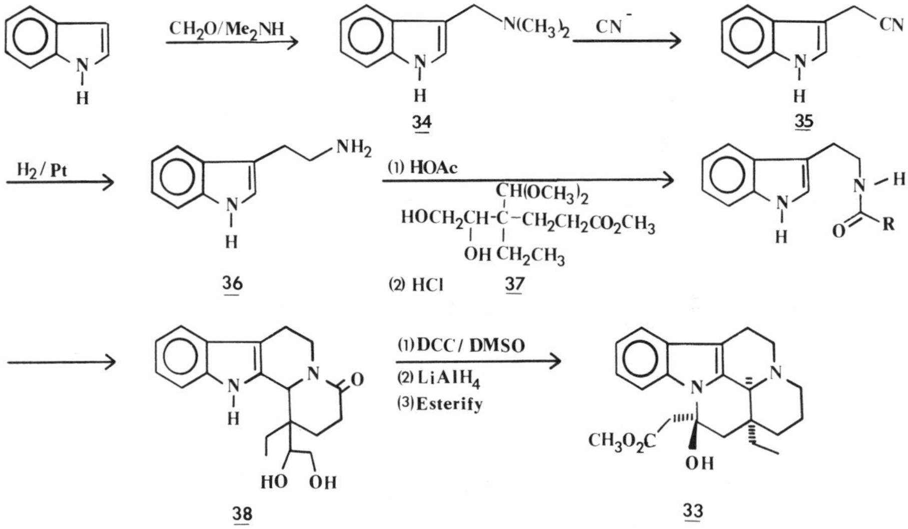

CH_2O/Me_2NH → 34 → CN^- → 35 → H_2/Pt → 36 → (1) HOAc, $HOCH_2CH(OH)-C(CH(OCH_3)_2)(CH_2CH_3)-CH_2CH_2CO_2CH_3$ 37, (2) HCl → 38 → (1) DCC/DMSO (2) $LiAlH_4$ (3) Esterify → 33

Scheme 13.2

The synthesis of the alkaloid *skopine* (**39**)[12] illustrates the reactivity of the two α-positions of an *N*-substituted pyrrole in a $[4 + 2]$ cycloaddition. The desired adducts **40** and **41** (2 : 1 ratio) are smoothly dehalogenated, then the *N*-carbomethoxy and ketone functions are reduced simultaneously with diisobutylaluminum hydride (DIBAH) to give **43**, as an isomeric (93 : 7) mixture of alcohols. Protection of the alcohol group, epoxidation, and deprotection affords **39**, as well as other C-6 and C-7 related oxygenated alkaloids.

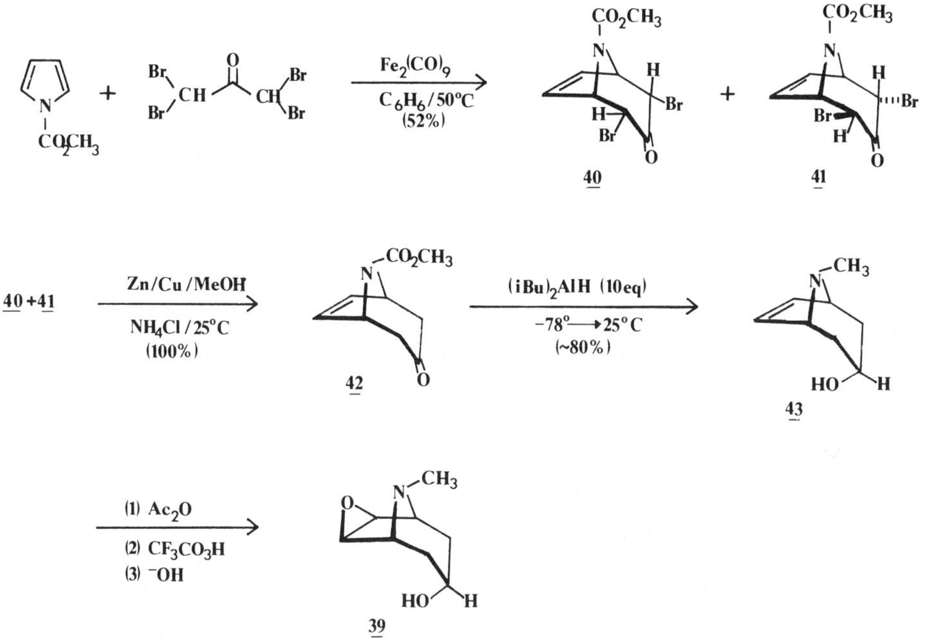

$Fe_2(CO)_9$, $C_6H_6/50°C$ (52%) → **40** + **41**

40 + **41** — Zn/Cu/MeOH, $NH_4Cl/25°C$ (100%) → **42** — $(iBu)_2AlH$ (10eq), $-78° \rightarrow 25°C$ (~80%) → **43**

(1) Ac_2O (2) CF_3CO_3H (3) ^{-}OH → **39**

The reactive species in the formation of **40** and **41** is the iron(II) salt **44**.

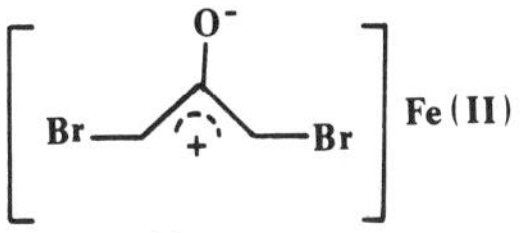

Fe(II)

44

13.3. NATURALLY OCCURRING FURANS

The number of naturally occurring furans is considerably less than the naturally occurring pyrroles. Yet a number of these have been isolated from a variety of sources. The formicine ant *Dendrolosius fuliginosus* produces a defensive secretion which contains the sesquiterpenoid furan *dendrolasin* (**45**). Interestingly, the same compound is also found in a plant source produced by fermentation of the sweet potato.[13] The starting material, geraniol (**46**), is converted to the cadmium reagent **47**, which reacts with acid chloride **48** to afford ketone **49**. Wolff–Kishner reduction of **49** readily yields dendrolasin.

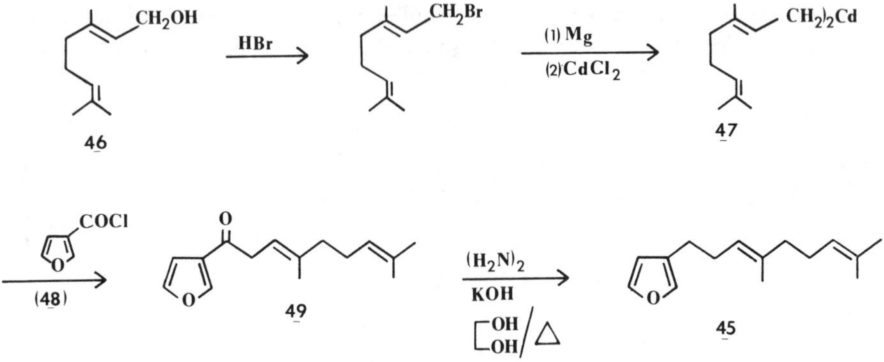

CH_2OH — HBr → CH_2Br — (1) Mg, (2) $CdCl_2$ → $CH_2)_2Cd$

46 47

COCl (48) → 49 — $(H_2N)_2$, KOH, $\begin{matrix}OH\\OH\end{matrix}$ / △ → 45

Ipomeamarone (**50**) has been isolated from infected tissue of the sweet potato. This same tissue, when infected by black rot fungus, also contains *ipomeanine* (**51**), batatic acid (**52**), and 3-furoic acid.[13]

O, CH_3, O — 51

O, CO_2H, CH_3 — 52

CO_2Et — EtOAc, Na° → O, CH_2CO_2Et — (1) Na°, (2) $BrCH_2C(CH_3){=}CHCO_2Et$ →

EtO_2C, $CH_2C(CH_3){=}CHCO_2Et$, O — 53 — H_3O^+, △ → O, CH_3, $CHCO_2H$ — (1) CH_2N_2, (2) $Al(O\underline{i}Pr)_3/\underline{i}PrOH$ →

$CHCO_2Me$, OH CH_3 — 54 — NaOH → CH_3, CH_2CO_2H — 55

(1) $SOCl_2$, (2) $[Me_2CHCH_2]_2Cd$ → CH_3, $CH_2COCH_2CHMe_2$ — 50

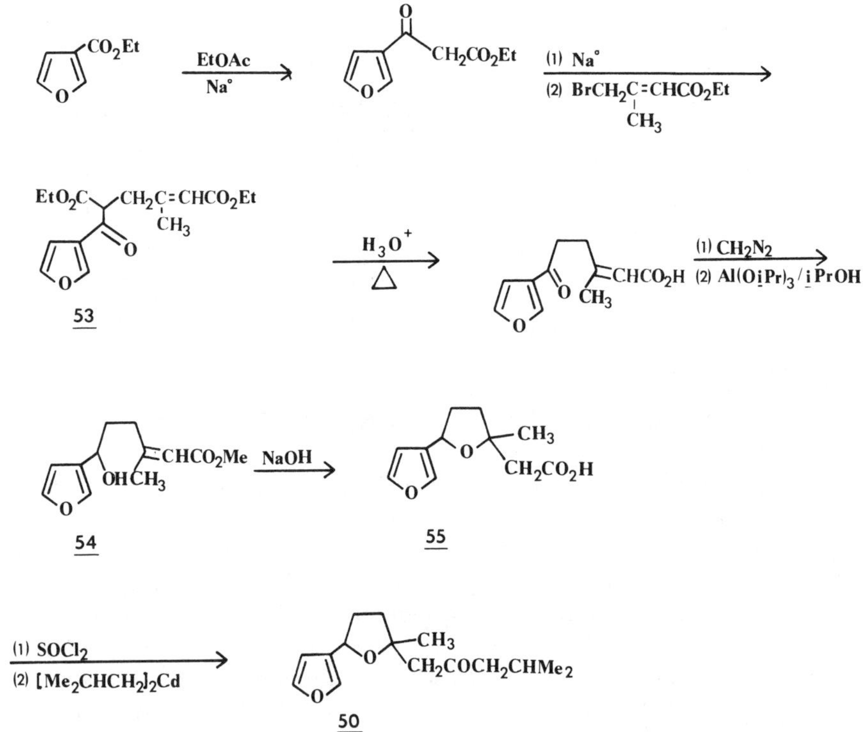

3-Carbethoxyfuran is transformed to **53** via a Claisen condensation followed by alkylation of the resultant β-ketoester. Hydrolysis of **53** causes decarboxylation; esterification and reduction gives alcohol **54,** which cyclizes under basic conditions by a Michael addition to the α,β-unsaturated ester to give the tetrahydrofuran **55.** Subsequent conversion to **50** is realized by addition of the appropriate cadmium reagent to the acid chloride of **55.**

One of the odor constitutents of rose oil is the disubstituted furan **56,** *rosefuran.* The synthesis of **56** utilizes the reactivity of the salt of furan-2-carboxylic acid (**57**) and involves an interesting transmetalation of the mercuri compound **58** to the 2-lithio-3-methylfuran (**59**). Condensation of **59** with an allyl bromide affords rosefuran in excellent yield.[14] The related *dehydroelsholtzione* (naginata ketone)[15] (**60**) is prepared by the following simple transformations[14] from 2-carbomethoxy-3-methylfuran:

CH_3, $CO_2^- Na^+$ (**57**) $\xrightarrow{HgCl_2}$ CH_3, $HgCl$ (**58**) $\xrightarrow[97\%]{Na_2S_2O_3}$

CH_3 CH_3, Hg $\xrightarrow[\text{ether}]{Li}$ [CH_3, Li] (**59**) $\xrightarrow{Br}$ CH_3, CH_3, CH_3 (**56**)

CH_3, CO_2Me $\xrightarrow[B^-]{CH_3COCH_3}$ CH_3, O, O, CH_3 $\xrightarrow{CH_3MgI}$

CH_3, O, CH_3, CH_3, OH + CH_3, O, CH_3, CH_3 (**60**)

The synthesis of the sesquiterpenes *dehydrofuropelargones* (**61** and **62**)[16] utilizes an interesting application of a Darzens glycidic ester condensation involving an epoxyacetal **63.** An enamine condensation, alkylation, and selective addition of a Grignard reagent to the isolated ketone gives a β-hydroxyketone, which undergoes a reverse aldol condensation, then recyclizes to give the isomeric ketones **61** and **62.**

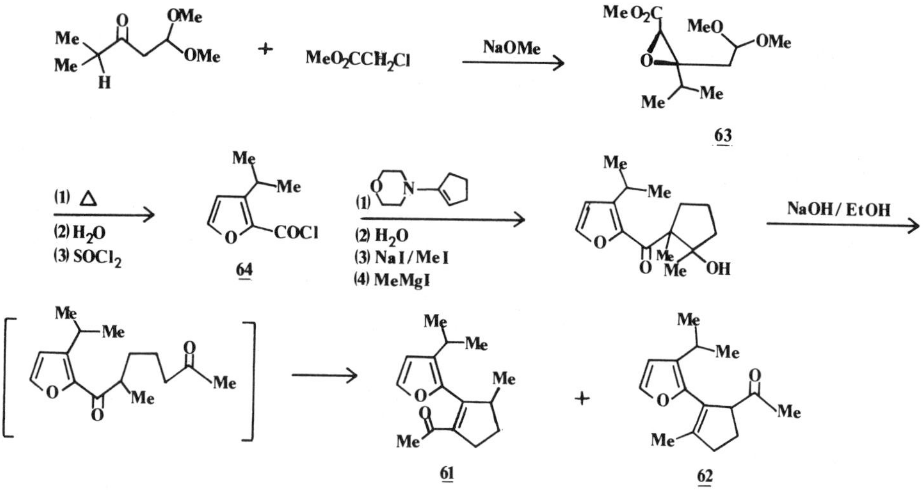

The initial conversion of the epoxide **63** to the furan **64a** is best rationalized by the following sequence:

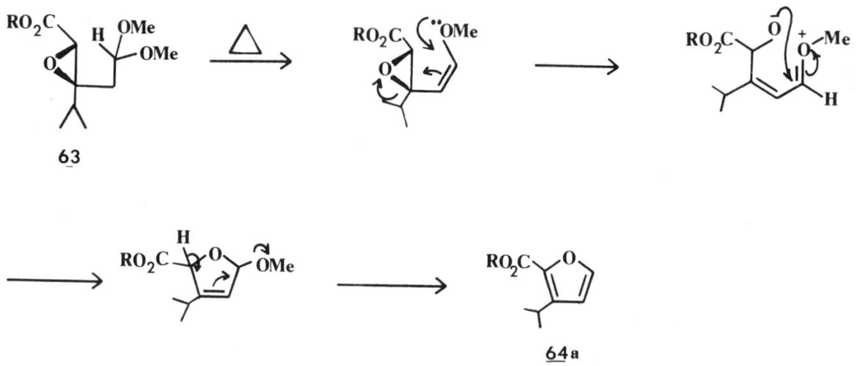

Dibenzofurans are a rather rare group of naturally occurring furans and are essentially found only in lichens.[17] Among these is *di-O-methylstrepcilin* (**65**). Its synthesis is accomplished as delineated in Scheme 13.3.[18] Formation of the benzofuran **66** follows the general synthesis already described. The Wittig reaction (**66** to **67a** and **67b**) followed by a [4 + 2] cycloaddition with methyl acetylenedicarboxylate affords, after prototropic rearrangement (**68** → **69**), the dihydro derivative **69**. Aromatization to **70**, followed by ester reduction with lithium aluminum hydride, gives **71**. Selective oxidative cyclization of **71** with dichromate generates the 2-oxodihydrofuran ring of **65**.

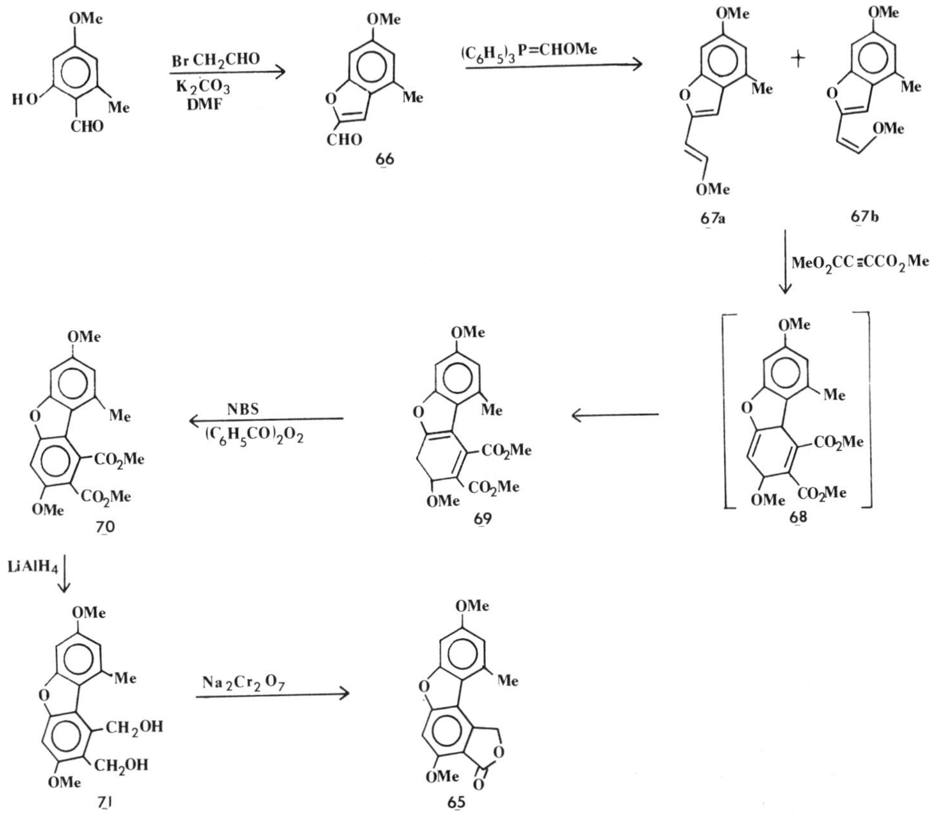

OMe HO Me CHO

Br CH_2CHO / K_2CO_3 / DMF

OMe O Me CHO **66**

$(C_6H_5)_3P{=}CHOMe$

OMe O Me OMe **67a** + OMe O Me OMe **67b**

$MeO_2CC{\equiv}CCO_2Me$

OMe O Me CO_2Me OMe CO_2Me **68**

OMe O Me CO_2Me OMe CO_2Me **69**

NBS / $(C_6H_5CO)_2O_2$

OMe O Me CO_2Me OMe CO_2Me **70**

$LiAlH_4$

OMe O Me CH_2OH CH_2OH OMe **71**

$Na_2Cr_2O_7$

OMe O Me O OMe O **65**

Scheme 13.3

Volatile constitutents present in nanogram quantities have been isolated from the body and glands of the Pine sawfly of the species *Neodiprion sertifer* Geoffr., and one has been identified as *transperillenal* (**72**).[19] The synthesis[20] of **72** starts from 3-furanmethanol, which by transetherification affords **73.** Upon pyrolysis, **73** undergoes a thermal Claisen–Cope type rearrangement to give **72.**

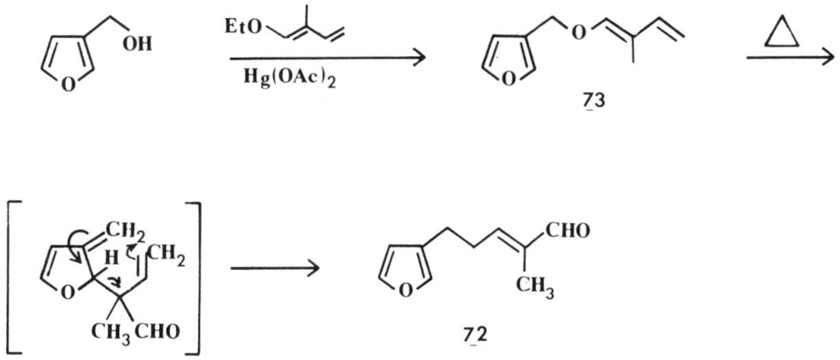

O OH

EtO / $Hg(OAc)_2$

O O **73**

Δ

CH_2 H CH_2 O CH_3 CHO

O CHO CH_3 **72**

The furan nucleus can be easily constructed by the acid-catalyzed rearrangement of γ-hydroxy-β-tosylacetals.[21] Thus *dl-menthofuran* (**74**) is successfully prepared by treatment of 3-methylcyclohexanone with lithium diisopropylamide in THF followed by gaseous acetaldehyde at −78°C to give **75,** which is dehydrated under mild conditions. Treatment of **77** with butyllithium and formaldehyde, followed by treatment with acid at −78°C, generates **74.**

(1) LDA; (2) CH_3CHO, −78 °C → **75** → Al_2O_3, molecular sieves, RT → **76** (57%)

(1) Tol SO_3H/THF; (2) ethylene glycol / H^+ (ref. 21) → **77** (67%) → (1) n-buLi; (2) HCHO; (3) $H^+/C_6H_6/\Delta$ → **74** (40%)

Generally, cycloaddition of furan with monoactivated dienophiles proceeds slowly and nonstereoselectively in low yields[22a]; however, 3-methoxyfuran (**78**) reacts with typical dienophiles to give the endo adducts with high stereoselectivity. Thus **78** with octyl vinyl ketone gives **79,** which is reduced with lithium tri-*t*-butoxyaluminum hydride to afford **80** in excellent overall yield. This intermediate is transformed into **81,** which is a key intermediate for the synthesis of *dl-avenaciobale.*[23]

78 → **79** (100%) → $LiAl(OtBu)_3H$, THF/0°C → **80** (90%) → **81**

13.4. NATURALLY OCCURRING THIOPHENES

Members of the Compositae family and some other higher plants frequently contain some thiophene derivatives along with polyacetylenes. Thus it appears reasonable to suggest that the thiophenes may well arise from addition of hydrogen sulfide (or its biogenetic equivalent) to 1,3-diacetylenes.[24] α,α-Terthienyl (**82**) has been isolated from the flowers of the Indian marigold. Although the temptation is great to suggest that this compound is formed from an appropriate polyacetylene, this has been disproved.[25]

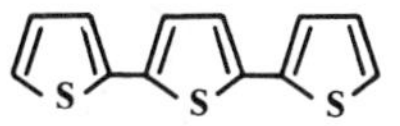

82

The tetrahydrothiophene *biotin* (**83**) is the most important naturally occurring thiophene so far known. The compound, also known as coenzyme R, is an important growth factor.[26] One of several syntheses of this compound is delineated in Scheme 13.4. Transformation of **84** to **85** utilizes the already discussed methylene group activity of β-oxothiophenes, while the reaction with hydroxylamine (**85** → **86**) employs the carbonyl group activity of this dihydrothiophene. The remaining steps follow synthetic transformations established in carbocyclic chemistry.

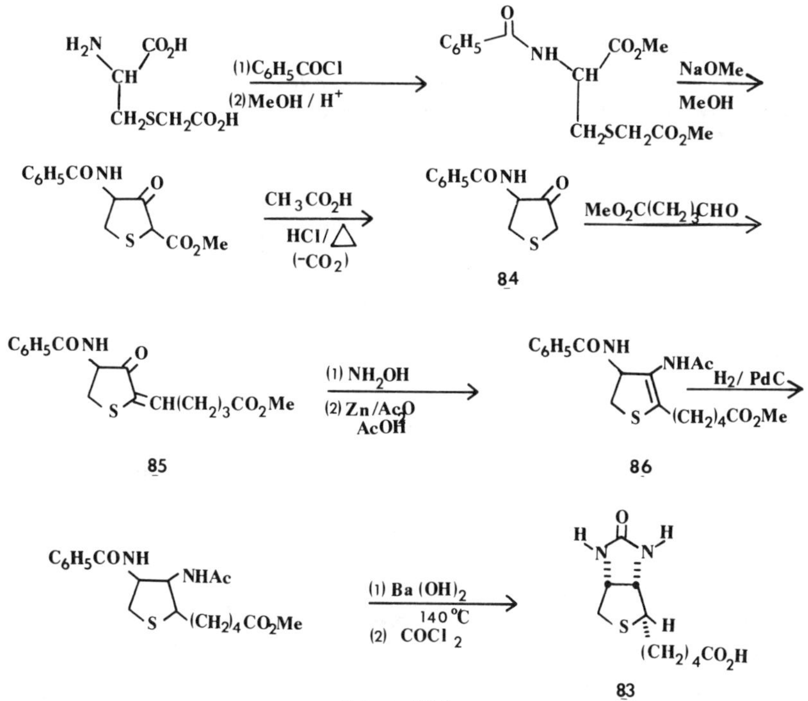

H_2N, CO_2H, CH, $CH_2SCH_2CO_2H$ —(1) C_6H_5COCl, (2) MeOH / H^+→ $C_6H_5C(O)NH$–CH(CO_2Me)–$CH_2SCH_2CO_2Me$ —NaOMe, MeOH→

C_6H_5CONH, O, S, CO_2Me —CH_3CO_2H, HCl/Δ, (−CO_2)→ C_6H_5CONH, O, S (84) —$MeO_2C(CH_2)_3CHO$→

C_6H_5CONH, O, S, $CH(CH_2)_3CO_2Me$ (85) —(1) NH_2OH, (2) Zn/AcO, AcOH→ C_6H_5CONH, NHAc, S, $(CH_2)_4CO_2Me$ (86) —H_2/ PdC→

C_6H_5CONH, NHAc, S, $(CH_2)_4CO_2Me$ —(1) Ba $(OH)_2$, 140 °C, (2) $COCl_2$→ H, N, O, N, H, S, H, $(CH_2)_4CO_2H$ (83)

Scheme 13.4

13.5. SYNTHESES OF SOME SELECTED PHARMACEUTICALS

13.5.1. PYRROLES

The pyrrolidines (totally reduced pyrroles) generally behave, in a biological sense, as do the corresponding open-chain amines.[27] A series of succinimide anticonvulsants has been developed for the treatment of petit mal. The simple synthesis of *phensuximide* (**87**) demonstrates the synthetic principles generally employed.

$$C_6H_5CH{=}C(CN)CO_2Et \xrightarrow{CN^-} C_6H_5(CN)CHCH(CN)CO_2Et \xrightarrow{HCl} C_6H_5CH(CO_2H){-}CH_2CO_2H$$

$$\xrightarrow{CH_3NH_2} \textbf{87}$$

87

13.5.2. FURANS

Derivatives of 2-nitrofuran have shown bacteriostatic as well as bacteriocidal activity and are especially useful for topical application. The synthesis of one of the active nitrofurans (*nitroxyzone;* **88**) is as follows[28]:

$$HOCH_2CH_2NHNH_2 \longrightarrow HOCH_2CH_2N(CONH_2)NH_2 \xrightarrow{O_2N\text{-furan-}CHO} O_2N\text{-furan-}CH{=}N{-}N(CONH_2)CH_2CH_2OH$$

88

A potent diuretic, *furosemide* (**89**),[29] includes the furan ring. The presence of two electron-withdrawing groups in **90** greatly activates the nucleophilic displacement of chloride by the furanylamine **91.**

$$\text{2,4-}Cl_2C_6H_3CO_2H \xrightarrow{ClSO_3H} \text{(}SO_2Cl\text{ derivative)} \xrightarrow[(2)\,H_3O^+]{(1)\,NH_3} \text{(}SO_2NH_2\text{ derivative)}$$

90

90 → (91, furfurylamine CH_2NH_2) → 89

A series of benzofurans have recently been developed for use as antispasmodic agents. *Benziodrane* (**92**) is one example of this type of active compound.[30]

$ClCH_2COCH_3$; $H_2N{-}NH_2$, $^{-}OH/\Delta$; p-$CH_3OC_6H_4COCl$, $AlCl_3$; (1) Pyr·HCl (2) I_2, OH^-, KI → 92

13.5.3. THIOPHENES

One of many interesting changes in biological behavior occurs when the phenyl rings in **93** are replaced by thiophene rings. The former compound has significant endocrine activity that is absent in its thiophene analogue (**94**),[31] which does, however, possess analgesic activity and is used in veterinary medicine.

$(CH_3)_2N{-}CH(CH_3)CH_2CO_2Et$ → (2-thienyllithium) → $-H_2O$ → 94

93: $(C_6H_5)_2C{=}CHCH(CH_3)N(CH_3)_2$

Some anthelminthic agents, in use in veterinary medicine, are represented by the synthesis of *pyrantel* (**95**).[32]

CH_3 CHO S — $NCCH_2CO_2H$ → CH_3 S CN — CH_3OH, H^+ →

CH_3 S NH OCH_3 — H N N CH_3 → CH_3 S N N CH_3

95

REFERENCES

1. R. G. Westall, *Nature* **170,** 614 (1952).
2. G. W. Kenner, K. M. Smith, and J. F. Unsworth, *J. Chem. Soc. Chem. Commun.* **1973,** 43.
3. H. Plieninger, P. Hess, and J. Ruppert, *Chem. Ber.* **101,** 240 (1968).
4. D. G. Davies and P. Hodge, *Tetrahedron Lett.* **1970,** 1673.
5. G. R. Birchall, C. G. Hughes, and A. H. Rees, *Tetrahedron Lett.* **1970,** 4879.
6. K. Bailey and A. H. Rees, *Chem. Commun.* **1969,** 1284.
7. H. Rapoport and K. G. Holden, *J. Am. Chem. Soc.* **84,** 635 (1962).
8. M. E. Flaugh and H. Rapoport, *J. Am. Chem. Soc.* **90,** 6877 (1968).
9. A. D. Adler, F. R. Longo, J. D. Finarelli, J. Goldmacher, J. Assour, and L. Korsakoff, *J. Org. Chem.* **32,** 476 (1967); R. E. Bozak and C. L. Hill, *J. Chem. Ed.* **59,** 36 (1982).
10. J. H. Brewster and E. L. Eliel, *Org. Reactions* **7,** 99 (1953).

10a. M. Kuehne, *Lloydia* **27,** 435 (1964).

11. J. Harley-Mason, L. Castedo, and T. J. Leeney, *Chem. Commun.* **1968,** 1186.
12. R. Noyori, Y. Baba, and Y. Hayakawa, *J. Am. Chem. Soc.* **96,** 3336 (1974).
13. T. A. Geissman and D. H. G. Crout, *Organic Chemistry of Secondary Plant Metabolism,* Freeman, Cooper, San Francisco, 1969, p. 271.
14. G. Büchi, E. sz. Kovats, P. Enggist, and G. Uhde, *J. Org. Chem.* **33,** 1227 (1968).
15. Y. R. Naves and P. Ochsner, *Helv. Chim. Acta* **43,** 568 (1960).
16. G. Buchi and H. Wuest, *Tetrahedron* **24,** 2049 (1968).
17. F. M. Dean, *Naturally Occurring Oxygen Ring Compounds,* Butterworths, London, 1963, pp. 144–160.
18. J. D. Brewer and J. A. Elix, *Tetrahedron Lett.* **1969,** 4139.
19. G. Ahlgren, G. Bergström, J. Löfqvist, A. Jansson, and T. Norin, *J. Chem. Ecol.* **5,** 309 (1979).
20. A. F. Thomas, *Chem. Commun.* **1968,** 1657; A. F. Thomas and M. Ozainne, *J. Chem. Soc. (C)* **1970,** 220.
21. K. Inomata, Y. Nakayama, M. Tsutsumi, and H. Kotake, *Heterocycles* **12,** 1467 (1979).

22. G. K. Cooper and L. J. Dolby, *Tetrahedron Lett.* **1976,** 4675.

22a. See: K. Matsumoto, T. Uchida, and R. M. Acheson, *Heterocycles,* **16,** 1369 (1981).

23. A. Murai, K. Takahashi, H. Taketsuru, and T. Masamune, *J. Chem. Soc. Chem. Commun.* **1981,** 221.

24. F. Bohlmann, T. Burkhardt, and C. Zdero, *Naturally Occurring Acetylenes,* Academic Press, London, 1973.

25. S. Challenges, *Sci. Progr.* (*London*) **41,** 593 (1953).

26. J. Knappe, K. Biederbick, and W. Brummer, *Angew. Chem. Int. Ed.* **1,** 401 (1962).

27. D. Lednicer and L. A. Mitscher, *Organic Chemistry of Drug Synthesis,* Wiley, New York, 1977, p. 226.

28. W. B. Stillman and A. B. Scott, U. S. Pat. 2,416,234 (1947) [*Chem. Abstr.* **41,** 3488 (1974)].

29. K. Sturm, W. Siedel, and R. Weyer, Ger. Pat. 1,122,541 (1962) [*Chem. Abstr.* **56,** 14032 (1962)].

30. N. P. Buu Hoi and C. Beaudet, U. S. Pat. 3,012,042 (1961) [*Chem. Abstr.* **57,** 11168 (1962)].

31. D. W. Adamson, *J. Chem. Soc.* **1950,** 885.

32. Pfizer Corp., Belg. Pat., 658,987 (1965) [*Chem. Abstr.* **64,** 8192 (1966)].

Chapter 14

SYNTHETIC ROUTES TO SELECTED COMMON SIX-MEMBERED π-DEFICIENT HETEROAROMATIC COMPOUNDS

14.1 GENERAL SYNTHETIC ROUTES

Up to this point in the text, the combinations possible to create small rings are rather limited. Thus when the need arises to synthesize any one of the millions of possible larger-ring organic compounds, a reasonable approach is to "subdivide the target molecule into fragments that will ultimately fit together to generate the desired structure." Scheme 14.1 shows several of the possible

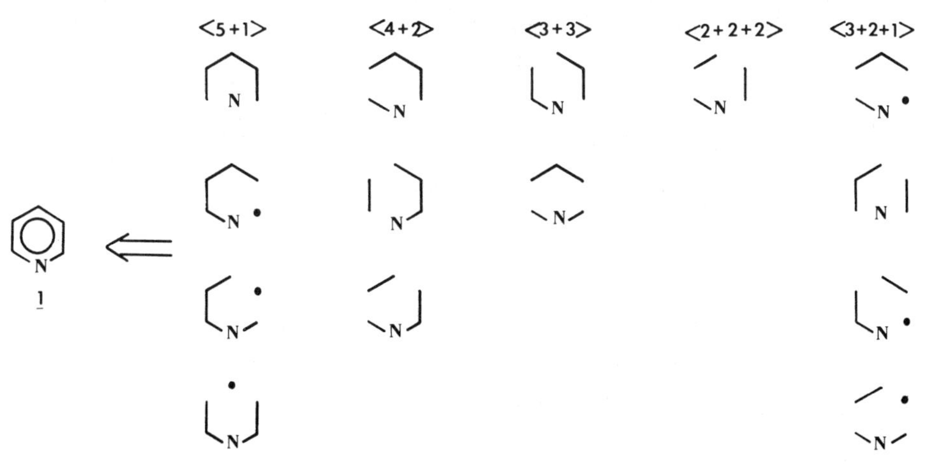

Scheme 14.1. Possible synthetic approaches to pyridine.

modes of combination to give the simplest π-deficient, six-membered, heteroaromatic molecule, the pyridine nucleus (**1**). A general "rule of thumb" is *to choose the largest readily available units for the construction* of the heteroaromatic nucleus. For example, ⟨4 + 2⟩ combination could be accomplished by a typical Diels–Alder reaction,[1] and ⟨5 + 1⟩ could be effected by the condensation of **2** with hydroxylamine.[2]

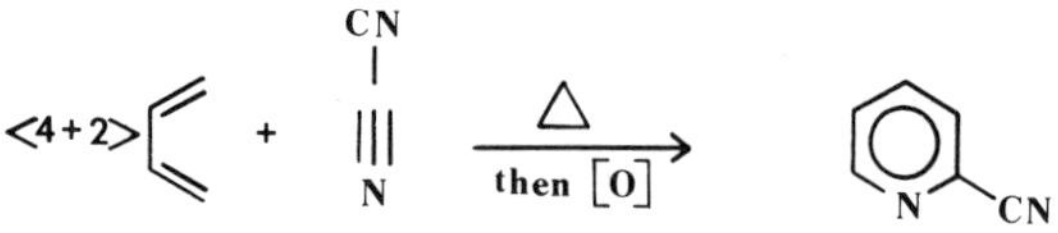

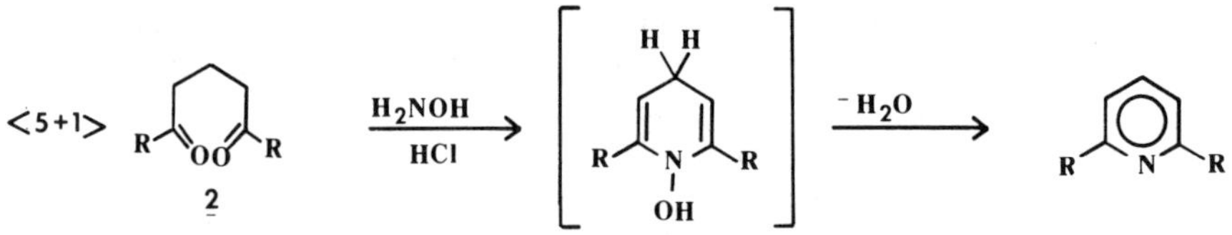

Schiff base chemistry is probably the most commonly adopted synthetic approach to π-deficient heteroaromatics. Thus diazines such as pyridazine (**3**) are prepared by the condensation of 1,4-diones (**4**) with hydrazine.[3] However, numerous other routes to **3** are available and are considered in this chapter.

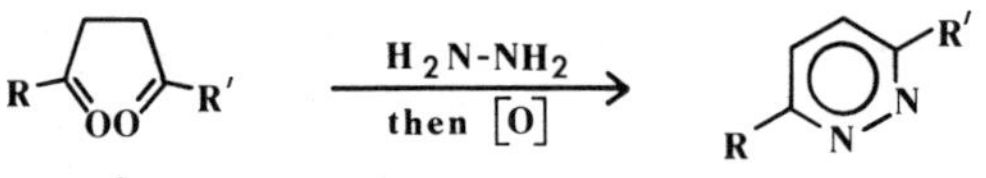

Similarly, pyrimidine (**5**)[4] is easily constructed from two three-membered fragments, such as **6** and an amidine derivative (**7**). Pyrazine (**8**),[5] on the other hand, is most readily formed by a ⟨4 + 2⟩ condensation of α-dione **9** and a bisamine **10,** followed by oxidation. In each of these simplified examples, the intermediate dihydro compounds are readily oxidized, often by air alone, to the aromatized products. In the simple condensation reactions, isomeric mixtures can be obtained when the R groups are not identical.

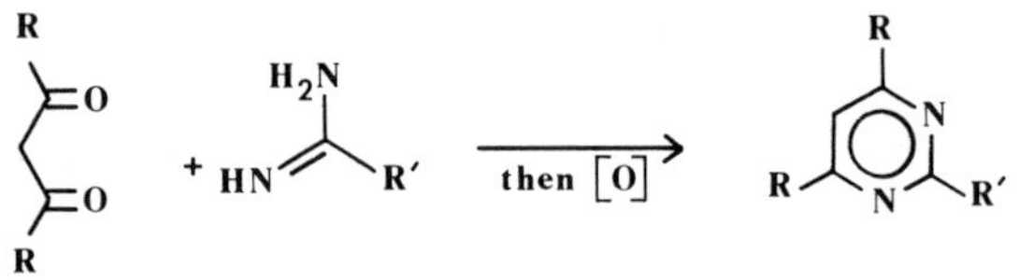

9 + 10 → [ ] —(O)→ 8

14.2. COMMON SYNTHETIC APPROACHES TO PYRIDINE AND ITS DERIVATIVES

14.2.1. ⟨5 + 1⟩ AND RELATED COMBINATIONS

The most direct synthesis of pyridine derivatives is by condensation of the appropriate unsaturated δ-dicarbonyl compounds with ammonia or a suitable nitrogen source. Thus reaction of glutacondialdehyde **(11)** with ammonium acetate affords **1.**[6a] Since unsaturated δ-diones are not readily available, the saturated analogues **12** are employed and condensed with hydroxylamine[6b] to give rise to an *N*-hydroxy intermediate, which readily dehydrates under the reaction conditions.

$OHCCH{=}CHCH_2CHO$ (**11**) $\xrightarrow[\Delta]{NH_4OAc}$ **1**

$OHC(CH_2)_3CHO$ (**12**) $\xrightarrow[\Delta]{H_2NOH}$ [N–OH intermediate] → **1**

Modifications of this ⟨5 + 1⟩ approach utilize a five-carbon unit generated *in situ* by means of a ⟨3 + 2 + (1)⟩ or a ⟨2 + 1 + 2 + (1)⟩ procedure. The latter represents the *Hantzsch synthesis,* which combines two molecules of an α-methylenic carbonyl compound, an aldehyde, and an ammonia source. Classically, ethyl acetoacetate, acetaldehyde, and ammonia give the dihydropyridine **13,** which is oxidized to **14.**[7] At pH 6, the yield of **13** is 3%, whereas at pH 3.25–5.0, a yield of 53% or better is realized. Thus the yield of the Hantzsch dihydro intermediate depends on both the pH and the aldehyde used.[8] The exact mechanistic sequence is unknown, but the Schiff base **15,** enamine **16,** and tetrahydropyridine **17** have been identified as intermediates in certain instances[9]; one logical sequence is shown in Scheme 14.2.

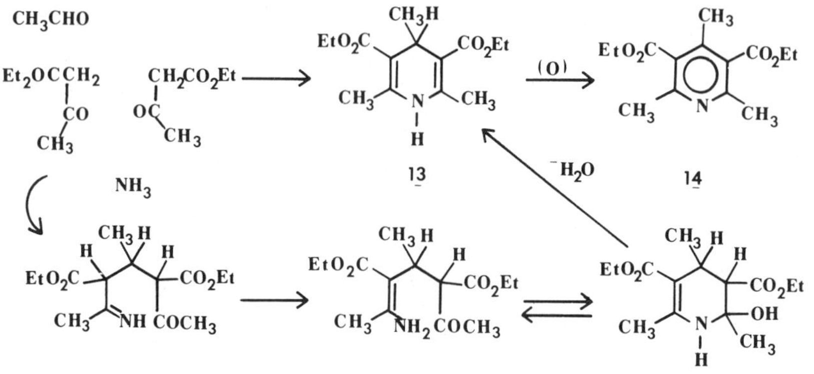

15 16 17

Scheme 14.2

The five-carbon cyanoacid chloride **18** reacts with hydrochloric acid to give an immonyl chloride which cyclizes by imine attack at the activated acyl halide carbon to afford **19.**[10] These five-carbon units can also be synthesized by condensation of C-2 and C-3 units.

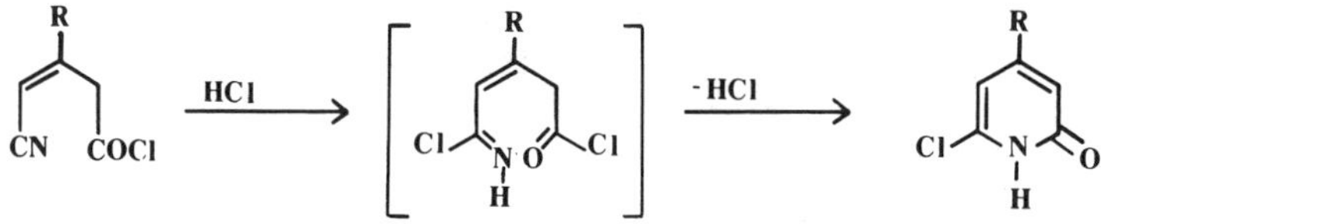

18

19 [R=Me(37%), R=H(63%)]

Pyrylium salts **20** react with ammonia to give excellent yields of substituted pyridines,[11] but the procedure is limited by the availability of the starting salts.[12]

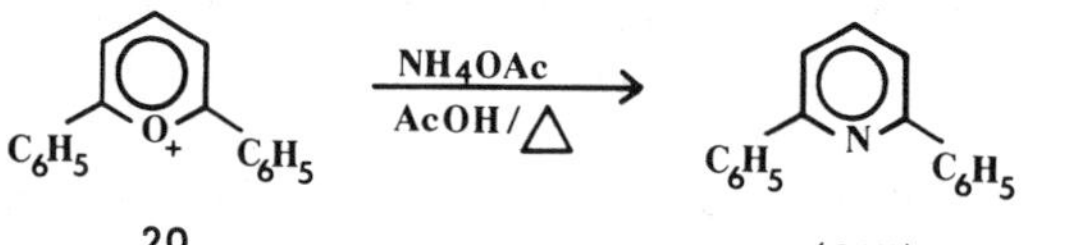

20

(96%)

Kröhnke[13a] described the facile *in situ* synthesis of the five-carbon unit **23** by the normal Michael addition of pyridinium salt **21** (C-2) to an α,β-unsaturated moiety (for example, **22**) (C-3). Subsequent treatment of **23** with ammonium acetate in glacial acetic acid results in the formation of the substituted pyridine in generally excellent yield. This has been demonstrated to be a convenient route to the pyridine nucleus.[13b]

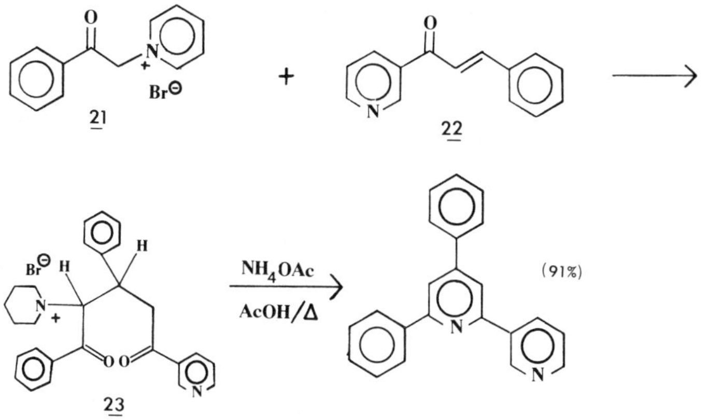

14.2.2. ⟨4 + 2⟩ AND ⟨2 + 2 + 2⟩ COMBINATIONS

Although the Diels–Alder reaction has been commonly employed to prepare substituted pyridines, the recent use of metallocyclopentadiene intermediates has greatly enhanced the effectiveness of the ⟨4 + 2⟩ route. Thus **24** can be generated from two acetylene molecules and a nitrile via intermediate **25.**[14] Specifically, co-oligomerization of α,ω-diynes (for example, **26**) with acetonitrile in the presence of η^5-cyclopentadienylcobalt dicarbonyl [$CpCo(CO)_2$] gives 3-methyl-5,6,7,8-tetrahydroisoquinoline (**27**).[15] Other cobalt catalysts have been used successfully in this procedure[16] as well. These metallocyclopentadiene intermediates react with isocyanates to give pyridinones.[17] This cobalt-catalyzed cotrimerization of acetylenes and "heteroacetylenes" is a novel, high-yield route to numerous electron-deficient systems, but to date limited work has been reported in the area.

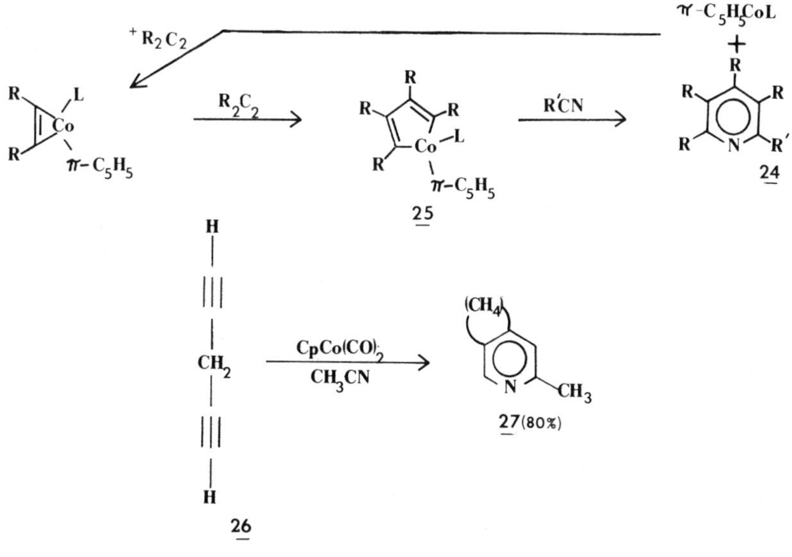

Diels–Alder reactions of diverse electron-poor heterocycles with electron-rich dienophiles give rise to bicyclic intermediates, which undergo retrocycloaddition to generate the heteroaromatic nucleus. Thus, oxazole **28** has been transformed to pyridoxine (**29**).[18] The α-pyrone **30** react with benzonitrile to give **31**,[19] and oxazinones (**32**) react with ynamines to give **33**.[20]

EtO, CH3 oxazole 28 + maleic anhydride ⟶ [bicyclic adduct: EtO, CH3] —HCl/EtOH→ HO, CH3, CO_2Et, CO_2Et pyridine —$LiAlH_4$→ HO, CH3, OH, OH pyridine **29**

EtO_2C, CH_3, H, H_3C pyrone **30** + benzonitrile (CN) —Δ→ [bicyclic adduct: CH_3, H, CO_2Et, CH_3] —$-CO_2$→ EtO_2C, CH_3, H_3C, phenyl pyridine **31**

R, CH_3, H, OEt —Δ, R=t-bu→ [H_3C, H, R iminoketene] ⟶ H_3C, H, R oxazinone **32** —Et_2N-C≡C-CH_3, Δ, $-CO_2$→ NEt_2, CH_3, H_3C, R pyridine **33** (94%)

14.2.3. ⟨6⟩ COMBINATION, THERMAL REARRANGEMENTS

More recently, inclusion of all six atoms of a potential pyridine nucleus in a single precursor molecule has been reported. Iminoketene **35** is generated from **34** by thermolysis at 400°C; a 1,5-sigmatropic rearrangement gives **36** which smoothly undergoes an electrocyclization and loss of hydrogen to form the substituted nicotinaldehyde **37**.[21] Similarly, **38** is alkylated to afford **39**, which, upon pyrolysis, expels carbon dioxide and cyclizes to give **40**.[22a] The rearrangement of unsaturated six-atom units is quite common and has been invoked in numerous mechanistic presentations.

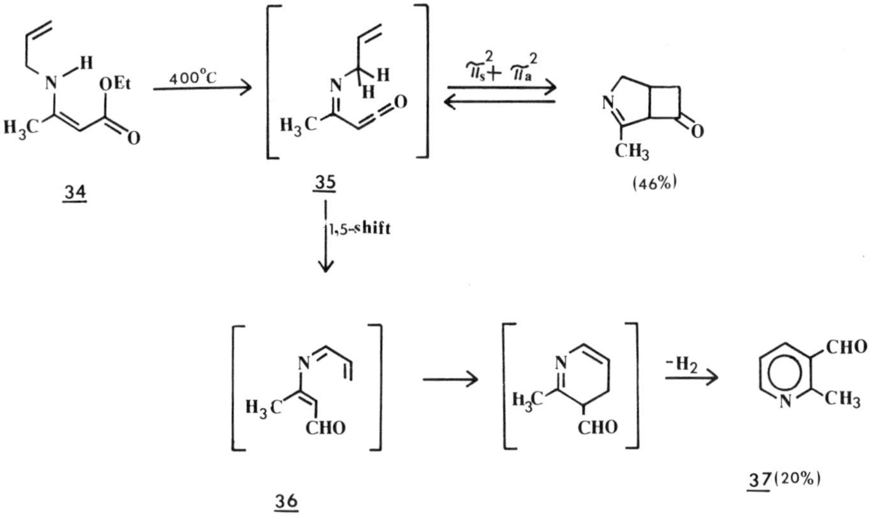

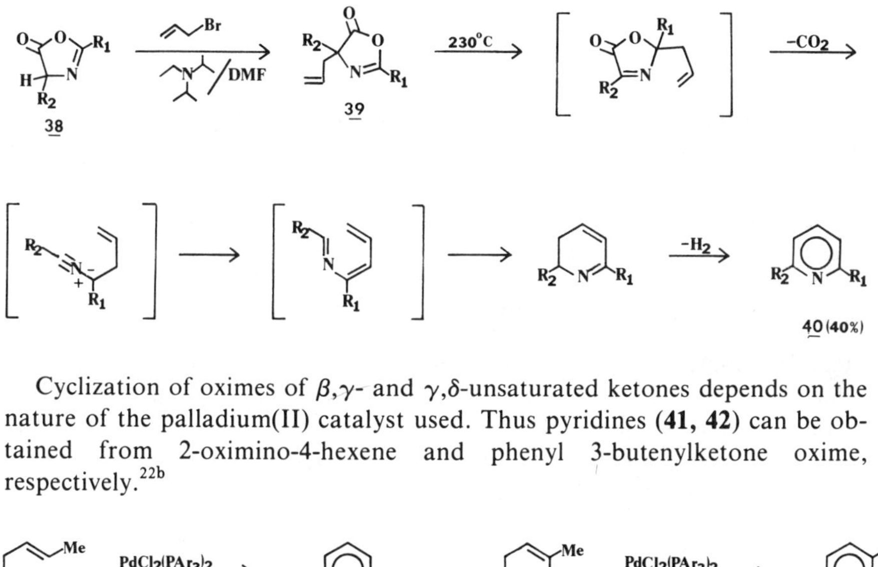

Cyclization of oximes of β,γ- and γ,δ-unsaturated ketones depends on the nature of the palladium(II) catalyst used. Thus pyridines (**41, 42**) can be obtained from 2-oximino-4-hexene and phenyl 3-butenylketone oxime, respectively.[22b]

14.2.4. ⟨3 + 3⟩ COMBINATIONS

2,3,5,6-Tetraphenylpyridine (**43**) is formed exclusively from 2,3-diphenyl-1-azirine (**44**) and diphenylcyclopropenyl bromide (**45**). This reaction represents a ⟨3 + 3⟩ combination and has been suggested to proceed by the mechanism shown in Scheme 14.3.[23]

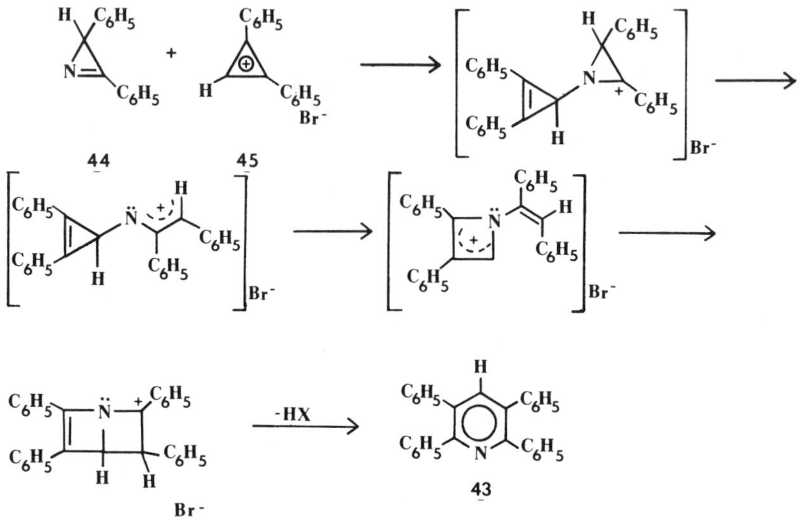

Scheme 14.3

14.3. COMMON SYNTHETIC APPROACHES TO PYRIDAZINE AND ITS DERIVATIVES

For a more detailed review of the synthetic routes to substituted pyridazines, *Recent Advances in Pyridazine Chemistry* should be consulted.[24]

14.3.1. ⟨4 + 2⟩ COMBINATIONS

Most commonly, unsaturated γ-diketones react with hydrazine to give the pyridazine nucleus. The cis isomer of **46** reacts with hydrazine to give **47,** whereas the trans isomer forms the monohydrazone.[25] Upon treatment with hydrazine, saturated γ-diketones afford the corresponding dihydropyridazine, which is easily aromatized by mild oxidation.[3a]

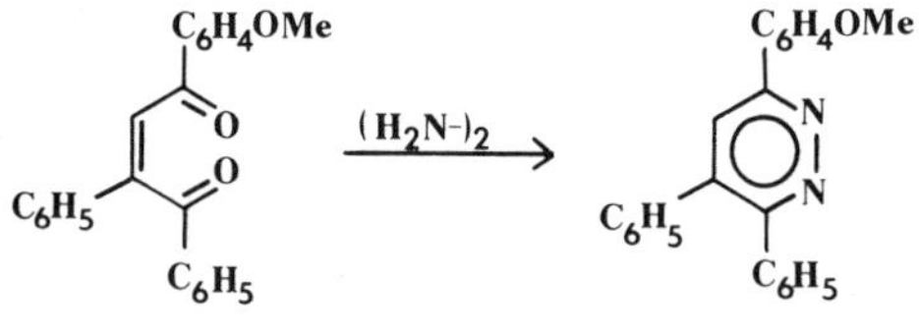

Benzil monohydrazone (**48**) reacts with dimethyl acetylenedicarboxylate to give, along with disproportionation products, the desired pyridazine **49.**[26]

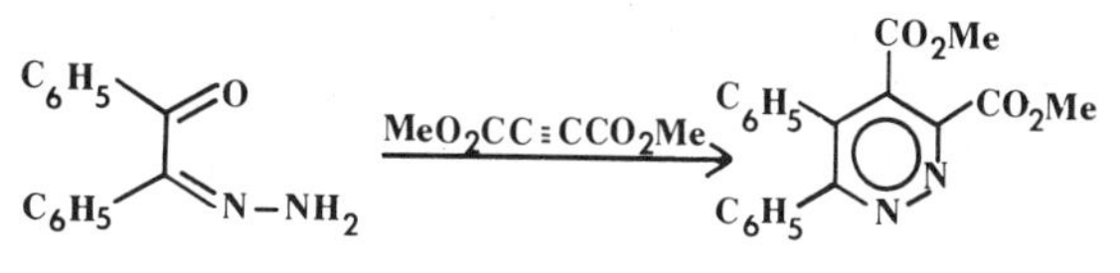

48 49

A [4 + 2] cycloaddition reaction usually occurs between electron-rich dienes and electron-poor dienophiles. Sommer reported the synthesis of **50** from electron-deficient azoalkenes (for example, **51**) and alkenes,[27] from which a mixture of regioisomers is obtained. *s*-Tetrazines (for example, **52**) react with alkenes and acetylenes to give the corresponding pyridazines.[28] These tetrazines (**53**) behave uniquely, in that reaction with thiophene gives a bicyclic product, whereas in the reaction with *N*-methylpyrrole, a second cycloaddition occurs to give **54**,[29] and with 2,5-dimethylfuran, subsequent ring opening affords **55**.[30]

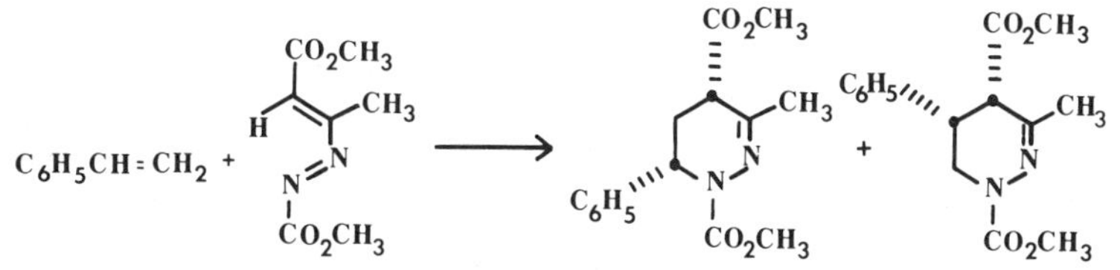

51 50a 50b

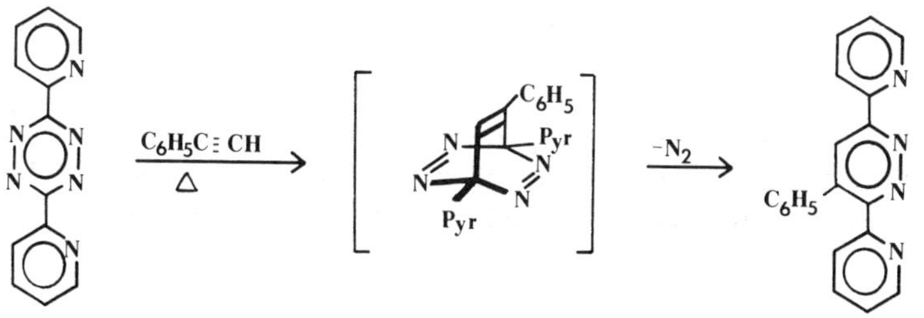

52

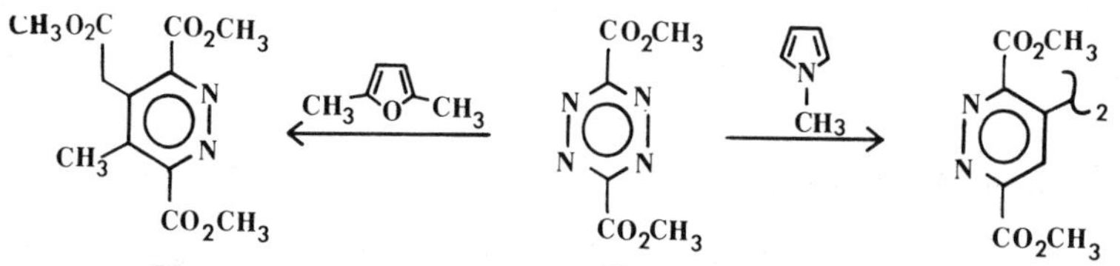

55 53 54

14.3.2. ⟨3 + 3⟩ COMBINATIONS

Arylazirine **56** is cleaved in the presence of titanium tetrachloride at −78°C to give **57**.[31] It has been proposed that formation of a nitrogen-titanium bond is followed by either carbon–nitrogen cleavage by chloride ion or by a polarization to produce zwitterionic species which combine to give **57.**

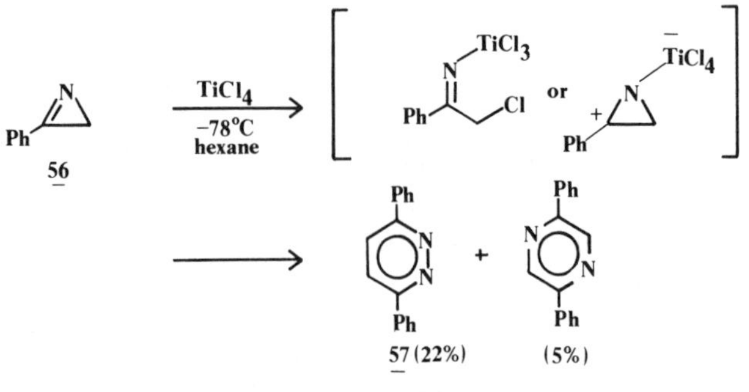

Cycloaddition of diazoalkanes to methylene cyclopropenes forms **58** via a sequence of rearrangements. However, since **59** is isolated in the case of diazoethene, a portion of this complex transformation can be established.[32]

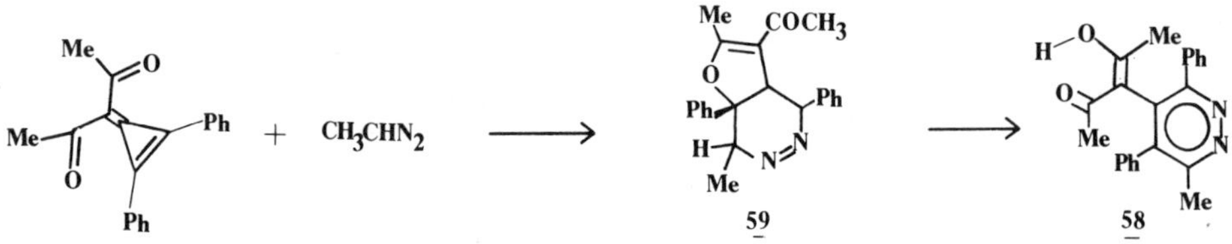

14.3.3. ⟨5 + 1⟩ COMBINATION

The carbene ring expansion of pyrazoles has been accomplished under neutral conditions. Thus 3,4,5-trimethylpyrazole **60** when treated with dichlorocarbene gives **61** under neutral and **62** under basic conditions.[33] Subsequent treatment of **62** with sodium ethoxide affords the pyridazine **63.**

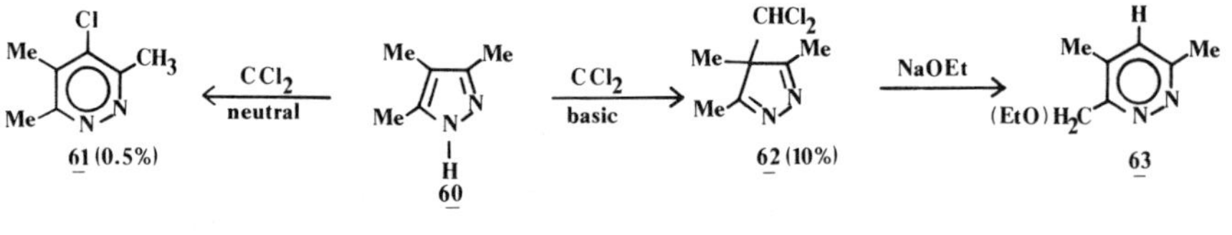

14.3.4. ⟨6⟩ COMBINATIONS

The production of a tetrahydropyridazine from the ketazine dianion **64,** generated from the azine on treatment with lithium diisopropylamide in THF, confirms the formation of a C—C bond that leads to **65.**[34] A similar reaction is the slow thermal decomposition of α-styrylazide (**66**) to give 2-phenylazirine (**56**), 2,5-diphenylpyrrole, and 3,6-diphenylpyridazine (**57**).[35] It has been proposed that the azide decomposes to an azo intermediate (**67**), which cyclizes and is aromatized.

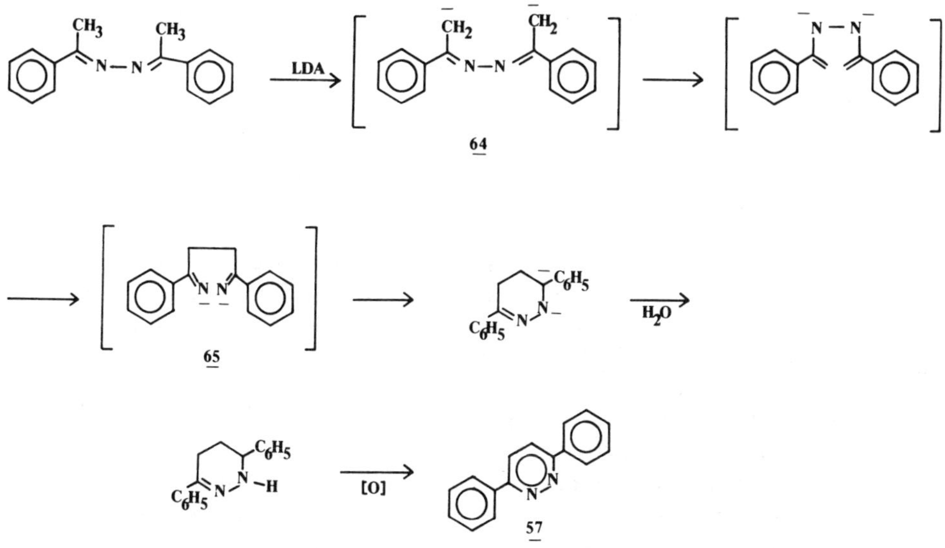

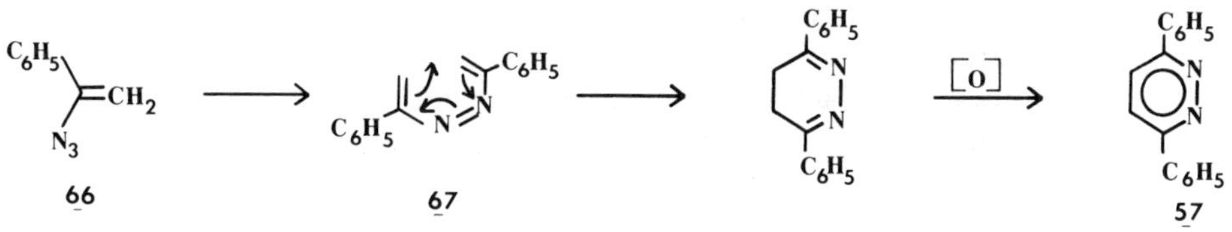

64

65

57

66

67

57

14.3.5. ⟨7 → 6⟩ RING CONTRACTIONS

Ketone **68** upon thermolysis at 475°C is converted into **69** by valence isomerization through a diazotropone.[35] 1,2-Diazepine **70** upon halogenation forms pyridazine **71** via a norcaradiene intermediate.[36] Numerous examples of ring contractions have recently been reported.

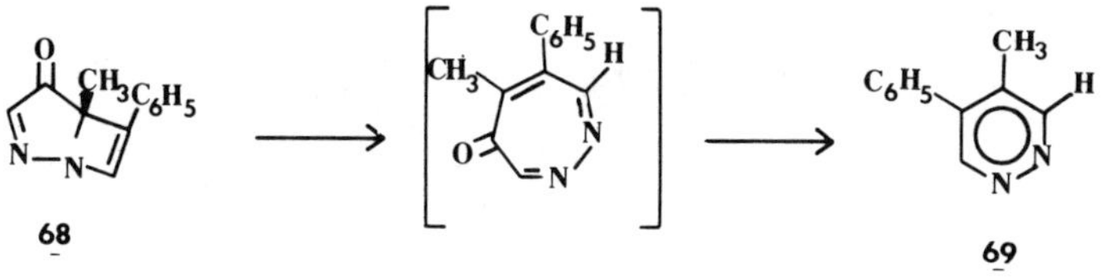

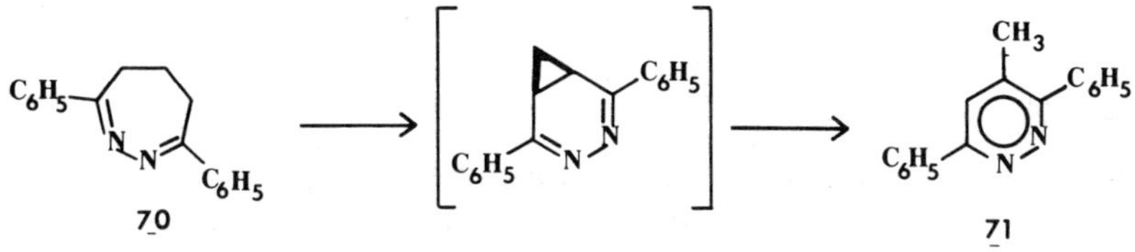

68

69

70

71

14.4. COMMON SYNTHETIC APPROACHES TO PYRIMIDINE AND ITS DERIVATIVES

Because of the biological and pharmaceutical importance of these compounds,[37] several excellent reviews treat the synthetic aspects.[38]

14.4.1. ⟨3 + 3⟩ COMBINATIONS

The most common and versatile route to substituted pyrimidines is the condensation of two three-atom units. For example, acetylacetone and guanidine carbonate react with or without solvent or a catalyst at 100°C to give 2-amino-4,6-dimethylpyrimidine (**72**).[39] A solvent is generally used and, depending on the reagents, an acidic or basic catalyst may facilitate the condensation. Thus acetylacetone reacts cleanly with formamidine to give **73**,[40] with urea to give **74**,[41] and with thiourea to give **75**.[40] This procedure to generate the pyrimidine nucleus is extremely flexible since one three-atom unit can have any combination of an aldehyde, ketone, ester, or nitrile group(s) and the nitrogen fragment can be an amidine, urea, thiourea, or guanidine moiety. For example, uracil (2,4-dihydroxypyrimidine) can be derived from a formylacetate and urea, thymine (2,4-dihydroxy-5-methylpyrimidine) can be obtained from ethyl acetoacetate and urea, and barbituric acid (2,4,6-trihydroxypyrimidine) can be produced from malonic acid and urea, in the presence of ethoxide. Numerous barbiturate drugs have been synthesized by this procedure.[38]

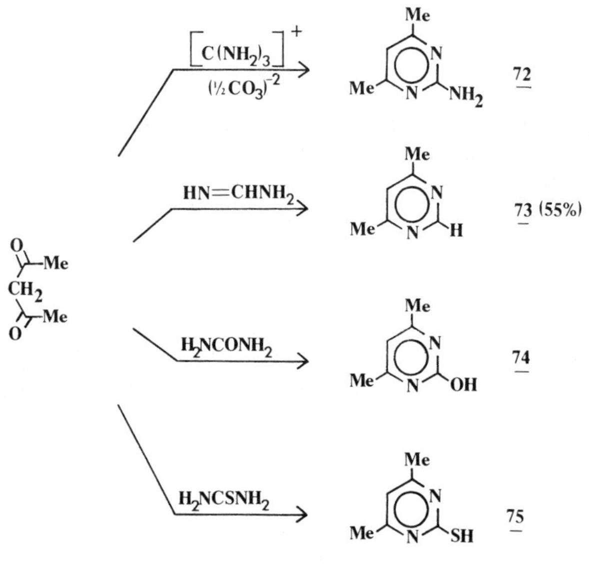

14.4.2. ⟨4 + 2⟩ COMBINATIONS

Reaction of a ethyl β-aminocrotonate with phenylisocyanate gives an ureido derivative which upon treatment with base cyclizes to afford the desired pyrimidine derivative **76**.[42]

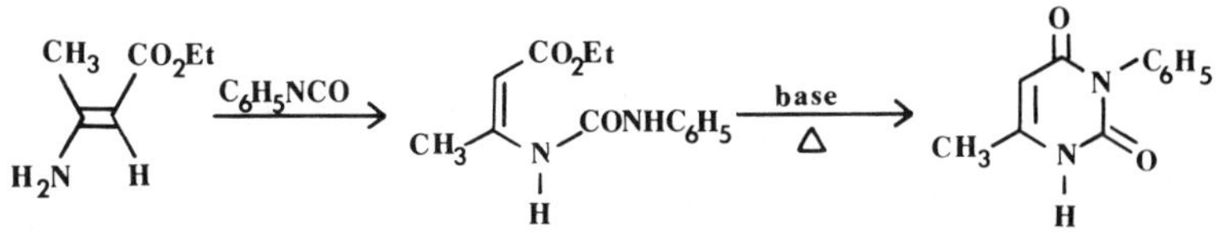

CH_3 CO_2Et / H_2N H $\xrightarrow{C_6H_5NCO}$ CO_2Et / CH_3 N(H) $CONHC_6H_5$ $\xrightarrow[\Delta]{base}$ O / N C_6H_5 / CH_3 N(H) O

76

14.4.3. ⟨5 + 1⟩ COMBINATIONS

Typical of this procedure, malondiamide reacts with an activated methylene group, as in a malonate, to give **77**.[43] Numerous other modifications of this procedure can be envisioned, such as the synthesis of diethylbarbituric acid (**78**) ⟨3 + 2 + 1⟩.[44]

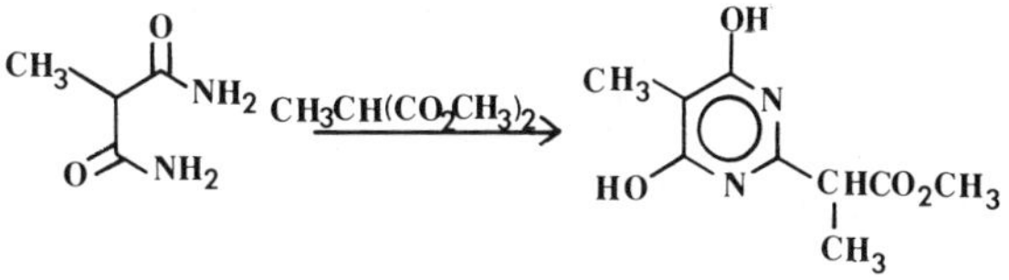

CH_3 O NH_2 / O NH_2 $\xrightarrow{CH_3CH(CO_2CH_3)_2}$ OH / CH_3 N / HO N $CHCO_2CH_3$ / CH_3

77

Et Et $COCl$ $COCl$ $\xrightarrow{H_2NCO_2Et}$ Et Et $NHCO_2Et$ $NHCO_2Et$ $\longrightarrow$ Et Et O N CO_2Et O N(H) O $\xrightarrow[H_2O]{OH^-}$ Et Et O N H O N(H) O

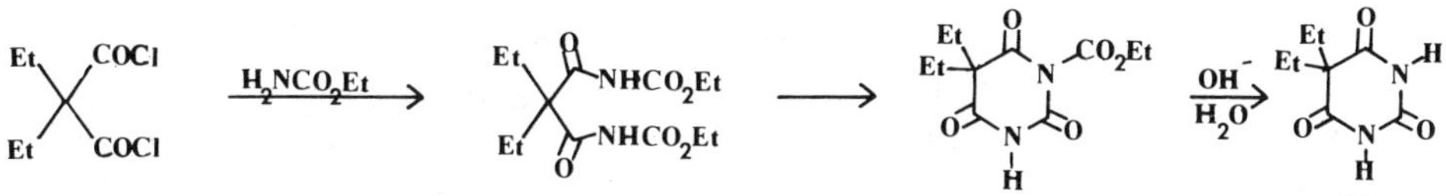

78

14.4.4. ⟨3 + 1 + 1 + 1⟩ COMBINATIONS

A new synthesis of dihydropyrimidines has been devised, in which substituted pyrimidines are directly isolated under oxidative conditions. Hence the reaction of dibenzoylmethane, benzaldehyde, and anhydrous ammonium acetate in dry dimethyl sulfoxide gives **79**.[45]

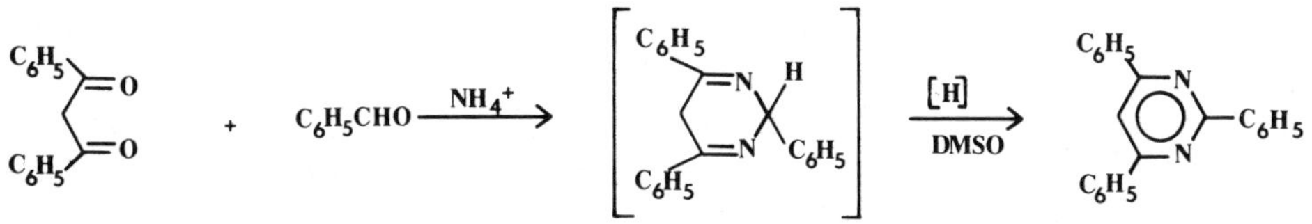

C_6H_5 =O / C_6H_5 =O + C_6H_5CHO $\xrightarrow{NH_4^+}$ [C_6H_5 N H N C_6H_5 C_6H_5] $\xrightarrow[DMSO]{[H]}$ C_6H_5 N N C_6H_5 C_6H_5

79

14.5. COMMON SYNTHETIC APPROACHES TO PYRAZINE AND ITS DERIVATIVES

Because substituted pyrazines also occur in nature, many synthetic routes have been devised and reviewed.[46] Alkylpyrazines such as 2-isopropyl-5-

methylpyrazines (in coffee) are important flavor constitutents in roasted foods such as coffee, cocoa, and peanuts.

14.5.1. ⟨4 + 2⟩ COMBINATIONS

The classical route to alkyl- and arylpyrazines utilizes Schiff base chemistry, in which α-diketones are condensed with 1,2-diamines. The intermediary dihydropyrazine **80** is isolated in good yield from the reaction of **81** and **82.**[47] Conditions for the dehydrogenation of the dihydro compounds are quite variable, ranging from copper chromate at 300°C to simple atmospheric oxidation.

81 + 82 → 80 —(O)→ (58%)

14.5.2. ⟨3 + 3⟩ COMBINATIONS

Two molecules of an α-aminoketone can self-condense to give a 2,5-dihydropyrazine (**83**), which is readily air-oxidized to the substituted pyrazine **84.**[48] α-Aminoketones are not easily synthesized, but may be prepared by treatment of α-hydroxyketones with ammonium acetate.[49] The reaction of α-azidoketones with triphenylphosphine gives a *P-N*-ylide **85,** which undergoes an intermolecular condensation to afford dihydro intermediates.[50]

2 —Base→ 83 —(O)→ 84

$P(C_6H_5)_3$ → 85 → → 86

A variant of this combination is the ⟨2 + 1 + 2 + 1⟩ process in which α-haloketones are condensed with ammonia. When ω-chloroacetophenone (**87**) is heated with alcoholic ammonia, **86** and **88** are formed in a 1 : 1 ratio.[51]

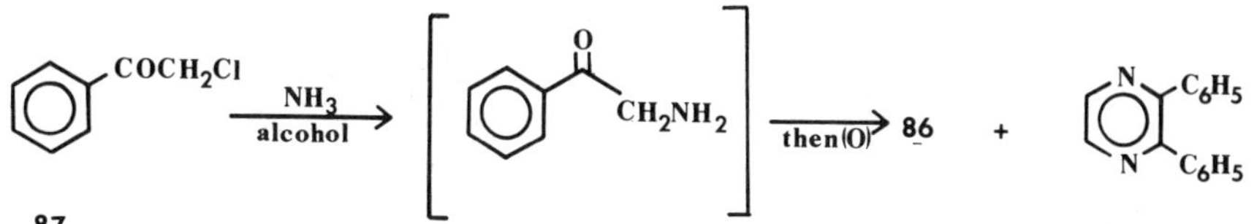

2,3,5,6-Tetrachloro-1,4-diformylpiperazine (**89**) is prepared from glyoxal and formamide in the presence of chlorine. Thermal decomposition of **89** at 185–200°C gives 2-chloropyrazine (**90**).[52]

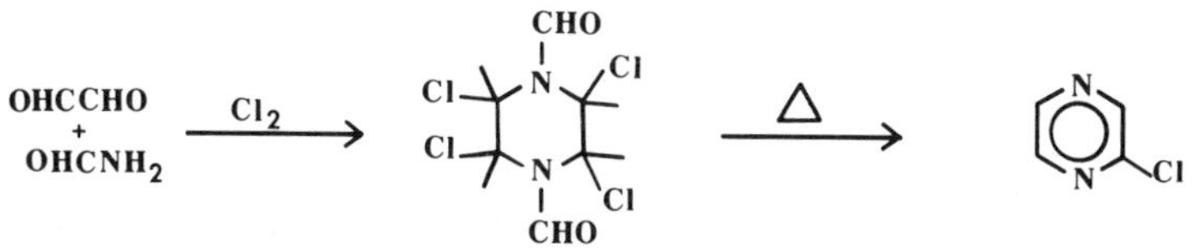

α-Aminoesters readily dimerize to give the 2,5-diketopiperazines. The Meerwein reagent converts these bis-lactones into the corresponding 3,6-dihydro-2,5-dialkoxypyrazines, which can be easily oxidized with dichlorodicyanobenzoquinone (DDQ).[53] Circumvention of normal C—N bond formation is demonstrated by the novel ⟨3 + 3⟩ approach, where **91** is treated with cupric bromide to give **92,** which arises from a dimerization and generation of two C—C bonds.[54]

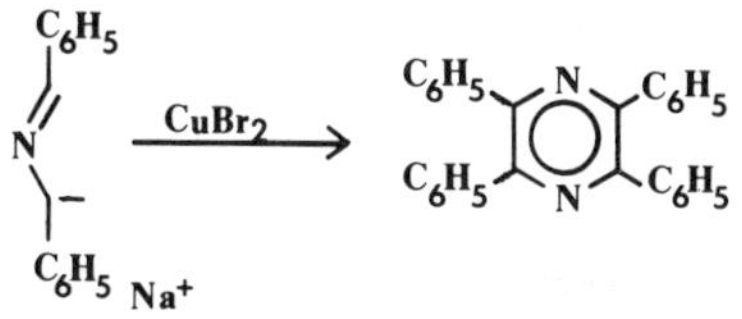

14.5.3. ⟨6⟶6⟩ INTERCONVERSIONS. TRANSPOSITIONS[55] OR METATHESIS[56]

Irradiation of **93** with a medium-pressure mercury lamp gives an almost quantitative yield of tetrafluoropyrazine (**94**).[57] However, thermolysis of **93** affords only traces of **94;** the major product is the corresponding pyrimidine (**95**). Scheme 14.4 illustrates this reaction sequence.

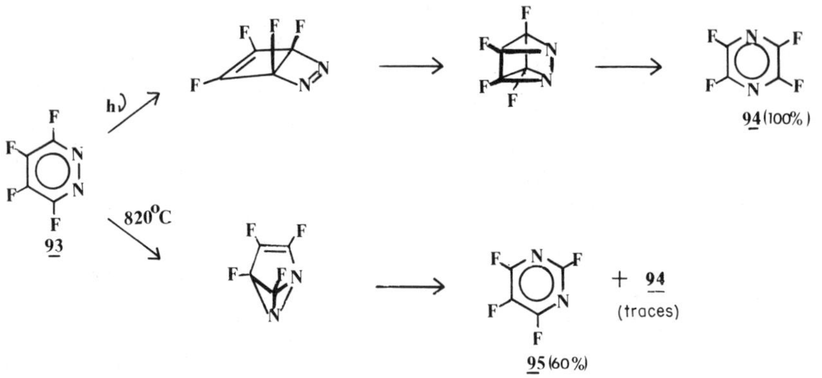

Scheme 14.4

14.6. COMMON SYNTHETIC APPROACHES TO THE TRI- AND TETRAZINES

14.6.1. 1,2,3-TRIAZINES

The mild thermolysis of 1,2,3-triphenylcyclopropenyl azide (**96**) affords convenient access to 4,5,6-triphenyl-1,2,3-triazine (**97**).[58] Little synthetic effort has been expended on this series.

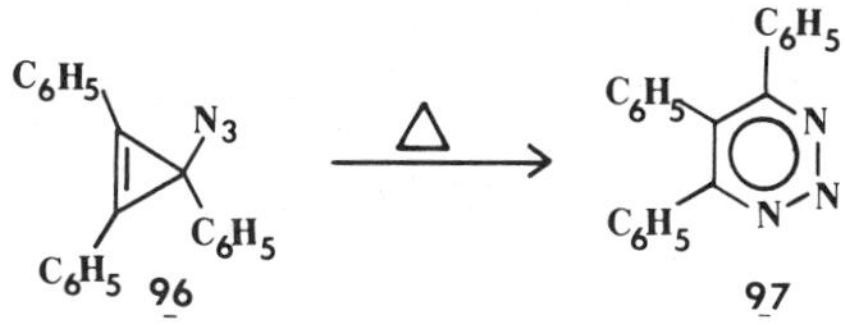

14.6.2. 1,2,4-TRIAZINES[59]

14.6.2.1. ⟨4 + 2⟩ Combinations. In a process analogous to the common synthesis of pyrazine, the condensation of α-diketones (for example, **98**) with semicarbazides, thiosemicarbazides, or aminoguanidine yields the initial hydrazone **99,** which is cyclized by treatment with base or by heating in solution to afford the 1,2,4-triazine derivative **100.**[60] Isomers generally result from this procedure. The parent 1,2,4-triazine was originally prepared from glyoxal and ethyl oxalamidrazonate via the carbethoxy derivative **101** ($R_1 = R_2 = H$; $X = CO_2Et$), which is readily hydrolyzed and the resulting carboxylic acid is converted to 1,2,4-triazine.[60a] In large scale syntheses, this compound is prepared directly (**101,** $X = R = H$)[60b] from 3-hydrazino-1,2,4-triazine by oxidation.[60c]

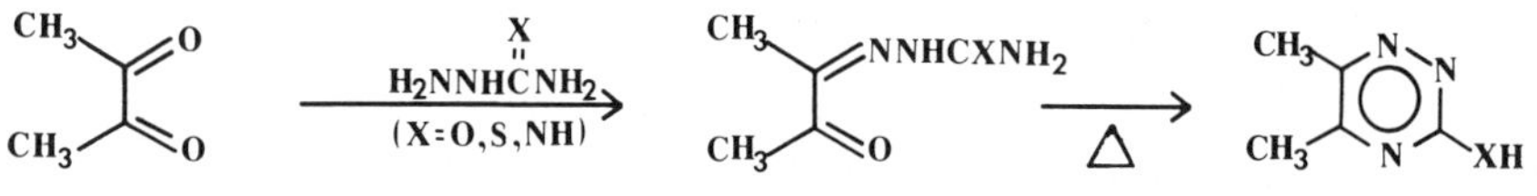

14.6.2.2. ⟨5 + 1⟩ Combinations. α-Diketones react with acrylhydrazides to generate an intermediate hydrazone, which undergoes a Hantzsch-type cyclization to the desired triazine **102.**[61] When the α-diketone is unsymmetrical, the most reactive carbonyl group gives the hydrazone intermediate. If both carbonyl groups are of equal reactivity, a mixture of products results. Greater regioselectivity can be realized if α-haloketones are used in place of the α-diketones.[62] Thus α-haloketones react with two equivalents of acylhydrazide. The initial hydrazone reacts with the second equivalent of hydrazide and subsequent cyclization generates a variety of 3,5,6-, 3,6-, and 6-substituted 1,2,4-triazines.

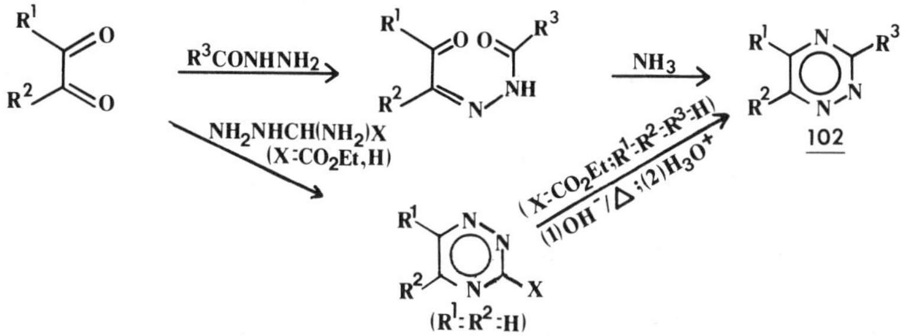

14.6.3. 1,3,5-TRIAZINES

The most common route to the 1,3,5-triazines **103** is via cyclotrimerization of nitriles, in which R can be hydrogen, alkyl, aryl, halo, amino, hydroxy, or other substituents. If different substituents are necessary on the ring, this trimerization process is not particularly effective since polymerization occurs and isolation of mixtures is required. Copolymerization of mixed nitriles has met with limited success.

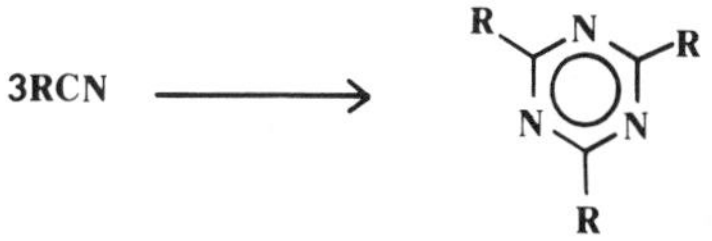

14.6.4. TETRAZINES AND PENTAZINES

No unambiguous examples of 1,2,3,4-tetrazines (**104**) are known, although dihydro derivatives (**105**) are reported.[63a] No synthesis of 1,2,3,5-tetrazine (**106**) has as yet been described.[63b]

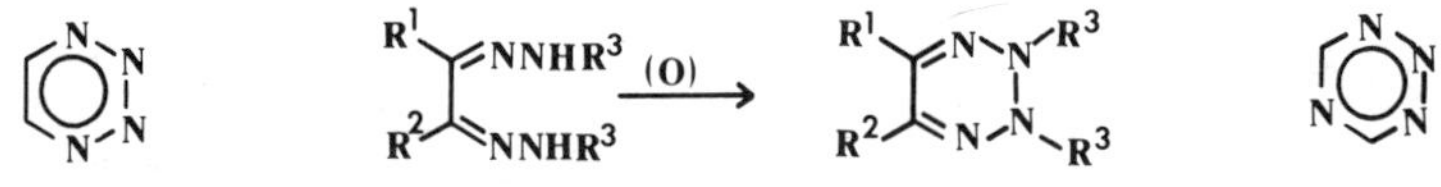

104 105 106

1,2,4,5-Tetrazine **107** is obtained by the oxidation of the dihydro intermediate **108**, which is produced by the dimerization of diazoacetate in the presence of base.[64a] Other derivatives that possess electron-withdrawing or stabilizing substituents can be prepared by this procedure.[64b]

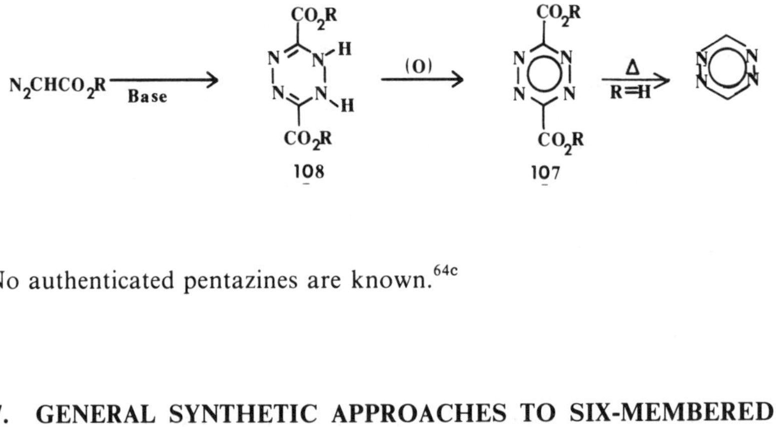

N_2CHCO_2R → (Base) **108** → (O) **107** → (Δ, R≡H)

No authenticated pentazines are known.[64c]

14.7. GENERAL SYNTHETIC APPROACHES TO SIX-MEMBERED HETEROAROMATIC RINGS CONTAINING ONE PHOSPHORUS, ARSENIC, ANTIMONY, OR BISMUTH

14.7.1. GENERAL PROCEDURES

Phosphorin (**110**),[65] arsenin (**111**),[65] and antimonin (**112**), are each prepared in one step by the treatment of 1,4-dihydro-1,1-di-*n*-butylstannabenzene (**109**) with phosphorus tribromide, arsenic trichloride, and antimony trichloride, respectively, followed by aromatization with 1,5-diazabicyclo[5.4.0]undec-5-ene (DBU). This procedure has also been applied to the synthesis of bismin (**113**). However, because of the transient nature of **113**, an adduct **114** was offered as evidence for its intermediacy.[66]

$HC{\equiv}C{-}CH_2{-}C{\equiv}CH$ $\xrightarrow{Bu_2SnH_2}$ **109** (1,1-dibutyl-1,4-dihydrostannin, $H\ H$, Sn, $Bu\ Bu$)

109 $\xrightarrow{(1)\ PBr_3,\ (2)\ DBU}$ **110** (P)

109 $\xrightarrow{(1)\ AsCl_3,\ (2)\ DBU}$ **111** (As)

109 $\xrightarrow{(1)\ SbCl_3,\ (2)\ DBU}$ **112** (Sb)

109 $\xrightarrow{BiCl_3}$ (Bi–Cl) $\xrightarrow{DBU}$ [**113** (Bi)] $\xrightarrow{CF_3C{\equiv}CCF_3}$ **114** (CF_3, CF_3, Bi)

14.7.2. PHOSPHORIN

In a manner similar to the preparation of pyridines, 2,4,6-triphenylpyrylium tetrafluoroborate (**115**) is treated with tris(hydroxymethyl)phosphine in refluxing pyridine to give 2,4,6-triphenylphosphorin (**116**).[67]

115 (C_6H_5, C_6H_5, C_6H_5, O^+, BF_4^-) $\xrightarrow[C_5H_5N/\Delta]{P(CH_2OH)_3}$ **116** (C_6H_5, C_6H_5, C_6H_5, P)

14.7.3. ARSENIN

Lithium arsenide (**117**), when treated with sodium phenyldichloroacetate, gives a carbene precursor **118** which decomposes slightly above its melting point to afford 2,3,6-triphenylarsenin (**119**).[68] An intramolecular carbene insertion has been proposed to account for arsacyclopropane betaine **120,** which rearranges to **119.**[69]

117 (C_6H_5, As, C_6H_5, Li) $\xrightarrow{C_6H_5CCl_2CO_2Na}$ (C_6H_5, As, C_6H_5; $C_6H_5{-}C(Cl){-}CO_2Na$) $\longrightarrow$ **119** (C_6H_5, As, C_6H_5, C_6H_5)

↓ [**118** (C_6H_5, As, C_6H_5, $\ddot{C}C_6H_5$)] $\longrightarrow$ [**120** (C_6H_5, As$^+$, $C^-{-}C_6H_5$, C_6H_5)] ↑ **119**

REFERENCES

1. G. J. Janz and S. C. Wait, Jr., *J. Am. Chem. Soc.* **76,** 6377 (1954).
2. Yu. I. Chumakov and V. P. Sherstyuk, *Tetrahedron Lett.* **1965,** 129; G. Caccia, G. Chelucci, and C. Botteghi, *Syn. Commun.* **11,** 71 (1981).
3. (a) F. Schon, L. Jung, and P. Cordier, *C. R. Acad. Sci. Ser. C* **267,** 490 (1968); (b) K. N. Zelenin and J. Dumpis, *Zh. Org. Khim.* **6,** 1349 (1970); **9,** 1295 (1973).
4. P. N. Evans, *J. Prakt. Chem.* **46,** 352 (1892); **48,** 489 (1893).
5. J. P. Marion, *Chimia* **21,** 510 (1967).
6. (a) Z. Arnold, *Experientia* **15,** 415 (1959); (b) J. Epsztajn, A. Bieniek, and J. Z. Brzezinski, *Pol. J. Chem.* **54,** 341 (1980), and refs. cited therein.
7. C. A. C. Haley and P. Maitland, *J. Chem. Soc.* **1951,** 3155.
8. A. Ehsan and Karimullah, *Pakistan J. Sci. Ind. Res.* **11,** 5 (1968).
9. K. L. Marsi and K. Torre, *J. Org. Chem.* **29,** 3102 (1964).
10. G. Simchen and G. Entenmann, *Angew. Chem. Int. Ed.* **12,** 119 (1973).
11. M. Simalty-Siemiatycki and R. Fugnitto, *Bull. Soc. Chim. Fr.* **1965,** 1944.
12. A. T. Balaban, W. Schroth, and G. Fischer, *Adv. Heterocycl. Chem.* **10,** 241 (1969); A. R. Katritzky, *Tetrahedron* **36,** 679 (1980).
13. (a) Review: F. Kröhnke, *Synthesis* **1976,** 1; R. S. Tewari, S. C. Chaturvedi, and D. K. Nagpal, *J. Chem. Eng. Data* **25,** 293 (1980); (b) R. S. Tewari, N. K. Misra, and A. K. Dubey, *J. Heterocycl. Chem.* **17,** 953 (1980).
14. Y. Wakatsuki and H. Yamazaki, *Tetrahedron Lett.* **1973,** 3383; *J. Chem. Soc. Dalton* **1978,** 1278.
15. J. R. Fritch and K. P. C. Vollhardt, *J. Am. Chem. Soc.* **100,** 3643 (1978); A. Naiman, *Diss. Abstr. Int. B* **41,** 202 (1980) [*Chem. Abstr.* **93,** 186121 (1980)]; D. J. Brien, A. Naiman, K. P. C. Vollhardt, *J. Chem. Soc. Chem. Commun.* **1982,** 133.
16. H. Bönnemann, R. Brinkmann, and H. Schenkluhn, *Synthesis* **1974,** 575; H. Bönnemann and R. Brinkmann, *Synthesis* **1975,** 600; Y. Wakatsuki, and H. Yamazaki, *J. Chem. Soc. Chem. Commun.* **1973,** 280; *Synthesis* **1976,** 26; *J. Chem. Soc. Dalton* **1978,** 1278; H. Bönnemann, *Angew. Chem. Int. Ed.* **17,** 505 (1978).
17. P. Hong and H. Yamazaki, *Synthesis* **1977,** 50; *Nippon Kagaku Kaishi* **1978,** 730.
18. R. A. Firestone, E. E. Harris, and W. Reuter, *Tetrahedron* **23,** 943 (1967).
19. T. Jaworski and S. Kwiatkowski, *Rocz. Chem.* **44,** 555 (1970).
20. W. Steglich, E. Buschmann, and O. Hollitzer, *Angew. Chem. Int. Ed.* **13,** 533 (1974).
21. A. Maujean, G. Marcy, and J. Chuche, *J. Chem. Soc. Chem. Commun.* **1980,** 92.
22. (a) S. Gotze, B. Kubel, and W. Stelich, *Chem. Ber.* **109,** 2331 (1976); (b) T. Hosokawa, N. Shimo, K. Maeda, A. Sonoda, and S.-I. Murahashi, *Tetrahedron Lett.* **1976,** 383.
23. R. E. Moerck and M. E. Battiste, *Tetrahedron Lett.* **1973,** 4421.
24. M. Tisler and B. Stanovnik, *Adv. Heterocycl. Chem.* **9,** 211 (1968); **24,** 363 (1979); R. N. Castle, Ed., *Pyridazines,* Wiley, New York, 1973.
25. J. C. Trisler, J. K. Doty, and J. M. Robinson, *J. Org. Chem.* **34,** 3421 (1969).
26. R. K. Gupta and M. V. George, *Indian J. Chem.* **10,** 875 (1972).
27. S. Sommer, *Tetrahedron Lett.* **1977,** 117; *Chem. Lett.* **1977,** 583.
28. F. H. Case, *J. Heterocycl. Chem.* **5,** 431 (1968).
29. G. Seitz and T. Kaempchem, *Chem. Ztg.* **99,** 292 (1975).
30. G. Seitz and T. Kampchen, *Arch. Pharm.* (*Weinheim*) **308,** 237 (1975).
31. H. Alper, J. E. Prickett, and S. Wollowitz, *J. Am. Chem. Soc.* **99,** 4330 (1977).

32. T. Eicher and E. von Angerer, *Chem. Ber.* **103,** 339 (1970).
33. R. L. Jones and C. W. Rees, *J. Chem. Soc.* (*C*) **1969,** 2251, 2255.
34. Z. Yoshida, T. Harada, and Y. Tamaru, *Tetrahedron Lett.* **1976,** 3823.
35. E. J. Volker, M. G. Pleiss, and J. A. Moore, *J. Org. Chem.* **35,** 3615 (1970).
36. R. G. Amiet and R. B. Johns, *Aust. J. Chem.* **21,** 1279 (1968); O. Tsuge and K. Kamata, *Heterocycles* **3,** 15 (1975).
37. D. T. Hurst, *An Introduction to the Chemistry and Biochemistry of Pyrimidines, Purines, and Pteridines,* Wiley, New York, 1980.
38. D. J. Brown, *The Pyrimidines,* Wiley-Interscience, New York, 1962; Suppl. I, 1970.
39. A. Combes and C. Combes, *Bull. Chim. Soc. Fr.* **7,** 788 (1892).
40. R. R. Hunt, J. F. M. McOmie, and E. R. Sayer, *J. Chem. Soc.* **1959,** 525.
41. P. N. Evans, *J. Prakt. Chem.* **46,** 352 (1892); **48,** 489 (1893).
42. R. Behrend, H. Meyer, and O. Buckholz, *Ann. Chem.* **304,** 200 (1901).
43. F. G. P. Remfry, *J. Chem. Soc.* **1911,** 610.
44. W. Traube, Ger. Pat. 180,424 (1905) [*Chem. Abstr.* **1,** 1654 (1907)].
45. A. L. Weis and V. Rosenbach, *Tetrahedron Lett.* **22,** 1453 (1981).
46. G. W. H. Cheeseman and E. S. G. Werstiuk, *Adv. Heterocycl. Chem.* **14,** 99 (1972).
47. J. P. Marion, *Chimia* **21,** 510 (1967).
48. D. G. Farnum and G. R. Carlson, *Synthesis* **1972,** 191.
49. J. Wiemann, N. Vinot, and M. Villadary, *Bull. Soc. Chim. Fr.* **1965,** 3476; N. Vinot and J. Pinson, *Bull. Soc. Chim. Fr.* **1968,** 4970.
50. E. Zbiral and J. Stroh, *Ann. Chem.* **727,** 231 (1969).
51. F. Tutin, *J. Chem. Soc.* **97,** 2495 (1910).
52. G. Fort, U.S. Pat. 3,356,679 (1967) [*Chem. Abstr.* **69,** 10471 (1968)].
53. K. W. Blake, A. E. A. Porter, and P. G. Sammes, *J. Chem. Soc. Perkin I* **1972,** 2494.
54. Th. Kaufmann, G. Beissner, H. Berg, E. Koppelmann, J. Legler, and M. Schonfelder, *Angew. Chem. Int. Ed.* **7,** 540 (1968).
55. R. D. Chambers, R. Middleton, and R. P. Corbally, *J. Chem. Soc. Chem. Commun.* **1975,** 731.
56. W. Mahler and T. Fukunaga, *J. Chem. Soc. Chem. Commun.* **1977,** 307.
57. C. G. Allison, R. D. Chambers, Yu. A. Cheburkov, J. A. H. MacBride, and W. K. R. Musgrave, *J. Chem. Soc. Chem. Commun.* **1969,** 1200.
58. E. A. Chandros and G. Smolinsky, *Tetrahedron Lett.* **1960,** 19; R. Gompper and K. Schonafinger, *Chem. Ber.* **112,** 1514, 1529, 1535 (1979) and refs. cited therein.
59. Reviewed by H. Neunhoeffer, in *Chemistry of 1,2,3-Triazines and 1,2,4-Triazines, Tetrazines, and Pentazines,* Wiley, New York, 1978, pp. 194–200.
60. (a) W. W. Paudler and J. Barton, *J. Org. Chem.* **31,** 1295 (1966); (b) H. Naunhoeffer, *Chem. Ber.* **102,** 847 (1968); (c) W. W. Paudler and D. Kress, *Synthesis* **1974,** 351.
61. R. Metz et al., *Chem. Ber.* **88,** 772 (1955); **89,** 2056 (1956); **90,** 481 (1957); **91,** 422, 1863 (1958).
62. T. V. Saraswathi and V. R. Srinivasan, *Tetrahedron* **33,** 1043 (1977).
63. Reviewed by P. F. Wiley, in *Chemistry of 1,2,3-Triazines and 1,2,4-Triazines, Tetrazines, and Pentazines,* Wiley, New York, (a) pp. 1287–1895; (b) 1296–1297.
64. (a) C. Fridh, L. Asbrink, B. O. Jonsson, and E. Lindholm, *Int. J. Mass Spectrom. Ion Phys.* **9,** 485 (1972) [*Chem. Abstr.* **77,** 113264 (1972)]; G. H. Spencer, Jr., P. C. Cross, and K. B. Wiberg, *J. Chem. Phys.* **35,** 1939 (1961); (b) reviewed in Ref. 63, pp. 1075–1283; (c) reviewed in Ref. 63, pp. 1298–1300.

65. A. J. Ashe, *J. Am. Chem. Soc.* **93,** 3293 (1971).
66. A. J. Ashe and M. D. Gordon, *J. Am. Chem. Soc.* **94,** 7596 (1972).
67. G. Märkl, *Angew. Chem. Int. Ed.* **5,** 846 (1966).
68. G. Märkl, H. Hauptmann, and J. Advena, *Angew. Chem. Int. Ed.* **11,** 441 (1972).
69. G. Märkl and R. Liebl, *Ann. Chem.* **1980,** 2095, and refs. cited therein.

Chapter 15

SYNTHETIC ROUTES TO THE BENZO-FUSED SIX-MEMBERED π-DEFICIENT HETEROAROMATICS AND COMMON POLYAZANAPHTHALENES

15.1. SPECIFIC ROUTES TO THE BENZO-FUSED PYRIDINES AND RELATED DIAZANAPHTHALENES

15.1.1. GENERAL COMMENTS

There are three benzo-fused pyridines: quinoline (**1**; 1-azanaphthalene or benzo[*b*]pyridine), isoquinoline (**2**; 2-azanaphthalene or benzo[*c*]pyridine), and the quinolizinium cation (**3**; 4*a*-azanaphthalene or benzo[*a*]pyridine). In view of the abundance of these ring systems in natural products, tremendous synthetic efforts have been devoted to their preparation and substituent manipulation. In many cases, the synthetic routes to the monoazanaphthalenes are adaptable to the construction of the related diazanaphthalenes (naphthyridines). There are 15 isomeric diazanaphthalenes; Scheme 15.1 gives some of their common or representative structure and accepted names.

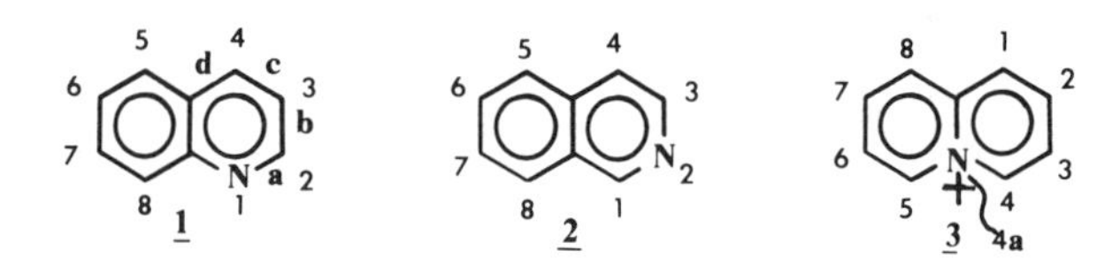

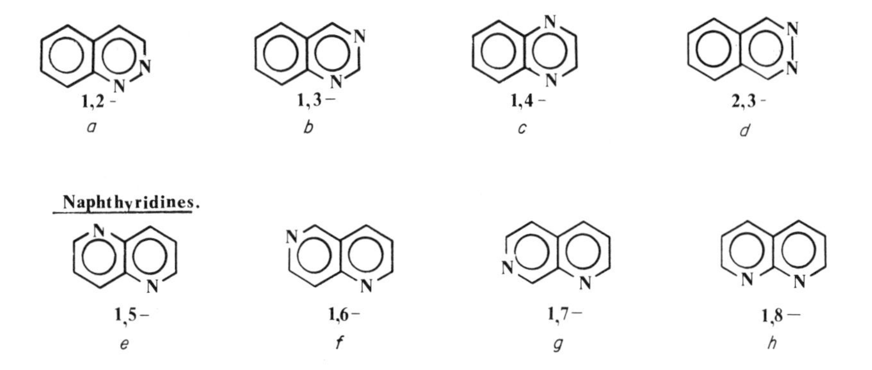

Scheme 15.1. Diazanaphthalenes. Common benzo-fused diazines: (*a*) cinnoline (benzo[*c*]pyridazine); (*b*) quinazoline (benzo[*d*]pyrimidine); (*c*) quinoxaline (benzo[*b*]pyrazine); (*d*) phthalazine (benzo[*d*]pyridazine). Naphthyridines: (*e*)–(*h*).

15.1.2. QUINOLINES, NAPHTHYRIDINES

There are four common routes to the synthesis of the quinoline nucleus, all of which generate the pyridine ring and are based on the use of a substituted aniline moiety. Because of the importance of this ring system, numerous classical name reactions are associated with its synthesis. The chemistry of quinoline and its derivatives have been reviewed in detail elsewhere.[1]

15.1.2.1. ⟨3 + 3⟩ Combinations. One of the classic heterocyclic syntheses is the *Skraup synthesis,*[2] in which quinoline is "magically" derived in generally high yield from a mixture of starting materials. Thus a mixture of aniline, nitrobenzene, glycerol (an *in situ* source of acrolein), and concentrated sulfuric acid gives quinoline. The Skraup reaction has often been regarded as a modification of the more general *Doebner–von Miller synthesis,*[3] in which aniline, nitrobenzene, paraformaldehyde, and sulfuric acid affords 2-methylquinoline (via the crotonaldehyde intermediate).

R^1-aniline + α,β-unsaturated carbonyl (R_2, R_3, R_4) $\xrightarrow[C_6H_5NO_2]{H_2SO_4}$ quinoline (R^1, R^2, R^3, R^4)

Mechanistically, the reaction pathway is shown in Scheme 15.2. The use of crotonaldehyde in this reaction gives rise to 2-methylquinoline[4] rather than to the 4-isomer. This is indicative of a Michael addition of the amine to the α-position of the unsaturated carbonyl compound. Elaboration of the remaining steps has been meticulously studied[5] and numerous key intermediates (**4–6**) have been isolated. Nitrobenzene acts exclusively as an oxidant in the later stages of the pathway and has been shown *not* to be the source of the ring nitrogen atom. Formation of the 3,4-bond is envisioned as an electrophilic substitution. However, even 5-nitro-2-carboxyaniline undergoes the Skraup synthesis with apparent ease (70%).[6]

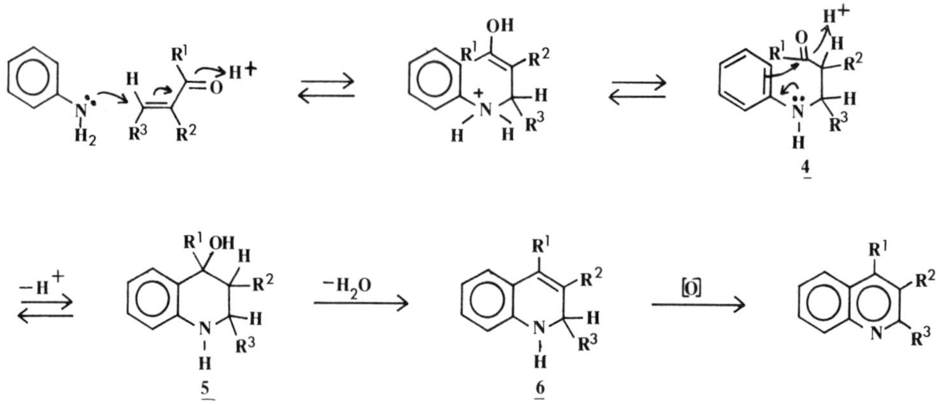

Scheme 15.2

Application of this procedure to the synthesis of naphthyridines has been reported.[7] Again, electrophilic cyclization does not seem reasonable into a pyridine nucleus; however, the favored preparation of **8** from 3-aminopyridine (**9**) must follow a similar mechanistic pathway.

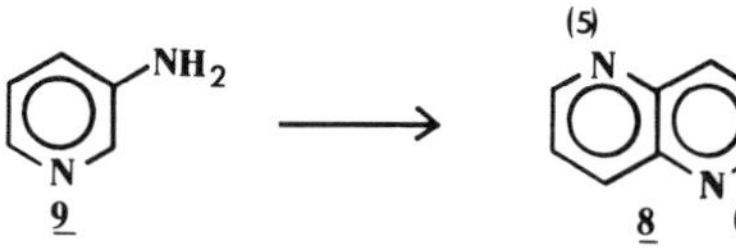

(1)

A wide range of substituted anilines and aminopyridines can be utilized. The ortho or para amines generally give a single product, whereas use of meta-substituted anilines gives rise to mixtures of 5- and 7-substituted quinolines, in which meta-deactivating (e.g., NO_2) groups favor the 5-substituted quinolines and o,p-activating (e.g., OCH_3) groups favor 7-substitution.[8] Numerous variations have been devised over the past century of work on this reaction; minor alterations in solvent and oxidants have remarkable effects on product yields.

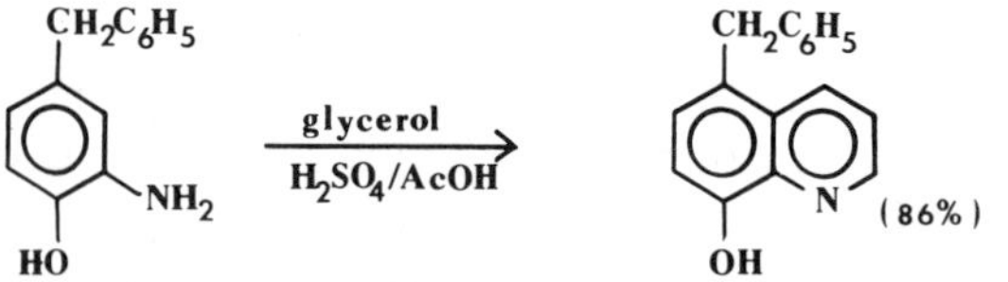

Several years after the introduction of the above reactions, Combes[9] condensed anilines with β-diketones to give imine intermediates (**10**), which were subsequently cyclized in the presence of acid to give the related 2,4-disubstituted quinolines. This reaction now carries his name (the *Combes synthesis*). The increased ease of cyclization of **10** with meta-electron-donating substituents supports the fact that the cyclization probably does proceed by electrophilic aromatic substitution.[10]

$(CH_3CO)_2CH_2$, H_2SO_4; 10a; 10b; $+H^+$; $-H^+$, $-H_2O$; (X=Cl; 100%)

The *Conrad–Limpach synthesis*[11] is the easiest route to 4-substituted quinolones (e.g., **11**) whereas the *Knorr synthesis*[12] gives access to the 2-substituted quinolones. Thus aniline reacts with diethyl oxosuccinate to give **12b,** which can undergo an electrocyclization and subsequent elimination of ethanol to afford quinoline **11.**

$EtO_2CCOCH_2CO_2Et$; 12a; 12b; $-EtOH$; 11

In the Knorr synthesis, β-ketoesters are converted into β-ketoanilides, followed by cyclization in the presence of acid to give the desired α-quinolone. Treatment of 2-methoxyaniline with acetoacetyl chloride affords **13,** which cyclizes in the presence of phosphoric acid to give **14.**[14] A logical extension of the above quinolone synthesis was devised by Rugheimer and involves intermediates, such as **15,** which upon action with phosphorus pentachloride are converted into the related 2,6-dichloroquinolines (**16**).[15] Generally, a stepwise sequence is preferred, as is shown by the conversion of **17** to **18.**[16]

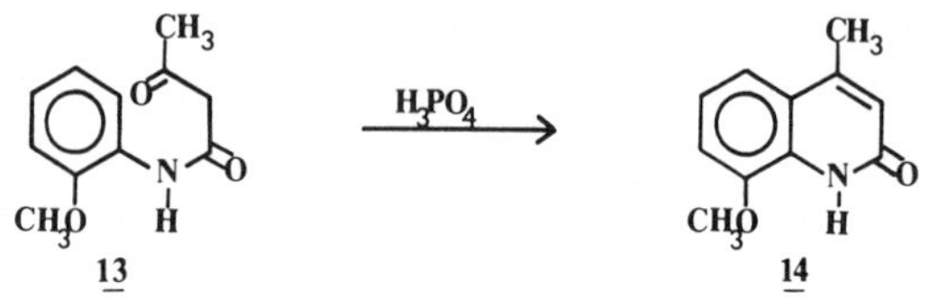

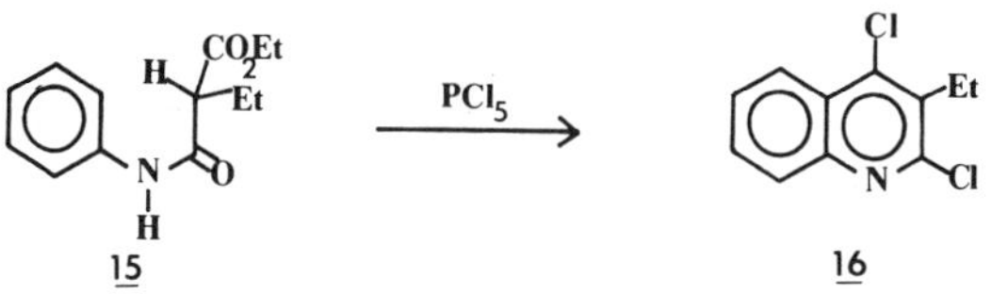

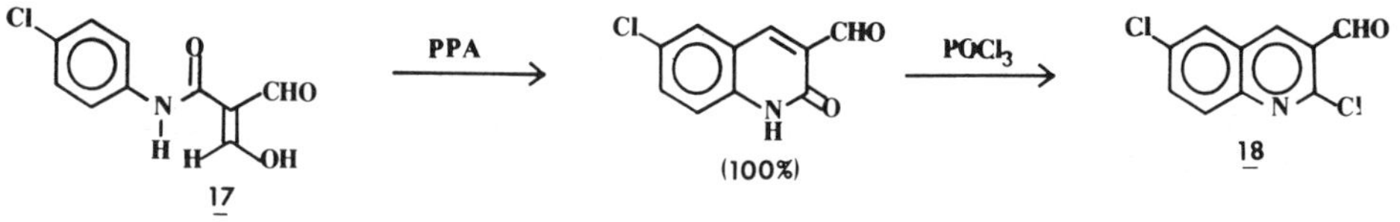

15.1.2.2. ⟨4 + 2⟩ Combinations. Condensation of 2-aminobenzaldehyde or acetophenones with α-methylenic aldehydes, ketones, or related compounds in the presence of base affords the quinoline nucleus; this reaction is known as the *Friedländer synthesis.*[17] The reaction of **19** with ethyl oxosuccinate gives **20;** initially the imine intermediate **21** is probably generated, followed by a Knoevenagel cyclocondensation.[18] In view of the limited mechanistic considerations given to this reaction, the reverse condensation sequence cannot be ruled out in all cases.

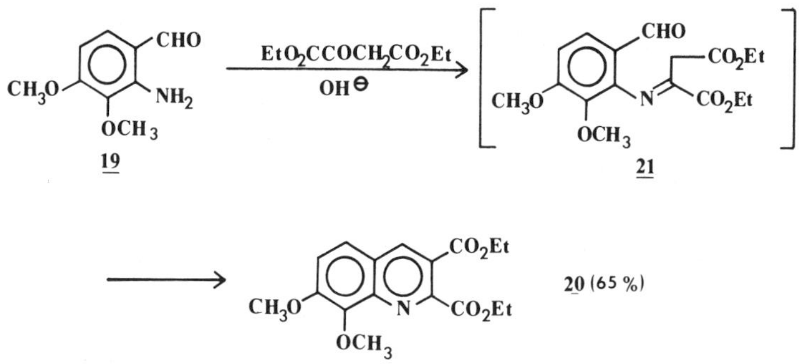

The facile self-condensation of aminoaldehydes poses a problem that can be circumvented by the *Pfitzinger synthesis.*[19] Condensation of aldehydes or ketones with isatin (**22**) in basic media forms a five-membered lactam, which ring-opens in base to give the 2-aminophenylglyoxylate (**23**). This compound presumably acts as an 2-aminophenyl ketone in a normal Friedländer reaction, although little mechanistic detail is known about the Pfitzinger synthesis.

O; O; N; H; 22; $C_6H_5COCH_3$; Base; H; COC_6H_5; O; N; H

CO_2^-; H; C_6H_5; N; H_2; O; 23; $-H_2O$; $(+H^+)$; CO_2H; N; C_6H_5

15.1.2.3. ⟨6⟩ Combinations. One of the original syntheses of the quinoline nucleus[20] involved the simple reduction of 2-nitro-*cis*-cinnamic acid (**24**) with tin or zinc in hydrochloric acid.

H; H; CO_2H; NO_2; 24; HCl; Zn; H; H; CO_2H; NH_2; $-H_2O$; Δ; N; O; H

Photocyclization of **25** gives a thietan **26** via an allowed [2 + 2] cycloaddition reaction. Cleavage of the four-membered ring affords thiol **27**, which subsequently eliminates hydrogen sulfide to give quinoline **28**.[21] The initial photochemical reaction probably proceeds through a singlet $\pi \rightarrow \pi^*$ state.

CH_3; NH; 25; S; CH_3; hν; CH_3CN; CH_3; S; N; H; CH_3; 26

H SH; CH_3; N; CH_3; 27; CH_3; N; CH_3; 28 69%

Other electrocyclizations have been utilized to create the pyridine nucleus in quinoline syntheses. Photoextrusion of nitrogen from **29** gives **30**, which ring-opens under the reaction conditions, and then **31** cyclizes thermally to give **32**.[22a] Aromatization occurs by dehydrogenation. An alternative route to

31 is the pyrolysis of **33** at 750°C (10^{-2} torr). Quinoline **34** is prepared similarly.[22b]

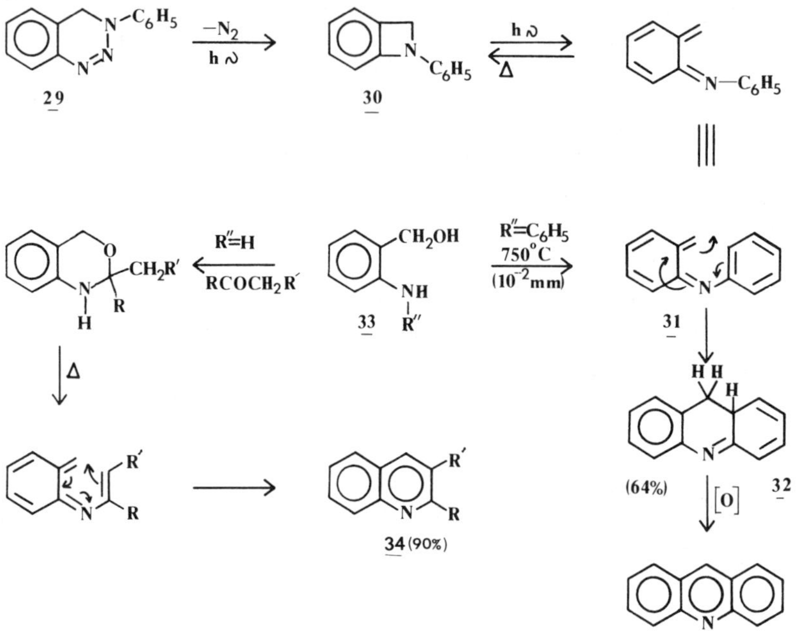

Uncharged "vinamidines," for example, 3-dimethylaminoallylidenearylamine (**36**), undergo an electrocyclic ring closure,[23] followed by elimination of dimethylamine on heating, to give the substituted quinoline **35.** Compounds of structure **36** are readily synthesized from an aniline and **37,** an adduct derived from 3-dimethylaminoacraldehydes and dimethyl sulfate. This reaction sequence is quite versatile and, in many cases, quantitative yields are realized.

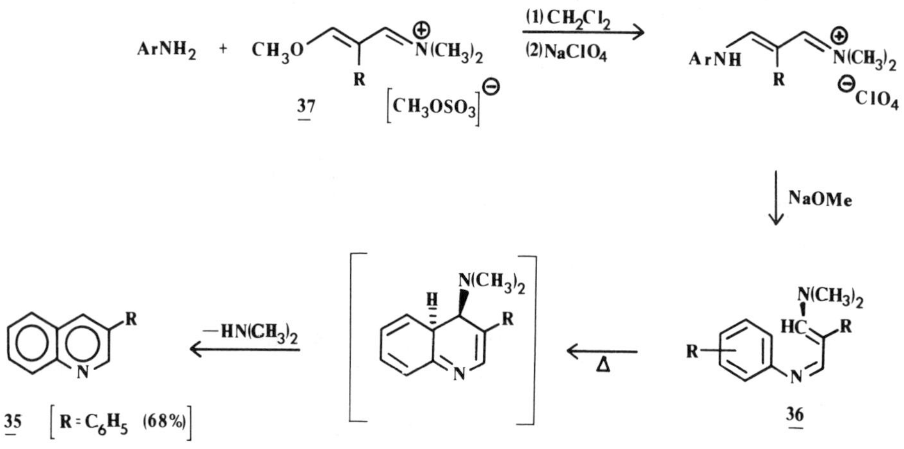

Although far less common, the generation of the benzene ring can be accomplished by the flash pyrolysis of either 2- or 4-(butadienyl)pyridine (**38**) to give quinoline[24] or isoquinoline. An interesting mechanistic profile has been proposed.[24]

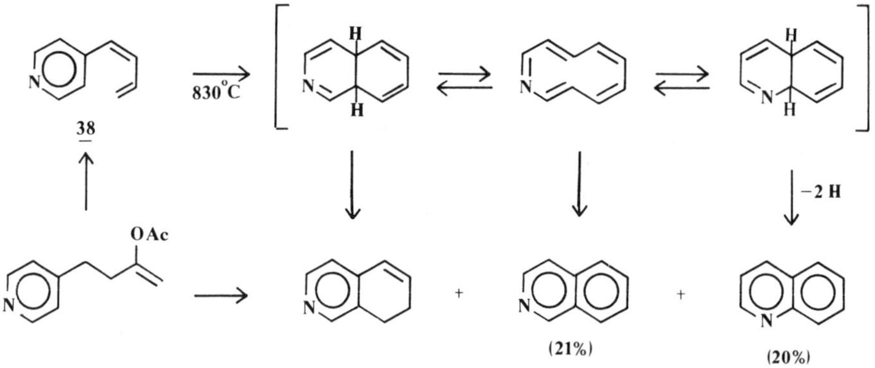

830°C

38

OAc

−2 H

(21%) (20%)

15.1.2.4. ⟨5 + 1⟩ Combinations and Ring Enlargements. The ring expansion of a five-membered ring generally involves either cyclic ketones (Beckmann, Schmidt, or diazo reactions) or indoles (carbene, azide, or oxygen insertion). Disubstituted indenones **39** have been treated with sodium azide and sulfuric acid to give **40,**[25] **41** being suggested as the intermediate.[26] The Schmidt reaction on **42** affords **43;** this reaction is also envisioned to proceed through indene intermediate **44.**[26] Indenes, in this latter case, can be isolated immediately after **39** is subjected to acid but prior to the addition of sodium azide.[27]

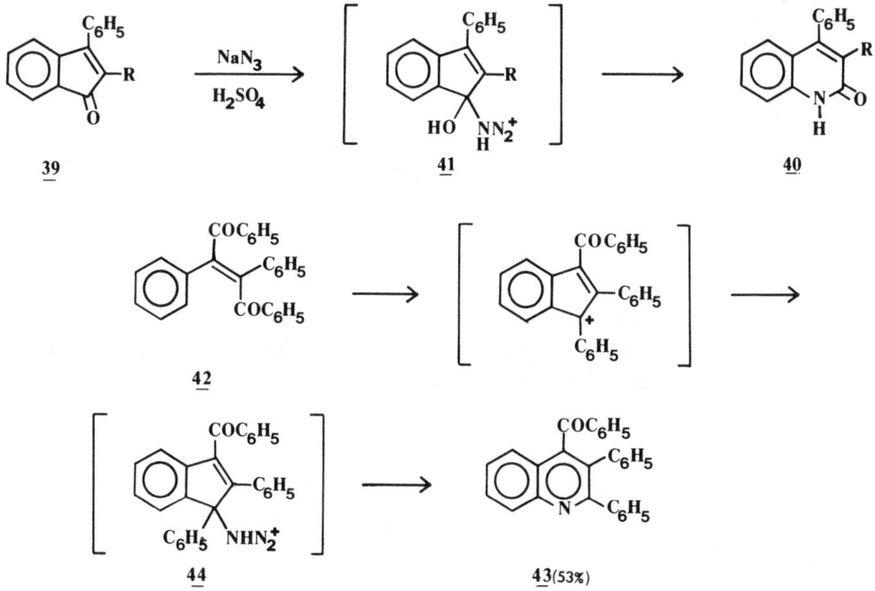

C_6H_5 R O

NaN_3 / H_2SO_4

HO NN_2^+ H

39 41 40

COC_6H_5 C_6H_5 COC_6H_5

42

C_6H_5 NHN_2^+

44 43(53%)

Since pyrroles are well-known to undergo carbene expansion to give pyridines, it is not surprising that indoles, when subjected to haloforms and strong base, give indolenines and quinolines. Preparation of these mixtures has been suggested[28] to proceed via a cyclopropane intermediate **45,** which subsequently has two modes for fragmentation. Improved yields of quinoline products can be realized by the use of phase-transfer catalysts[29] or crown ethers.

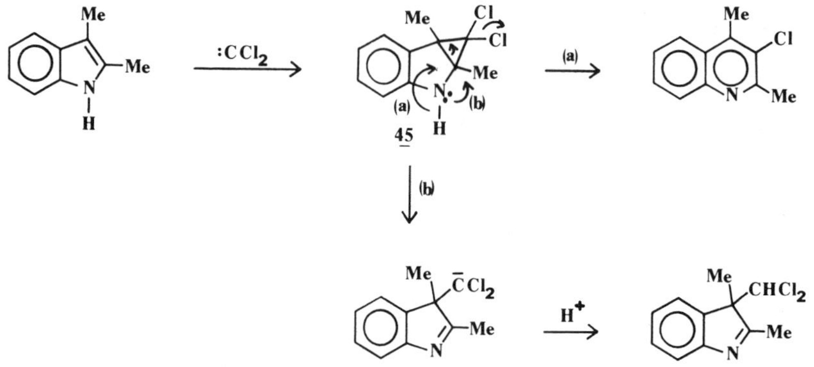

:CCl$_2$ (a) (b) 45 H$^+$

One of the most novel ring expansions in this series is the conversion of **46** to **47.** The pyrolysis has been suggested to occur through the electrocyclization of ketone **48.**[30a]

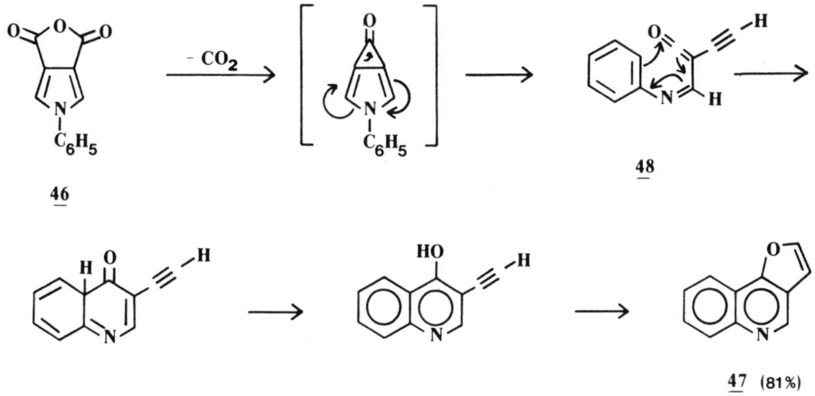

− CO$_2$ 46 48 47 (81%)

The Vilsmeier formylation of activated methylenes has also been used to transform *N*-(ω-chloroacyl)anilines (**49**) into **50,** which cyclizes to give excellent yields of the *N*-aryl-2-quinoline (**51**) upon hydrolysis.[30b]

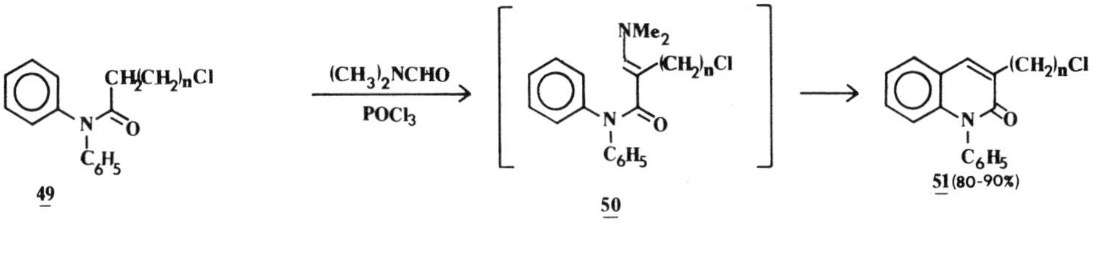

(CH$_3$)$_2$NCHO / POCl$_3$ 49 50 51 (80-90%)

15.1.2.5. ⟨7 → 6⟩ Combinations; Ring Contractions. Most simply, tetrahydro-1*H*-benzazepine (**52**), when heated with palladized charcoal and a hydrogen acceptor such as ethyl cinnamate, is transformed to a mixture of quinoline and methylquinolines.[31] Dioxobenzazepines (**53**) undergo ring fragmentation in the presence of base, then cyclodehydrate to give quinoline-2-carboxylic acids.[32]

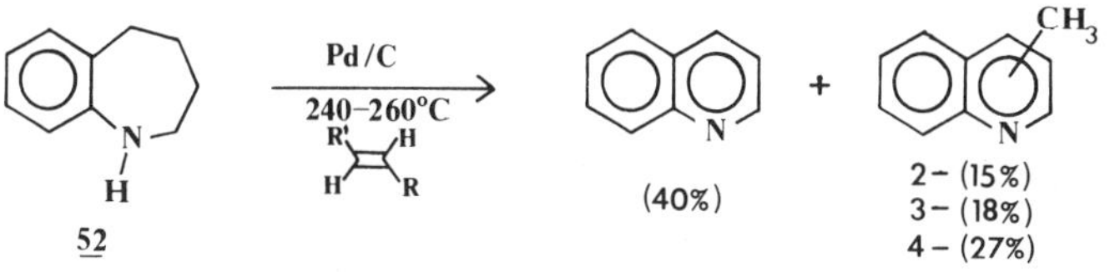

Unusual rearrangements which generate the quinoline nucleus include the molecular rearrangement of the bridged azocine **54** to give all of the methylquinolines *except* the 2-isomer, and that of the isomeric azasemibullvalene **55,** which affords *predominantly* 2-methylquinoline[33]!

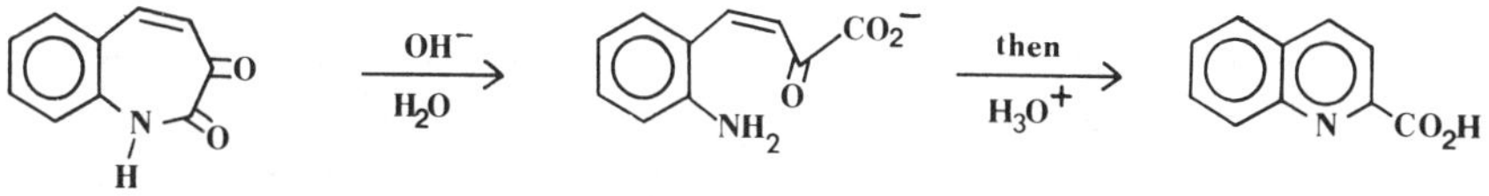

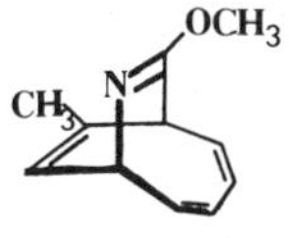

15.1.3. ISOQUINOLINES

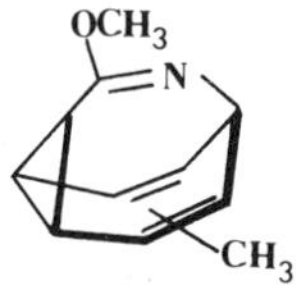

In view of the tremendous synthetic effort expended toward the total syntheses of isoquinoline alkaloids over the past century, few routes to the substituted aromatic isoquinoline nucleus have been devised.[34] Most routes give rise to the partially reduced heteroaromatic compounds in which the hetero ring is at the di- or tetrahydro stage. If aromatization is desired, it is accomplished in a separate dehydrogenation step. Both steps are presented, even though the aromatization step is quite facile.

15.1.3.1. ⟨5 + 1⟩ Combinations. *The Bischler–Napieralski cyclization*[35] is one of the better routes to the simple[36] dihydroisoquinolines. In this reaction the amide derived from a substituted phenylethylamine is cyclized under dehydrative acidic conditions: P_2O_5, PCl_5, or $POCl_3$ in boiling xylene or decalin, or the potent new acid catalyst P_2O_5 in methanesulfonic acid.[36] Most phenethylamines are derived from the reduction of β-nitrostyrenes or from the reduction of an arylacetonitrile with lithium aluminum hydride in the presence of aluminum chloride. As expected, aromatic electron-releasing substituents facilitate the electrophilic cyclization, whereas electron-withdrawing groups retard the process. Fortunately, most isoquinoline alkaloids have one or more electron-releasing substituent.[34] Thus amide **56** when treated with phosphorus pentachloride gives the iminoyl chloride **57,** which via the nitrilium ion[37] **58** cyclizes to give 1-phenyl-3,4-dihydroisoquinoline **59.**

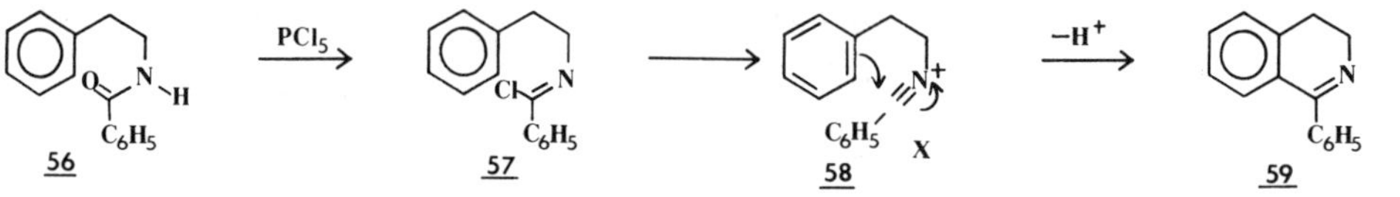

A novel modification that gives rise to the aromatic quinoline uses ω-phenylisonitrosopropiophenone (**60**), which is reduced catalytically, *N*-formylated, and cyclized with $POCl_3$ at elevated temperatures.[38] This modification has been termed the *Pictet–Gams synthesis.*[39]

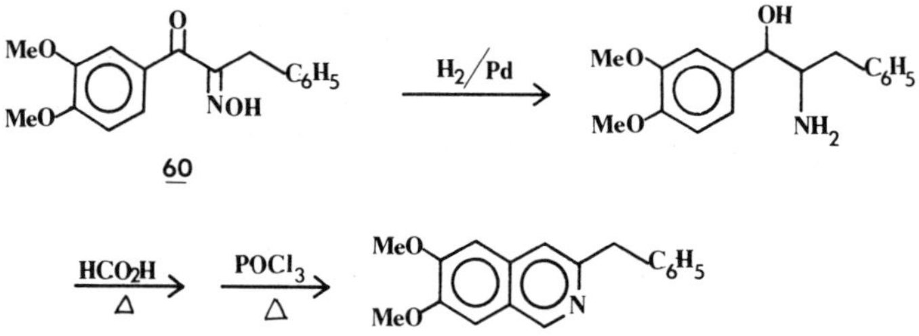

The *Pictet–Spengler synthesis*[40] is a further alteration by which the tetrahydroquinoline nucleus is formed. The procedure can be conducted in one or two stages, as desired. Thus the aldehyde is condensed with the appropriate β-phenethylamine, then cyclized by treatment with acid. To exemplify this procedure, mescaline hydrochloride (**61**) is condensed with an activated carbonyl compound **62,** then cyclized to give *O*-methylpeyoxylic acid (63).[41]

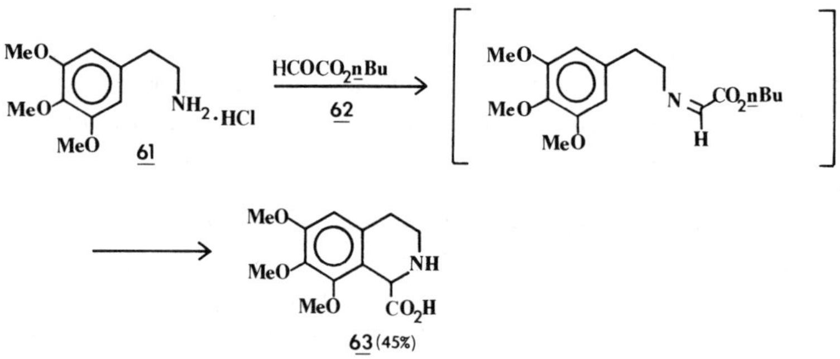

Utilization of the carbanion generated from 2-tolunitrile (**64**) offers an easy procedure to prepare an aminoketone **65**, which readily cyclizes in the presence of acid.[42]

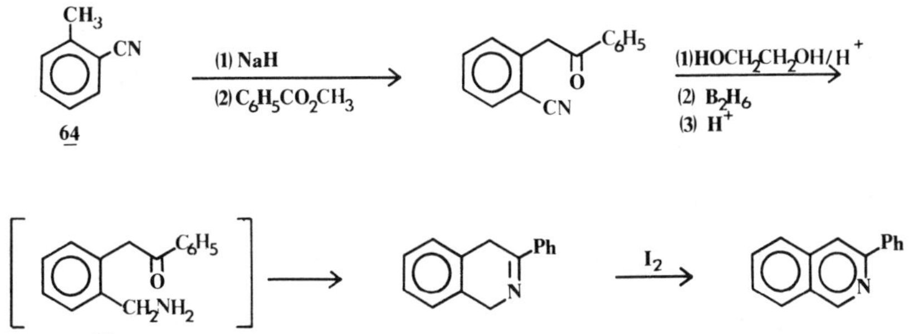

15.1.3.2. ⟨6⟩ Combinations. The *Pomeranz–Fritsch synthesis*[43a] can be envisioned as a simple electrophilic cyclization to generate the 4,4*a*-bond, hence the designation ⟨6⟩. In consideration of the actual starting materials, this reaction may be considered as either a ⟨4 + 2⟩ or a ⟨3 + 3⟩ combination. Substituted quinolines (**66**) can be formed from aldehyde **67** or from amine **68**, as shown.

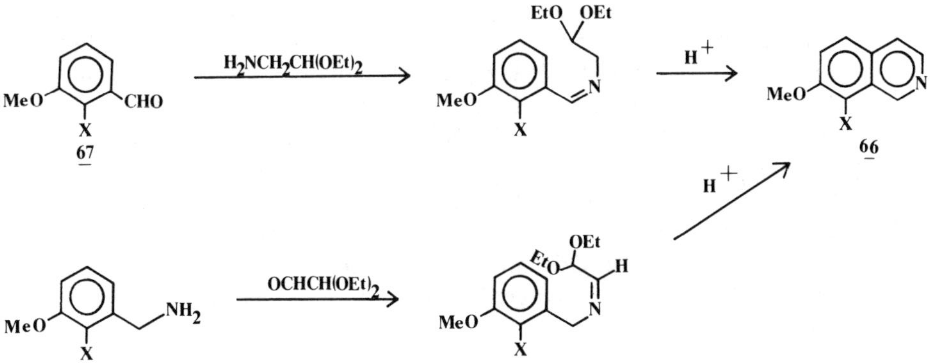

Photocyclization of α-chloroamides, such as **69**, has proved to be an excellent route to the isoquinoline lactam **70**.[43]

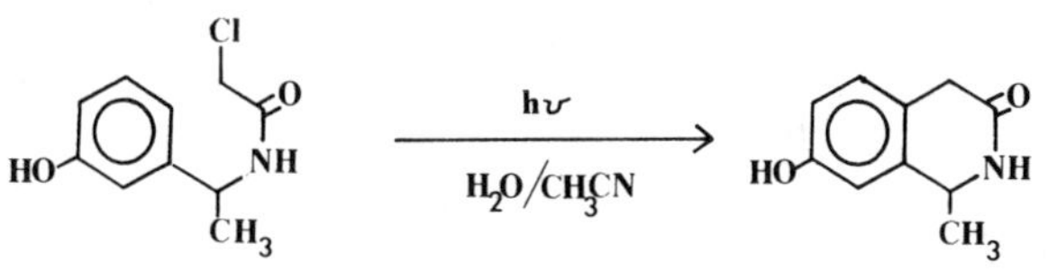

Since isoquinoline is simply a benzo-*c*-fused pyridine ring, conversion of the appropriate pyrylium salt **71** to **72** occurs readily in the presence of ammonia.[44]

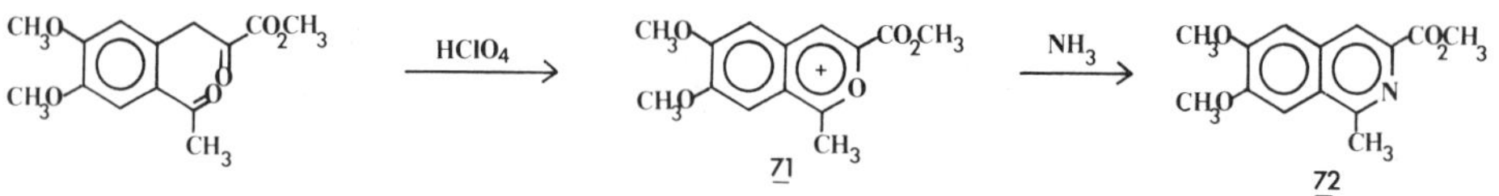

15.1.4. QUINOLIZINIUM CATION

Although the quinolizine ring system occurs in numerous alkaloids, it is normally in the reduced form. However, the aromatic quinolizinium ring system can be isolated from natural sources as shown by sempervirin (**73**), present in *Gelsemium sempervirens.* The chemistry of quinolizine has been reviewed[45] and the relationship to alkaloids has been considered.[46]

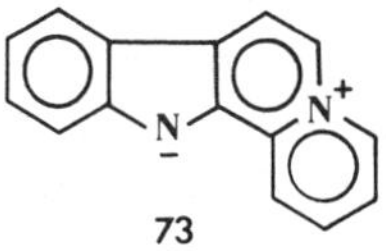

15.1.4.1. ⟨3 + 3⟩ Combinations. Synthesis of the parent quinolizinium cation (**74**) is accomplished in poor yield by the condensation of 2-picolyllithium with β-isopropoxyacrolein (**75**). When β-ethoxypropionaldehyde (**76**) is substituted for **75, 77** is prepared. Subsequent treatment of **77** with hydrogen iodide gives rise to **78,** which is dehydrated quantitatively with acetic anhydride and dehydrogenated with platinum catalyst in nitrobenzene.[47] The weak link in this sequence is the dehydrogenation stage. Numerous modifications have been attempted, but with limited success. Circumvention of this problem is accomplished by addition of the Grignard reagent **79** to 2-cyanopicoline (**80**), followed by cyclization with hydrogen bromide, then dehydration with acetic anhydride.[48]

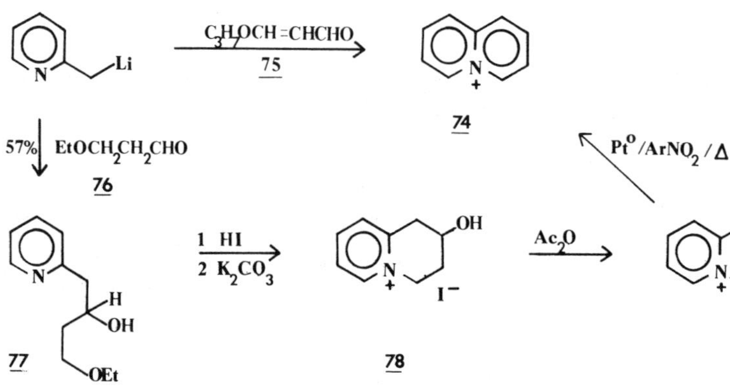

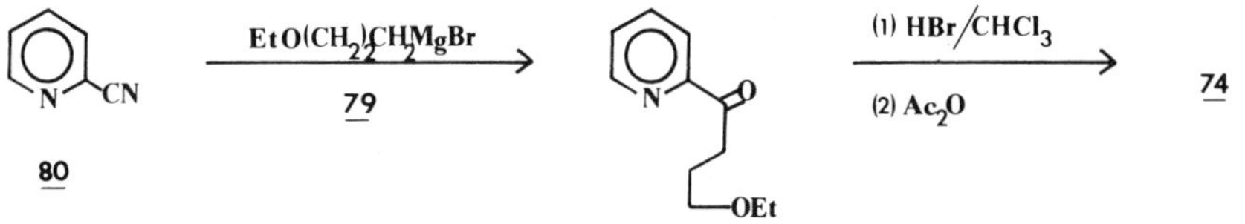

15.1.4.2. **⟨4 + 2⟩ Combinations.** Condensation of diketones with α-picolinium salts possessing an *N*-activated methylene can best be demonstrated by the reaction of **81** with benzil in the presence of a weak base. Spontaneous hydrolysis and decarboxylation affords **82** in very good yields. If strong base is used in the condensation stage, the related indolizine is formed.[49]

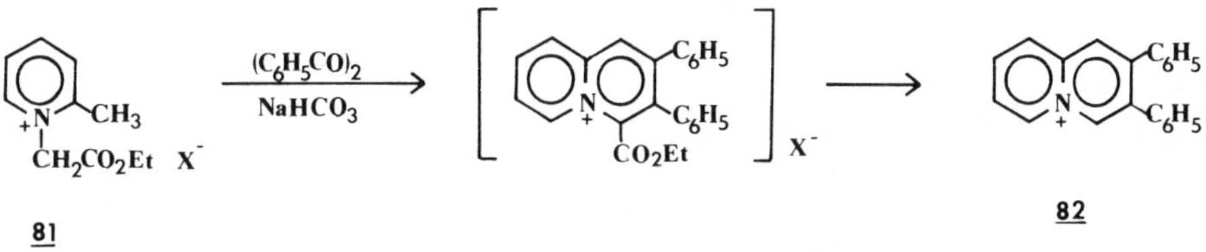

Slight variations of this procedure have been accomplished, as shown below in the stepwise transformation of 2-pyridinecarboxaldehyde (**83**) to **82.**[50]

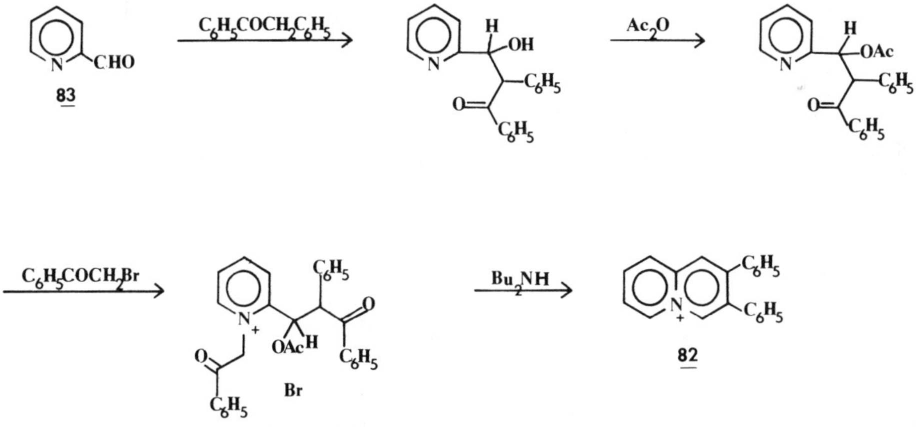

15.1.4.3. ⟨2 + 2 + 2⟩ Combinations. The Diels–Alder reaction is conveniently applied to the synthesis of dihydroquinolizines. Treatment of pyridine and substituted pyridines with dimethyl acetylenedicarboxylate affords three adducts **84–86:** the first, **84,** is quite labile and the last two are stable.[51] Several older reviews delineate the chemistry of this reaction and the subsequent products,[52] a more modern instrumental approach to structure elucidation would be of interest, in view of the tentative nature of many of the structures.

$MeO_2CC{\equiv}CCO_2Me$, RT/ether → **84** →

85 + **86**

15.1.5. CINNOLINES (1,2-DIAZANAPHTHALENES)[53]

15.1.5.1. ⟨6⟩ Combinations. The most widely used method to prepare cinnolines which possess alkyl, aryl, or heteroaryl groups in either the 3- and/or 4-positions is called the *Widman–Stoermer synthesis.*[54] Generally, a diazotized *o*-aminoarylethylene cyclizes upon standing to generate the cinnoline. Since all attempts to prepare an unsubstituted heteroaryl ring by this route have failed, it is important that the α-carbon of the ethylene group has an alkyl or (hetero)-aryl group. Diazotization of the amine **87** generates **88,** which undergoes slow decomposition and concomitant cyclization to give **89.**

87 —($NaNO_2$, H^+)→ [**88**] —(H^+, Δ)→ **89**

Similarly, the *Borsche synthesis*[55] utilizes the thermal decomposition of a diazonium salt; the corresponding 4-hydroxycinnoline (**90**) is prepared from the substituted 2-aminoacetophenone (**91**).[56] A further modification of these diazonium salt decompositions is referred to as the *Richter synthesis.*[57] Preparation of 4-hydroxycinnolines is exemplified by the reaction of *o*-aminophenylacetylene **92** with sodium nitrite in acid.[58] This procedure has only limited application in view of the better routes available.

91 —($NaNO_2$, H^+, (90%))→ **90**

NH_2

92

$\xrightarrow[H^+]{NaNO_2}$ [$\overset{+}{N}_2$] $\xrightarrow{(59\%)}$ 90

The *Neber–Bossel synthesis* also utilizes diazonium salts, which are reduced chemically, cyclized, and dehydrated. Sodium *o*-aminomandelate (**93**) is transformed to 3-hydroxycinnoline (**94**).[54]

$CH(OH)CO_2Na$, NH_2

93

$\xrightarrow[H^+]{NaNO_2}$ [$CH(OH)CO_2H$, $\overset{+}{N}_2$] $\xrightarrow{SnCl_2}$

$CH(OH)CO_2H$, $NHNH_2$ $\xrightarrow[H^+]{-H_2O}$ [H, OH, O, N–H, N, H] $\xrightarrow{-H_2O}$ OH, N, N

94

H, OH, O, N, H

95

The free acid of **93** upon direct diazotization in acidic medium rapidly generates the dioxindole **95.**

Hydrazones are convenient starting materials in cinnoline syntheses as shown by the *Friedel–Crafts* cyclization of mesoxalyl chloride phenylhydrazones. Hydrazone **96** is hydrolyzed carefully to the diacid, transformed to the corresponding acid chloride, then cyclized in the presence of a Lewis acid.[60] Decarboxylation of **97** at elevated temperature affords the 4-hydroxycinnoline in good overall yield.

EtO_2C, CO_2Et, N, N, H

96

$\xrightarrow[(2)\ SOCl_2]{(1)\ EtOH/NaOH}$ ClOC, COCl, N, N, H $\xrightarrow[C_6H_5NO_2,\ 100°c]{TiCl_4}$ OH, CO_2H, N, N

97

15.1.5.2. ⟨5 → 6⟩ Ring Expansions. The ring-expansion sequence in which 1-aminooxindole (**98**) is oxidized with lead tetraacetate[61] gives **94.** Utilization of *t*-butyl hypochlorite[62] as the oxidant in this reaction affords a quantitative yield of the same compound. A nitrene (route *a*) mechanism is proposed[61,62]; however, an alternative pathway would involve the *N*-chloro intermediate (route *b*) rather than the nitrene.

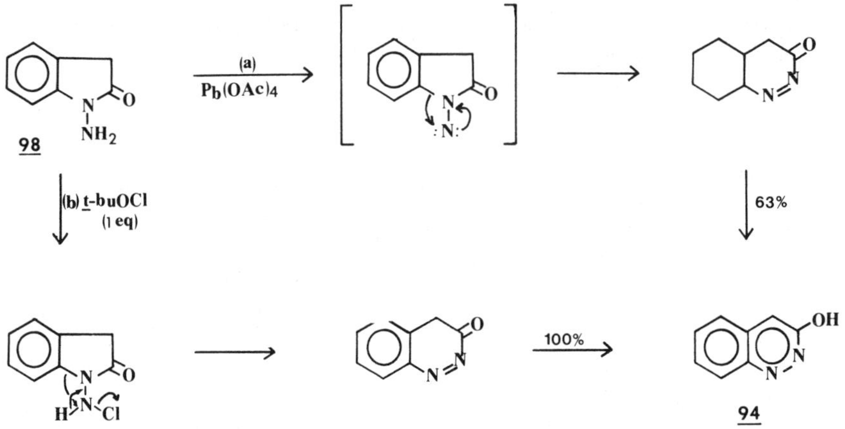

(a) $Pb(OAc)_4$

(b) t-buOCl (1 eq)

63%

100%

98

94

Ring-expansion of 2-phenylisatogen (**99**) by treatment with ammonia in ethanol at 140–145°C gives **100.**[63] A mechanism, that has been suggested, proceeds via the addition product **101,** which ring-cleaves and then recyclizes to give the dihydrocinnoline **102.** The mechanistic aspects of this ring expansion probably require reevaluation.

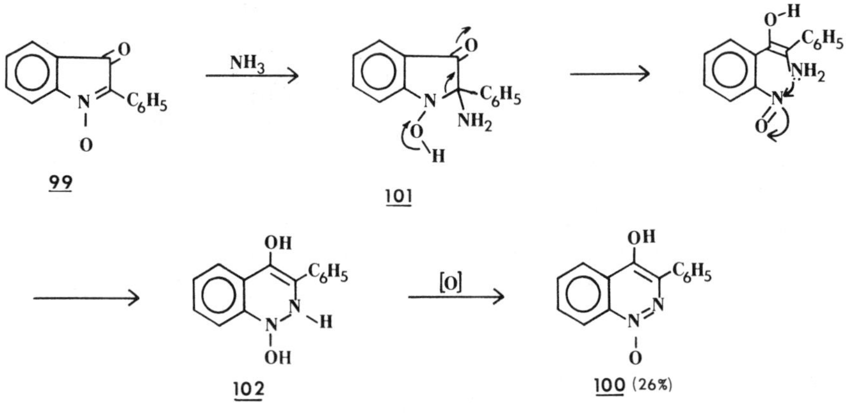

NH_3

[O]

99

101

102

100 (26%)

Benzo[*c*]cinnoline (**103**) has been synthesized by cyclization of the hetero ring via (*a*) formation of the biphenyl central bond or (*b*) formation of either 4*a*–5 (CN) or 5–6 (NN) bonds. Cyclization reactions of 2,2′-disubstituted biphenyls have been reviewed[64] and can be best illustrated by the reduction of 2,2′-dinitrobiphenyl (**104**) to generate **103.** As anticipated, numerous other partially reduced products can be isolated via this route.

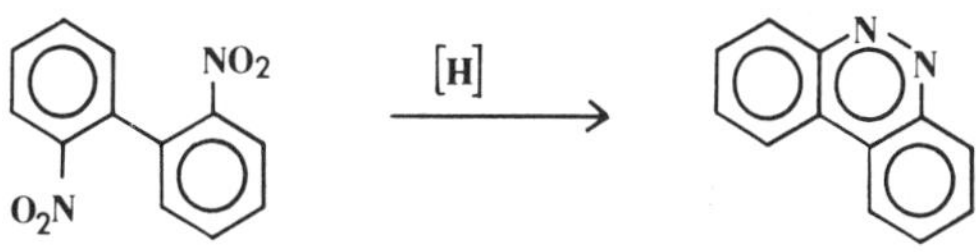

Upon irradiation in acidic medium, azobenzene (**105**) undergoes an initial rapid *cis–trans* equilibration to monoprotonated azobenzene (**105** ⇌ **106**), followed by cyclization of the excited state (lowest transition $\pi^* \rightarrow \pi$) of **106** and dehydrogenation via reduction of a second equivalent of azobenzene to give hydroazobenzene (**107**). This cyclization process is greatly affected by substituents, especially in the para position. The proposed[65] mechanism is shown in Scheme 15.3. Other routes to substituted **103** have been reviewed.[65a]

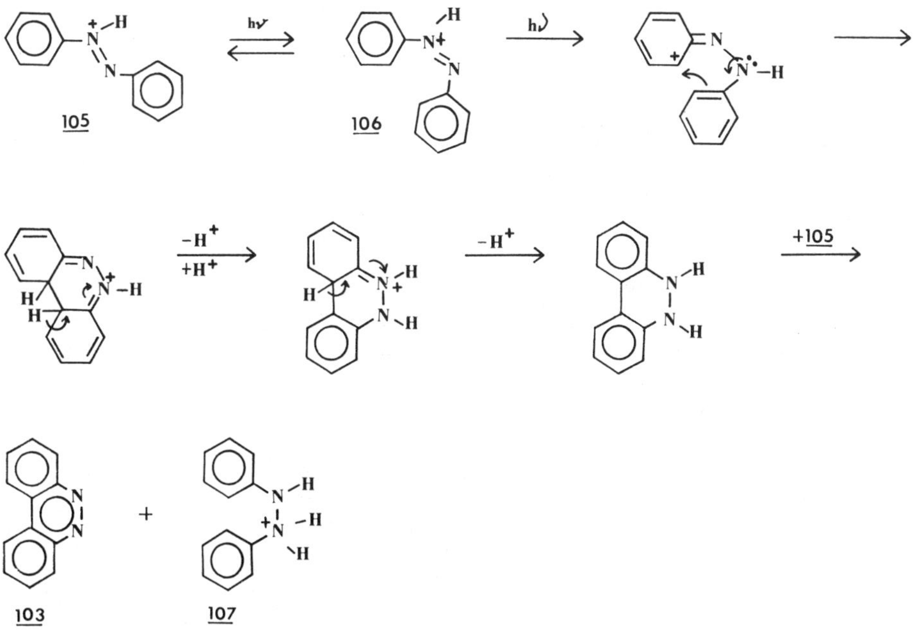

Scheme 15.3

15.1.6. PHTHALAZINES (2,3-DIAZANAPHTHALENES)

15.1.6.1. ⟨4 + 2⟩ Combinations. Phthalazine **108** was initially prepared by reaction of $\alpha,\alpha,\alpha',\alpha'$-tetrachloro-*o*-xylene **109** with hydrazine.[66] More recently, it has been determined that execution of the reaction in >90% sulfuric acid increases the yields to about 90%.[67] Reaction of the readily available *o*-phthaldehyde (**110**) with hydrazine permits the synthesis of **108** in 96% yield.[68] In each of these cases, substituted phthalazines can be readily prepared.[69]

$CHCl_2$ / $CHCl_2$ —N_2H_4, 90% H_2SO_4 (~90%)→ [phthalazine] ←N_2H_4 (96%)— CHO / CHO

109 108 110

A more novel, but limited, procedure to produce substituted phthalazines involves the Diels–Alder reaction. Treatment of 1,4-diphenyl-*sym*-tetrazine (**111**) with benzyne (generated from anthranilic acid diazonium betaine) gives **112.**[70]

111 → [intermediate] —$-N_2$→ 112 (67%)

CO_2H / NH_2 → O / O^- / $N{\equiv}N^+$

15.1.6.2. ⟨6⟩ Combination. A new synthesis of 1-arylphthalazines has recently been reported in which aromatic aldazines **113** are cyclized upon treatment with aluminum chloride and triethylamine (2 : 1) at about 200°C.[71] Without triethylamine, dearylation occurs.

[phthalazine] ←$AlCl_3$— 113 —$AlCl_3$, Et_3N→ [1-phenylphthalazine]

113

15.1.7. QUINAZOLINES (1,3-DIAZANAPHTHALENES)

Generally, the heterocyclic ring is synthesized from a derivative of an *o*-substituted benzaldehyde or benzoic acid. It is beyond the scope of this text to delineate the numerous routes available; more details can be obtained from several reviews.[72]

15.1.7.1. ⟨5 + 1⟩ Combinations. One of the more versatile routes to substituted quinazolines is the *Bischler synthesis,* which is illustrated by the reaction of an *N*-acyl derivative of *o*-aminobenzophenone (**114**)[73] and ammonia at elevated temperatures. This synthesis is quite diversified in that a wide range of substituents can be placed on either ring.

NH_3 (alc), 170°

114

The reductive cyclization (*Reidel synthesis*) of bisformamide derivatives of *o*-nitrobenzaldehyde (**115**) may become a successful route to **116,**[74] since the starting material is readily available.[75]

H_2CNOH

115 116

o-(*N′*,*N′*-Dimethylaminomethyleneamino)benzonitriles (**117**), derived from the *β*-oximes of isatins and the Vilsmeier reagent [$(CH_3)_2NCHO$ and $POCl_3$], are readily cyclized to the 4-aminoquinazoline (**118**) with ammonium acetate.[76]

NH_4OAc

117 118 (75–80%)

15.1.7.2. ⟨4 + 2⟩ Combinations. The facile *Neimentowski synthesis* of 4-hydroxyquinazolines (3,4-dihydro-4-oxoquinazoline) is exemplified by condensation of anthranilic acid (**119**) with an amide at elevated temperatures. Because this procedure is limited by the steric bulk of the amide substituents, the synthesis is most effective for 4-oxoquinazolines in which the 2-substituent is a hydrogen. Fusion of substituted anthranilic acids with formamide affords excellent yields of **120.**[77]

$HCONH_2$, 150–200°C

119 120 (80–90%)

15.1.7.3. ⟨6⟩ Combinations. Reaction of *o*-amidobenzonitriles (**121**) with alkaline hydrogen peroxide at about 40°C affords excellent yields of the 3,4-dihydro-4-oxoquinazoline **122.**[78] The reaction proceeds via the intermediate amide, which can be easily cyclized by base. This is a superior route to these compounds in that the yields are high and there are few side reactions.[79]

CN NHCOMe $\xrightarrow[NaOH]{H_2O_2}$ [$CONH_2$ NHCOMe] $\xrightarrow{OH^-}$ O N H N Me

121 122

Aromatic pyrylium salts, such as 2-substituted 4-phenyl-3,1-benzoxazin-3-ium perchlorates (**123**), are easily transformed into the corresponding quinazolines by treatment with ammonia.[80]

Ar ClO_4^- O^+ N R $\xrightarrow{NH_3}$ Ar N N R

123

15.1.7.4. ⟨2 + 2 + 2⟩ Combinations. Aryldiazonium borofluorides (**124**) react with two equivalents of a nitrile at 20–50°C to give variable yields of substituted quinazolines (**125**) in which the 2- and 4-positions bear similar substituents.[81]

N_2^+ BF_4^- $\xrightarrow{ArCN}$ [$\overset{+}{N}$ Ar] $\xrightarrow{ArCN}$ [Ar + N N Ar] $\longrightarrow$ Ar N N Ar

124 125 40–80%

15.1.7.5. ⟨3 + 1 + 1 + 1⟩ Combinations. Treatment of *p*-toluidine (**126**) with formaldehyde in acid was reported by Tröger in 1887 to give a molecule later to be known as "Tröger's base" (**127**).[82] Two key intermediates were subsequently retrieved from the reaction mixture and shown to be dihydro (**128**) and tetrahydro (**129**) quinazolines. The reaction has been proposed to proceed by the stepwise condensation sequence shown in Scheme 15.4. Prelog and Wieland[83] resolved **127** by column chromatography through activated *d*-lactose hydrate. Studies performed with chiral **127** have added considerably to our stereochemical knowledge of the trivalent nitrogen atom.

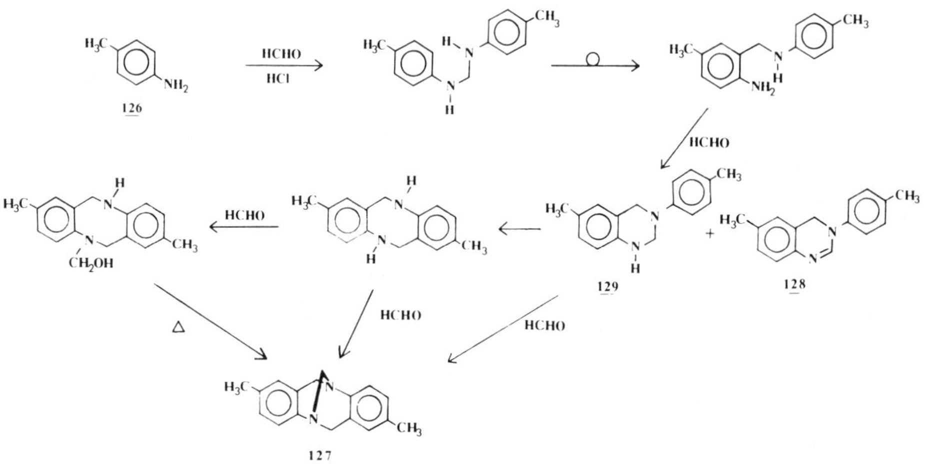

Scheme 15.4

15.1.8. QUINOXALINES (1,4-DIAZANAPHTHALENES)

Recently, an excellent review of this benzo-fused pyrazine ring system has appeared.[84]

15.1.8.1. ⟨4 + 2⟩ Combination. The most common route to the substituted quinoxalines is the condensation of an *o*-disubstituted benzene with a suitable two-carbon synthon. Classically, *o*-phenylenediamine (**130**) is condensed with glyoxal sodium bisulfite to give the parent quinoxaline (**131**).[85] In general, judicious selection of starting materials permits the preparation of almost all substituted quinoxalines by this procedure.

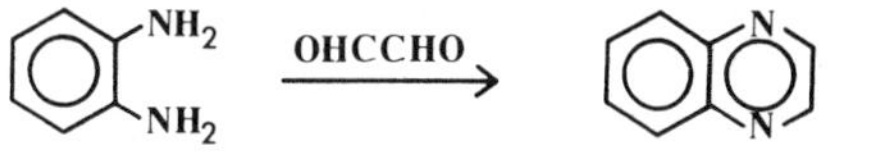

130 → (OHCCHO) → 131 (85–90%)

15.2. SELECTED COMMON TRIAZANAPHTHALENES

Although many of the more than 60 isomers of triazanapthalene have been synthesized, a considerable number remain to be prepared. Impetus for the synthesis of these compounds is their close biological and physical relationship to the quinazolines and especially to the pteridines. In Japan, the pyrido[2,3-*d*]pyrimidines (**132**) have recently been shown to possess anti-inflammatory, analgesic, and CNS depressant properties. Because it is beyond the scope of this book[86] to be all inclusive, the preparation of a family of pyridopyrimidines (**132–135**) is presented to depict the synthetic logic that leads to these triazaheterocycles.

-(2,3-d)- 132 -(3,2-d)- 133 -(3,4-d)- 134 -(4,3-d)- 135

15.2.1. ⟨4 + 2⟩ COMBINATIONS

Scheme 15.5 illustrates the synthetic concept that converts either pyridines or pyrimidines to triazanaphthalenes.

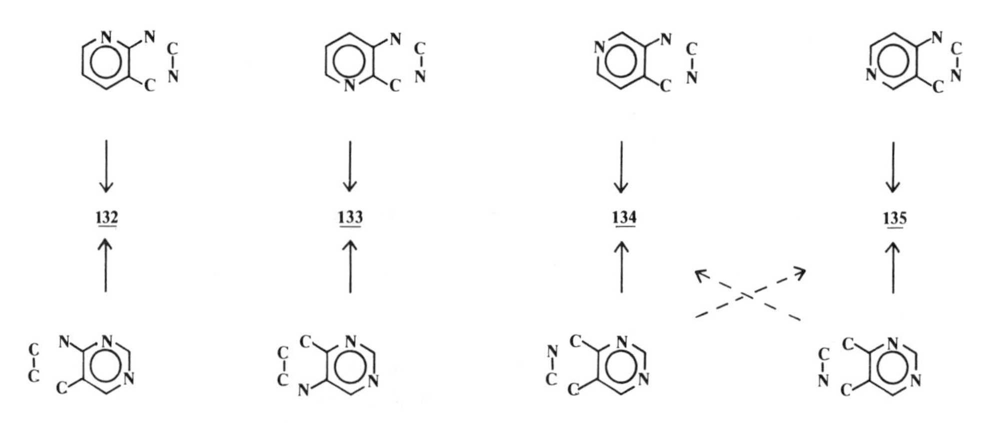

Scheme 15.5. ⟨4 + 2⟩ combinations.

As an example of this route, 4-amino-2-(3′-pyridyl)pyrido[2,3-*d*]pyrimidine (**136**) is prepared from 2-aminonicotinonitrile and nicotinonitrile.[87] Similarly, the *o*-aminonitrile **137,** when treated with formamide, is converted in good yield to **138.**[88] This procedure is merely an extension of the Niementowski quinazolone synthesis: 3-aminopicolinic acid with formamide affords pyrido[3,2-*d*]-pyrimidin-4(3*H*)-one (**139**).[89]

136

$HCONH_2$

137 138

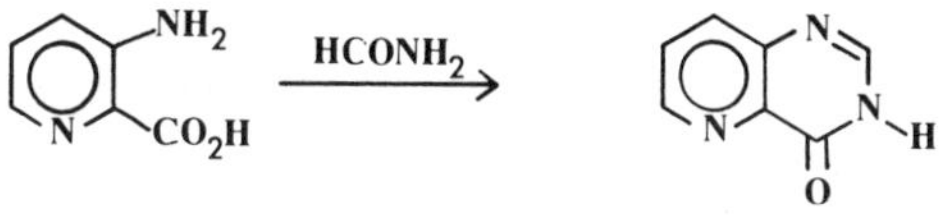

NH_2 N CO_2H $\xrightarrow{HCONH_2}$ N N N H O

139

15.2.2. ⟨3 + 3⟩ COMBINATION

A major synthetic route to this family of compounds uses electrophilic substitution onto an electron-deficient ring which possesses electron-rich substituents. Thus 4-aminopyrimidine (**140**) is condensed with a β-dicarbonyl compound in the presence of acid, such as phosphoric or sulfuric acid.[90]

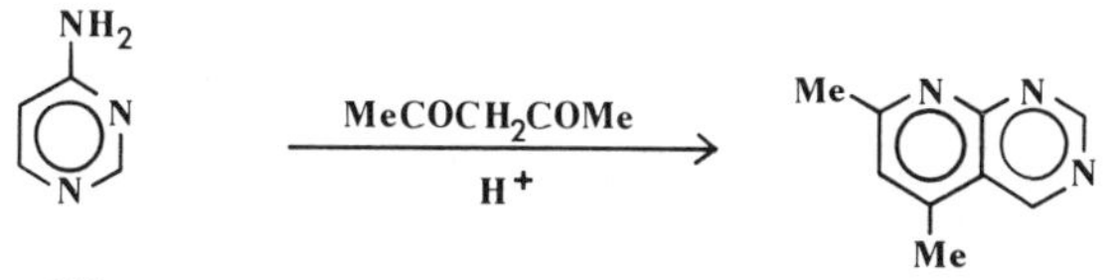

NH_2 N N $\xrightarrow[H^+]{MeCOCH_2COMe}$ Me N N N Me

140

From these limited examples, it is quite apparent that the combinations in question can be derived by application of the previously considered synthetic routes to the mono- and diazanaphthalenes.

15.3. SELECTED COMMON TETRAZANAPHTHALENES

The most common member of this family is the pteridine (**141**) ring system. Derivatives of **141** such as xanthopterin (**142**) and leucopterin (**143**) are isolated as pigments from butterflies and other insects. Folic acid (**144:** pteroylglutamic acid) is another well-known derivative of this ring system. One of the outstanding, currently popular anticancer drugs, methotrexate (**145**), is also a member of this class. In view of the numerous (>60) possible isomers of the tetrazanaphthalene group,[91] only the best-known examples are considered.

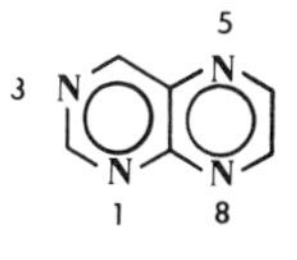

5 3 N N N N 1 8 — OH N N OH H_2N N N — OH N N OH H_2N N N OH

141 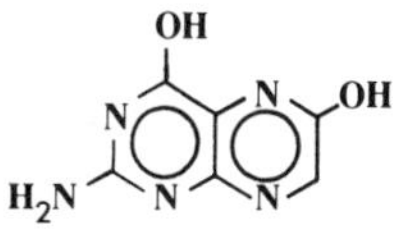 142 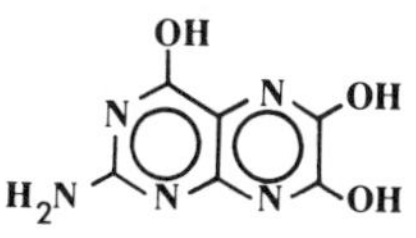 143

CO_2H
$CONHCHCH(CH_2)_2CO_2H$
X
H_2N CH_2 N R

144 X=OH; R=H

145 $X=NH_2$; R=Me

15.3.1. PTERIDINES[92]

One general route to the parent compound (**141**) is via the treatment of 4,5-diaminopyrimidine (**146**) with glyoxal (the *Isay reaction*),[93] a procedure that can be readily modified to prepare simple derivatives such as **143.** The oxidation state of the pyridazine ring is controlled by the α-dicarbonyl source.

NH_2 NH_2 —OHCCHO→

146 141

OH NH_2 H_2N NH_2 —$HO_2C)_2$→ OH OH H_2N OH

143

Mixed dicarbonyl compounds generally lead to formation of isomers; however, by control of the pH a preferential condensation occurs. Treatment of **146** with **147** in mildly acidic or neutral media gives the 7-hydroxy isomer **148,** whereas under highly acidic conditions the 6-hydroxy isomer **149** is isolated.[94] Although a rationale that hinges on the base strength of the 5- versus the 6-amino group in the condensation stage has been proposed,[95] further work is still necessary.

OH Me ←147, H^+— NH_2 NH_2 —$MeCOCO_2H$ (147)→ Me OH

149 146 148

Another procedure to control the regiochemistry utilizes a nitroso compound (e.g., **150**), which can be readily condensed with an activated methylenic carbonyl compound.[96]

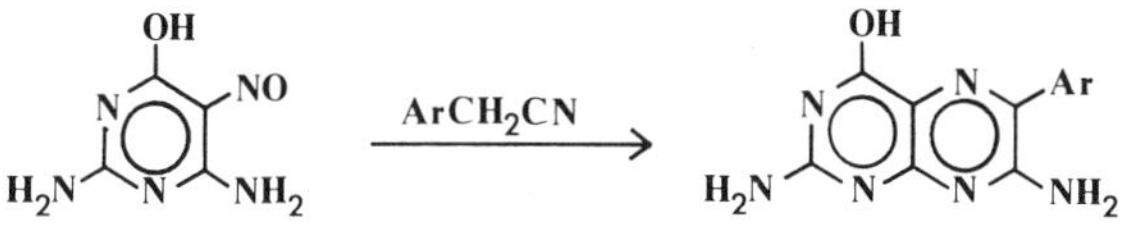

Riboflavin (**151**), or vitamin B_2, is a benzo-fused substituted pteridine that occurs in animal and plant tissues. The simplest member of this group is alloxazine (**152**), which is readily synthesized from alloxan (**153**). Riboflavin can be synthesized by this same procedure except for the initial attachment of the D-ribose side chain. Care must be exercised with **151** since it is sensitive to light and, upon irradiation, decomposes to give lumiflavin (**154**) and lumichrome (**155**).[97]

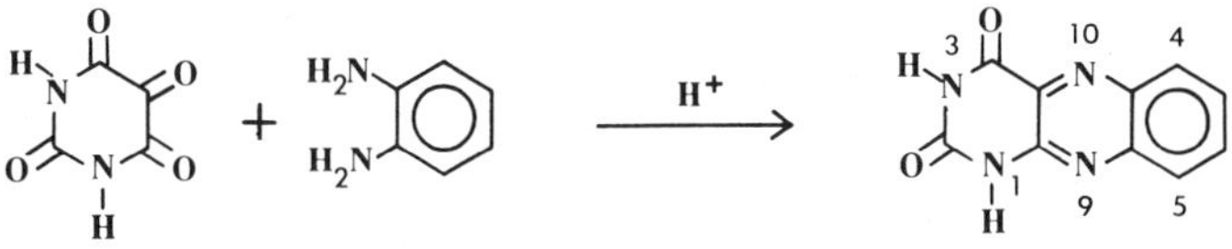

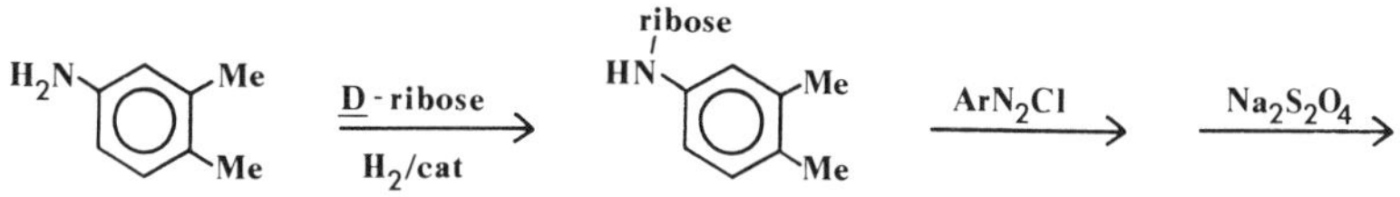

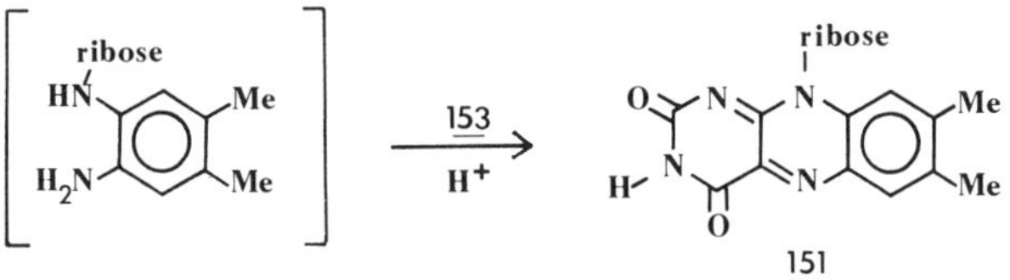

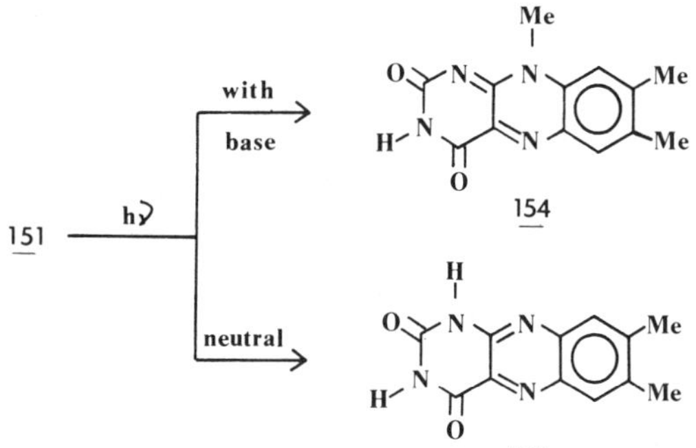

15.3.2. PYRAZINO[2,3-*b*]PYRAZINES (1,4,5,8-TETRAZANAPHTHALENES)

These compounds (**156**) have been synthesized by condensation of 2,3-diaminopyrazine with an appropriate α-dicarbonyl compound. The monooxo (**157**) and dioxo (**158**) derivatives are prepared from ethyl diethoxyacetate or diethyloxalate, respectively.[98] Numerous derivatives have been patented as potential pharmaceuticals.[99]

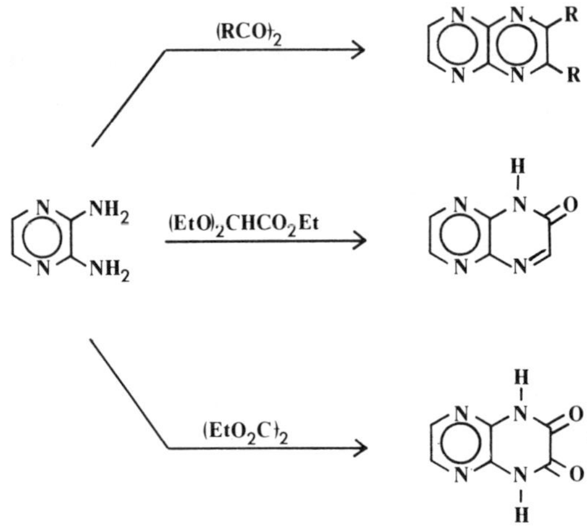

$(RCO)_2$ → 156

$(EtO)_2CHCO_2Et$ → 157

$(EtO_2C)_2$ → 158

15.3.3. PYRIDAZINO[4,5-*d*]PYRIDAZINES (2,3,6,7-TETRAZANAPHTHALENES; (159))

Problems plagued the original researchers who attempted the condensation of tetracarbonyls with hydrazine since several isomers are possible and their structures are difficult to assign.[100] For example, **160** (R = C_6H_5) condenses with hydrazine to give both **159** and **161.** The original structural assignments are based on UV data[101] and should be considered as tentative.

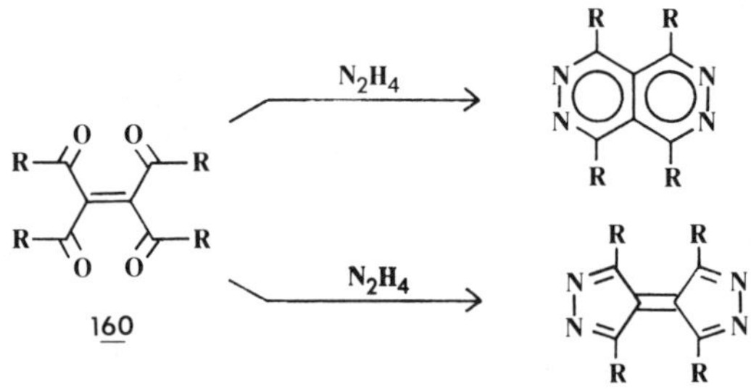

160 —N_2H_4→ 159

160 —N_2H_4→ 161

Unequivocal routes to **159** use the appropriately substituted pyridazine precursors. For example, **159** (R = H) is prepared from diethyl pyridazine-

4,5-dicarboxylate (**162**) by reduction with lithium aluminum hydride at −70°C and trapping of the diformyl intermediate **163** with hydrazine.[102]

$$\text{162} \xrightarrow[-70^\circ C]{LiAlH_4} [\text{163}] \xrightarrow{N_2H_4} \text{159}$$

15.3.4. PYRIMIDOPYRIDAZINES

Pyrimidopyridazines have been reviewed.[103] The synthetic routes to these tetrazanaphthalenes, as well as to the other members of this group, generally utilize previously described procedures.

REFERENCES

1. R. H. Manske, *Chem. Rev.* **30,** 113 (1942); F. W. Bergstrom, *Chem. Rev.* **35,** 77 (1944); R. C. Elderfield, *Heterocyclic Compounds,* Vol. 4, Wiley, New York, 1952, Chap. 1; N. Campbell in *Rodd's Chemistry of Carbon Compounds,* Vol. 4F, 2nd ed., S. Coffey, Ed., Elsevier, Amsterdam, 1976, pp. 231–256; G. Jones, Ed., *Quinolines,* Wiley, New York, 1977.
2. Z. H. Skraup, *Chem. Ber.* **13,** 2086 (1880); reviewed by R. H. F. Manske and M. Kulka, *Org. Reactions* **7,** 59 (1953).
3. O. Doebner and W. von Miller, *Chem. Ber.* **14,** 2812 (1881).
4. C. M. Leir, *J. Org. Chem.* **42,** 911 (1977).
5. G. M. Badger, H. P. Crocker, B. C. Ennis, J. A. Gayler, W. E. Matthews, W. G. C. Raper, E. L. Samuel, and T. M. Spotswood, *Aust. J. Chem.* **16,** 814 (1963); H. O. Jones and P. E. Evans, *J. Chem. Soc.* **99,** 334 (1911); M. G. Edwards, R. E. Garrod, and H. O. Jones, *J. Chem. Soc.* **101,** 1376 (1912).
6. L. Bradford, T. J. Elliott, and F. M. Rowe, *J. Chem. Soc.* **1947,** 437.
7. H. Rapoport and A. D. Batcho, *J. Org. Chem.* **28,** 1753 (1963); W. W. Paudler and T. J. Kress, *J. Heterocycl. Chem.* **2,** 393 (1965); *J. Org. Chem.* **23,** 832 (1967), and refs. cited therein.
8. M. H. Palmer, *J. Chem. Soc.* **1962,** 3645.
9. A. Combes, *Bull. Soc. Chim. Fr.* **49,** 89 (1888).
10. S. Tamura and E. Yabe, *Chem. Pharm. Bull. Tokyo* **21,** 2105 (1973).
11. M. Conrad and L. Limpach, *Chem. Ber.* **21,** 523 (1888).
12. L. Knorr, *Ann. Chem.* **236,** 69 (1886).
13. J. C. Jutz and R. M. Wagner, *Angew. Chem. Int. Ed.* **11,** 315 (1972).
14. S. Nakano, *Yakugaku Zasshi* **80,** 1510 (1960); **82,** 492 (1962).
15. L. Rügheimer and C. G. Schramm, *Chem. Ber.* **20,** 1235 (1887).
16. O. Meth-Cohn, B. Narine, and B. Tarnowski, *Tetrahedron Lett.* **1979,** 3111.
17. P. Friedlander, *Chem. Ber.* **15,** 2572 (1882).
18. W. Ried, A. Berg, and G. Schmidt, *Chem. Ber.* **85,** 204 (1952).
19. W. Pfitzinger, *J. Prakt. Chem.* **33,** 100 (1886).
20. L. Chiozza, *Ann. Chem.* **83,** 117 (1852).

21. P. deMayo, L. J. Sydnes, and G. Wenska, *J. Chem. Soc. Chem. Commun.* **1979,** 499.

22. (a) E. M. Burgess and L. McCullagh, *J. Am. Chem. Soc.* **88,** 1580 (1966); (b) Y. -l. Mao and V. Boekelheide, *J. Org. Chem.* **45,** 1547 (1980).

23. J. C. Jutz, *Topics Org. Chem.* **73,** 125 (1978).

24. M. Yoshida, H. Sugihara, S. Tsushima, and T. Miki, *Chem. Commun.* **1969,** 1223; M. Yoshida, H. Sugihara, S. Tsushima, E. Mizuta, and T. Miki, *Yakugaku Zasshi* **94,** 304 (1974); B. I. Rosen and W. P. Weber, *J. Org. Chem.* **42,** 47 (1977).

25. A. Marsili, *Ann. Chim. (Rome)* **52,** 3 (1962); *Tetrahedron Lett.* **1963,** 1143.

26. A. Marsili, *Tetrahedron* **24,** 4981 (1968).

27. R. E. Pratt, W. J. Welstead, Jr., and R. E. Lutz, *J. Heterocycl. Chem.* **7,** 1051 (1970).

28. M. Nakazaki, *Nippon Kagaku Zasshi* **76,** 1169 (1955).

29. S. Kwon, Y. Nishimura, M. Ikeda, and Y. Tamura, *Synthesis* **1976,** 249.

30. (a) M. P. Cava and L. Bravo, *Tetrahedron Lett.* **1970,** 4631; (b) R. Hayes, O. Meth-Cohn, and B. Tarnowski, *J. Chem. Res.* **1980,** 414.

31. S. Uyeo, T. Shingu, and H. Harda, *Yakugaku Zasshi* **85,** 314 (1965).

32. C. G. Hughes and A. H. Rees, *Chem. Ind. (London)* **1971,** 1439.

33. L. A. Paquette, G. R. Krow, and J. R. Malpass, *J. Am. Chem. Soc.* **91,** 5522 (1969).

34. K. W. Bentley, *The Isoquinoline Alkaloids,* Pergamon Press, New York, 1965; M. Shamma, *The Isoquinoline Alkaloids,* Academic Press, New York, 1972; M. Shamma and J. L. Moniot, *Isoquinoline Alkaloids Research 1972–1977,* Plenum Press, New York, 1978; *Isoquinolines,* Vol. 38, G. Grethe, Ed., Wiley-Interscience, 1981.

35. W. M. Whaley and T. R. Govindachari, *Org. Reactions,* **6,** 74 (1951).

36. P. E. Eaton, G. R. Carlson, and J. T. Lee, *J. Org. Chem.* **38,** 4071 (1973).

37. G. Fodor, J. Gal, and B. A. Phillips, *Angew. Chem. Int. Ed.* **11,** 919 (1972).

38. L. Simon and G. Talpas, *Pharmazie* **29,** 314 (1974).

39. A. Pictet and A. Gams, *Chem. Ber.* **42,** 2943 (1909).

40. A. Pictet and T. Spengler, *Chem. Ber.* **44,** 2030 (1911).

41. G. J. Kapadia, G. S. Rao, M. H. Hussain, and B. K. Chowdhury, *J. Heterocycl. Chem.* **10,** 135 (1973).

42. C. K. Bradsher and T. G. Wallis, *Tetrahedron Lett.* **1972,** 3149.

43. (a) J. M. Bobbitt, *Adv. Heterocycl. Chem.* **15,** 99 (1973); W. J. Gensler, *Org. Reactions* **6,** 191 (1951); (b) M. Ikeda, K. Hirao, Y. Okuno, and O. Yonemitsu, *Tetrahedron Lett.* **1974,** 1181.

44. G. N. Dorofeenko, S. V. Krivun, and V. G. Korobkova, *Khim. Geterotsikl. Soedin.* **1973,** 1458.

45. B. S. Thyagarajan, *Chem. Rev.* **54,** 1019 (1954); *Adv. Heterocycl. Chem.* **5,** 291 (1965); W. L. Mosby, *Heterocyclic Systems with Bridgehead Nitrogen Atoms,* Part Two, A. Weissberger, Ed., Interscience, New York, 1961, pp. 1001–1044.

46. N. J. Leonard, in *The Alkaloids: Chemistry and Physiology,* Vol. 3, R. F. Manske and H. L. Holmes, Eds., Academic Press, New York, 1953, p. 120; T. A. Henry, *The Plant Alkaloids,* 4th ed., Blakiston, Philadelphia, 1949, pp. 116–153 and 661–672.

47. V. Boekelheide and W. G. Gall, *J. Am. Chem. Soc.* **76,** 1832 (1954).

48. E. Glover and G. Jones, *Chem. Ind.* **1956,** 1456.

49. O. Westphal, K. Jann, and W. Heffe, *Arch. Pharm.* **294,** 37 (1961).

50. O. Westphal and G. Feix, *Angew. Chem. Int. Ed.* **2,** 96 (1963).

51. O. Diels and H. Pistor, *Ann. Chem.* **530,** 87 (1937).

52. V. Boekelheide and A. Sieg, *J. Org. Chem.* **19,** 587 (1954); S. Sugasawa and N. Sugimoto, *Chem. Ber.* **72,** 977 (1939).

53. G. M. Singerman, in *Condensed Pyridazines including Cinnolines and Phthalazines,* R. N. Castle, Ed., Wiley, New York, 1973, pp. 1–322.

54. J. M. Bruce, *J. Chem. Soc.* **1959,** 2366; L. S. Besford and J. M. Bruce, *J. Chem. Soc.* **1964,** 4037.

55. W. Borsche and A. Herbert, *Ann. Chem.* **546,** 293 (1952).

56. D. W. Ockender and K. Schofield, *J. Chem. Soc.* **1953,** 3706.

57. V. von Richter, *Chem. Ber.* **16,** 677 (1883); reviewed in J. C. E. Simpson, in *The Chemistry of Heterocyclic Compounds,* Vol. 5, A. Weissberger, Ed., Interscience, New York, 1953; T. L. Jacobs, in *Heterocyclic Compounds,* Vol. 6, R. C. Elderfield, Ed., Wiley, New York, 1957, p. 136.

58. K. Schofield and T. Swain, *J. Chem. Soc.* **1949,** 2393.

59. E. J. Alford and K. Schofield, *J. Chem. Soc.* **1952,** 2102.

60. H. J. Barber, K. Washbourn, W. R. Wragg, and E. Lunt, *J. Chem. Soc.* **1961,** 2828.

61. H. E. Baumgarten, P. L. Creger, and R. L. Zey, *J. Am. Chem. Soc.* **82,** 3977 (1960).

62. H. E. Baumgarten, W. F. Whittman, and G. J. Lehmann, *J. Heterocycl. Chem.* **6,** 333 (1969).

63. W. E. Noland and D. A. Jones, *J. Org. Chem.* **27,** 341 (1962).

64. R. E. Buntrock and E. C. Taylor, *Chem. Rev.* **68,** 209 (1968).

65. (a) G. M. Badger, R. J. Drewer, and G. E. Lewis, *Aust. J. Chem.* **19,** 643 (1966); (b) J. W. Barton, in *Advances in Heterocyclic Chemistry,* Vol. 24, A. R. Katritzky and A. J. Boulton, Eds., Academic Press, New York, 1979, Chap. 4, pp. 152–185.

66. S. Gabriel and G. Pinkus, *Chem. Ber.* **26,** 2210 (1893).

67. French Pat. 1,438,827 (1966) [*Chem. Abstr.* **66,** 95069 (1967)].

68. A. Hirsch and D. Orphanos, *J. Heterocycl. Chem.* **2,** 206 (1965).

69. N. R. Patel, in *Condensed Pyridazines including Cinnolines and Phthalazines,* Vol. 27, R. N. Castle, Ed., Wiley, New York, 1973, pp. 323–760.

70. J. Sauer and G. Heinrichs, *Tetrahedron Lett.* **1966,** 4979.

71. S. K. Robev, *Tetrahedron Lett.* **22,** 345 (1981).

72. W. L. F. Armarego, *Adv. Heterocycl. Chem.* **1,** 253 (1963); **24,** 1 (1979); in *Fused Pyrimidines,* Part 1—*Quinazolines,* D. J. Brown, Ed., Wiley, New York, 1967.

73. A. Bischler and D. Barad, *Chem. Ber.* **25,** 3080 (1892).

74. G. S. Sidhu, G. Thyagarajan, and N. Rao, *Indian J. Chem.* **1,** 346 (1963).

75. W. F. Beech, *J. Chem. Soc.* **1954,** 1297.

76. M. N. Deshpoude and S. Seshadri, *Indian J. Chem.* **11,** 538 (1973).

77. W. L. F. Armarego, *J. Appl. Chem.* **11,** 70 (1961).

78. E. C. Taylor, R. J. Knopf, and A. L. Borror, *J. Am. Chem. Soc.* **82,** 3152 (1960).

79. M. T. Bogert and W. F. Hand, *J. Am. Chem. Soc.* **28,** 94 (1906).

80. V. I. Dulenko, N. N. Alekseev, V. M. Golyak, and Yu. A. Nikolyukin, *Khim. Geterotsikl. Soedin.* **1976,** 1286.

81. H. Meerwein, P. Laasch, R. Mersch, and J. Nentwig, *Chem. Ber.* **89,** 224 (1956).

82. Tröger, *J. Prakt. Chem.* **36,** 225 (1887); M. A. Spielman, *J. Am. Chem. Soc.* **57,** 583 (1935).

83. V. Prelog and P. Wieland, *Helv. Chim. Acta* **27,** 1127 (1944).

84. G. W. H. Cheeseman and R. F. Cookson, in *Condensed Pyrazines,* A. Weissberger and E. C. Taylor, Eds., Wiley, New York, 1979, pp. 1–290.

85. R. G. Jones and K. G. McLaughlin, *Org. Syn.* **30,** 86 (1950); A. Zmujdzin, Pol. Pat. 69,644 (1974) [*Chem. Abstr.* **81,** 77966Z (1974)].

86. W. J. Irwin and D. G. Wibberley, *Adv. Heterocycl. Chem.* **10,** 149 (1969).

87. E. C. Taylor and A. L. Borror, *J. Org. Chem.* **26,** 4967 (1961).
88. D. M. Mulvey, S. G. Cottis, and H. Tieckelmann, *J. Org. Chem.* **29,** 2903 (1964).
89. A. Albert and A. Hampton, *J. Chem. Soc.* **1954,** 505.
90. B. Lythgoe, A. R. Todd, and A. Topham, *J. Chem. Soc.* **1944,** 315.
91. D. Barton and W. D. Ollis, *Comprehensive Organic Chemistry,* Vol. 4, P. G. Sammes, Ed., Pergamon Press, New York, 1979.
92. D. T. Hurst, *An Introduction to the Chemistry and Biochemistry of Pyrimidines, Purines, and Pteridines,* Wiley, New York, 1980, pp. 86–103; W. Pfleiderer, *Chemistry and Biology of Pteridines,* de Gryuter, New York, 1975; A. Albert, *Quart. Rev.* **6,** 197 (1952); *Fortschr. Chem. Org. Naturst.* **11,** 350 (1954).
93. O. Isay, *Chem. Ber.* **39,** 250 (1906).
94. G. B. Elion and G. H. Hitchings, *J. Am. Chem. Soc.* **69,** 2553 (1947).
95. G. B. Elion, G. H. Hitchings, and P. B. Russell, *J. Am. Chem. Soc.* **72,** 78 (1950).
96. R. G. W. Spickett and G. M. Timmis, *J. Chem. Soc.* **1954,** 2887.
97. R. M. Acheson, *An Introduction to the Chemistry of Heterocyclic Compounds,* 2nd ed., Wiley, New York, Chap. 7, pp. 366–367.
98. W. L. F. Armarego, *J. Chem. Soc.* **1963,** 4304.
99. Brit. Pat. 1,145,730 [*Chem. Abstr.* **71,** 30505 (1969)]; Brit. Pat. 1,159,412 [*Chem. Abstr.* **71,** 91535 (1969)]; Fr. M. 6,196 [*Chem. Abstr.* **72,** 12767 (1970)].
100. W. L. Mosby, *J. Chem. Soc.* **1957,** 3997, and refs. cited therein.
101. H. Keller and H. Halban, *Helv. Chim. Acta* **27,** 1253 (1944).
102. G. Adembri, F. DeSio, R. Nesi, and M. Scotton, *Chem. Commun.* **1967,** 1006.
103. M. Tisler and B. Stanovnik, in *Condensed Pyridazines Including Cinnolines and Phthalazines,* R. N. Castle, Ed., Wiley, New York, 1973, Chap. 3, pp. 761–1056.

Chapter 16

AMINO, OXO, AND THIO SIX-MEMBERED π-DEFICIENT HETEROCYCLES

16.1. GENERAL COMMENTS

As is the case for π-excessive heteroaromatics that are functionalized with amino or hydroxyl groups, the possibility of tautomerism (**1** ⇄ **2**; **3** ⇄ **4**) also exist for π-deficient systems.[1,2] Tautomeric behavior of this type is not possible for the corresponding 3-substituted isomers since the substituents are neither "ortho" nor "para" to the heteroatom. However, the possible existence of zwitterionic structures (**5**) must be considered.

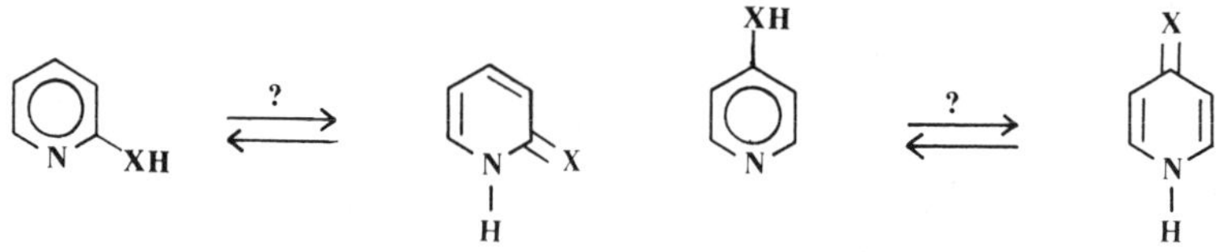
XH
?
X
H
XH
?
X
H

1 2 3 4

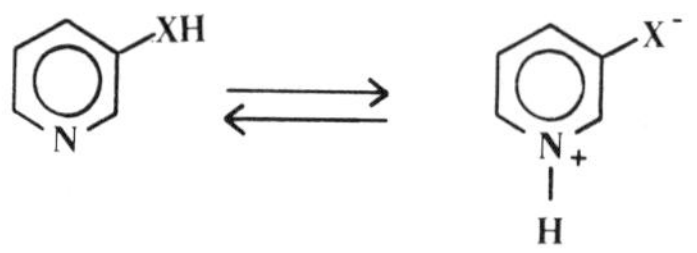
XH
X⁻
H

5

16.2. AMINOPYRIDINES AND AMINODIAZINES

The three possible monoaminopyridines can generally be obtained from the appropriate halopyridines by treatment with ammonia, or from the corresponding carboxamido derivatives via a Hofmann reaction. Based upon the availability of starting materials and yields, the following reactions are preferred for the syntheses of 2-, 3-, and 4-aminopyridine (**6, 7, 8,** respectively).[1-3]

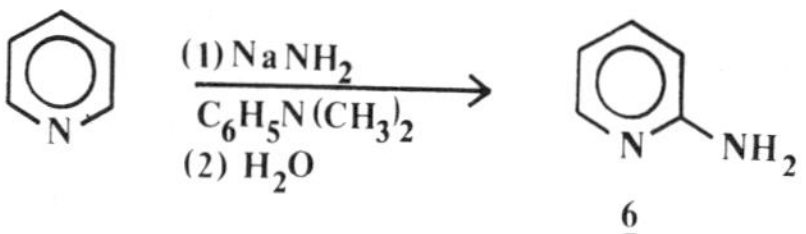
(1) NaNH2
C6H5N(CH3)2
(2) H2O
NH2

6

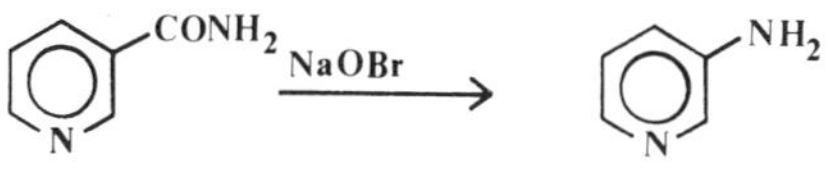
CONH2
NaOBr
NH2

7

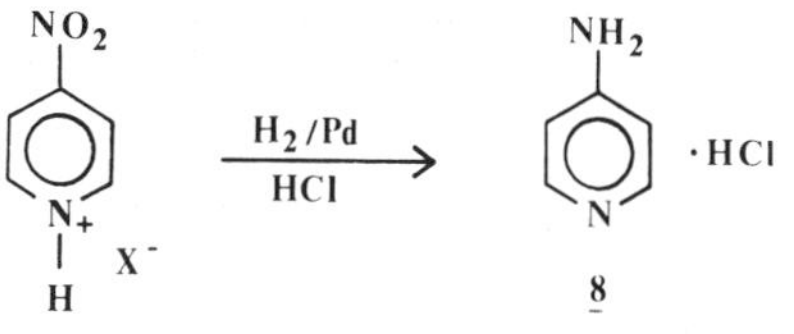
NO2
X⁻
H
H2/Pd
HCl
NH2
·HCl

8

A comparison of the ultraviolet spectra of the pyridine derivatives **9** and **10** with the spectrum of 2-aminopyridine (**6**) shows its identity with **6** and **9**. Thus

2-aminopyridine exists in the amino rather than the imino form. A similar comparison of the ultraviolet spectra of **11** and **12** with that of 4-aminopyridine (**8**) establishes, by the identity of the spectra of **11** and 4-aminopyridine, that 4-aminopyridine exists as such rather than as its imino tautomer.[2] Similar conclusions have been drawn from studies of the acid dissociation constants of 2- and 4-aminopyridine.[1]

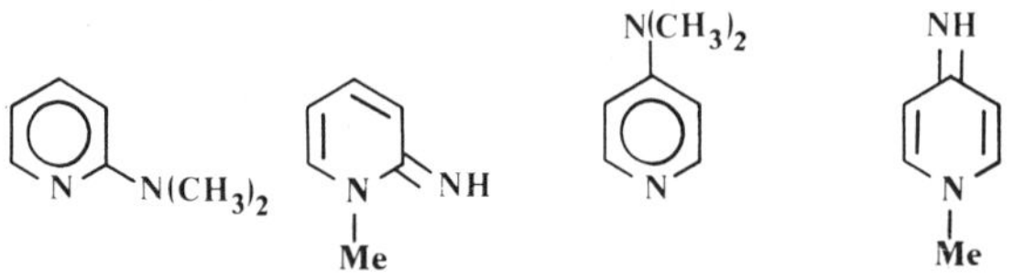

$N(CH_3)_2$ NH N $N(CH_3)_2$ N NH Me N N Me

9 10 11 12

Since the dipole moments of 2- and 4-aminopyridine are appreciably greater than the values calculated from an appropriate combination of the pyridine and aniline dipole moments,[1] resonance-contributing structures **13** and **14** must make a significant contribution to the ground-state of these aminopyridines.

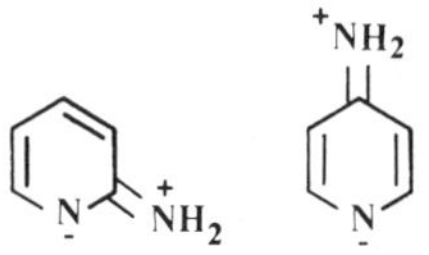

$^{+}NH_2$ N $\overset{+}{N}H_2$ N

13 14

The stabilization of ring-protonated 2- and 4-aminopyridines, gained by the presence of additional resonance structures (**15** and **16**, respectively), accounts for their high basicities. 3-Aminopyridine cannot be stabilized by this type of resonance and is thus a considerably weaker base than the other two isomers.

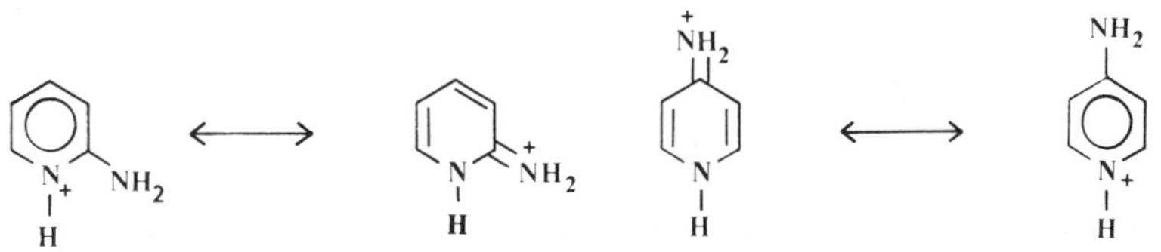

$\overset{+}{N}H_2$ NH_2 N_+ NH_2 ⟷ N $\overset{+}{N}H_2$ N ⟷ N_+ H H H H

15 16

A significant chemical difference between the 2- and 4-aminopyridines and the 3-amino isomer is exemplified by the behavior of the corresponding diazonium salts. Unlike diazonium salt **18,** compounds **17** and **19** are very reactive and cannot be trapped with β-naphthol. Instead, they form the corresponding oxo derivatives **20** and **21,** respectively.

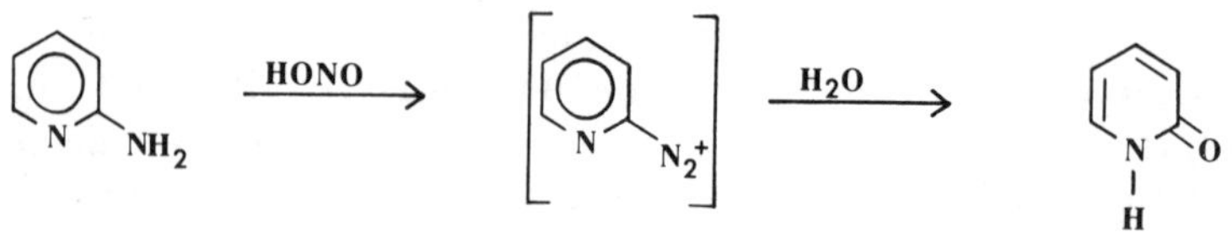

HONO → [17] H_2O → 20

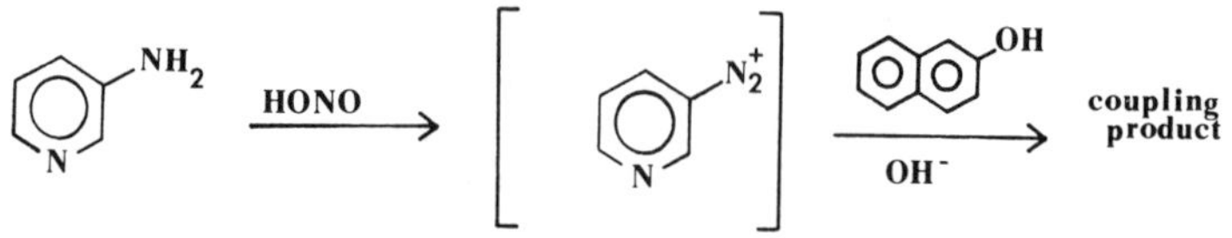

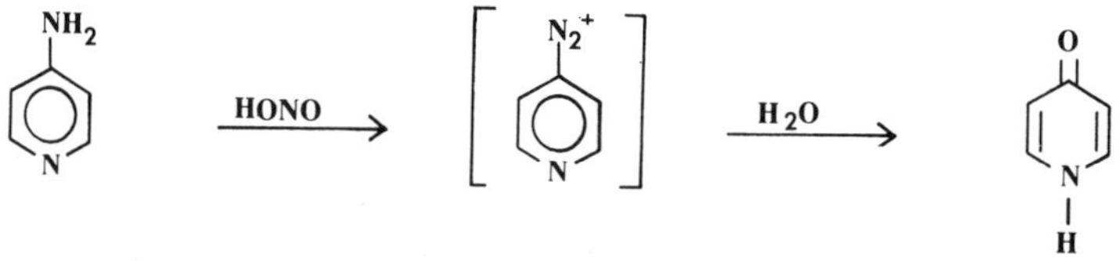

HONO → [18] OH⁻ → coupling product

HONO → [19] H_2O → 21

The sodium diazotate obtained by reaction of 2-aminopyridine with boiling sodium ethoxide and amyl nitrite is sufficiently stable to couple with 2-naphthol. Thus it appears that the reactivity of **17** and **19** is due to the actual existence of these compounds in their protonated forms **22** and **23,** respectively, in the acidic diazotization medium.[4]

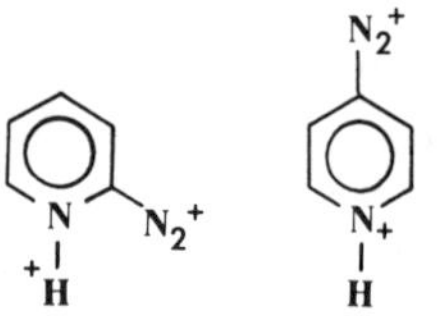

22 23

Interestingly, the 2- and 4-aminopyridine *N*-oxides can be diazotized to form stable diazonium salts (**24** and **25**) which react with β-naphthol, as expected.[5] This increased stability may well be accounted for by the contribution of structures **24a** and **25a** to the ground state of these diazonium salts.

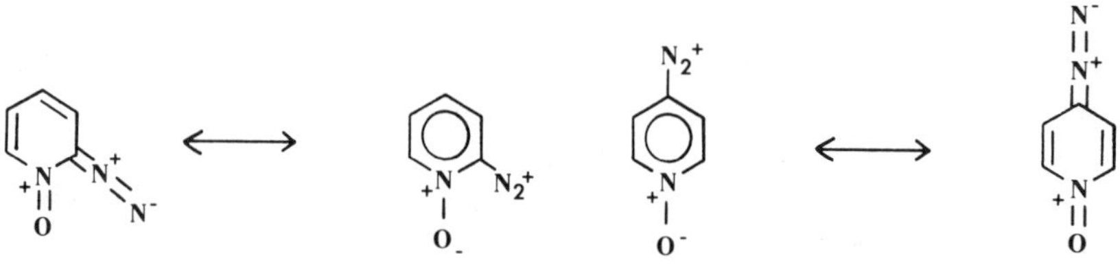

24a ⟷ 24 25 ⟷ 25a

In the pyrimidine series, 2-aminopyrimidines are directly available by cyclization of compounds derived from guanidine. The 4- and 6-aminopyrimidines are obtained from nitriles by the syntheses described in Chapter 14.[6–8] Finally, 5-aminopyrimidines (**26**) can be prepared by the following sequence of reactions:

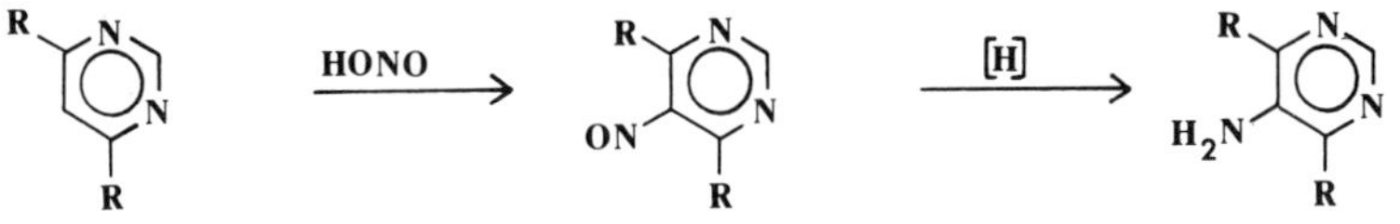

HONO → [H] → **26**

Aminopyrazine (**27**) is most readily generated from fluoropyrazine (**28**) on stirring with ammonia.[10] The other aminodiazines are synthesized by either direct nucleophilic substitution or Hofmann degradation of the corresponding amide by hypohalite.[11,12]

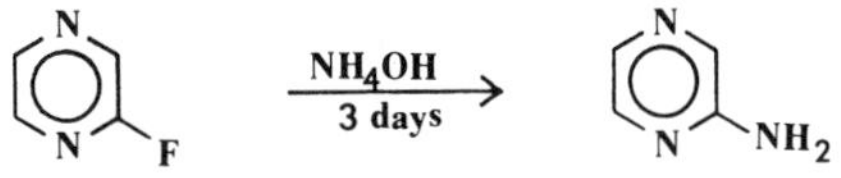

28 $\xrightarrow{NH_4OH,\ 3\ days}$ **27** (70%)

$\xrightarrow{NH_3,\ 100°C/6\ hrs.}$ $\xrightarrow{[H]}$

H_2NOC $\xrightarrow{NaOBr}$ H_2N

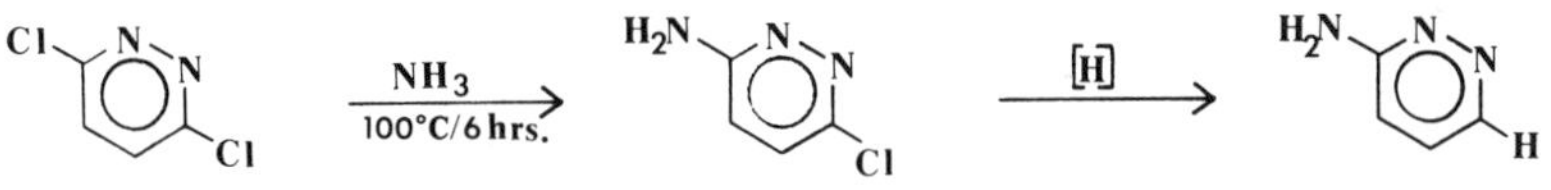

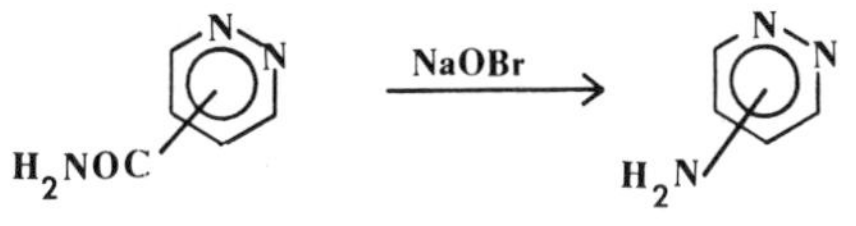

All these amino heteroaromatics exist in the amino rather than the imino forms,[9] and thus behave as do the aminopyridines. The diazonium salts obtained from these various amines vary considerably in their stabilities.[13] A diazo group that is either α or γ to a ring nitrogen is highly reactive, and in aqueous media is normally replaced by a "hydroxyl" group. On the other hand, when the diazonium group occupies some other position with respect to a ring nitrogen, as in **28**, it is more stable and can be coupled with β-naphthol.[7]

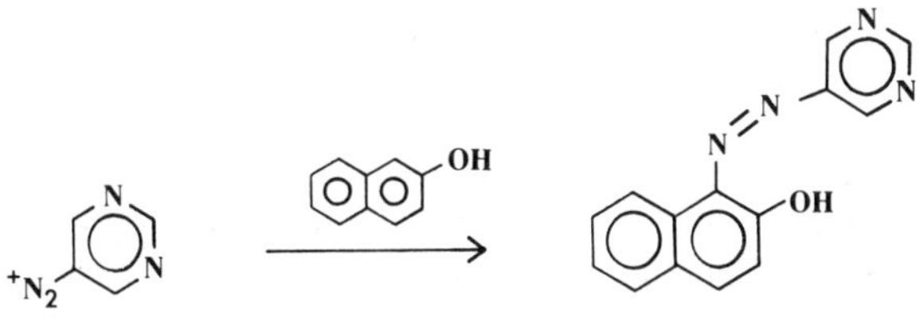

28

16.3. OXO- AND THIOPYRIDINES AND DIAZINES

The three possible oxopyridines (**20, 21, 29**) are most readily obtained by the following reactions:

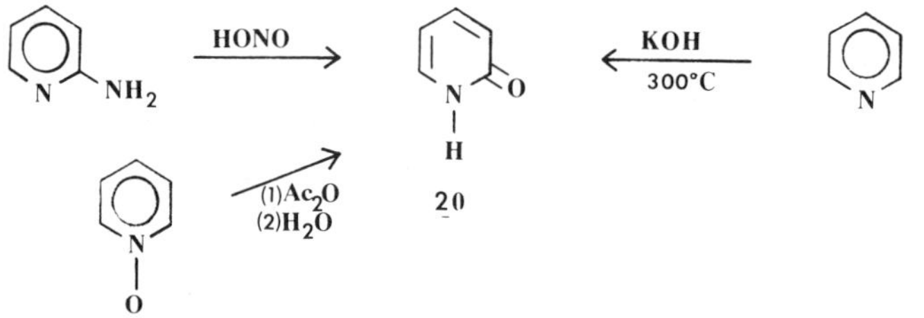

HONO; KOH, 300°C; (1)Ac_2O (2)H_2O; 20

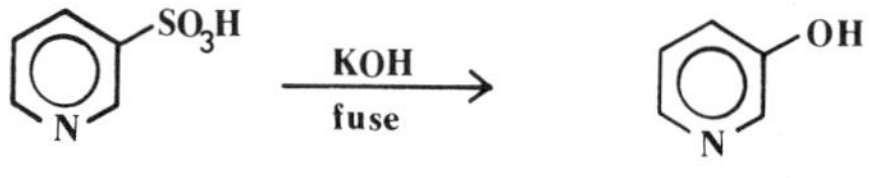

KOH fuse; 29

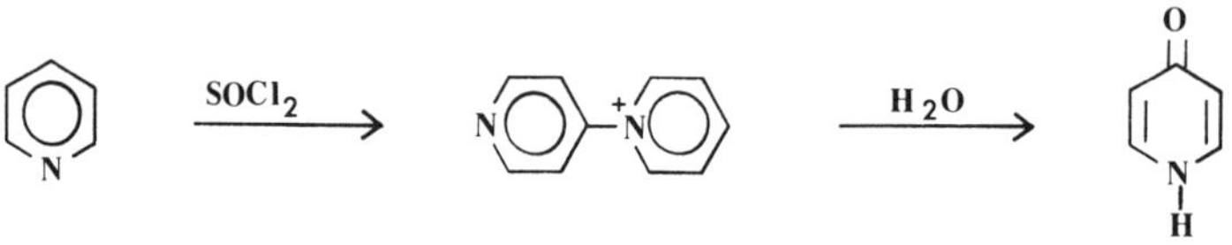

$SOCl_2$; H_2O; 21

As in the case of the aminopyridines, ultraviolet studies have been employed to examine the possible existence of tautomeric equilibria in these compounds.[9]

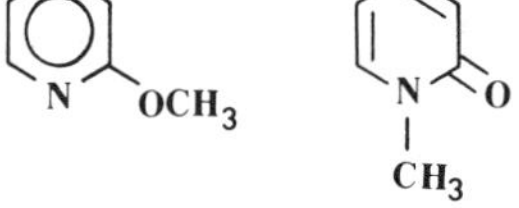

30 31

Comparison of the ultraviolet spectra of the reference compounds **30** and **31** with that of "2-hydroxypyridine" clearly establishes that, in polar solvents, the 2-oxo form (**20**) is preferred. Analogously, the 4-oxo isomer (**21**) is preferred over the 4-hydroxy form.[14] Similar conclusions are drawn for the corresponding thio compounds.[9] Significantly, ultraviolet studies in the gas phase[15] have shown that, under these conditions, equilibria *do* exist between the hydroxy and oxo forms on the one hand and the mercapto and thio forms on the other. These results are presented by the following:

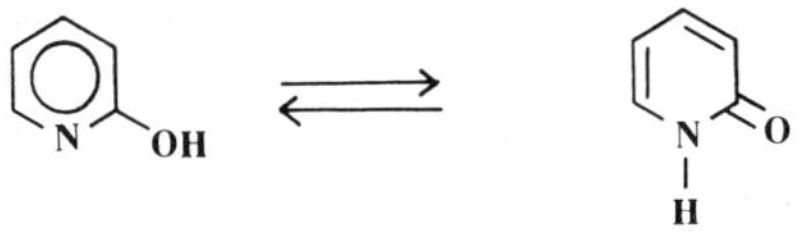

OH N ⇌ N O H

0% in solution
0% in solid state
72% in gas phase

100% in solution
100% in solid state
28% in gas phase

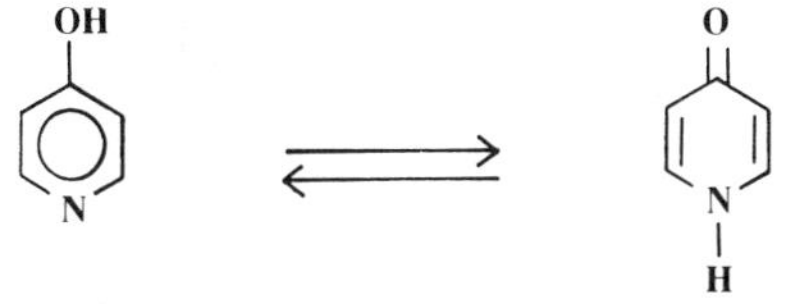

OH N ⇌ O N H

0% in solution
0% in solid state
Predominates in gas phase

100% in solution
100% in solid state
Minor tautomer in gas phase

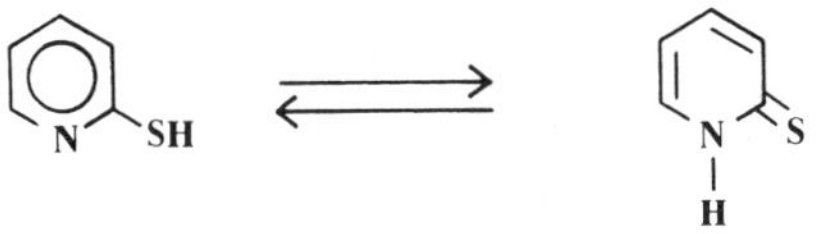

N SH ⇌ N S H

0% in solution
Predominates in gas phase

100% in solution
Minor tautomer in gas phase

One of the few examples of an α-hydroxypyridine that predominates in solution over the oxo form is found in the chloro compound **32**.[14]

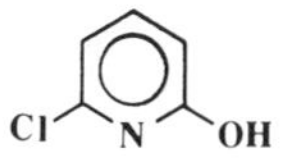

Cl N OH ⇌ Cl N O H

32

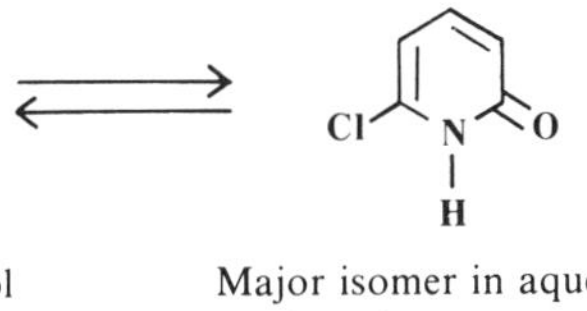

Major isomer in ethanol
Major isomer in gas phase

Major isomer in aqueous ethanol

Conversion of α- and γ-pyridinone to the *N*-oxide (**33** and **34,** respectively) does not alter the solution tautomeric behavior of these substances. They continued to exist in the oxo forms.[7,9] The corresponding sulfur analogues (**35** and **36**) behave similarly.[9]

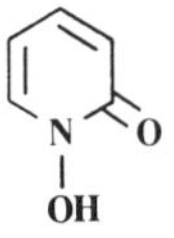

33 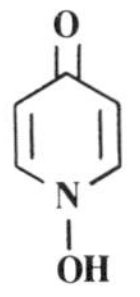 **34** 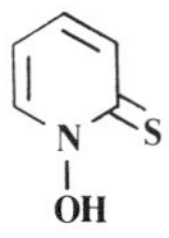 **35** 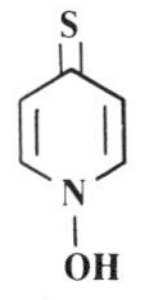 **36** 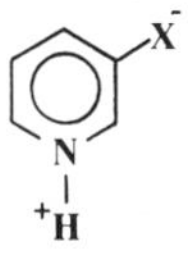 **37**

As indicated earlier, the 3-hydroxy- and 3-mercaptopyridines cannot exist as oxo and thio isomers, respectively. However, they do exist partially as betaines (**37**). 3-Hydroxypyridine behaves in much the manner expected of a phenolic compound. Thus it gives the "ferric chloride" test for phenols and reacts with formaldehyde (**29** → **38**). Depending on the reaction conditions, 3-hydroxypyridine (**29**) can be either *N*- (**39**) or *O*- (**40**) alkylated[16]:

38 ← 29

29 → 39 (CH_2N_2 or $(CH_3)_2SO_4$, OH^-)

29 → 40 (CH_2N_2, t-BuOH, −15°C)

The site of alkylation of the 2- (**20**) and 4-oxopyridines also depends on the reaction conditions. If alkali salts are the starting materials, *N*-alkylation takes place by an S_N2 mechanism. Silver salts yield *O*-alkylated products via an S_N1 type mechanism.[17]

20 → *N*-methyl-2-pyridone ($(CH_3)_2SO_4$, $NaOCH_3$, CH_3OH)

20 → *N*-methyl-2-pyridone + 2-methoxypyridine (CH_2N_2)

20 → 2-methoxypyridine ((1) Ag^+; (2) CH_3I)

The "hydroxydiazines," which generally serve as the starting materials for the syntheses of other diazines, are readily prepared[18–20]:

$OHCCH{=}CHCO_2H \xrightarrow{(H_2N-)_2}$ 3(2*H*)-pyridazinone

$H_2NOCNH_2 + ROCCH_2COR \longrightarrow$ 4,6-disubstituted 2(1*H*)-pyrimidinone

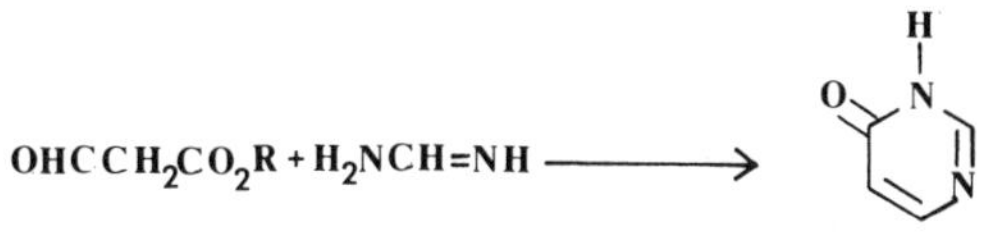

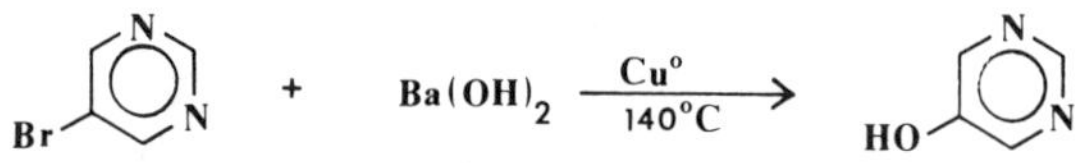

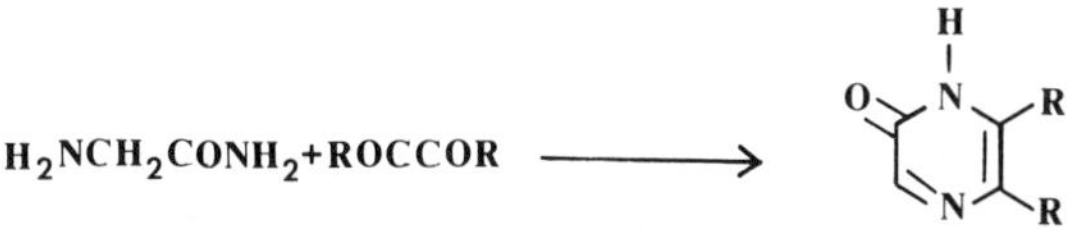

$OHCCH_2CO_2R + H_2NCH{=}NH \longrightarrow$

Br + $Ba(OH)_2 \xrightarrow[140°C]{Cu^0}$ HO

$H_2NCH_2CONH_2 + ROCCOR \longrightarrow$

In the case of "2-hydroxypyrimidine," the predominant tautomer in solution is the pyrimid-2(1*H*)-one (**41**); in the vapor phase the 2-hydroxy form (**42**) predominates. For "4-hydroxypyrimidine" (**43**), the oxo form (either **44** or **45**) is preferred both in solution and in the vapor phase.[14,15] 5-Hydroxypyrimidine (**46**), in which the hydroxyl group is neither α nor γ to a nitrogen, exists as such and the hydroxyl group is phenolic in character.

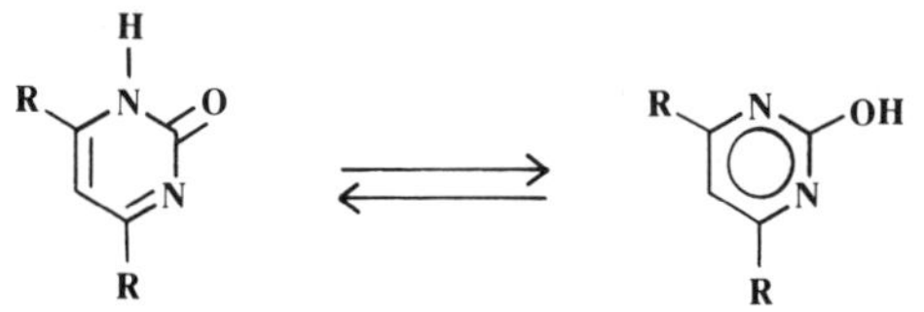

41 (R = H)
Predominates in solution

42 (R = H)
Predominates in vapor phase

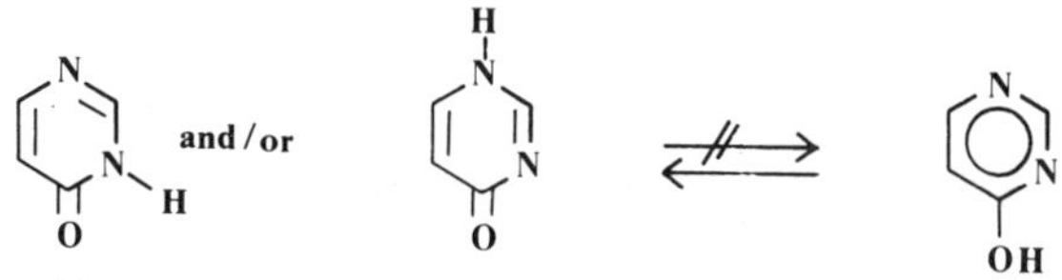

44 and/or 45 ⇄ 43

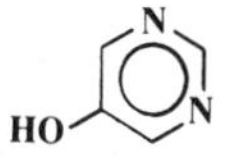

46

Pyridazines, insofar as they have been investigated,[21] exist in the oxo form (**47** and **48,** respectively).

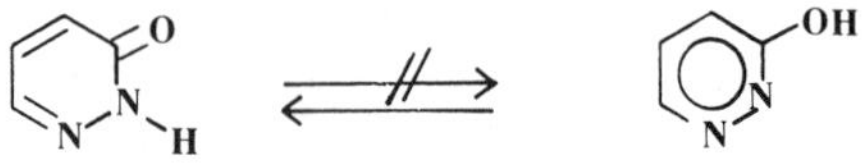

47

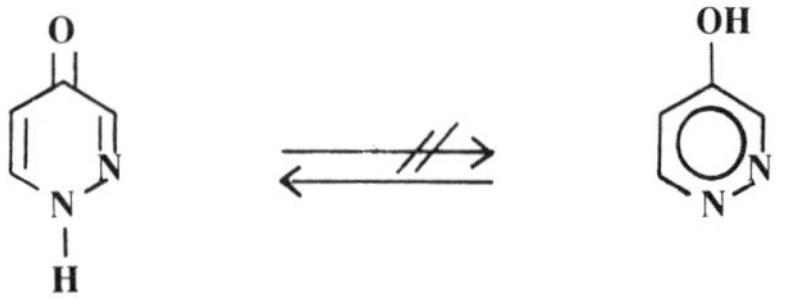

48

In the pyrazine family, the oxo isomer is also generally preferred, although substituent effects may become operative in some instances.[22] Thus 2-chloro-6-hydroxypyrazine (**49**) exists mainly as such:

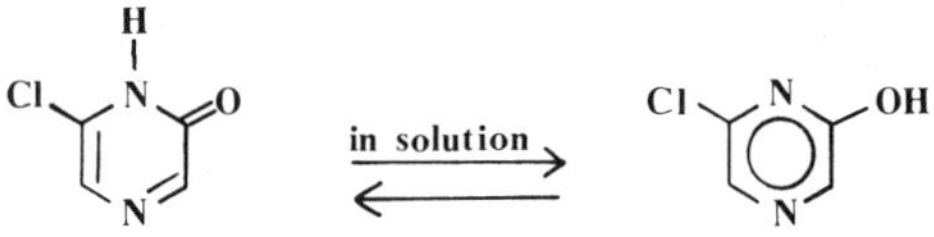

in solution

49

The few instances for which di- and trihydroxy derivatives have been examined are best summarized as follows:

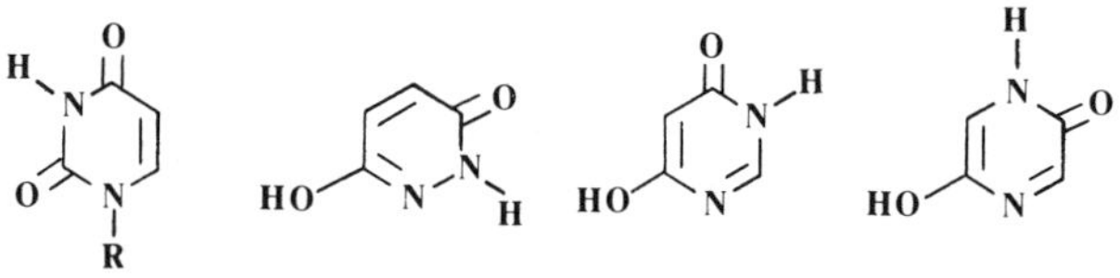

cytidine

In general, alkylations of these diazine derivatives by alkyl halides, alkaline dimethyl sulfate, or diazomethane take place on nitrogen[23]; in the case of diazomethane, *O*-alkylation occurs as well. Some of these transformations of **42** and **50** are illustrated[7]:

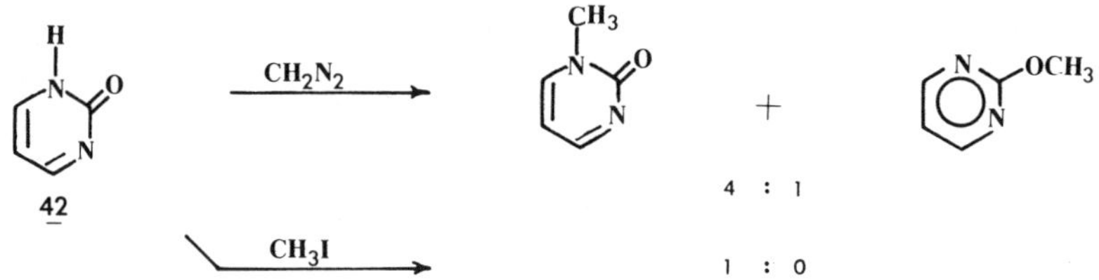

42

CH_2N_2 → 4 : 1

CH_3I → 1 : 0

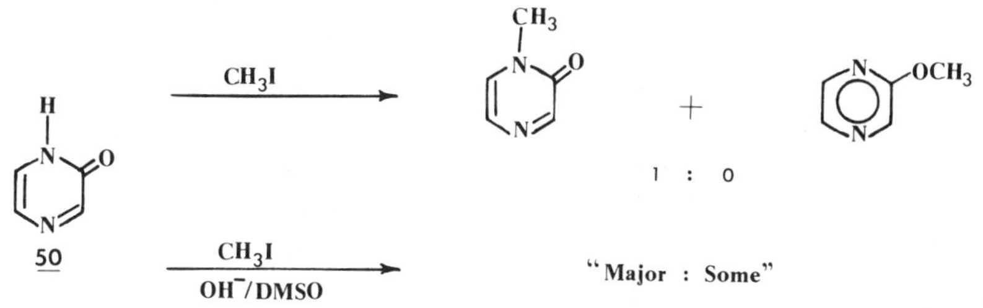

H N O N 50 — CH_3I → CH_3 N O N + N OCH_3 N

1 : 0

CH_3I / OH^-/DMSO → "Major : Some"

Since many naturally occurring pyrimidines are alkylated at N-1, a specific reaction for N-1 alkylation is of considerable importance. The following reaction sequence demonstrates an excellent path that accomplishes this transformation.[24]

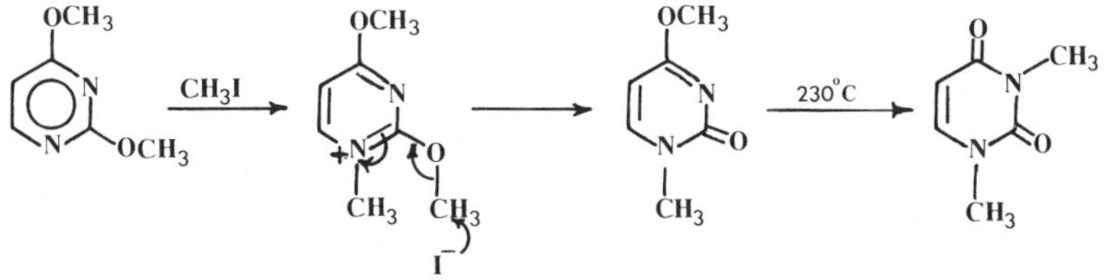

OCH_3 N N OCH_3 — CH_3I → OCH_3 N +N O CH_3 CH_3 I^- → OCH_3 N N O CH_3 — 230°C → O N CH_3 N O CH_3

16.4. *N*-OXO COMPOUNDS

16.4.1. GENERAL COMMENTS

A major group of oxo derivatives of pyridine and the diazines are the *N*-oxides (**51**). These compounds are of great synthetic utility because of the reactivity-modifying influence that the oxygen atom has on the ring systems. The presence of the oxygen atom can have, depending upon the reagent used in any particular transformation, either an electron-withdrawing (**52–54**) or an electron-donating effect (**55–57**).

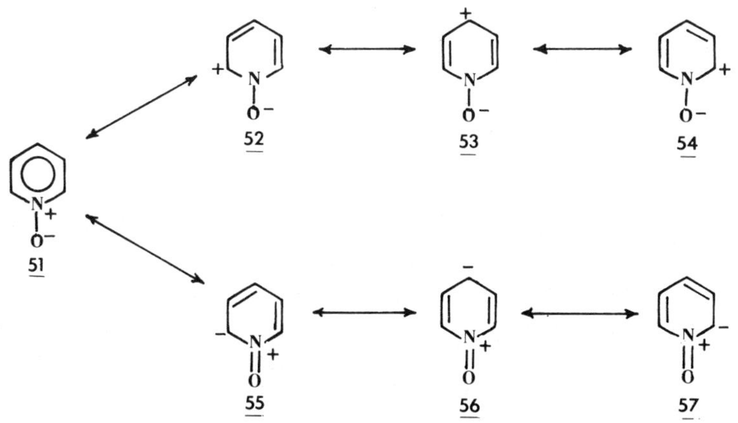

N+ O− 51

+ N O− 52 ⟷ + N O− 53 ⟷ N + O− 54

− N+ O 55 ⟷ − N+ O 56 ⟷ N+ − O 57

The fact that the dipole moment of pyridine *N*-oxide is 4.24D suggests that the electron-withdrawing resonance structures (**52–54**) are of similar importance to the electron-donating ones (**55–57**).[25] Based on these ground-state considerations, the presence of an *N*-oxide function in pyridine and the diazines should enhance nucleophilic substitution reactivities at C-2 and C-4 (**52–53**) as well as the electrophilic reactivity at the same positions (**55** and **56**).[26] This dichotomous behavior leads to some interesting reactivity patterns in *N*-oxide chemistry.

16.4.2. PYRIDINE AND DIAZINE *N*-OXIDES

N-Oxidation of pyridines and the diazines is generally accomplished by reaction with peracids. Peracetic acid is normally prepared *in situ* from 30% hydrogen peroxide in acetic acid and is the most commonly used reagent.[27] Aromatic peracids such as perbenzoic, *m*-chloroperbenzoic, and perphthalic acids are also frequently employed. Peroxytrifluoroacetic acid is occasionally the peracid of choice in difficult cases.[28] Generally, the weaker the base, the stronger the peracid that is required to cause *N*-oxidation. Some typical conversions are as follows:

CH_3CO_3H → **51**

Cl_5 CF_3CO_3H → Cl_5

$C_6H_5CO_3H$ →

58 (m-$ClC_6H_4CO_3H$) → "low yield"

59 $C_6H_5CO_3H$ → CF_3CO_3H → **60**

Pyrimidine (**58**) is the only one of these compounds that does not afford high yields by direct *N*-oxidation; substantial amounts of decomposition products are formed. Clearly, the diazines are candidates for conversion to di-*N*-oxides. Of this class of compounds, only pyrazine itself (**59**) has been directly converted to the di-*N*-oxide (**60**) by means of peroxytrifluoroacetic acid.[29] An indirect synthesis of pyrimidine di-*N*-oxide (**61**) has recently been described,[30] and pyridazine di-*N*-oxides (**62**) are afforded by a similar route.[31]

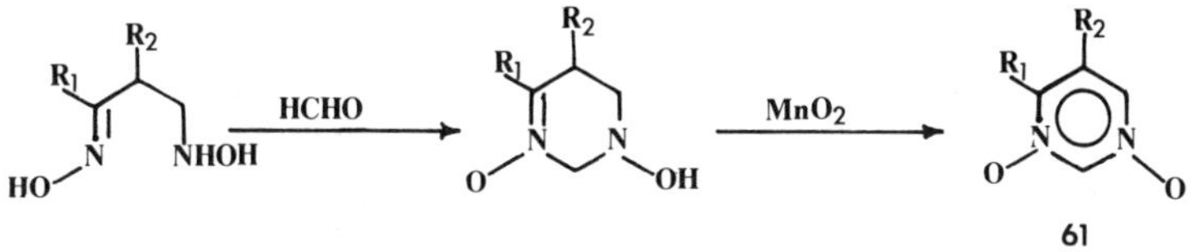

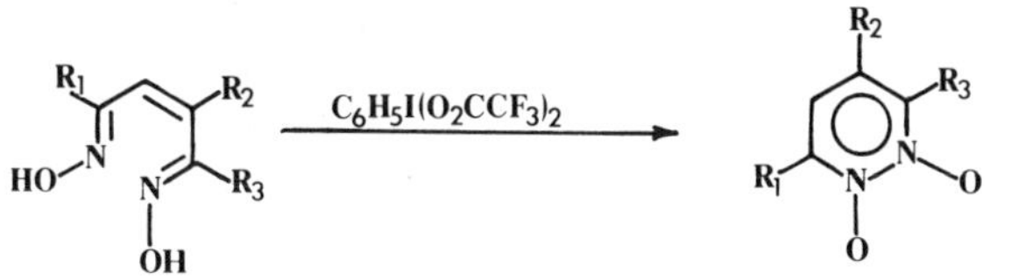

The pyridine *N*-oxides (**51**) can be deoxygenated by a variety of reducing agents,[32] such as $Pd/C/H_2$, $Fe^{\circ}/AcOH$, $NaBH_4/AlCl_3$, $PCl_3/CHCl_3$, $(C_6H_5)_3P$, and $(EtO)_3P$.

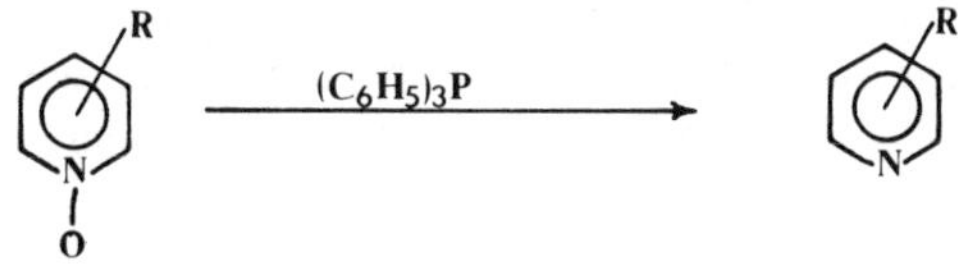

The diazine *N*-oxides, on the other hand, are considerably more stable toward deoxygenation; Pd/C and Raney Ni appear to be the reducing agents of choice in these instances. However, these agents reduce a number of other functional groups as well. Thus, depending upon reaction conditions, **64** can be debenzylated in a rapid step to form **65**. This *N*-oxide is then slowly deoxygenated with the same catalyst to afford **66**. The nitro group in **67** is reduced along with the *N*-oxide in going from **67** to **68**.[33]

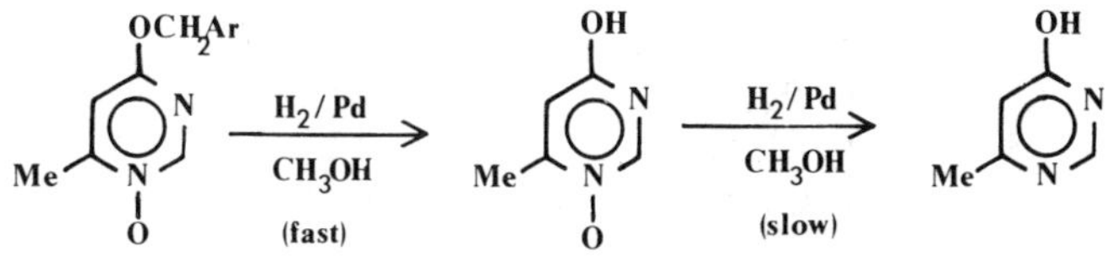

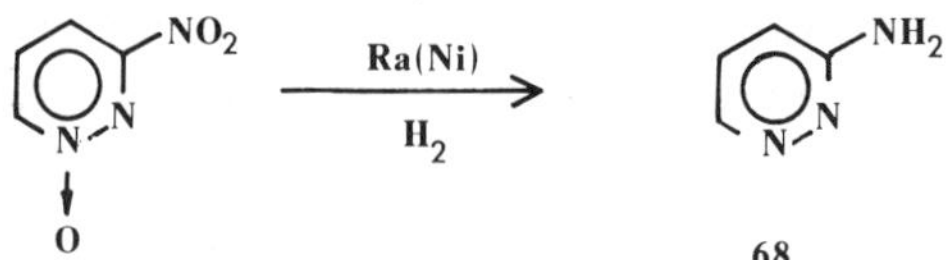

NO_2 Ra(Ni) H_2 NH_2

67 68

The ring reactivity, as modified by the presence of the *N*-oxide grouping, facilitates a number of reactions. Pyridine *N*-oxide (**51**) when treated with butyllithium forms the intermediate carbanion **69**, which reacts with a number of electrophiles.[34]

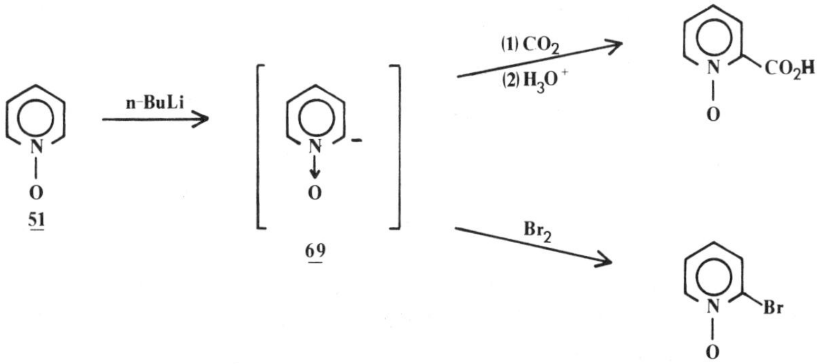

n-BuLi (1) CO_2 (2) H_3O^+ CO_2H Br_2 Br

51 69

Many of the transformations of these *N*-oxides are accompanied by concomitant loss of the *N*-oxide group. The pyridine *N*-oxides (for example, **51**) undergo many different AE_a reactions of notable synthetic utility. A number of these are shown below[35–37]:

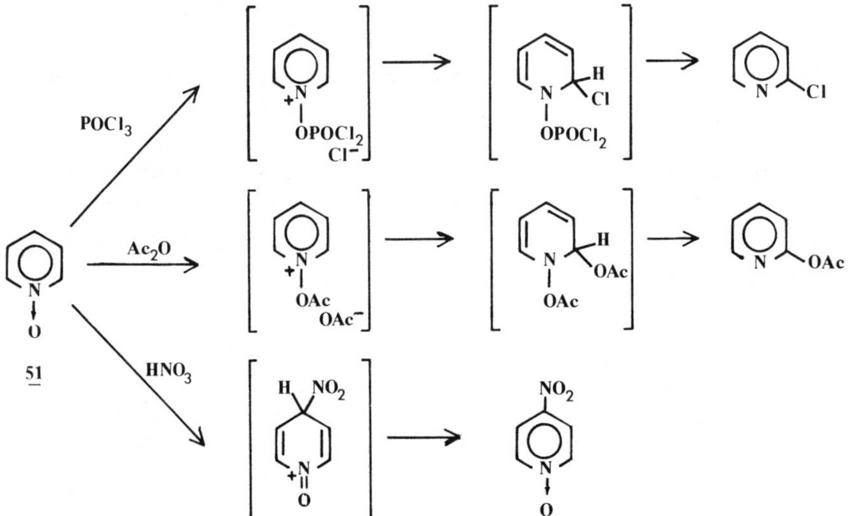

$POCl_3$ $OPOCl_2$ Cl^- H Cl $OPOCl_2$ Cl

Ac_2O OAc OAc^- H OAc OAc OAc

51 HNO_3 H NO_2 NO_2

70

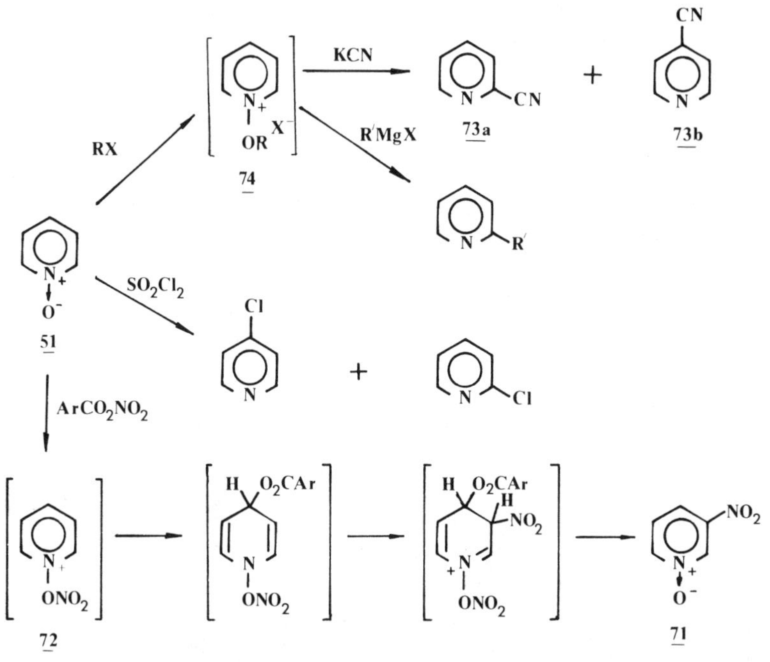

Reactions of pyridine *N*-oxide (**51**) with either $POCl_3$ or acetic anhydride are synthetically important since they permit facile introduction of reactive substituents into the 2-position. Direct nitration of **51,** on the other hand, forms the 4-nitropyridine *N*-oxide (**70**) in excellent yield. If benzoyl nitrate is the nitrating agent, the 3-nitropyridine *N*-oxide (**71**) is the sole product. It has been suggested that the 3-nitro compound is obtained in this reaction by the initial formation of an *O*-nitro intermediate (**72**), which is converted to **71** by the indicated transformations.

A mixture of 2- and 4-substituted deoxygenated products (**73**) is obtained when pyridine *N*-oxide (**51**) is initially *O*-alkylated to form **74,** then allowed to react with potassium cyanide. A mixture of 2- and 4-chloropyridines is obtained when sulfuryl chloride is the chlorinating agent. An interesting analogue of the benzoyl nitrate transformation is exemplified by treatment of **51** with bromine in the presence of acetate ion in acetic anhydride,[38] from which 3,5-dibromopyridine *N*-oxide (**75**) is obtained. This compound is presumably formed by the sequence of reactions shown below:

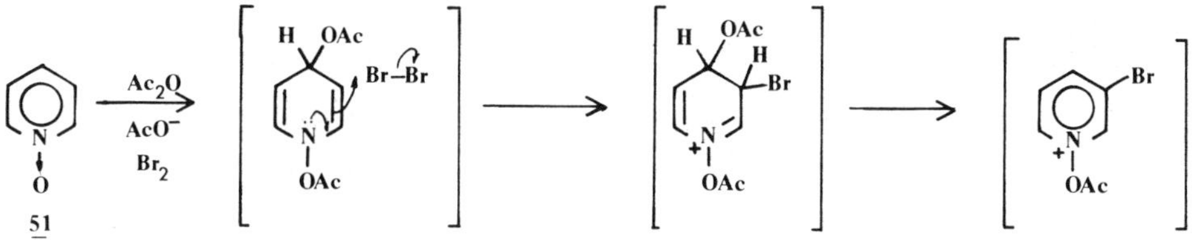

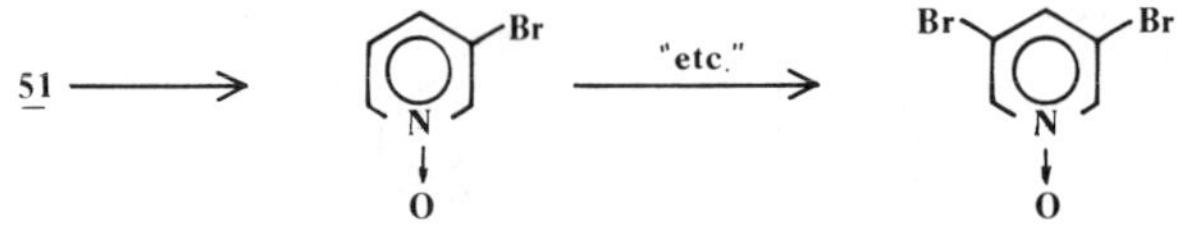

51 → "etc." → 75 (35%)

The *O*-alkylated compounds (**76**) can also undergo a "simple" deoxygenation with appropriate nucleophiles.[39]

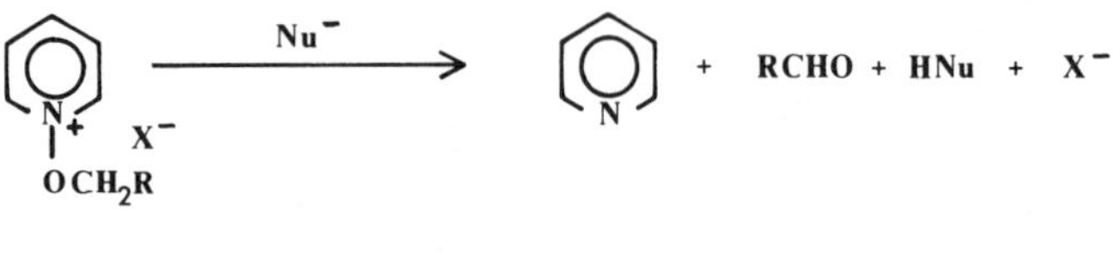

76 —Nu^-→ pyridine + RCHO + HNu + X^-

In a reaction analogous to that of pyridine *N*-oxide, the nitration of pyridazine 1-oxide (**77**) with mixed acids affords the 4-nitro derivative (**78**).[40] Although 3-methylpyridazine 1-oxide (**79**) cannot be nitrated, the 2-oxide (**80**) can, in excellent yield. Thus the reactivity of this ring system with NO_2^+ appears to be very susceptible to substituent effects. Interestingly, nitration with a mixture of silver nitrate and benzoyl chloride (benzoyl nitrate is the reactive species) affords the 3-nitro derivative (**81**). Again, this change in site of substitution follows the pyridine behavior.[40,41]

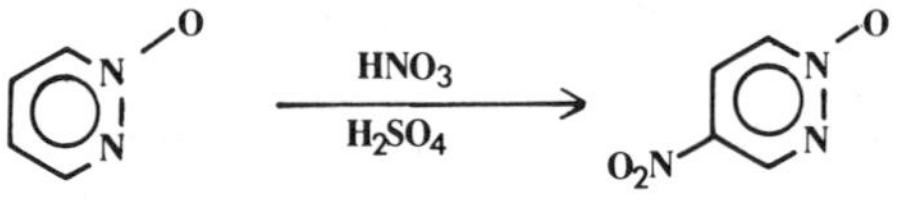

77 —HNO_3 / H_2SO_4→ 78

79 —HNO_3 / H_2SO_4→ NR

80 —HNO_3 / H_2SO_4→

77 —$AgNO_3$ / C_6H_5COCl→ [→] →

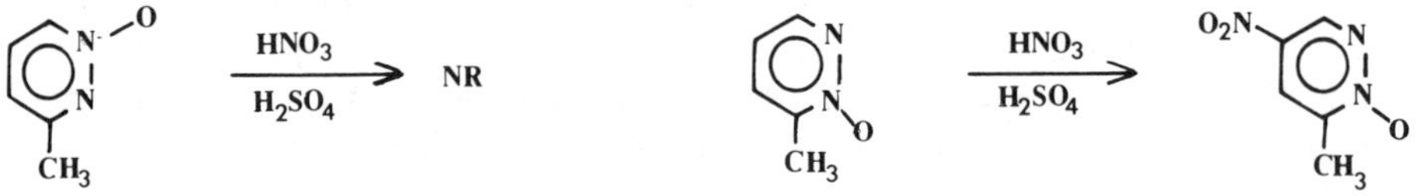

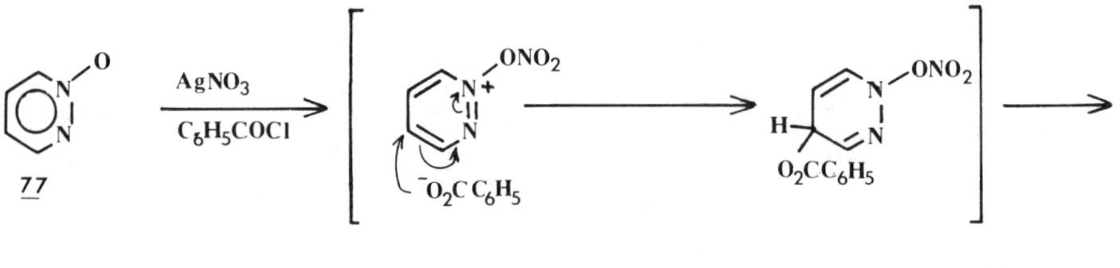

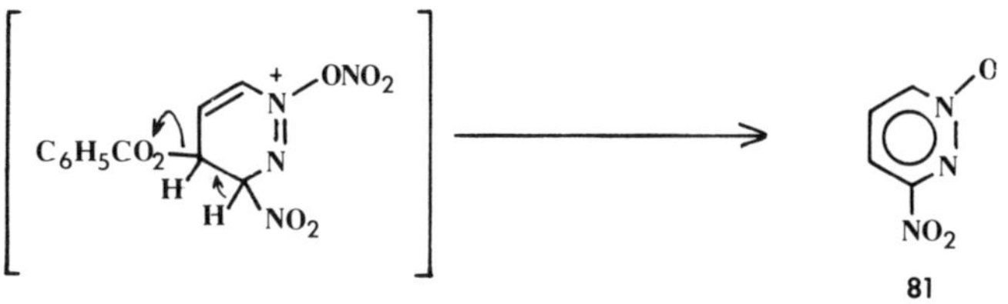

$C_6H_5CO_2$ H H $\overset{+}{N}$—ONO_2 N NO_2 ⟶ O N N NO_2

81

The nucleophilic reactions of diazine *N*-oxide have been extensively examined.[42]

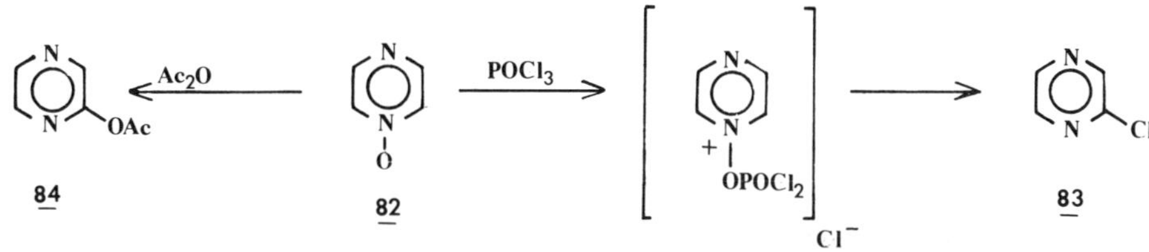

N N OAc ⟵(Ac_2O)— N N O —($POCl_3$)⟶ [N $\overset{+}{N}$ $OPOCl_2$] Cl^- ⟶ N N Cl

84 **82** **83**

N N OAc ⟵(Ac_2O)— N N O —(KCN, C_6H_5COCl, OH^-)⟶ [N $\overset{+}{N}$ $OCOC_6H_5$] CN^- ⟶ N N CN

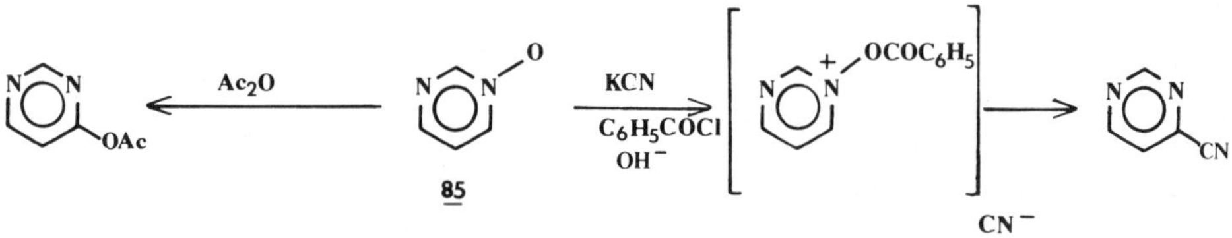

85

As is the case with the pyridine *N*-oxides, **82** with both acetic anhydride and phosphorus oxychloride affords the 2-substituted pyrazines (**83** and **84**, respectively).[43] In the case of pyrimidine *N*-oxide (**85**), acetylation occurs at C-4 (**86**) rather than at the other possible ortho position, C-2.[44] The Reissert reaction of pyrimidines, on the other hand, affords the 2-substituted products (**87**). Alkyl substitution at the α-carbon of pyridine (**88, 89**) and the diazine *N*-oxides (**90, 91**) significantly modifies the reactions of these compounds.[45–47]

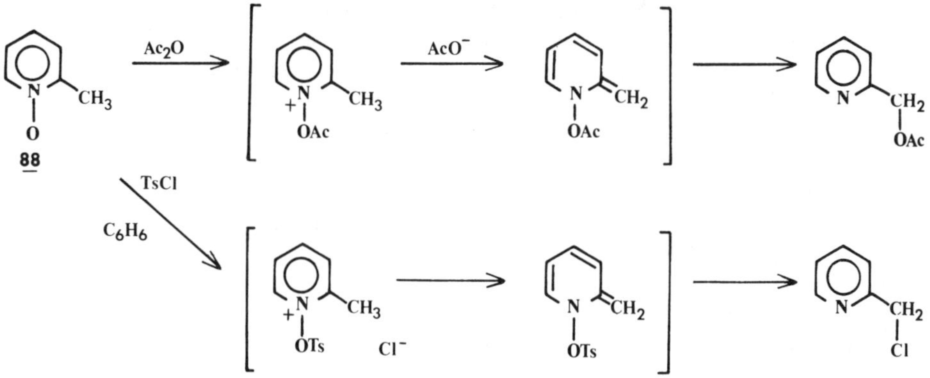

N CH_3 O —(Ac_2O)⟶ [$\overset{+}{N}$ CH_3 OAc —(AcO^-)⟶ N CH_2 OAc] ⟶ N CH_2 OAc

88

TsCl, C_6H_6 ⟶ [$\overset{+}{N}$ CH_3 OTs Cl^- ⟶ N CH_2 OTs] ⟶ N CH_2 Cl

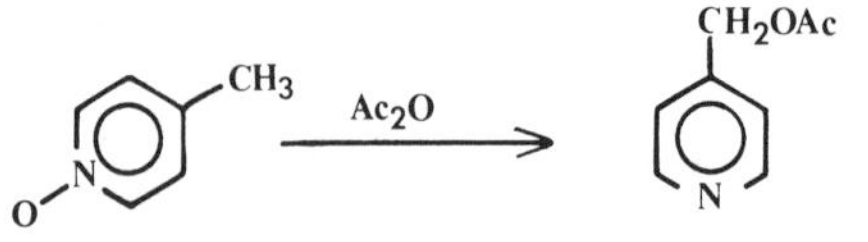

Clearly, in all of the instances illustrated, the alkyl group is more reactive toward substitution than the heterocyclic ring.

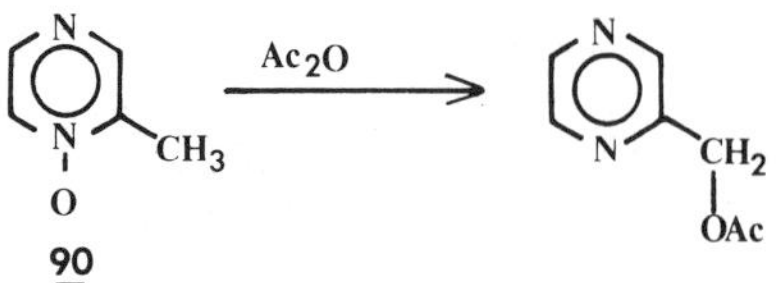

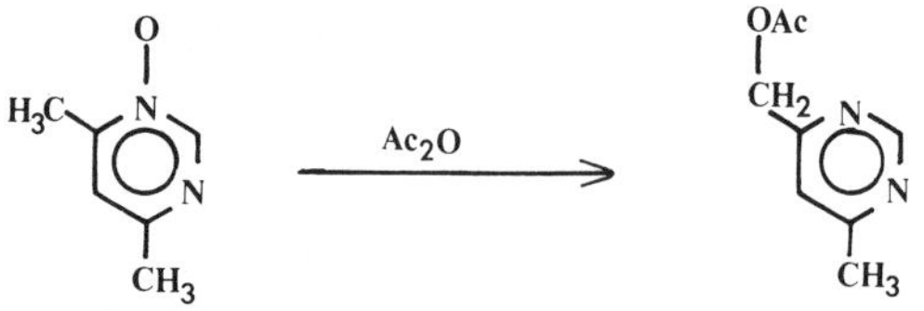

REFERENCES

1. H. Angyal, *J. Chem. Soc.* **1952,** 1461.
2. J. Anderson and I. Seeger, *J. Am. Chem. Soc.* **71,** 340 (1943).
3. R. M. Acheson, *An Introduction to the Chemistry of Heterocyclic Compounds*, Interscience, New York, 1967, p. 215.
4. A. S. Tomcutcik and L. N. Starkes, in *Pyridine and its Derivatives*, Part 3, E. Klingsberg, Ed., Interscience, New York, 1962, Chap. 9.
5. E. N. Shaw, in *Pyridine and its Derivatives*, Part 2, E. Klingsberg, Ed., Interscience, New York, 1961, Chap. 4.
6. M. Tisler and B. Stanovnik, *Adv. Heterocycl. Chem.* **9,** 211 (1968).
7. A. W. Kenner and A. Todd, in *Heterocyclic Compounds*, Vol. 6, R. C. Elderfield, Ed., Wiley, New York, 1957.
8. A. W. H. Cheeseman and E. S. G. Werstiuk, *Adv. Heterocycl. Chem.* **14,** 99 (1972).
9. A. R. Katritzky and J. M. Lagowski, *Adv. Heterocycl. Chem.* **1,** 131 (1962).
10. H. Rutner and P. E. Spoerri, *J. Heterocycl. Chem.* **3,** 435 (1966).
11. G. W. Anderson, H. E. Faith, H. W. Marson, P. S. Winnek, and R. O. Robin, *J. Am. Chem. Soc.* **64,** 2902 (1942).
12. C. M. Atkinson and R. E. Rodway, *J. Chem. Soc.* **1959,** 1.
13. R. A. Baxter, G. T. Newbold, and F. S. Spring, *J. Chem. Soc.* **1947,** 370.
14. J. Frank and A. R. Katritzky, *J. Chem. Soc. Perkin II* **1976,** 1428.
15. P. Beak, F. S. Fry, Jr., J. Lee, and F. Steele, *J. Am. Chem. Soc.* **98,** 171 (1976).
16. D. A. Prins, *Rec. Trav. Chim.* **76,** 58 (1957).
17. N. Kornblum, R. A. Smiley, R. K. Blackwood, and D. C. Iffland, *J. Am. Chem. Soc.* **77,** 6269 (1955).
18. D. T. Moury, *J. Am. Chem. Soc.* **75,** 1909 (1953).
19. P. A. Lavene and F. B. LaForge, *Chem. Ber.* **45,** 608 (1912).

20. Y. A. Tota and R. C. Elderfield, *J. Org. Chem.* **7,** 313 (1942).
21. P. Cucka, *Acta Cryst.* **16,** 318 (1963).
22. R. A. Godwin, Ph.D. Thesis, London University, 1970.
23. F. Arndt, *Angew. Chem.* **61,** 397 (1949).
24. G. E. Hilbert and T. B. Johnson, *J. Am. Chem. Soc.* **52,** 2001 (1930).
25. E. P. Linton, *J. Am. Chem. Soc.* **62,** 1945 (1940).
26. H. H. Jaffe and G. O. Doak, *J. Am. Chem. Soc.* **77,** 4441 (1955).
27. E. C. Taylor and A. J. Crovetti, *Org. Synth. Coll. Vol.* **4,** 704 (1963).
28. T. Kab, Y. Yamanaka, and T. Shibata, *J. Pharm. Soc. Japan* **87,** 1096 (1967).
29. K. W. Blake and P. G. Sammes, *J. Chem. Soc.* (*C*) **1970,** 1070.
30. A. Ya. Tikhonov and L. B. Volodarsky, *Tetrahedron Lett.* **1975,** 2721.
31. S. Spyroudis and A. Vargvoglis, *Synthesis* **1976,** 837.
32. A. R. Katritzky and J. M. Lagowski, *Chemistry of the Heterocyclic* N-*Oxides*, Academic Press, London, 1971.
33. H. Yamanaka, *Chem. Pharm. Bull.* (*Japan*) **1,** 158 (1959).
34. R. A. Abramovitch et al., *J. Org. Chem.* **37,** 1690, 3584 (1972).
35. J. Kavalek, A. Lycka, V. Machacek, and V. Sterba, *Coll. Czech. Chem. Commun.* **41,** 67 (1976), and refs cited therein.
36. Y. Kobayashi and I. Kumadaki, *Chem. Pharm. Bull.* (*Japan*) **17,** 510 (1969).
37. M. Ferles and M. Jankowsky, *Coll. Czech. Chem. Commun.* **33,** 3848 (1968).
38. M. Hamana and M. Yamazeka, *Chem. Pharm. Bull.* (*Japan*) **9,** 414 (1961).
39. A. R. Katritzky and E. Lund, *Tetrahedron* **25,** 4291 (1969).
40. T. Nakagome, *Yakugaku Zasshi* **82,** 253 (1963).
41. M. Ogata and H. Kono, *Chem. Pharm. Bull.* (*Japan*) **11,** 29 (1963).
42. E. Ochiai and C. Komeko, *Chem. Pharm. Bull.* (*Japan*) **5,** 56 (1957).
43. B. Klein, N. E. Hetman, and M. E. O'Donnell, *J. Org. Chem.* **28,** 1682 (1963).
44. H. Bredereck, R. Grompper, and H. Herlinger, *Chem. Ber.* **91,** 2832 (1958).
45. E. Ochiai and Y. Yamanoka, *Chem. Pharm. Bull.* (*Japan*) **3,** 175 (1955).
46. V. J. Traynelis and A. I. Gallagher, *J. Org. Chem.* **35,** 2792 (1970).
47. J. F. Vozza, *J. Org. Chem.* **27,** 3856 (1962).

Chapter 17

ELECTROPHILIC SUBSTITUTION. SIX-MEMBERED π-DEFICIENT HETEROAROMATICS

17.1. GENERAL COMMENTS

Since pyridine and the diazabenzenes are all π-deficient systems, it is not surprising that these compounds undergo electrophilic substitutions with great difficulty.[1-4] The partial rate factors for pyridine itself are on the order of 10^{-6}, that is, very similar to those for nitrobenzene. The reluctance of diazines to undergo electrophilic substitution is even greater. This behavior is also in accord with the predictions of quantum chemistry and concepts of traditional resonance theory (**1** → **2**). An additional difficulty is caused by the basicity of pyridine and the diazines: they may undergo electrophilic attack at the ring nitrogen(s) prior to electrophilic attack at a carbon. Thus two mechanisms must be considered.

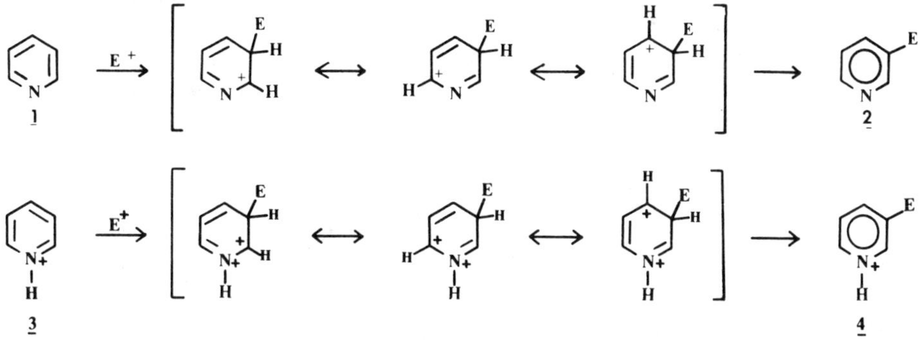

Clearly, the intermediate cation formed during the transformation **1** → **2** (path 1) is *not* stabilized by the presence of ring heteroatom(s). This lack of stability is even greater in the protonated intermediate cation of reaction **3** → **4** (path 2). Consequently, it is not surprising that the partial rate factors for electrophilic substitution by path 1 are of the order 10^{-6}, whereas those of pyridinium compounds are estimated[5] to be as low as 10^{-20}. These considerations also indicate that electrophilic substitution mechanisms may well depend on the substituents present on the π-deficient ring system. In fact, nitration of the more basic pyridines ($pK_a > 1$) occurs on the conjugate acids (path 2), whereas nitration of the less basic ones ($pK_a < -2.5$) normally takes place on the free bases (path 1).

17.2. NITRATIONS

Pyridine is not *C*-nitrated by nitronium fluoroborate; only *N*-nitropyridinium fluoroborate (**5**) is obtained. Extremely vigorous conditions ($NaNO_3$ in fuming H_2SO_4 at >300°C), are required to cause nitration, and even then only low yields of 3-nitropyridine (**6**) are produced.[4,6] The presence of two or three methyl groups at appropriate sites (**7, 8**) on the pyridine ring facilitates the nitration process.[7]

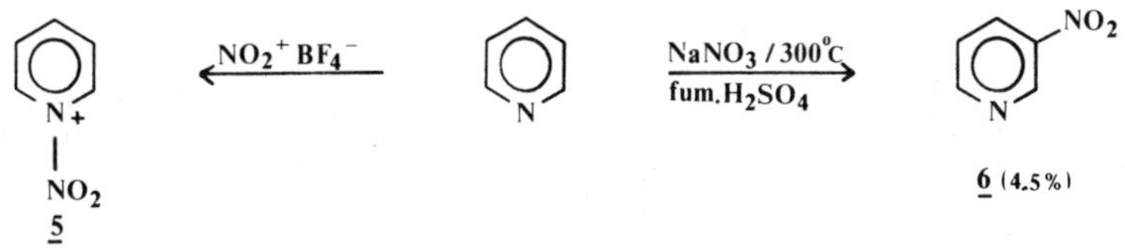

$NO_2^+ BF_4^-$; $NaNO_3$ / 300°C, fum. H_2SO_4

5 ; **6** (4.5%)

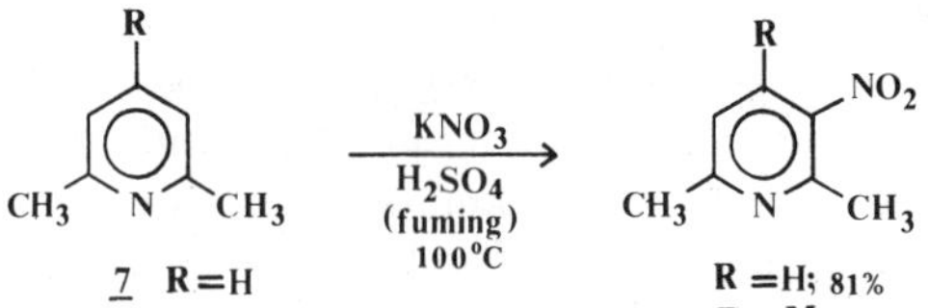

R | R
CH_3 N CH_3 | NO_2 CH_3 N CH_3

7 R=H
8 R=Me

KNO_3 / H_2SO_4 (fuming) / 100°C

R =H; 81%
R =Me; 94%

These nitrations proceed via attack on the conjugate acid (that is, path 2); however, the difference between electrophilic substitution on the conjugate acid and on the free base is clearly demonstrated by the reactivity difference between 3-bromopyridine (**9**), a fairly good base (pK_a 2.84), and 2,6-dichloropyridine (**10**), a very poor base (pK_a −2.86).[1]

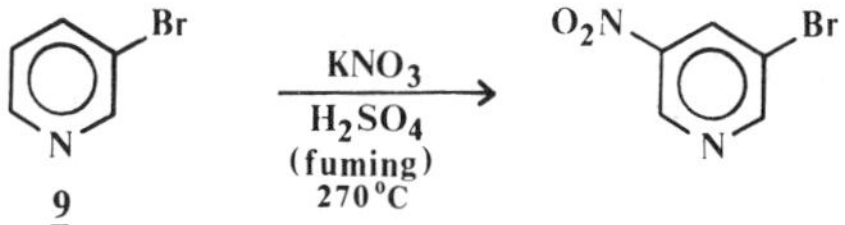

Br N 9 — KNO_3 / H_2SO_4 (fuming) / 270°C → O_2N Br N

Cl N Cl 10 — 95% HNO_3 / 90% H_2SO_4 / 115°C → NO_2 Cl N Cl

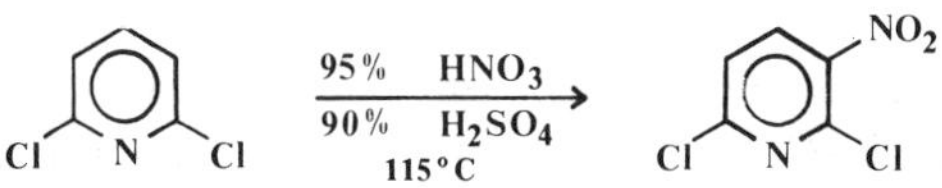

Strongly electron-donating substituents (for example, **11, 12**) facilitate electrophilic substitution and, in fact, control the site of substitution.[8] Nitration of aminopyridines initially produces nitroaminopyridines (**13**) which ultimately rearrange to *o*- and *p*-nitroaminopyridines (**14, 15**).[9]

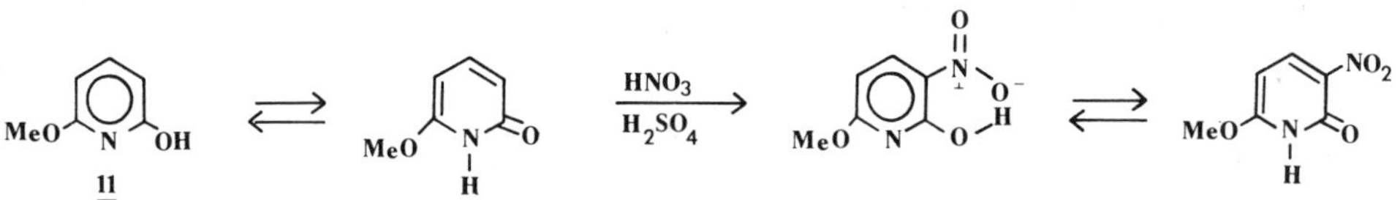

MeO N OH 11 ⇌ MeO N(H) O — HNO_3 / H_2SO_4 → MeO N O H O N O^- ⇌ MeO N(H) O NO_2

OR N 12 — HNO_3 / H_2SO_4 → OR N NO_2

(R=Me; 44%)

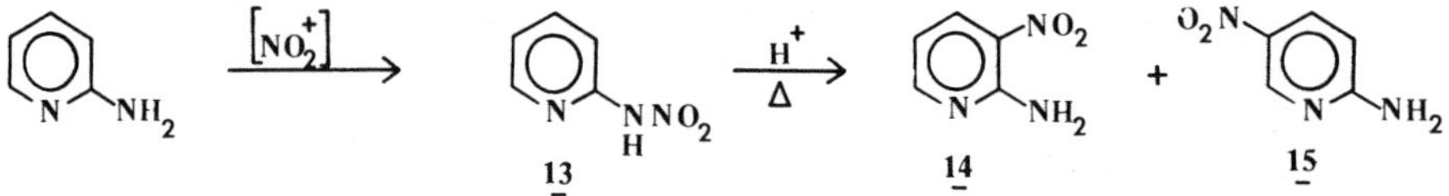

N NH_2 — [NO_2^+] → N NNO₂ H 13 — H^+ / Δ → NO_2 N NH_2 14 + O_2N N NH_2 15

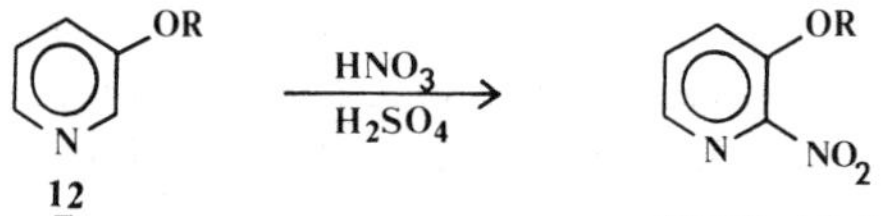

Chloro-substituted 3-aminopyridines follow the same initial pattern in that the 3-nitroaminopyridine is obtained. However, treatment of these compounds with hot acid forms a mixture of azo- and hydroxypyridine derivatives.[10] On the other hand, the *N*-methylaminopyridine **16** is nitrated, apparently directly, to form the 2-nitro compound **17**![11]

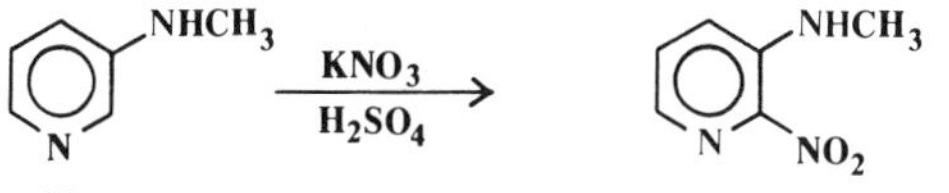

$$\text{16 (3-NHCH}_3\text{-pyridine)} \xrightarrow[\text{H}_2\text{SO}_4]{\text{KNO}_3} \text{17 (3-NHCH}_3\text{-2-NO}_2\text{-pyridine)}$$

Nitration of pyridine *N*-oxide (**18**) might be expected to follow the same path as that of a protonated pyridine. However, not only are these compounds nitrated easily, but the nitro group enters at the 4-position (**19**).[12] Nitration that occurs on the protonated *N*-oxide is rationalized by invoking back-donation of the lone pair of electrons on the oxygen atom (**20**). Interestingly, when **18** is treated with benzoyl nitrate, or *p*-nitrobenzoyl chloride, followed by silver nitrate, the 3- (**21**) and 3,5-nitrated *N*-oxides are formed.[13] This result has been attributed to initial formation of an *o*-nitro cation (**22**), which prevents back-donation.

As with pyridine, one or more electron-donating groups are required in order for the diazines to be successfully nitrated. As a rough comparison, it appears that two activating substituents, in the proper position, will make the diazines as reactive as benzene itself. The presence of three activating substituents brings some of these diazines into the reactivity range of phenol.

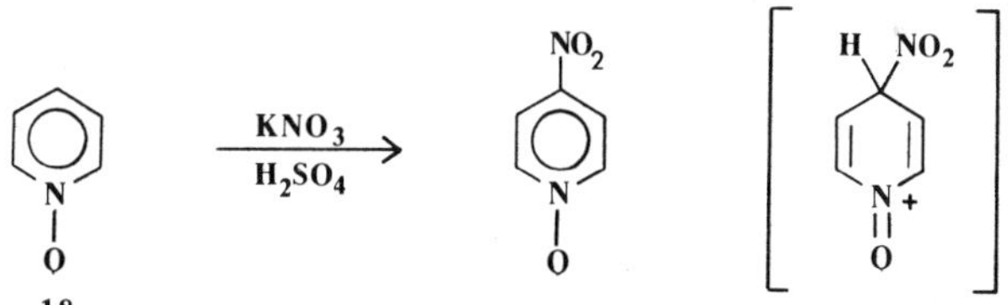

$$\text{18} \xrightarrow[\text{H}_2\text{SO}_4]{\text{KNO}_3} \text{19 (4-NO}_2\text{)} \qquad [\text{20}]$$

$$\text{22} \xrightarrow{C_6H_5CO_2^-} \left[\text{intermediate}, NO_2^+\right] \longrightarrow \text{21 (3-NO}_2\text{)}$$

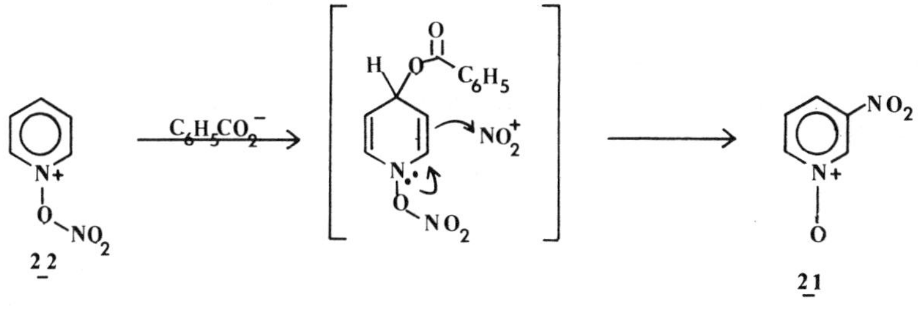

The low reactivity of the pyridazine ring is exemplified by the observation that phenylpyridazine (**23**) is nitrated on the phenyl rather than on the pyridazine ring.[14] When suitable activating groups are present on the pyridazine (**24** → **25**), nitration becomes feasible.

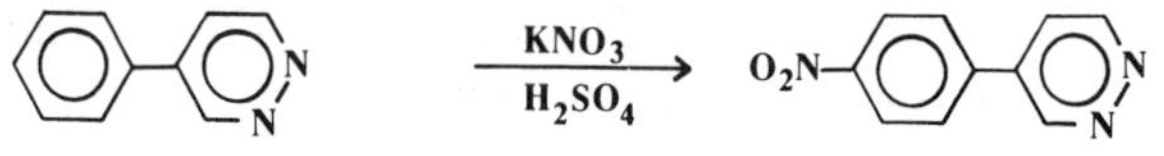

KNO_3 / H_2SO_4

23

KNO_3 / H_2SO_4

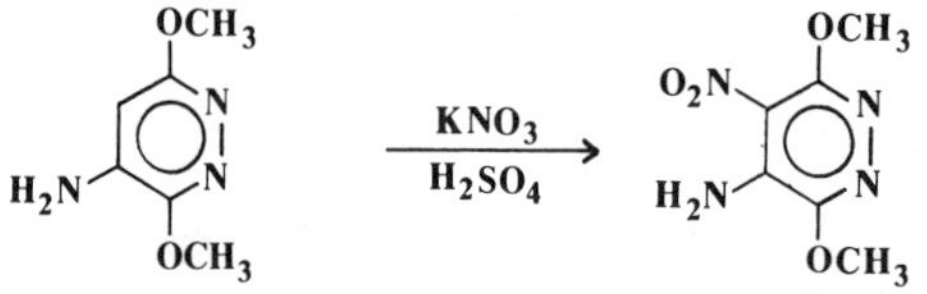

24 **25**

Electrophilic substitution of activated pyrimidines occurs, generally, at C-5; thus 4,6-diaminopyrimidine (**26**) yields the 5-nitro derivative (**27**). 2,4-Diaminopyrimidine (**28**) does not undergo nitration. Hence the position of the activating groups is of critical importance in these nitrations.[14] No nitrations on any pyrazine derivatives seem to have been reported.[15]

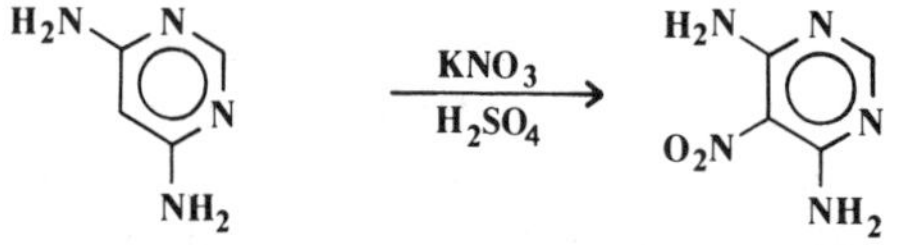

KNO_3 / H_2SO_4

26 **27**

KNO_3 / H_2SO_4 N. R.

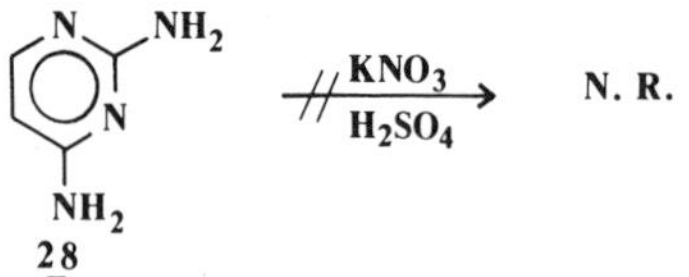

28

17.3. HALOGENATIONS

As is the case with the nitration reactions of pyridine and the diazines, electrophilic halogenation of the parent compounds requires, where the reaction proceeds at all, very severe conditions.[16]

Pyridine reacts with chlorine in the presence of a large excess of aluminum chloride to afford a low yield (~30%) of 3-chloropyridine (**29**).

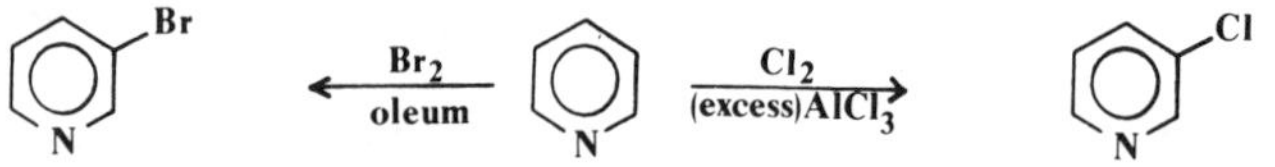

Br_2 / oleum; Cl_2 / (excess)$AlCl_3$

9 **29**

The use of bromine in oleum, however, affords 3-bromopyridine (**9**) in excellent yield (86%).[17] The methylpyridines are brominated with equal facility by this means. When the pyridine ring is activated by an amino group, for example, **30–32,** halogenation occurs with ease.[18–20] The amino group directs the halogen ortho and/or para to itself. These brominations apparently occur on the free bases.

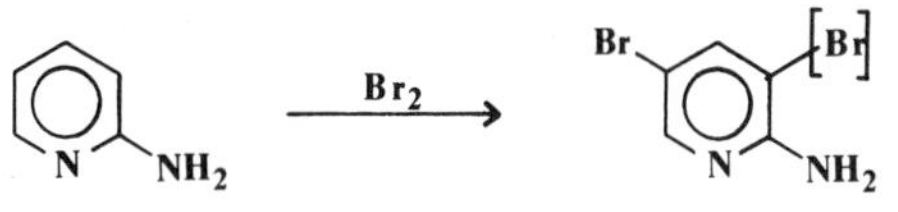

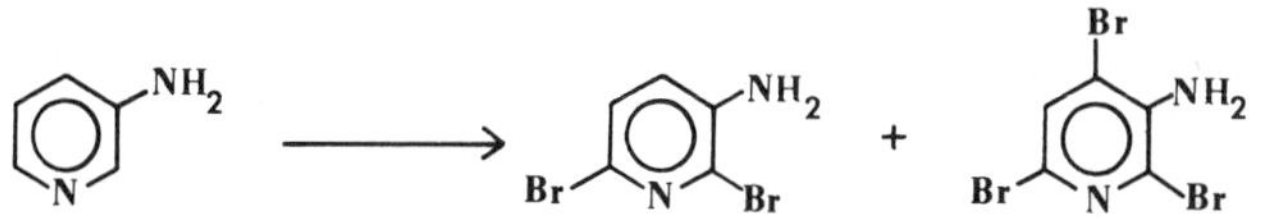

In contrast to the ease of nitration of pyridine *N*-oxides, bromination of these compounds occurs with difficulty. Bromination of **18** in acetic anhydride/sodium acetate affords the 3,5-dibromo compound **33** in poor yield.[21] Bromination in oleum at 120°C also does not afford good yields of the 3-bromo isomer (**34**).[22] It is worthwhile to reiterate that "normal" nitration forms the 4-substituted *N*-oxides, whereas bromination yields the 3-substituted products.

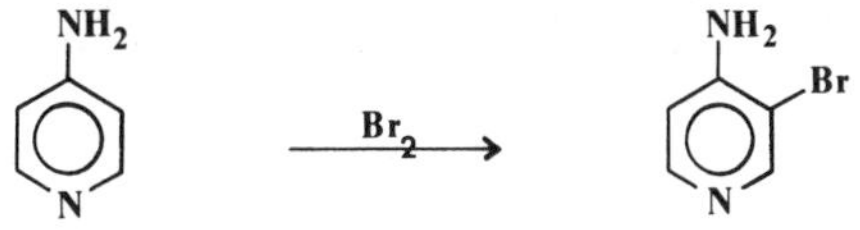

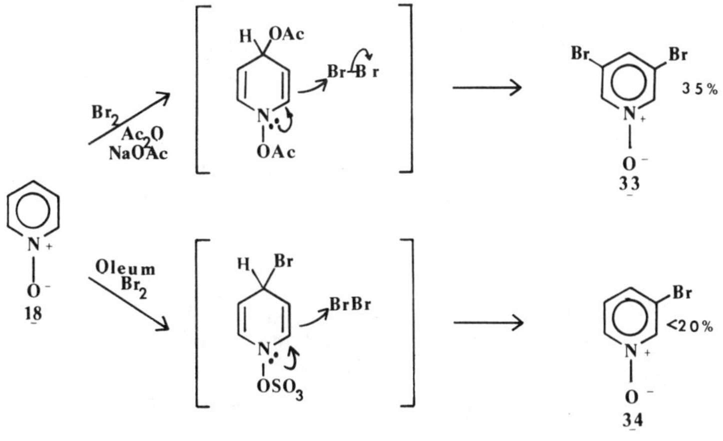

As in nitrations, the parent diazines do not undergo electrophilic halogenation. Even though pyrimidine reacts with bromine at 160°C to form 5-bromopyrimidine (**35**), it does not do so by a simple electrophilic substitution but rather via a perbromide intermediate.[23]

Similarly, even though 2-methylpyrazine (**36**) can be chlorinated to afford 2-chloro-3-methylpyrazine, this reaction proceeds via an addition-elimination mechanism.[24] Chlorination of 2-amino-3-carbomethoxypyrazine (**37**), an activated compound, results in ring chlorination to afford the 5-chloro derivative (**38**).[24]

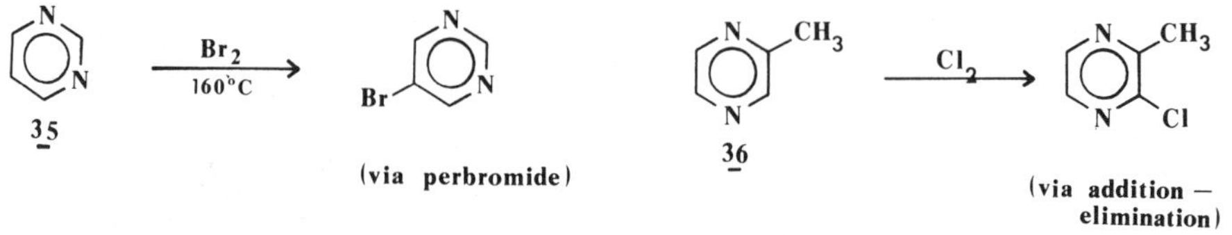

Br_2, 160°C; **35** → (via perbromide)

36 → Cl_2 → (via addition – elimination)

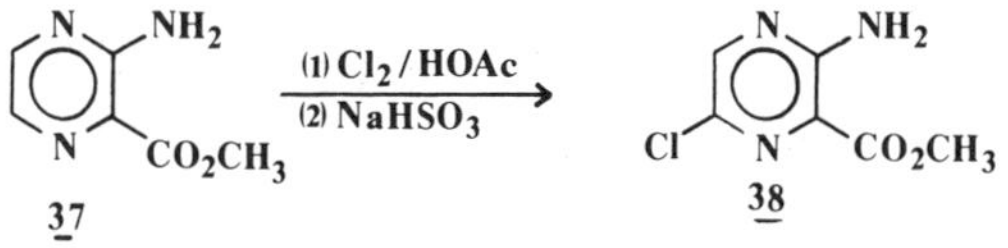

37 → (1) Cl_2/HOAc (2) $NaHSO_3$ → **38**

17.4. SULFONATIONS

At 220–270°C, pyridine in oleum can be sulfonated in the presence of mercuric sulfate to yield the 3-sulfonic acid (**39**).[25] At higher temperatures, pyridine-3-sulfonic acid is desulfonated and may "rearrange" to the 4-sulfonic acid. The three methylpyridines (**40–42**) are similarly sulfonated under these conditions.[26]

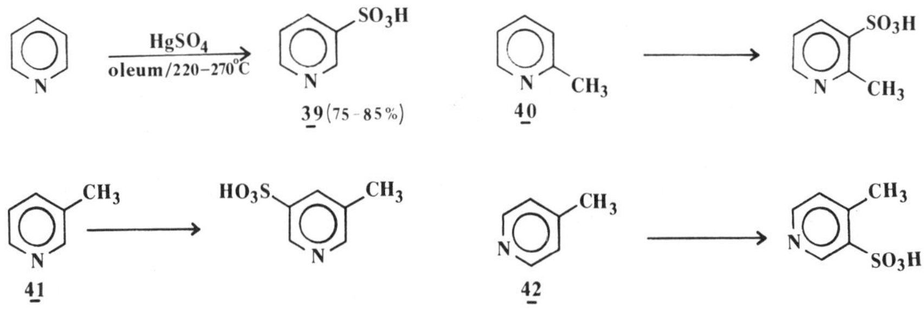

$HgSO_4$, oleum/220–270°C → **39** (75–85%)

40 →

41 →

42 →

A rather unusual "sulfonation" has been reported in the case of 2,6-di-*tert*-butylpyridine (**43**). At the amazingly low temperature of −10°C, the normal sulfonation product (**44**) is formed,[27] whereas at higher temperature a cyclic sulfone (**45**) is obtained.[28] The great ease with which this sulfonation takes place has been attributed to the steric block caused by the two *tert*-butyl groups which prevents the formation of the intermediate pyridine–SO_3 complex.

SO_3, liq. SO_2, −10° ; 240–250° ; Me ; Me ; (15–20%) ; 43 ; 44 ; 45

In the case of the diazines (for example, **26**), sulfonation occurs only where electron-donating groups (amino and/or alkoxy) are present to activate these π-deficient systems.[23]

H_2N ; NH_2 ; H_2SO_4 ; HO_3S ; 26

17.5. MERCURATIONS

When pyridine is heated with mercuric acetate at 170–180°C, 3-pyridyl mercuriacetate (**46**) is the principal product. Treatment of **46** with sodium chloride converts it to the mercurichloride (**47**) which, when treated with bromine, affords 3-bromopyridine (**9**) in good yield.[29] Similar reactions afford the corresponding bromomethylpyridines.[29] It is of some interest to note that pyridine *N*-oxide (**18**) is mercurated mainly at positions 2 and 6 (**48, 49**).[30]

$Hg(OAc)_2$, 170–180° ; HgOAc ; NaCl ; HgCl ; Br_2 ; Br ; 46 ; 47 ; 9

$Hg(OAc)_2$; HgOAc ; AcOHg ; 18 ; 48 ; 49

17.6. REACTIONS WITH WEAK ELECTROPHILES

As is the case with phenols and anilines, many amino- and hydroxyazines and diazines react with diazonium salts and can be nitrosated as well as thiocyanated. Some examples of these reactions with 2,6-diaminopyridine (**50**) are[31–33]:

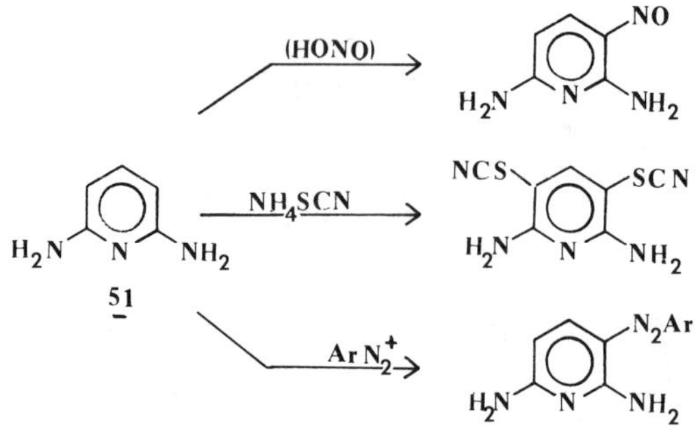

Interestingly, 3-amino-2-thiocyanatopyridines (**51**) cyclize spontaneously to thiazolo[5,4-*b*]pyridines (**52**).[34] Nitrosation of the activated pyrimidines (**26**) has been put to good synthetic use since nitroso compounds (**53**) can be readily reduced to their corresponding amino derivatives.[35]

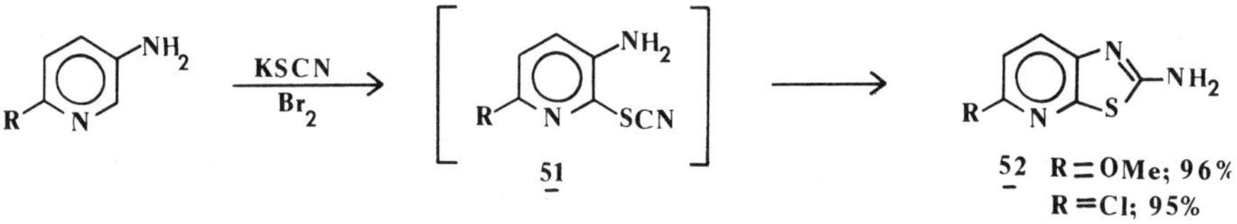

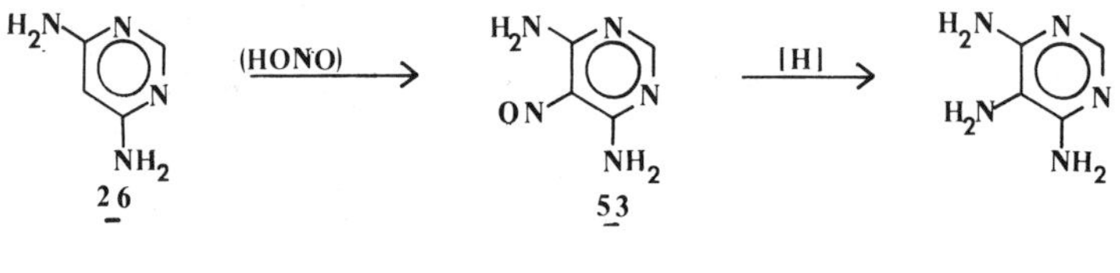

52 R=OMe; 96%
R=Cl; 95%

REFERENCES

1. A. R. Katritzky et al., *J. Chem. Soc.* (*B*) **1967,** 1204, 1211, 1213, 1219, 1222, 1226.
2. K. Schofield, *Heteroaromatic Nitrogen Compounds: Pyrroles and Pyridines*, Butterworths, London, 1967, p. 162.
3. R. A. Abramovitch and J. G. Saha, *Adv. Heterocycl. Chem.* **6,** 229 (1966).
4. Zh. I. Aksel'rod and V. M. Berezovskii, *Russ. Chem. Rev.* **39,** 627 (1970).
5. A. R. Katritzky and B. J. Ridgewell, *J. Chem. Soc.* **1963,** 3753, 3882.
6. P. Schorigin and A. Toptschiew, *Chem. Ber.* **69,** 1874 (1936).
7. E. V. Brown and R. H. Neil, *J. Org. Chem.* **26,** 3546 (1961).
8. A. P. Katritzky, H. O. Tarhan, and S. Tarhan, *J. Chem. Soc.* (*B*) **1970,** 114.
9. A. E. Chichibabin and B. A. Razorenov, *J. Russ. Phys. Chem. Soc.* **47,** 1286 (1915).

10. W. Czuba, *Rocz. Chem.* **34,** 905, 1639, 1647 (1960).
11. E. Plazek, A. Marcinkow, and C. Stammer, *Rocz. Chem.* **15,** 365 (1935).
12. E. Ochiai, *J. Org. Chem.* **18,** 534 (1953).
13. E. Ochiai and C. Kaneko, *Chem. Pharm. Bull. (Japan)* **8,** 28 (1960).
14. J. D. Mason and D. L. Aldous in, *Heterocyclic Compounds*, Vol. 28, R. V. Castle, Ed., Wiley, New York, 1973.
15. M. Tisler and B. Stanovnik, *Adv. Heterocycl. Chem.* **9,** 211 (1968).
16. D. E. Pearson, G. W. Habgrove, J. K. T. Chow, and B. R. Suthers, *J. Org. Chem.* **26,** 789 (1961).
17. H. J. den Hertog, L. van der Does, and C. A. Landheer, *Rec. Trav. Chim.* **81,** 864 (1962).
18. O. von Schickh, A. Binz, and A. Schulz, *Chem. Ber.* **69,** 2593 (1936).
19. E. Plazek and A. Marcinkow, *Rocz. Chim.* **14,** 326 (1934).
20. H. J. Hertog and J. G. Schogt, *Rec. Trav. Chim.* **40,** 353 (1951).
21. M. Hamana and M. Yamazaki, *Chem. Pharm. Bull. (Japan)* **9,** 414 (1961).
22. M. van Ammers, H. J. den Hertog, and B. Haase, *Tetrahedron* **18,** 227 (1962).
23. H. Bredereck, R. Gompper, and H. Herlinger, *Chem. Ber.* **91,** 2832 (1958).
24. A. Hirschberg and P. E. Spoerri, *J. Org. Chem.* **26,** 2356 (1961).
25. H. J. den Hertog, M. C. van der Plas, and D. J. Buurman, *Rec. Trav. Chim.* **77,** 963 (1958).
26. S. M. McElvain and M. A. Goese, *J. Am. Chem. Soc.* **65,** 2233 (1943).
27. H. C. van der Plas and H. J. den Hertog, *Rec. Trav. Chim.* **81,** 841 (1962).
28. H. C. van der Plas and T. H. Crawford, *J. Org. Chem.* **26,** 2611 (1961).
29. N. P. McCleland and R. H. Wilson, *J. Chem. Soc.* **1932,** 1263.
30. M. van Ammers and H. J. den Hertog, *Rec. Trav. Chim.* **77,** 340 (1958).
31. A. Chichibabin and O. Zeide, *J. Russ. Phys.-Chem. Soc.* **46,** 1216 (1914).
32. J. M. Cox, J. A. Elvidge, and D. E. H. Jones, *J. Chem. Soc.* **1964,** 1423.
33. J. A. Baker and S. A. Hill, *J. Chem. Soc.* **1962,** 3464.
34. C. A. Okafor, *J. Org. Chem.* **38,** 4383 (1973).
35. R. M. Evans, P. G. Jones, P. J. Palmer, and F. F. Stephens, *J. Chem. Soc.* **1956,** 4106.

Chapter 18

NUCLEOPHILIC SUBSTITUTION. SIX-MEMBERED π-DEFICIENT HETEROAROMATICS

18.1. GENERAL COMMENTS

As a consequence of their π-deficient nature, pyridine and the diazabenzenes should undergo nucleophilic substitution reactions more readily than benzene. For the carbocyclic compounds, two nucleophilic substitution mechanisms are operative: (1) addition followed by elimination, *AE*, and (2) elimination followed by addition, *EA*. Besides these "normal" mechanisms, a third possibility can also be operative: an addition-elimination mechanism which leads to "cine-substitution," AE_a. For the π-deficient heteroaromatic systems, a fourth mechanism is occasionally operative. This complex reaction—ANRORC—involves an *a*ddition *n*ucleophilic *r*ing-*o*pening *r*ing-*c*losing sequence.[1,2]

18.2. ADDITION-ELIMINATION MECHANISM—*AE*

The position of nucleophilic attack, correctly predicted by molecular orbital calculations (See Appendix A), on pyridine, pyridazine, pyrazine, and pyrimidine is always at the carbon that enables stabilization of the negative charge on the ring nitrogen(s) (**1–4**). In pyridines three possible anions (**1, 5, 6**) can arise from addition of a charged nucleophile to the three different carbon atoms. It has been established that the anions resulting from addition at the 2- (**1**) and 4-positions (**5**) are of lower energy than the one arising from addition at the 3-position (**6**).[3,4]

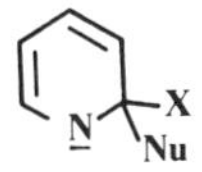

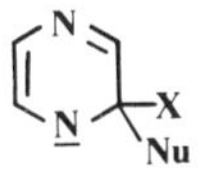

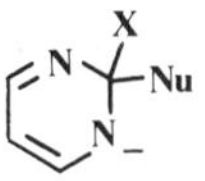

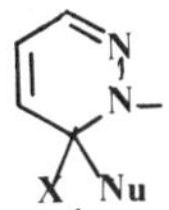

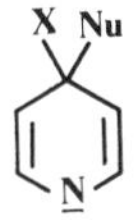

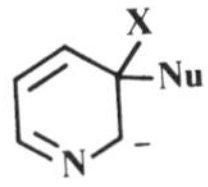

One may well predict that the corresponding azinium ions, like *N*-oxides, should be more susceptible toward nucleophilic attack because of the greater stability of the intermediates (**7** and **8**, respectively), and this is observed.[3] The *AE* mechanism is operative, by and large, in those instances where the leaving group is at an "active" position; that is, ortho or para to the heteroatom. These *AE* mechanisms can involve displacement of a hydrogen (X = H) or, more commonly, displacement of a halogen (X and/or Y = halogen).

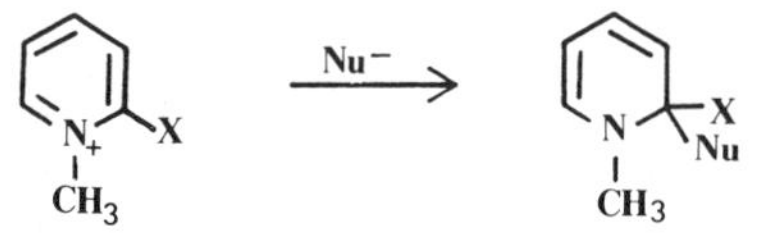

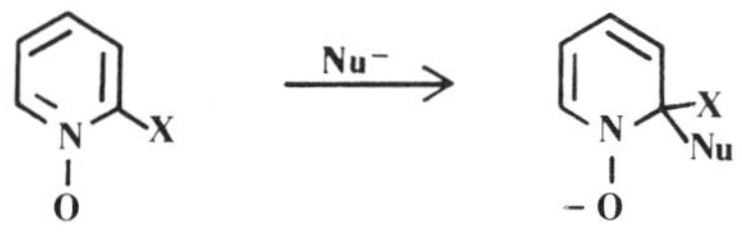

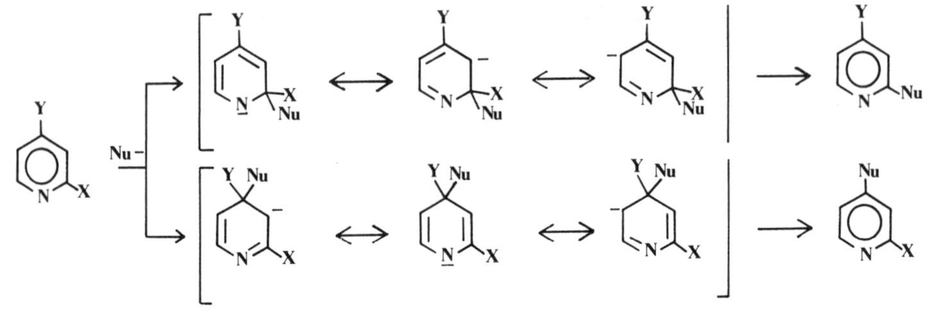

18.2.1. AMINATIONS

When pyridine or pyrazine is treated with $NaNH_2/NH_3$(liq.) (the *Tschichibabin amination*), the corresponding 2-amino compounds (**9** and **10**, respectively) are obtained.[5,6]

Pyrimidine itself does not appear to have been studied in this reaction. 4-Methylpyrimidine (**11**) forms the 2-amino compound (**12**) in low yield. The direct amination of pyridazine has not been reported.

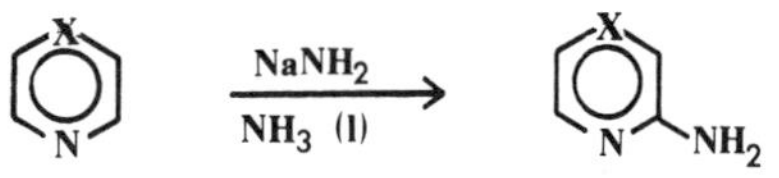

9 (X=CH) 70%

10 (X=N)

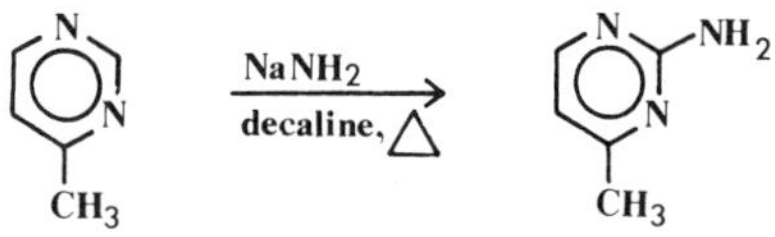

11

12 (low yield)

18.2.2. NUCLEOPHILIC SUBSTITUTION OF HALOGEN

Most nucleophilic substitutions on pyridines and diazines involve the displacement of substituents other than hydrogen, most commonly a halogen.

It is instructive to compare the displacement rates in the *AE* reactions of methoxide ion on 2-, 3-, and 4-chloropyridines (see Table 18.1).[7] Under these conditions, the relative reactivities are 4- > 2- ≫ 3-. The fact that 3-chloropyridine is 10,000 times less reactive than 2-chloropyridine and about 100,000 less so than 4-chloropyridine reflects the absence of resonance-stabilizing forms which involve a negative charge on the ring nitrogen atom in the intermediate formed from 3-chloropyridine. Reactivity differences between the 2- and 4-chloropyridines are relatively small, a factor of only 10. These minor differences are explained by an examination of the two Wheland intermediates (**13** and **14**) involved in these reactions.

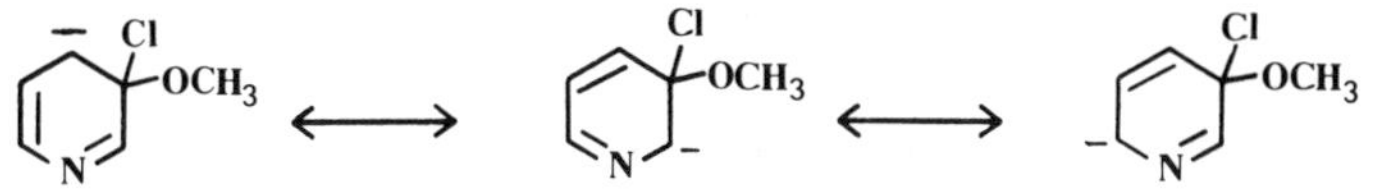

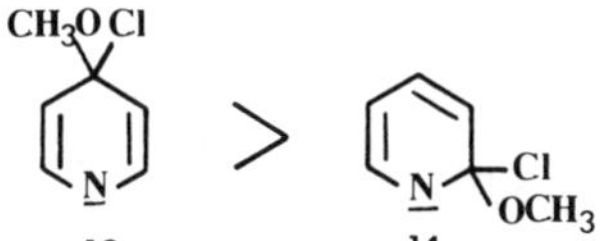

Cl OCH$_3$ N ⟷ Cl OCH$_3$ N ⟷ Cl OCH$_3$ N

CH$_3$O Cl N > N Cl OCH$_3$

13 14

The greater stabilization of the intermediates brought on by *N*-alkylation is reflected in the *AE* rates involving the *N*-methyl-2-, 3-, and 4-chloropyridinium ions (see Table 18.1). These compounds react faster than the nonalkylated

TABLE 18.1. Substituent Rate Factors of the Addition–Elimination Reaction of Methoxide Ion with Chloropyridines

K_2 (1 mol^{-1} sec^{-1})	Compound
2.76×10^{8}	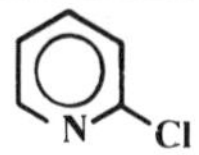 N, Cl
9.12×10^{4}	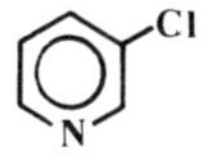 Cl, N
7.43×10^{9}	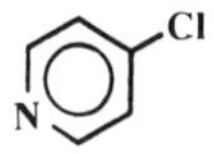 Cl, N
1.28×10^{21}	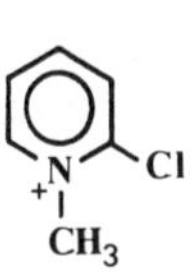 N$^+$, Cl, CH$_3$
2.62×10^{13}	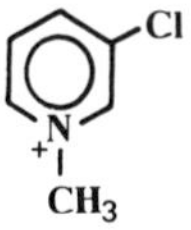 Cl, N$^+$, CH$_3$
4.23×10^{19}	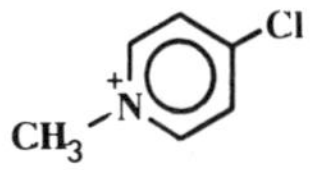 Cl, CH$_3$–N$^+$
1.0	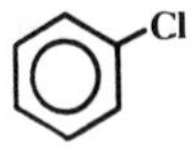 Cl

analogues by roughly a factor of 10^{10}! Again, the relative rates for the different positions are 4- > 2- ≫ 3-. The corresponding 2- (**15**) and 4-chloro (**16**) intermediates, which are neutral, reflect this increased reactivity.

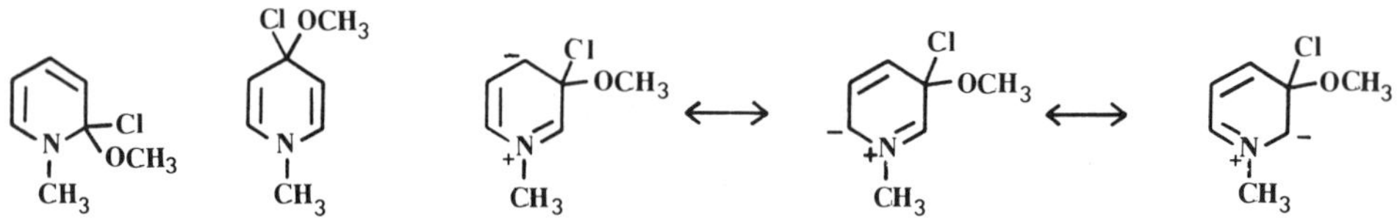

15 16 17a ⟷ 17b ⟷ 17c

For the *N*-methyl-3-chloropyridinium ion intermediate (**17a–c**), the increased stability is due to improved charge delocalization caused by the presence of the positive charge on the ring nitrogen. This increased reactivity of the halodiazines in *AE* reactions is also shown by the observation that 2-chloropyridine is stable at 0°C, whereas 3-chloropyridazine (**18**) decomposes spontaneously at that temperature.

The reaction of *p*-nitrophenoxide ion with a series of chlorodiazines has established the following relative reactivities in these reactions:[8]

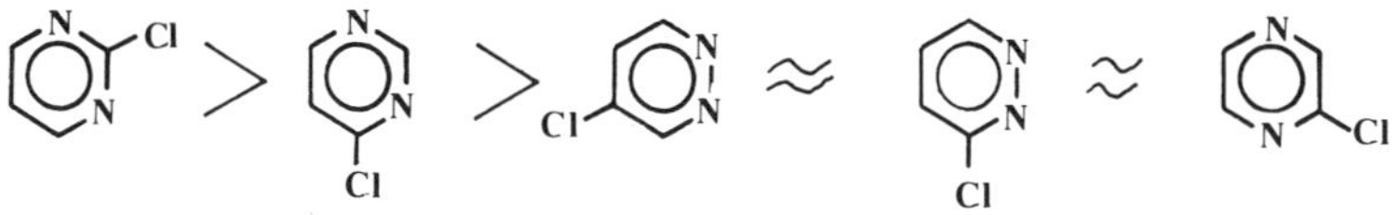

19 > 20 > 21 ≈ 18 ≈ 22

This sequence appears to be generally true for the *AE* reactions, except that 4-chloro- (**20**) is usually more reactive than 2-chloropyrimidine (**19**).[8] The relative reactivity differences between di- and polyhalogenated diazines become a major factor in determining the structures of the substitution products. Halogens located para to the ring nitrogen are generally more reactive than those in an ortho position.[9]

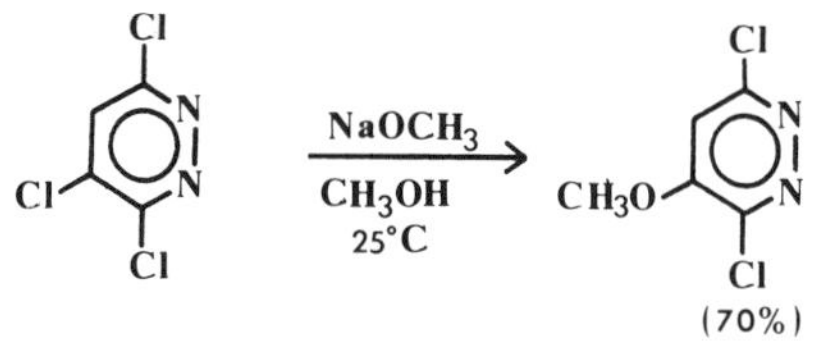

$\xrightarrow[\text{CH}_3\text{OH, 25°C}]{\text{NaOCH}_3}$ (70%)

18.3. ELIMINATION-ADDITION MECHANISM—*EA*

This mechanism is generally operative when the leaving group is situated at an unactivated position such as the 3-position in pyridine (**23**) and when the nucleophile is strongly basic.[10a] *A priori* one might expect to form either the 3,4- (**24**) and/or the 2,3-pyridyne (**25**). However, it has, been sufficiently well established that where a choice exists, the 3,4-pyridyne (**24**) is favored.[10b] The experimental support for this mechanism is illustrated by the reaction of 3-bromopyridine and potassium amide in liquid ammonia, in which a mixture of 3- and 4-aminopyridine is formed in a 1:2 ratio.[10–12]

Some other reactions involving alkoxypyridynes (**26–28**) as intermediates are listed below.[11,13]

The apparently lone example of a reaction that forms both a 2,3- (**29**) as well as 3,4-pyridyne (**30**) is the reaction of pentachloropyridine with *N*-butyllithium.[14]

Among the diazines, 3,6-dichloropyridazine (**31**), when treated with sodium amide, is reported to yield 4-amino-6-chloropyridazine (**32**). Although the structure has never been firmly established, the product is different from the known 3-amino-6-chloropyridazine (**33**). The intermediacy of a pyridazyne[10b] is a possibility.

Cl, Me, N, N, NH_2 — 33 ← NH_3 (R=Me) — Cl, R, N, N, Cl — 31 — $NaNH_2$ (R=H) → Cl, H_2N, N, N — 32

In the pyrimidine series, some evidence for the existence of a pyrimidyne has been obtained. 4-Phenyl-5-bromopyrimidine (**34**), when treated with potassium amide, affords as the lone product 6-amino-4-phenylpyrimidine (**35**). Similarly, 4-methoxy-5-bromopyrimidine (**36**) yields 4-methoxy-6-aminopyrimidine (**37**).[15]

Br, N, N, C_6H_5 — 34 — KNH_2 → H_2N, N, N, C_6H_5 — 35 (15%)

Br, N, N, OCH_3 — 36 — KNH_2 → H_2N, N, N, OCH_3 — 37 (70%)

18.4. ABNORMAL ADDITION-ELIMINATION MECHANISM—AE_a

The pyridinium salts, in which the *N*-substituent is a facile leaving group (such as an alkoxy or acyloxy), react with nucleophiles by addition at the 2- or 4-position, followed by subsequent loss of the *N*-substituent. The intermediate **38**, formed from the reaction of pyridine *N*-oxide with acetic anhydride, represents one such example.[16]

+N, OAc — AcO^- → N, H, OAc, OAc — 38 → N, OAc

N-Methoxypyridinium bromide (**39**; X = Br), when treated with methylmagnesium bromide, affords 2-picoline (**40**) as the major product along with some 4-picoline and pyridine. The "unexpected" pyridine is formed by attack of the nucleophile on the methyl group of the starting material. Nucleophiles such as CN^-, among others, have also been used in these reactions.[17] Thus *N*-methoxypyridinium iodide (**39**; X = I) affords the 2- (**41**) as well as 4-cyanopyridines (**42**).

$\xrightarrow[(X=Br)]{CH_3MgBr}$ → 40

39

$\xrightarrow[(X=I)]{KCN}$ 41 (47%) + 42 (4%)

18.5. ANRORC MECHANISM

An attempt by van der Plas to identify the presence of pyrimidynes in some displacement reactions led to the establishment of a new and very novel mechanism. The unique character of this mechanism makes a detailed discussion worthwhile.[2]

When 4-substituted 5-bromopyrimidine (**43**) is treated with potassium amide in liquid ammonia, the resulting pyrimidine is simply a 4-substituted 6-aminopyrimidine (**44**). That no 4-substituted 5-aminopyrimidine (**45**) is isolated indicates the reaction might well occur via the pyrimidyne (**46**).[15,18] When a 4-substituted 6-bromopyrimidine (**47**) is treated with potassium amide in liquid ammonia at −75°C, the exclusive product, again, is the corresponding 6-aminopyrimidine (**48**). In this instance, the intermediacy of the initial addition product **49** can be considered.

$\xrightarrow[liq.NH_3(-33°C)]{KNH_2}$

43 (R=t-bu, C_6H_5, CH_3, OCH_3, OH, NH_2) 44 No 45

46 47 $\xrightarrow[NH_3, -75°]{KNH_2}$ 49 → 48 No

In order to differentiate between these two possibilities, the corresponding 4-substituted 5-deuterio-6-bromopyrimidine (**50**) is allowed to react with potassium amide in liquid ammonia; the resulting aminopyrimidine (**51**) does *not* contain deuterium![2] Since the starting materials (**50, 47**) have been shown to be immune to H → D exchange, the conclusion can be drawn that this transformation involves a pyrimidyne intermediate. Such an occurrence would be rather surprising, since past experience has shown that a halogen at the indicated site is usually subject to an $S_N(AE)$ reaction. Superimposed on this curious set of results is the observation that a 4-substituted 6-bromopyrimidine (**52**), when treated with lithium piperidide in piperidine/ether, does not afford the 6-piperidyl derivative (**53**); instead, **54** is isolated. In this case none of the "usual" displacement reactions have taken place.[19] Why does the reaction with potassium amide in liquid ammonia proceed by an initial deprotonation at C-5 whereas reaction with lithium piperidide in piperidine occurs by an initial addition at C-2? Might the same sort of ring-cleaved intermediate (**56**) be formed even in the potassium amide/liquid ammonia reaction and, being unstable under prevailing conditions, cyclize to the aminopyrimidine (**57**)?

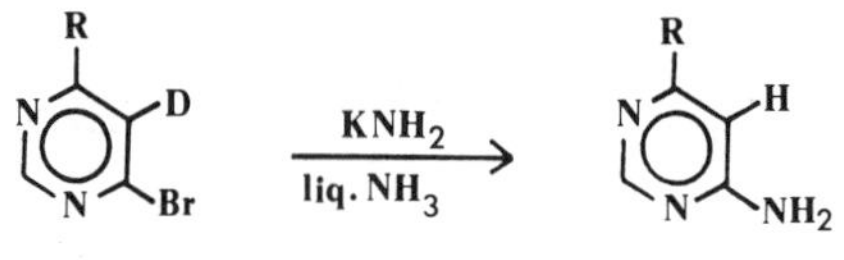

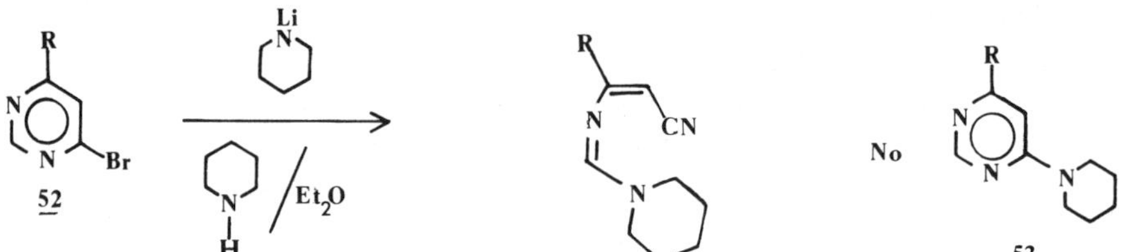

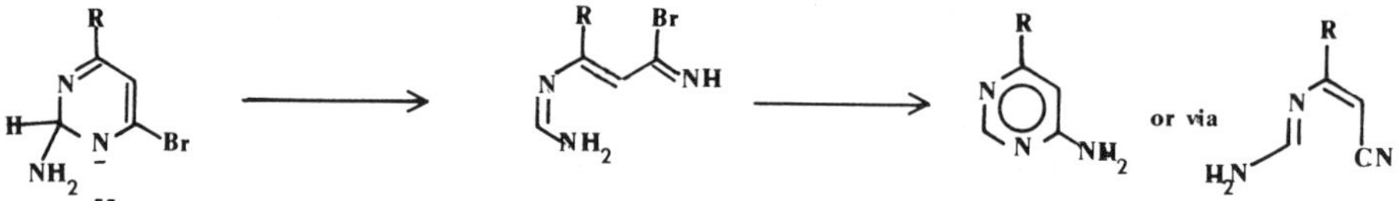

Treatment of the nitrogen-labeled 6-bromo-4-phenylpyrimidine (**58**) with potassium amide/liquid ammonia indicates that "scrambling" between the ring nitrogen and the exocyclic nitrogen does indeed take place.

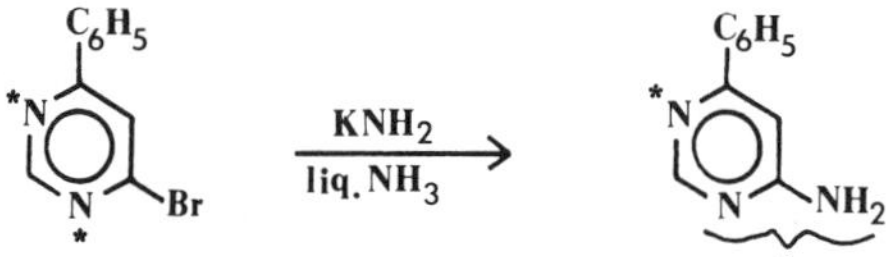

Thus it appears that both these reactions occur via a ring-opening/ring-closing process, for which the acronym ANRORC has been employed: *A*ddition *N*ucleophilic; *R*ing *O*pening; *R*ing *C*losing. This mechanism also explains the curious H → D exchange reaction observed in the deuteriated compound mentioned earlier. Determination of the generality of this reaction has been accomplished by syntheses of the corresponding 6-fluoro (**59**), 6-chloro (**60**), and 6-iodo (**61**) nitrogen-labeled pyrimidines and their subjection to the potassium amide/liquid ammonia reaction. Scheme 18.1 lists the percent of ANRORC mechanism operative for each of these compounds.[2]

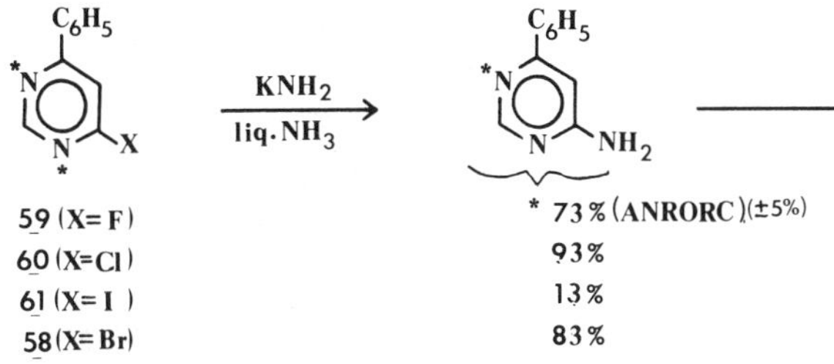

C_6H_5 KNH$_2$ liq. NH$_3$ NH$_2$ Br

59 (X= F) — * 73% (ANRORC)(±5%)
60 (X=Cl) — 93%
61 (X= I) — 13%
58 (X= Br) — 83%

Scheme 18.1

Since these ANRORC reactions suggest that the initial attack must occur at C-2 of the pyrimidine nucleus, an extension of this reaction to 2,4-diphenyl-6-bromopyrimidine (**62**) is a logical one.[20] Compound **62** does *not* show any nitrogen scrambling. Hence nucleophilic displacement in this substance occurs by the normal nucleophilic displacement reaction, and not by the ANRORC mechanism. For the 6-iodo-4-phenylpyrimidine (**63**), 13% of the displacement occurs by the ANRORC mechanism. By what process does the remainder of this reaction proceed? When 5-deuterio-**63** is subjected to this reaction, the corresponding 6-amino product (**64**) is free of any deuterium; thus an $S_N(EA)$ mechanism is "competing" with the ANRORC in this instance.[2]

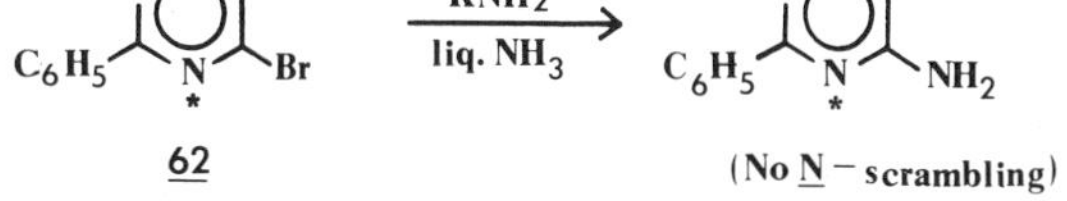

62 — KNH$_2$, liq. NH$_3$ → (No N – scrambling)

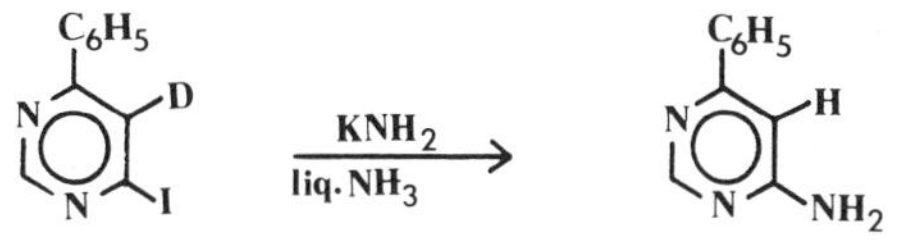

63 — KNH$_2$, liq. NH$_3$ → 64

Clearly, isolation of the open-ring intermediate from the potassium amide/liquid ammonia reactions on halopyrimidines would aid greatly in substantiation of the proposed mechanism. So far all attempts at doing so have failed. In the case of the 4-chloro-5-cyano-6-phenylpyrimidine (**65**) in which the cyano group is labeled, one might anticipate the following mechanism to be operative:

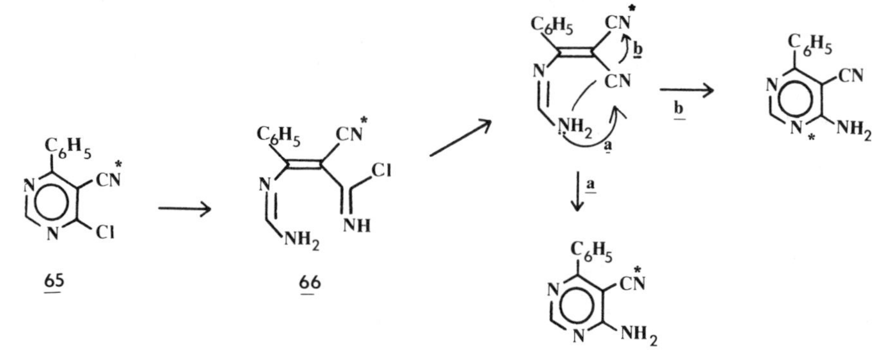

Two alternatives exist: path *a* or path *b*. If path *b* prevails, the cyano group carbon atom becomes part of the pyrimidine ring. If path *a* prevails, the carbon remains as an exocyclic nitrile carbon. No incorporation of the cyano group carbon into the ring is observed. It has been suggested that this may reflect the steric limitations around the carbon–carbon double bond, which prevent the labeled cyano group from coming sufficiently close to the attacking amino function. However, since tautomerism is a possibility, it appears that this is not a totally valid argument. The other alternative is that intermediate **66** is involved in this reaction.[21]

If 4-bromo-6-phenylpyrimidine (**67**) is treated with lithium isopropylamide in isopropylamine at 20°C, the only compound isolated is the direct substitution product (**70**).[22] If this same reaction is done at −75°C, **68** and **69** are obtained in about a 10:1 ratio. The cyano compound (**68**), on standing, is converted to the iminopyrimidine (**69**) at room temperature. If **69** is treated with lithium isopropylamide in isopropylamine at 20°C, a *Dimroth rearrangement* takes place and the *same* product (**70**) is obtained as from the direct amination at 20°C. The following, rather complex, mechanism is proposed to account for these facts[22]:

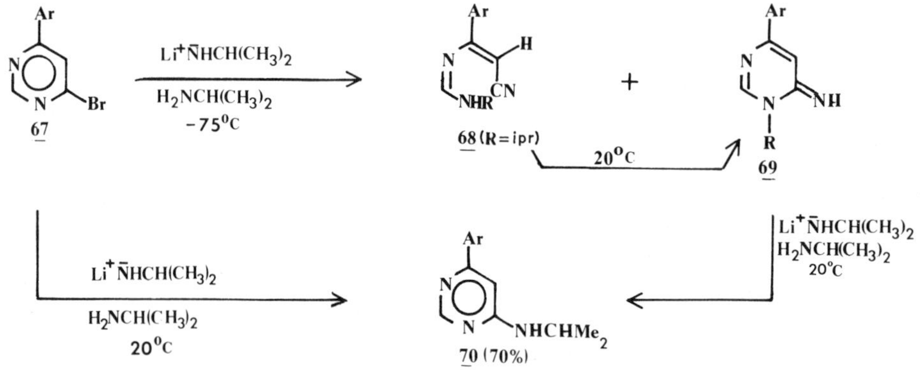

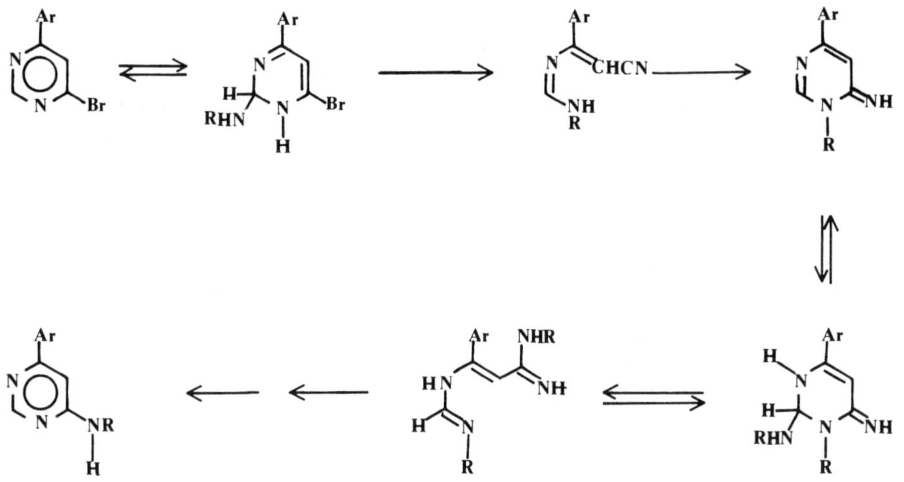

An extension of this study involves treatment of 5-bromo-4-*tert*-butylpyrimidine (**71**) with potassium amide/liquid ammonia, which only very slowly affords the corresponding amine. If two equivalents of potassium amide are used, no starting material is left after 10 minutes; NMR spectroscopy has established that the intermediate forms from the 1,2-addition anion **72.** Use of labeled nitrogen establishes that the reaction proceeds 50% via the ANRORC mechanism.[23] The problem that these results illuminate is obvious. The labeling experiment shows 50% ANRORC mechanism, whereas the NMR suggests total attack at the 6-carbon atom. It has been suggested that the discrepancy can be explained by invocation of intermediate **73,** which is presumed to be so reactive that it cannot be trapped.

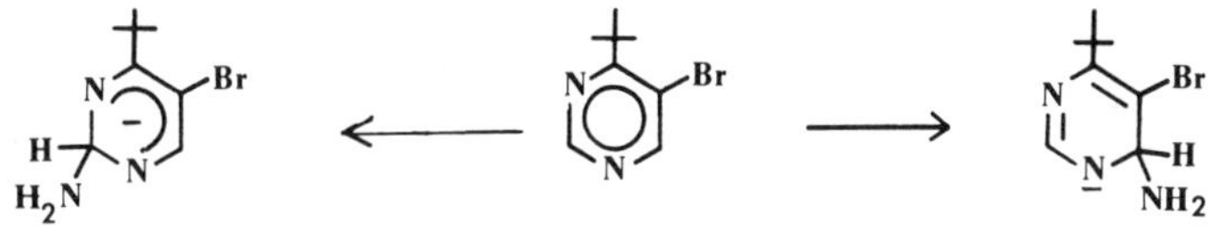

18.6. REACTIONS WITH ORGANOMETALLIC REAGENTS

The reaction of π-deficient heterocyclic compounds with Grignard reagents or organolithium compounds is basically the displacement of a hydride ion.[24] These reactions are akin to the Tschischibabin amination, previously considered.

Some of the reactions, which involve the addition of organolithium compounds to the —C=N— linkage of the π-deficient systems, do not, however, involve a direct elimination step; they result instead in the formation of dihydro derivatives. These dihydro derivatives are chemically reconverted to the aromatic systems by oxidation. Since the overall reaction is still a matter of "hydrogen" replacement, it is, superficially, the same as the "direct" alkylation-hydride displacement. Thus both of these reactions are best discussed under one heading.

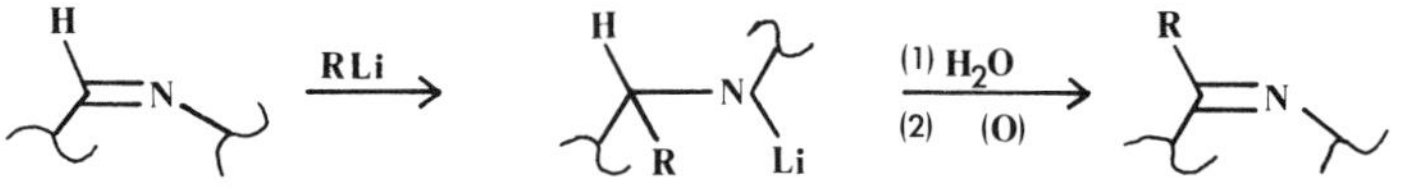

RLi; (1) H_2O (2) (O)

Alkylations of pyridines with Grignard reagents normally give low yields of the corresponding alkylpyridines. The one exception is the phenylation of pyridine which affords 2-phenylpyridine (**74**).[25]

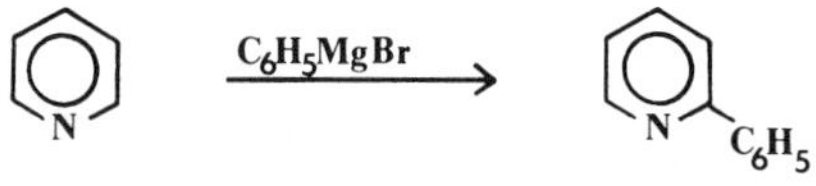

C_6H_5MgBr

74 (44%)

Pyridazine (**75**) and pyrimidine (**76**) are also phenylated with phenylmagnesium halide. In these cases, the intermediate dihydro derivatives are obtained and can be oxidized (often during the reaction workup) to the desired heteroaromatics (**77** and **78**, respectively).[26,27]

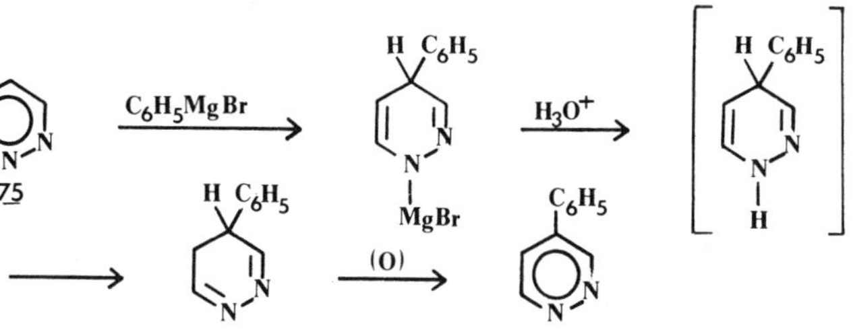

75 $\xrightarrow{C_6H_5MgBr}$ $\xrightarrow{H_3O^+}$ $\longrightarrow$ $\xrightarrow{(O)}$ 77 (7.3%)

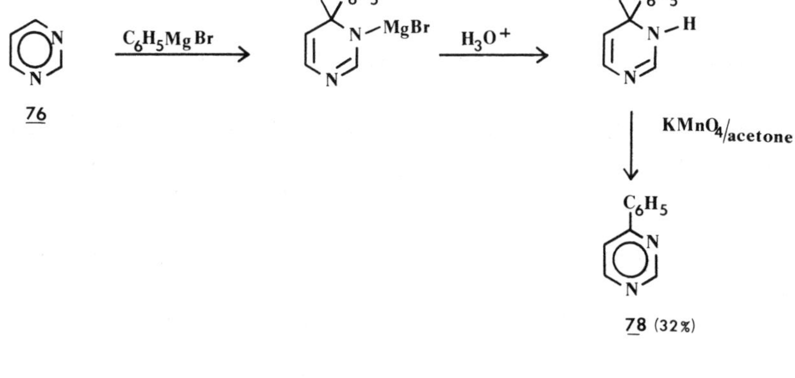

76 $\xrightarrow{C_6H_5MgBr}$ $\xrightarrow{H_3O^+}$ $\xrightarrow{KMnO_4/acetone}$ 78 (32%)

It is of some interest to note that, although the dihydropyridazine intermediate is oxidized by oxygen, the dihydropyrimidine derivative requires a much stronger oxidizing agent (e.g., permanganate) to accomplish this transformation.[24]

Treatment of pyridines with alkyl- or aryllithiums rather than with the alkyl Grignard reagents affords alkyl- and arylpyridines in moderate yields.[28]

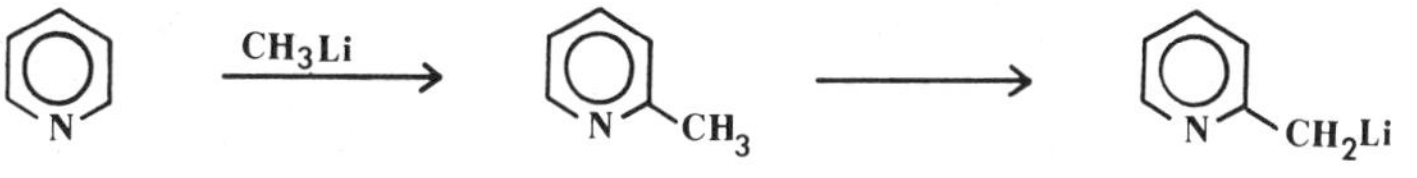

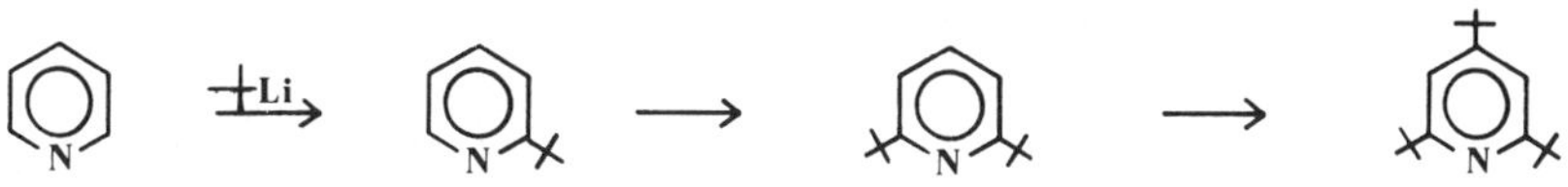

CH_3Li → CH_3 → CH_2Li

In those instances where the alkyl group introduced has no α-hydrogen (*tert*-butyllithium), and thus cannot form a carbanion, 2,6-di- and 2,4,6-trialkylpyridines can be obtained if an excess of the alkyllithium reagent is used.[29] As is the case with Grignard reagents, the diazines react with alkyl-and aryllithium compounds to form relatively stable dihydro derivatives.[26] In the case of pyridazine, alkyllithium compounds afford the 3-alkyl derivatives, whereas the 4-alkyl derivatives predominate with alkyl Grignard reagents.[26,29]

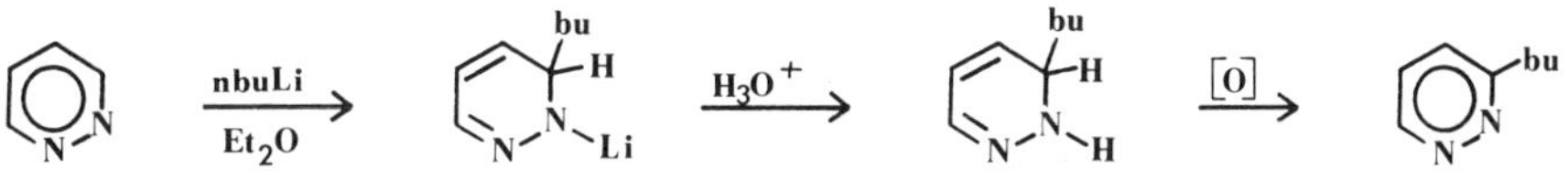

nbuLi, Et_2O → bu, H, N, Li → H_3O^+ → bu, H, N, H → [O] → bu

Alkyllithiums react with pyrimidines[30] to afford the dihydroalkyl derivatives in better yields than when the corresponding Grignard reagent is used.[31] If the 4-position of pyrimidine is blocked, substitution occurs at the 2- (**79**) and 6-positions (**80**).[32]

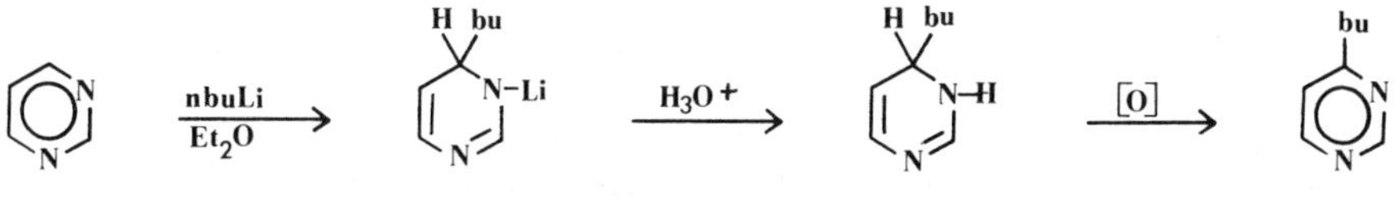

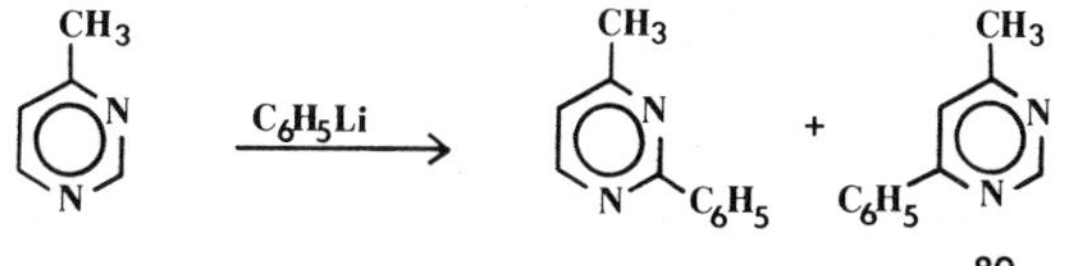

nbuLi, Et_2O → H bu, N–Li → H_3O^+ → H bu, N–H → [O] → bu

CH_3 → C_6H_5Li → CH_3, C_6H_5 (79) + CH_3, C_6H_5 (80)

In the pyrazine series, phenylation of a substituted pyrazine follows the pattern illustrated by the following reaction.[30]

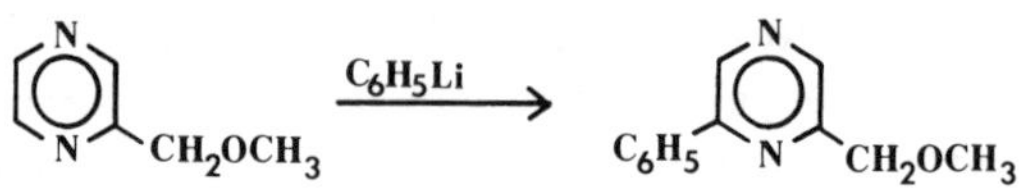

In the presence of the free metal (lithium or magnesium), alkylation of pyridines by RX occurs almost exclusively at the 4-position (**81**).[27] This type of reaction may not involve an *AE* process at all but proceed by a radical anion mechanism.[33] Hence, depending upon reaction conditions, alkylation of pyridines and diazines can take place at "preselected" locations. Consequently, these reactions have great utility in synthetic heterocyclic chemistry.

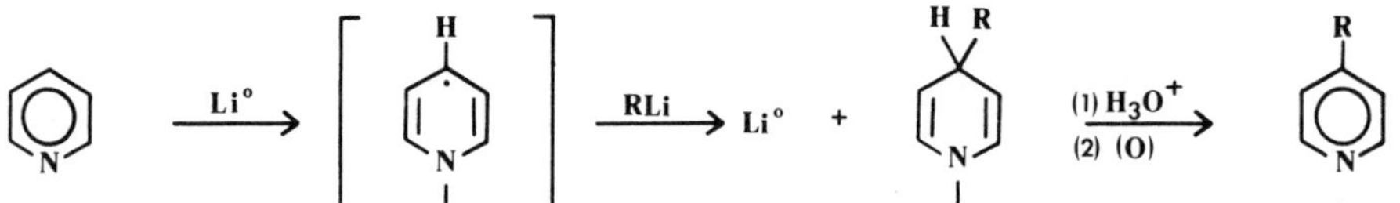

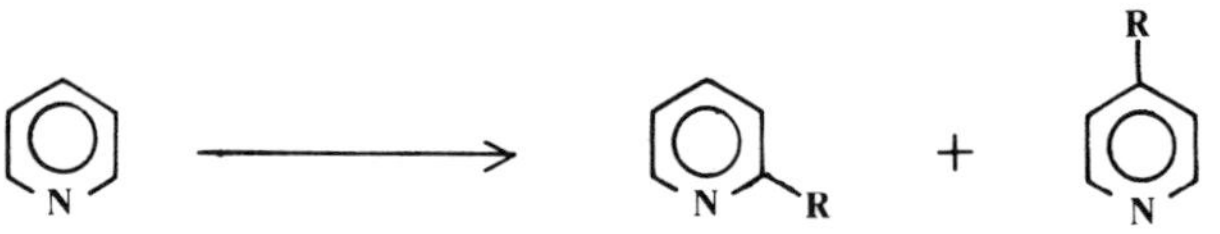

Reagents	82	(Ratio %)	81	Overall %Yield
n-BuLi	100		0	—
n-BuCl + Li	0		100	10
n-BuMgI	100		0	18
n-BuCl+ Mg	0		100	57

REFERENCES

1. K. Schofield, *Heteroaromatic Nitrogen Compounds: Pyrroles and Pyridines,* Butterworths, London, 1967, p. 277.
2. H. C. van der Plas and J. de Valk, *Rec. Trav. Chim.* **91,** 1414 (1972).
3. R. G. Shepherd and J. L. Fedrick, *Adv. Heterocycl. Chem.* **4,** 145 (1965).
4. G. Illuminati, *Adv. Heterocycl. Chem.* **3,** 285 (1964).
5. (a) M. L. Crossley and J. P. English, U.S. Pat., 2,394,963 (1946) [*Chem. Abstr.* **40,** 3143 (1946)]; (b) E. Ochiai and J. Karrii, *J. Pharm. Soc.* (*Japan*) **59,** 18 (1939).
6. (a) A. E. Tschitschibabin and O. A. Zeide, *J. Russ. Phys. Chem. Soc.* **46,** 1216 (1914) [*Chem. Abstr.* **9,** 1901 (1915)]; (b) F. W. Bergstrom and R. A. Ogg, Jr., *J. Am. Chem. Soc.* **53,** 245 (1931).
7. (a) M. Liveris and J. Miller, *J. Chem. Soc.* **1963,** 3486; (b) *Aust. J. Chem.* **11,** 297 (1958).
8. T. L. Chan and J. Miller, *Aust. J. Chem.* **20,** 1595 (1967).
9. K. Eichenberger, R. Rometsch, and J. Druey, *Helv. Chim. Acta* **39,** 1755 (1956).
10. T. Kauffmann and R. Wirthwein, *Angew. Chem. Int. Ed.* **10,** 20 (1971); (b) W. Adam, A. Grimison, and R. Hoffmann, *J. Am. Chem. Soc.* **91,** 2590 (1969); M. J. S. Dewar and G. P. Ford, *Chem. Commun.* **1977,** 539.

11. M. J. Pieterse and H. J. den Hertog, *Rec. Trav. Chim.* **80,** 1376 (1961).
12. H. J. den Hertog and H. C. van der Plas, *Adv. Heterocycl. Chem.* **4,** 121 (1965).
13. H. J. den Hertog, M. J. Pieterse, and D. J. Buurmon, *Rec. Trav. Chim.* **82,** 1173 (1963).
14. J. D. Cook, B. J. Wakefield, H. Heaney, and J. M. Jablonski, *J. Chem. Soc.* (*C*) **1968,** 2727.
15. H. C. van der Plas and G. Geurtsen, *Tetrahedron Lett.* **1964,** 2093.
16. J. H. Markgraf et al., *J. Am. Chem. Soc.* **85,** 958 (1965).
17. M. Ferles and M. Jankovsky, *Coll. Czech. Chem. Commun.* **33,** 3848 (1968).
18. H. C. van der Plas, *Tetrahedron Lett.* **1965,** 555.
19. H. C. van der Plas and A. Koudijs, *Rec. Trav. Chim.* **89,** 129 (1970).
20. J. de Valk and H. C. van der Plas, *Rec. Trav. Chim.* **92,** 145 (1973).
21. J. de Valk and H. C. van der Plas, *Rec. Trav. Chim.* **92,** 471 (1973).
22. H. C. van der Plas and A. Koudijs, *Rec. Trav. Chim.* **92,** 711 (1973).
23. J. P. Geerts et al., *Rec. Trav. Chim.* **93,** 231 (1974).
24. R. A. Abramovitch and J. S. Saha, *Adv. Heterocycl. Chem.* **6,** 229 (1966).
25. F. W. Bergstrom and S. H. McAllister, *J. Am. Chem. Soc.* **52,** 2845 (1930).
26. R. L. Letsinger and R. U. Lasco, *J. Org. Chem.* **21,** 812 (1956).
27. H. Bredereck, R. Gompper, and H. Herlinger, *Chem. Ber.* **91,** 2832 (1958).
28. F. V. Scalzi and N. F. Golob, *J. Org. Chem.* **36,** 2541 (1971).
29. I. Crossland and L. K. Ramussen, *Acta Chem. Scand.* **19,** 1652 (1965).
30. H. Bredereck, R. Gompper, and H. Herlinger, *Angew. Chem.* **70,** 571 (1958).
31. D. Bryce-Smith, P. J. Morris, and B. J. Wakefield, *Chem. Ind.* (*London*) **1964,** 495.
32. T. D. Heyes and J. C. Roberts, *J. Chem. Soc.* **1951,** 328.
33. G. R. Newkome and D. C. Hager, *J. Org. Chem.* **47,** 599 (1982).

Chapter 19

FREE RADICAL AND PHOTOCHEMICAL REACTIONS. SIX-MEMBERED π-DEFICIENT HETEROAROMATICS

19.1. FREE RADICAL REACTIONS

The addition-elimination (*AE*) reaction of free radicals with carbocyclic aromatics is a well-known process. It is only reasonable to expect that pyridines as well as the diazines will undergo this reaction. Among the heteroaromatic systems, the pyridines have been studied in greatest detail. Molecular orbital calculations on pyridine predict that free radical attack should occur preferentially at the 2-position.[1] The limited data available on homolytic heteroaromatic substitution of diazines suggest that positions α and γ to the ring nitrogens are most susceptible.[2] Some comparative reactions are given.

$C_6H_5N(NO)COCH_3$, Δ: 1 → 2 (46%) + 3 (43%) + 4 (11%)

$C_6H_5N(NO)COCH_3$, Δ: 5 → 6 (low yield!)

$C_6H_5N_2^+$: 7 → 8 (60%) + 9 (40%) (low overall yield)

Clearly, the diazines **5** and **7** give considerably lower yields in these arylation reactions than does pyridine (**1**). Free radical alkylation of pyridine has, however, been more extensively studied. The alkyl radicals have been generated by thermolysis of dialkyl peroxides or by oxidative decarboxylation of acids.[3] Respective percentages of the different alkylpyridines (**10–15**) obtained in various reactions are given below.

1, $(CH_3CO)_2O$, Δ → 10 (63%) + 11 (20%) + 12 (17%)

1, $(t\text{-BuO})_2$ → 13 (62%) + 14 (23%) + 15 (15%)

1, CH_3CO_2H, electrolysis → 10 (74%) + 11 (0%) + 12 (26%)

Even though all three alkylpyridines are formed in the reactions involving acetyl peroxide and *tert*-butyl peroxide (**10–12** and **13–15**, respectively), the 3-isomer (**11**) is not formed in the reaction that depends on oxidative decarboxylation to generate the methyl radicals. In this reaction, selectively larger amounts of 2- and 4-methylpyridines are obtained. Similarly, homolytic acylation of pyridine exclusively affords the 2- and 4-substituted compounds (**16** and **17**), with no 3-substituted product being detected. These acyl radicals are conveniently generated from aldehydes, a hydroperoxide, and a ferrous salt.[4–6]

$$\underset{1}{} \xrightarrow[\uparrow\ Fe^{2+} + R'CO_2H + R'CHO]{[R'CO\cdot]} \underset{16}{\text{2-}COR'\text{-pyridine}} + \underset{17}{\text{4-}COR'\text{-pyridine}}$$

Free radical amidation of pyridine (**18, 19**) and pyrazine (**20**) has been examined.[6] Again, no 3-substituted pyridines are formed in this reaction. The amido radicals are readily obtained by oxidation of formamide with hydrogen peroxide in the presence of a ferrous salt.

$$\underset{20}{\text{2-}CONH_2\text{-pyrazine}} \xleftarrow[{[\cdot CONH_2]}]{X=N} \text{(X-azine)} \xrightarrow[{[\cdot CONH_2]}]{X=CH} \underset{18}{\text{2-}CONH_2\text{-pyridine}} + \underset{19}{\text{4-}CONH_2\text{-pyridine}}$$

At sufficiently high temperatures, the vapor phase halogenation of pyridine yields mixtures of the 2- and 2,6-dihalopyridines (**21, 22**). This is a free radical reaction, since electrophilic substitution on pyridine has been demonstrated to take place at C-3.[7]

$$\underset{1}{\text{pyridine}} \xrightarrow[300\text{-}350°C]{X_2} \underset{21}{\text{2-X-pyridine}} + \underset{22}{\text{2,6-}X_2\text{-pyridine}}$$

Free radical arylation and alkylation of a number of substituted pyridines reveals that, as expected, the isomer distribution depends on both the nature of the substituent and the means by which the free radicals are generated.[8,9] When phenyl radicals are generated by thermolysis of benzoyl peroxide, isomer distribution A is obtained. Isomer distribution B prevails when benzoyl peroxide is heated in a mixture of acetic and hydrochloric acids. The isomer ratio obtained in this latter instance is very similar to that obtained for the various dimethylpyridines (**27–30**) when 2-methylpyridine is allowed to react with methyl radicals that are generated by thermolysis of methyl iodide.

$$\text{2-}CH_3\text{-pyridine} \xrightarrow{C_6H_5\cdot} \underset{23}{\text{3-}C_6H_5\text{-2-}CH_3} + \underset{24}{\text{4-}C_6H_5\text{-2-}CH_3} + \underset{25}{\text{5-}C_6H_5\text{-2-}CH_3} + \underset{26}{\text{6-}C_6H_5\text{-2-}CH_3}$$

	23	24	25	26
A:	31	15	20	34
B:	8	41	4.5	46

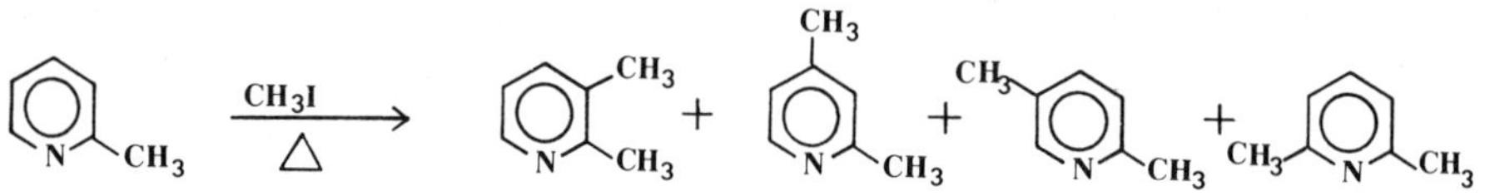

CH_3I, Δ

Isomer Ratio: 27: 8.3; 28: 32.4; 29: 2.1; 30: 57.2

Arylation of 2-methoxypyridine (**31**)[9] affords the 3-phenyl isomer **33** as the major product along with minor amounts of the 6-phenyl isomer. In **32,** the presence of the cyano group deactivates the 3-position relative to the influence of a 2-methoxy group and yields equal amounts of the 3- and 5-phenyl isomers (**33, 34**).

Phenylation of 3-methoxy- and 3-cyanopyridine affords the 2-, 4-, and 6-arylated products in the given ratio, along with traces of the 5-isomer.

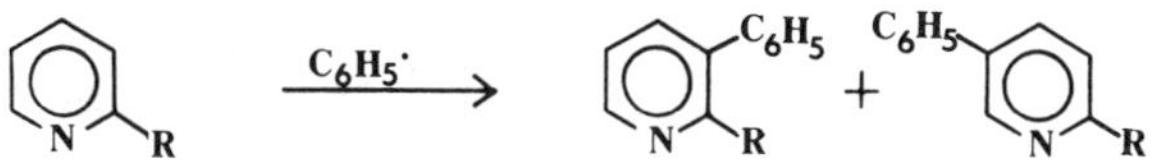

$C_6H_5\cdot$

32 R=C≡N (R=CN: Isomer Ratio: 33 : 41)
31 R=OMe

Arylation of a number of 4-substituted pyridines (**35–37**) yields the arylated isomers in the indicated ratios.

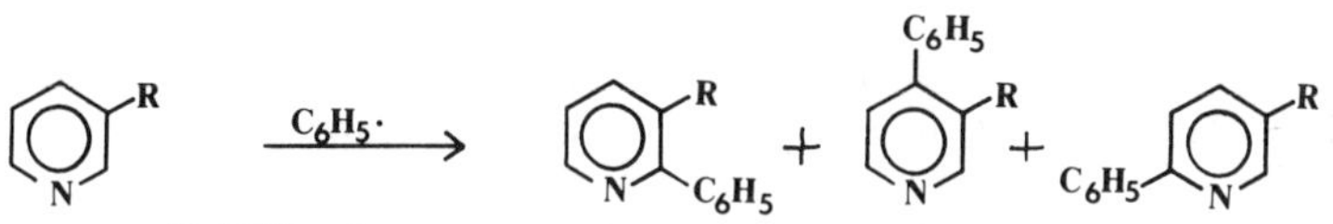

$C_6H_5\cdot$

	Isomer Ratio:		
$R=OCH_3$	59	19	14
R=CN	33	25	36

$C_6H_5\cdot$

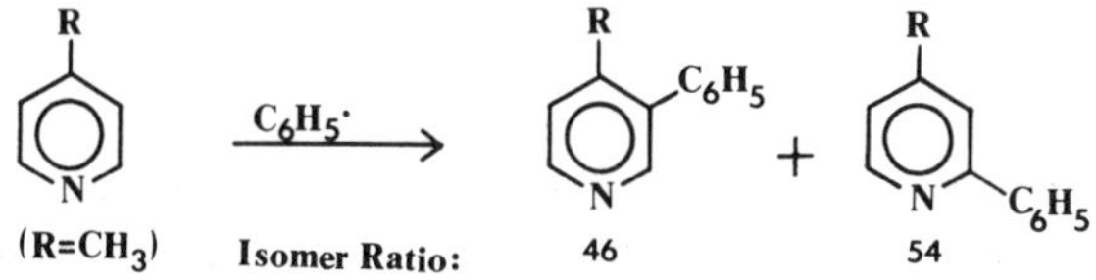

	Isomer Ratio:	
35 ($R=CH_3$)	46	54
36 ($R=OCH_3$)	66	34
37 (R=CN)	76	24

19.2. PHOTOCHEMICAL TRANSFORMATIONS

Irradiation of pyridine in solution at 253.7 nm gives a Dewar pyridine (**38**) which is extremely unstable: it reverts to pyridine unless it is "trapped" by conversion to a dihydro (**39**) derivative or by hydrolysis to **40.**[10] The alternative "Dewar pyridine" (**41**) is not detected after irradiation of simple pyridines, but irradiation of **42** gives the 1,4-bonded Dewar pyridine (**43**), which thermally reverts to **42.**[11] When **43** is treated with $NaOCD_3$ in CD_3OD, deuterium exchange occurs at the 2- and 6-methyl groups, but *not* at the 4-methyl group; under identical exchange conditions, *all* three methyl groups of **42** undergo exchange. These data suggest that a "push–pull" electronic interaction contributes to the enhanced stability of **43.**[11] Palladium(II) and platinum(II) complexes of **43** have been isolated and upon standing they slowly isomerize to complexes of **42.**[12]

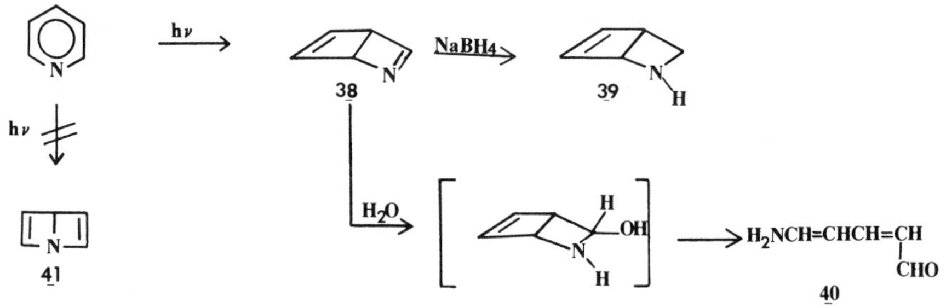

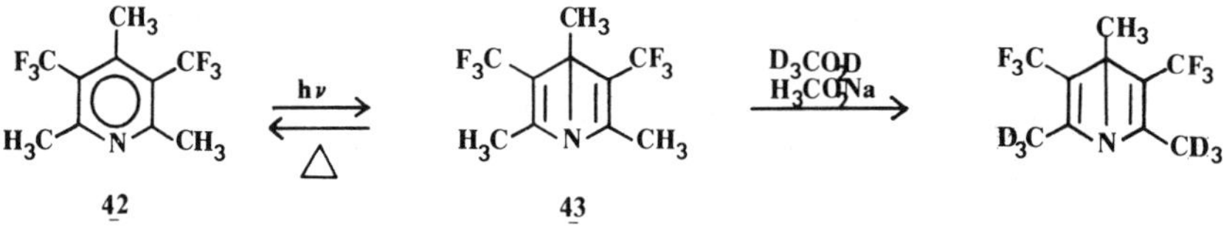

When pyridazine **44** is irradiated, it is converted to pyrimidine **46.** This transformation is envisioned to occur via the tetrafluorodiazabenzvalene **45.**[13]

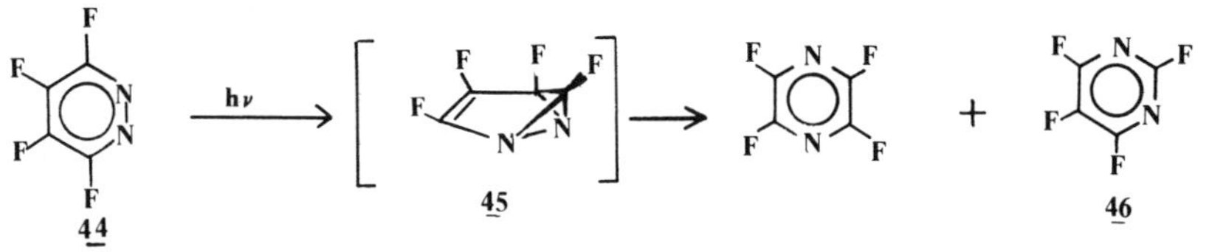

2,6-Dialkylpyridines when irradiated undergo an equilibrium transformation that is best explained by the intermediacy of an azaprismane **47.**[11] Similar transformations have been observed for 2-methyl- (**48**), 2,6-dimethyl- (**49**), and 2,5-dimethylpyrazine (**50**).[14] The probable intermediate in all of these transformations is the appropriate valence isomer: azaprismane **47** or azabenzvalene **45.**

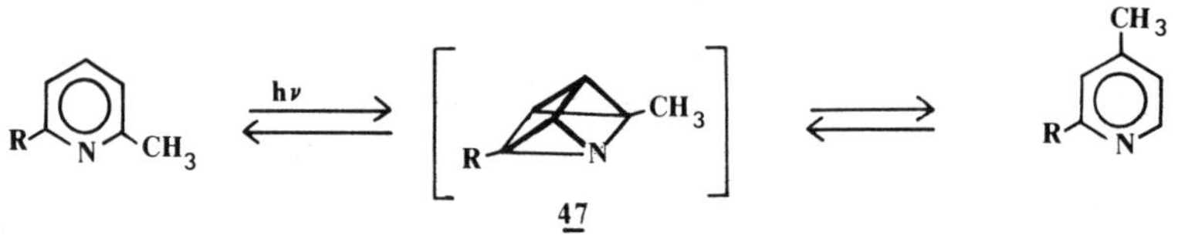

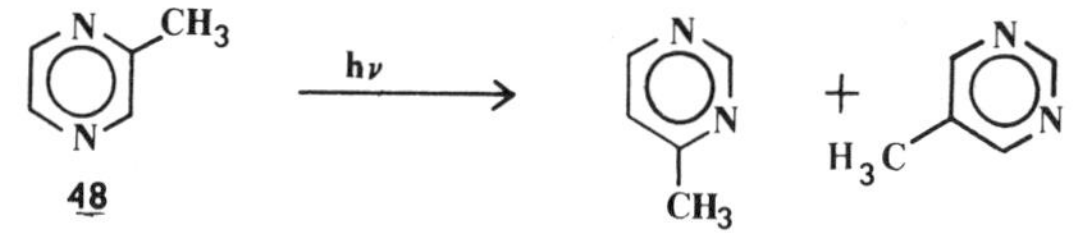

48

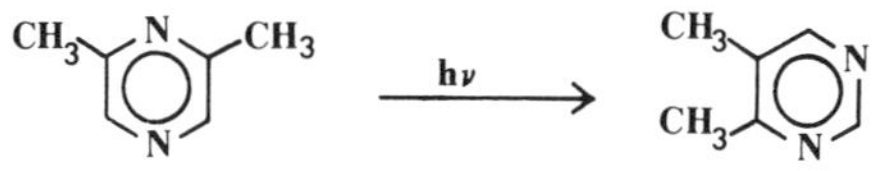

49

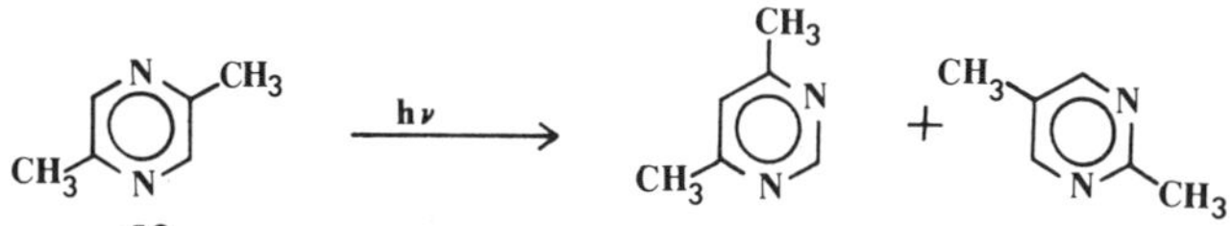

50

Photolysis of α-pyridinones (**51**)[15] and 2-aminopyridine hydrochloride (**52**)[16] gives rise to a dimer (**53a** or **53b**, respectively). When **54** is photolyzed under nitrogen at 20–25°C for 15 hours with a mercury arc lamp, anthranilinonitrile (**55**) is isolated in good yield. This transformation of **54** (or the methide **56**) may proceed through the 3-substituted 2-azabicyclo[2.2.0]hexa-2,5-diene (**57**) or **58**, which rearranges either by a radical or thermal [3,3]-sigmatropic rearrangement to **59**, followed by rearomatization to **55**.[17]

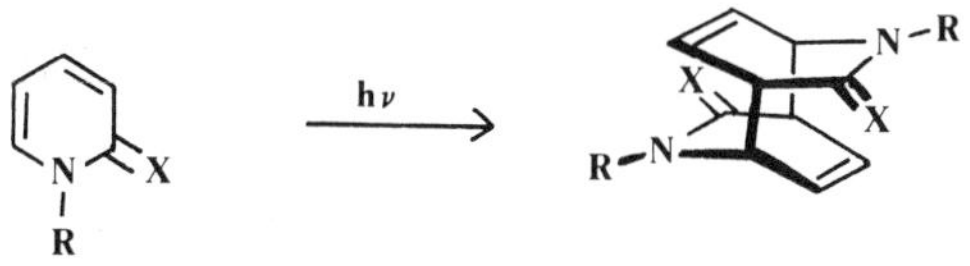

51 (X= O)

52 (X=NH; R = H·HCl)

53a (X=O)

53b ($X=\overset{+}{N}H_2$)

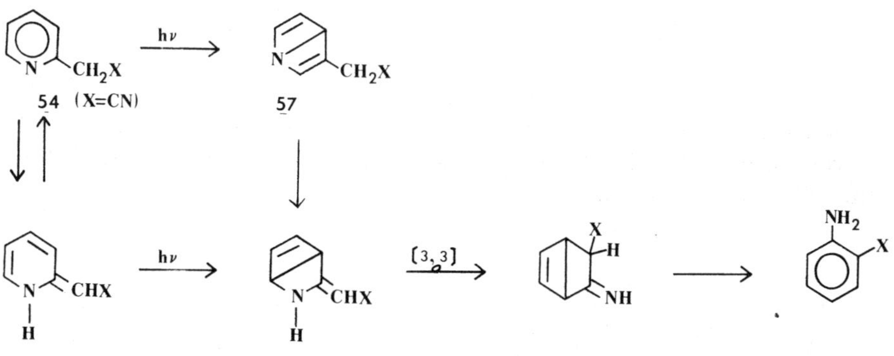

54 (X=CN)

57

56

58

59

55 (44%) (X=CN)

Azabenzvalene **60**, yet another possible valence isomer of pyridine, has been proposed in the photochemical transformation of *N*-methylpyridinium chloride.[18] The photochemical transformation of zwitterion **61** probably occurs via a similar intermediate.[19]

N^+ CH_3 Cl^- — $h\nu$, H_2O → [60] → N–CH_3, H OH

61 (C_6H_5, +N, O^-) — $h\nu$ → N–C_6H_5, O

N-Substituted pyridines undergo a variety of interesting photochemical transformations[20–23]; some of these are given below.

R +N R, –N CO_2Et — $h\nu$, acetone → R, R, N, N–CO_2Et

— $h\nu$, benzene → N–CO_2Et + $C_6H_5NHCO_2Et$

+N–$C(CN)_2$ — $h\nu$, benzene → CN, CN + N–H, CN, CN, H

Photosubstitution of the cyano group in **32** by ethanol[24] and of the substituent in 4-cyano-(or chloro-)pyridine by 1,3-dioxolane (**62**) (an excellent source of carboxaldehyde)[25] has been reported. The latter is suggested to proceed by the following mechanism:

X, +N–H (X=CN; Cl) — $h\nu$, [Hg^0] → [X, +N–H]* + **62** → [X, H, N+, H; H–dioxolanyl] **63**

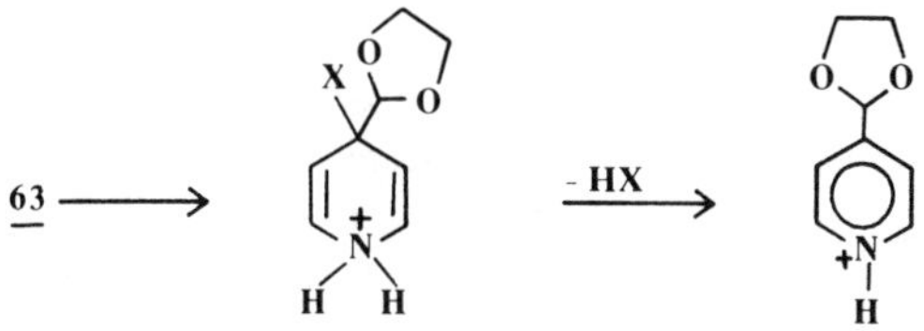

Heteroaromatic *N*-oxides have been used as oxygen sources.[26] When *N,N'*-dimethyltryptamine (**64**) is irradiated (253.7 nm) with pyridine *N*-oxide or benzo[*c*]cinnoline *N*-oxide, the intermediate epoxide **65** is trapped by an intramolecular cyclization to give **66** in low but reasonable yield.[27]

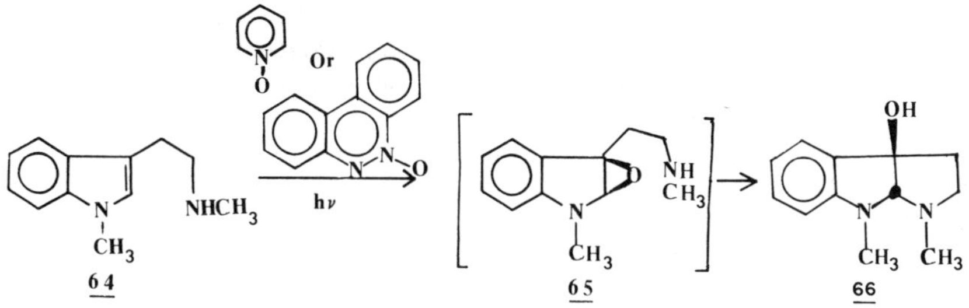

REFERENCES

1. J. Palecek, V. Skala, and J. Kuthan, *Coll. Czech. Chem. Commun.* **34,** 1110 (1969).
2. H. J. M. Dou, G. Vernin, and J. Metzger, *Bull. Soc. Chim. Fr.* **1971,** 4593; G. Vernin, H. J. M. Dou, and J. Metzger, *Bull. Soc. Chim. Fr.* **1972,** 1173; F. Minisci and O. Porta, *Adv. Hetercycl. Chem.* **16,** 123 (1974); G. Vernin, *Bull. Soc. Chim. Fr.* **1976,** 1257.
3. K. C. Bass and P. Nababsing, *J. Chem. Soc.* (*C*) **1970,** 2169.
4. (a) F. Minisci, R. Bernardi, F. Bertini, R. Galli, and M. Perchinunno, *Tetrahedron* **27,** 3575 (1971); (b) W. Buratti, G. P. Gardini, F. Minisci, F. Bertini, R. Galli, and M. Perchinunno, *Tetrahedron* **27,** 3655 (1971).
5. T. Caronna, G. Fronza, F. Minisci, and O. Porta, *J. Chem. Soc. Perkin II* **1972,** 2035.
6. R. Bernardi, T. Caronna, R. Galli, F. Minisci, and M. Perchinunno, *Tetrahedron Lett.* **1973,** 645.
7. M. M. Boudakian, F. F. Frulla, D. F. Gavin, and J. A. Zaslowsky, *J. Heterocycl. Chem.* **4,** 375, 377 (1967).
8. J. M. Bounier and J. Courd, *Compt. Rend.* (*C*) **265,** 133 (1967).
9. R. Dufournet, J. Court, and J. M. Bonnier, *Bull. Soc. Chim. Fr.* **1974,** 1112.
10. K. E. Wilzbach and D. J. Rausch, *J. Am. Chem. Soc.* **92,** 2178 (1970).
11. Y. Kobayahsi, A. Ohsawa, M. Baba, T. Sato, and I. Kumadaki, *Chem. Pharm. Bull.* (*Japan*), **24,** 2219 (1976); R. D. Chambers, R. Middleton, and R. P. Corbally, *J. Chem. Soc. Chem. Commun.* **1975,** 731.
12. Y. Kobayahi and A. Ohsawa, *Chem. Pharm. Bull.* (*Japan*) **24,** 2225 (1976).
13. C. G. Allison, R. D. Chambers, Yu. A. Cheburkov, J. A. H. MacBride, and W. K. R. Musgrave, *Chem. Commun.* **1969,** 1200.

14. F. Lahmani and N. Ivanoff, *Tetrahedron Lett.* **1967,** 3913.
15. R. C. De-Selms and W. R. Schleigh, *Tetrahedron Lett.* **1972,** 3563, and refs. cited therein.
16. L. A. Paquette and G. Slomp, *J. Am. Chem. Soc.* **85,** 765 (1963).
17. Y. Ogata and K. Takagi, *J. Am. Chem. Soc.* **96,** 5933 (1974).
18. L. Kaplan, J. W. Paulik, and K. E. Wilzbach, *J. Am. Chem. Soc.* **94,** 3283 (1972).
19. A. R. Katritzky and H. Wilde, *J. Chem. Soc. Chem. Commun.* **1975,** 770.
20. T. Sasaki, K. Kanematsu, A. Kakehi, I. Ichikawa, and K. Hayakawa, *J. Org. Chem.* **35,** 426 (1970).
21. R. A. Abramovitch and T. Takaya, *J. Org. Chem.* **38,** 3311 (1973).
22. V. Snieckus and G. Kan, *Chem. Commun.* **1970,** 172.
23. J. Streith, A. Blind, J. M. Cassal, and C. Sigwalt, *Bull. Soc. Chim. Fr.* **1969,** 948.
24. T. Furihata and A. Sugimori, *J. Chem. Soc. Chem. Commun.* **1975,** 241.
25. T. Caronna, S. Morrocchi, and B. Vittimberga, *Chem. Ind.* **58,** (1976).
26. J. Streith, B. Danner, and C. Sigwalt, *Chem. Commun.* **1967,** 979; H. Igeta, T. Tsuchiya, M. Yamada, and H. Arai, *Chem. Pharm. Bull.* (*Japan*) **16,** 767 (1968); H. Igeta, and T. Tsuchiya, *J. Syn. Org. Chem. Japan* **31,** 867 (1973).
27. M. Nakagawa, T. Kaneko, H. Yamaguchi, T. Kawashima, and T. Hino, *Tetrahedron* **30,** 2591 (1974).

Chapter 20

REDUCED SIX-MEMBERED π-DEFICIENT HETEROCYCLIC COMPOUNDS

20.1. GENERAL COMMENTS

Because π-deficient heteroaromatic compounds react with nucleophiles, it is not surprising that they can be readily reduced by nucleophilic reducing agents such as complex metal hydrides.[1,2] The more electron-deficient pyridine derivatives, such as pyridinium salts, can even be reduced by mild reducing agents like sodium dithionite. Reduction with sodium metal transforms many of these compounds to their perhydro derivatives. The dihydro and tetrahydro derivatives that are possible for pyridine (**1**) are given on the next page:

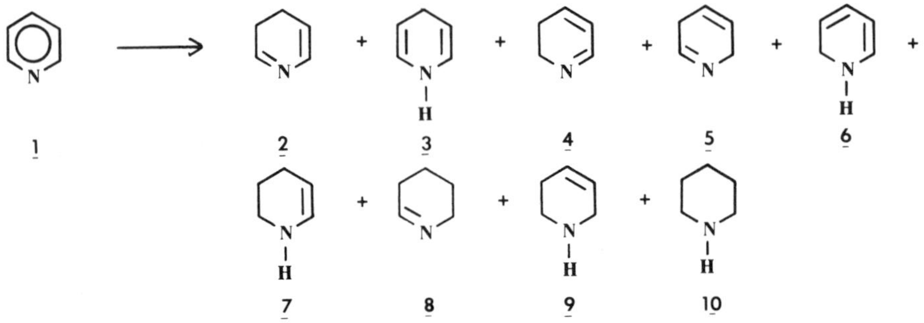

1 2 3 4 5 6

7 8 9 10

Two dihydro isomers (**2** and **4**) result from reduction of a C=C bond, and one isomer (**6**) from reduction of the C=N bond. The 1,4-dihydropyridines (**3** and **5**) result from a 1,4-reduction and are the only dihydro compounds among these which lack conjugated double bonds. The three tetrahydro isomers (**7–9**) generally react in accordance with their structures: an enamine (**7**), an imine (**8**), and an allylamine (**9**). The perhydropyridine, known as piperidine (**10**), is a "normal" secondary amine, analogous to diethylamine.

For pyridazine, five dihydro and four tetrahydro derivatives are possible. Pyrimidine can give rise to five dihydro and three tetrahydro compounds, and pyrazine can, in principle, afford four dihydro and two tetrahydro isomers. Since the relative position of the two nitrogens in these compounds changes from 1,2 to 1,3 to 1,4, significant chemical differences between them can be expected. This is especially true for pyridazine and pyrimidine derivatives, for which reduction affords hydrazino and amidino derivatives, respectively.

20.2. SYNTHESES

Reduction of pyridine (**1**) with lithium aluminum hydride forms a complex (**11**) which contains both 1,2- and 1,4-dihydropyridines.[3] Although it has not yet been possible to isolate the dihydropyridines from this complex, it is nevertheless a very useful selective reducing agent for ketones. The reduction of pyridine and some of its derivatives with $LiAlH_4$ in the presence of aluminum chloride[4] affords a mixture of 1,2,3,6-tetrahydropyridines (**9**) and piperidine (**10**).

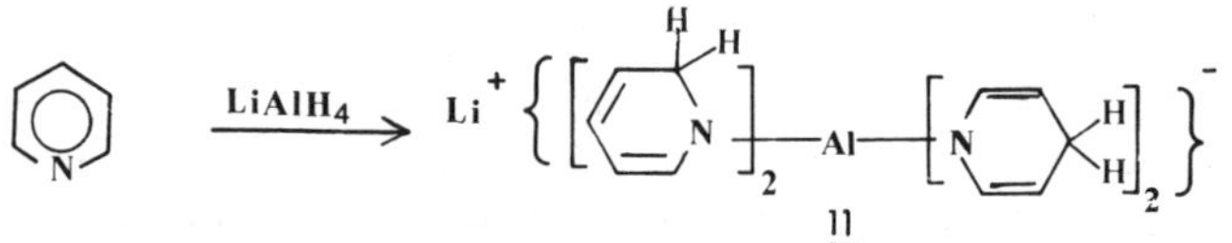

$LiAlH_4$ → Li^+ 11

1 $LiAlH_4$ / $AlCl_3$ → 9 + 10

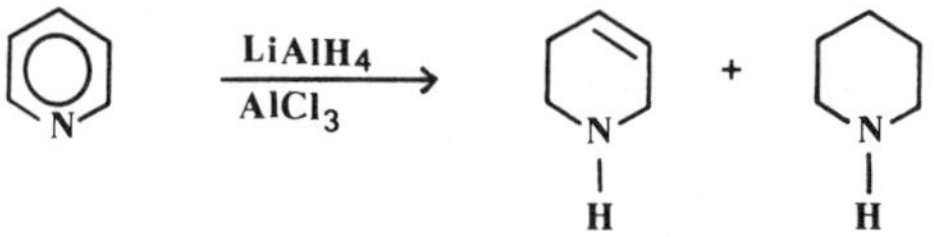

Even though $NaBH_4$ does not appear to reduce pyridine, itself, it does form 1,2- (**12**) and 1,4-dihydro (**13**) derivatives when strongly electron-withdrawing groups are present.[5–7] A few examples are known in which reduction to tetrahydro derivatives occurs in the presence of protic solvents.[8]

$NaBH_4$ / EtOH (→ **12**); $NaBH_4$ / pyridine (→ **13**)

12 **13**

The quaternized 3-cyanopyridine (**14**), when subjected to $NaBH_4$ reduction in protic solvents, affords the 1,2- (**15**) and 1,6-dihydro (**16**) and the 1,2,5,6-tetrahydro (**17**) compounds. This contrasts with the formation of the 1,2,3,4-tetrahydro derivative formed when 3-cyanopyridine is reduced under these conditions.

14 —$NaBH_4$ / MeOH→ **15** + **16** + **17**

Among the pyridazines, 4,5-dihydro-3(2*H*)-pyridazinone (**18**) has been reduced with $LiAlH_4$ to the 1,4,5,6-tetrahydro derivative (**19**). An excess of $LiAlH_4$ reduces **19** to the hexahydro (**20**) derivative.

18 —$LiAlH_4$→ **19** —$LiAlH_4$ excess→ **20**

The use of a metal, such as sodium, in the reduction of these ring systems is a procedure of long standing. Generally, the hexahydro derivatives are obtained (**10, 20, 22, 25**). Metal-induced reduction of pyrimidines (**21**) affords low yields of hexahydropyrimidines which exist in equilibrium with the unstable imine **23.**

Na/ROH → **10**

→ **20** (low yields)

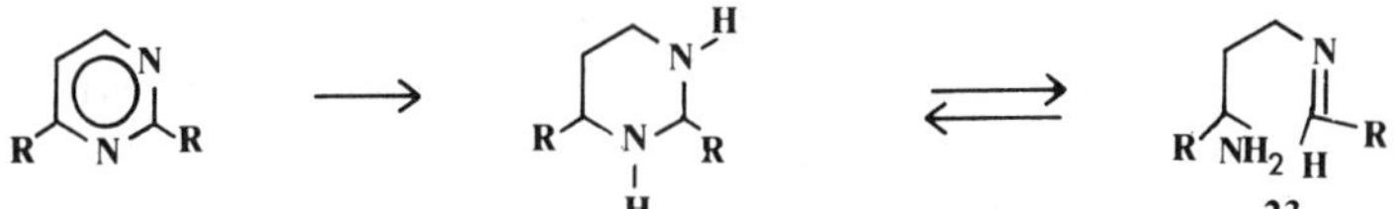

21 → 22 ⇄ 23

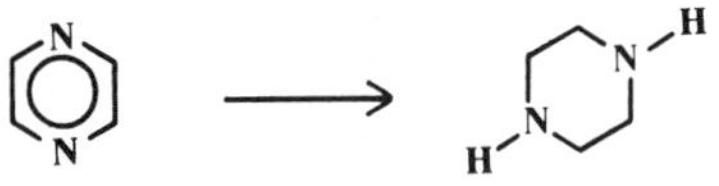

24 → 25

When the sodium metal reduction of pyridine or some of its derivatives is done in aprotic solvents,[9] tetrahydrobipyridyls (**26**) are obtained that are readily air-oxidized to bipyridyl (**27**). 3,6-Diphenylpyridazine (**28**) when treated with sodium in alcohol affords the dihydro derivative **29.**[2] These 1,2-dihydro derivatives are easily oxidized to the aromatic starting materials.

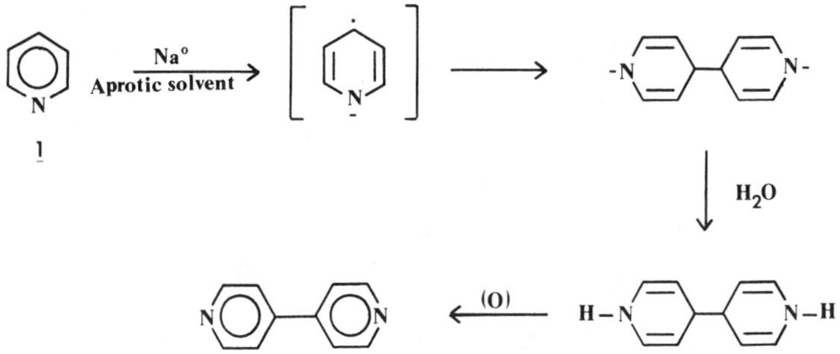

Na° / Aprotic solvent

H_2O

(O)

1 27 26

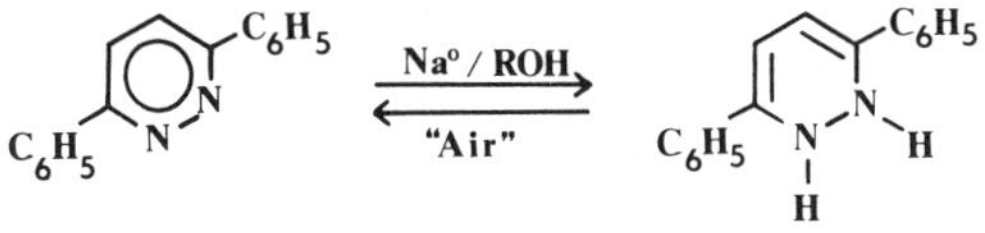

Na° / ROH

"Air"

28 29

Although either catalytic or $LiAlH_4$ reduction of 3-cyanopyridine affords the same tetrahydro derivative (**13**), a 3-carboxylic acid substituent alters this reaction[10] and affords the 2,3,4,5-tetrahydro derivative **8** as the result of a facile reductive decarboxylation.

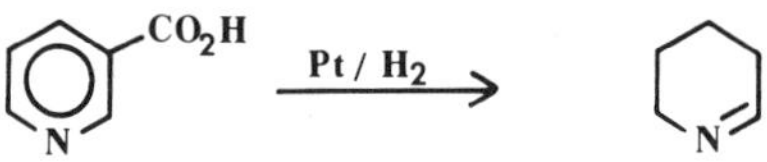

Pt / H_2

8

Catalytic reduction of the double bond in 1,2,3,6-tetrahydropyridazines (**30**) has been employed to prepare hexahydro derivatives (**20**).[2]

N–H, N–H **30** —(Pt / H_2)→ N–H, N–H **20**

The catalytic reduction of uridine (**31**) over Adam's catalyst represents an example of the formation of a 5,6-dihydropyrimidine (**32**).

O, N–H, N, O, $HOCH_2$, O, OH OH **31** —(Pt/H_2)→ O, N–H, N, O, $HOCH_2$, O, OH OH **32**

Addition of organometallic reagents to pyridines, diazines, or their quaternary salts gives rise to relatively stable 1,2-dihydro derivatives. The following examples will serve to demonstrate these reactions.[12–16] All these 1,2-dihydro derivatives (as well as the 1,4-dihydro compound **33**, which is an exception) are readily oxidized to the substituted heteroaromatic compounds.

N, H, $CH_2C_6H_5$, H (63%) + H $CH_2C_6H_5$, N, H **33** ←(1) $C_6H_5CH_2MgCl$ (2) H_3O^+— N —(1) C_6H_5Li (2) H_3O^+→ N, H, C_6H_5, H

NC, N^+, CH_3, I —(1) C_6H_5MgBr (2) H_3O^+→ NC, N, C_6H_5, H, CH_3

N, N —(1) nbuLi / Et_2O (2) H_3O^+→ nbu, H, N, N–H

R, N–H, N, H, bu + R, N, H, N, bu, H ←1. n buLi 2. H_3O^+— R, N, N —(R=H) (1) C_6H_5MgBr (2) H_3O^+→ H, C_6H_5, N–H, N

A rather unusual reduction occurs when pyrazine is reduced with lithium in the presence of trimethylsilyl chloride.[17] The resultant 1,4-disubstituted pyrazine **34** spontaneously decomposes in air.

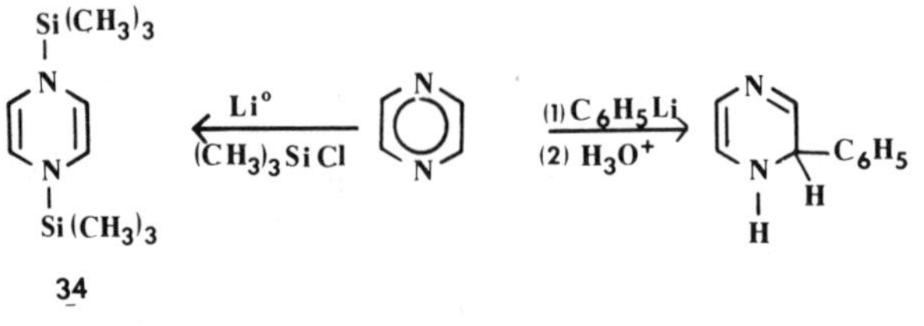

Si(CH3)3 Si(CH3)3 Li° (CH3)3SiCl (1) C6H5Li (2) H3O+ C6H5 H H

34

A considerable amount of information exists on the direct syntheses of these reduced ring systems. The Hantzsch pyridine synthesis lends itself ideally to the preparation of 1,4-dihydropyridines (**35**).[6,18,19]

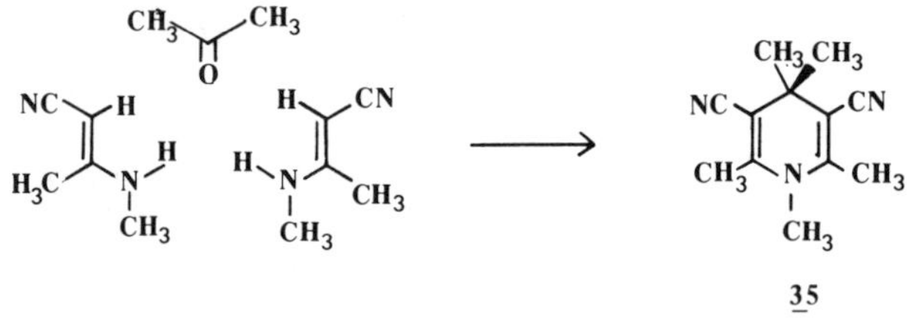

35

Direct synthesis of 1,2-dihydropyridines (**36**) can also be accomplished, where some cycloaddition reactions[6] have been employed in this quest.

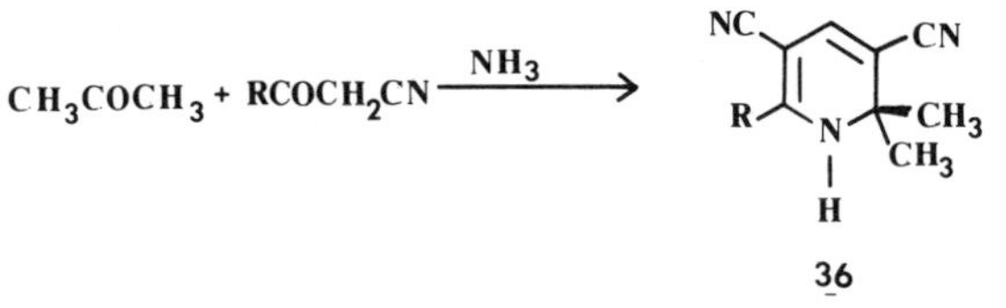

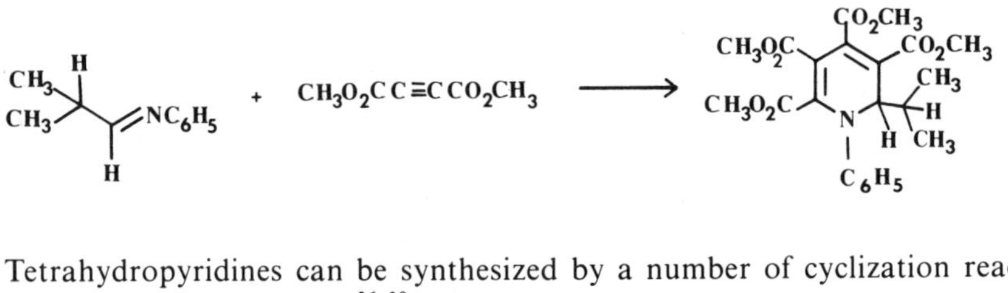

$CH_3COCH_3 + RCOCH_2CN \xrightarrow{NH_3}$

36

Tetrahydropyridines can be synthesized by a number of cyclization reactions. Some examples follow[26–30]:

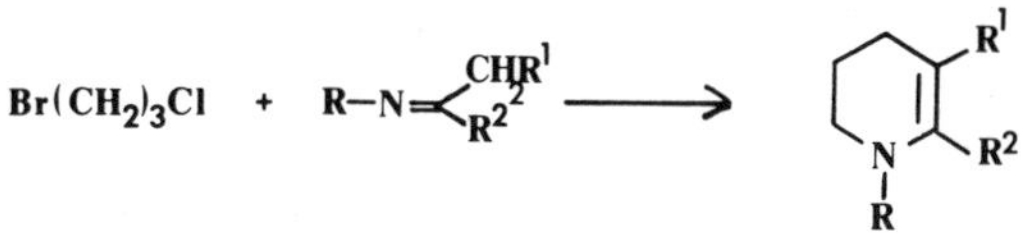

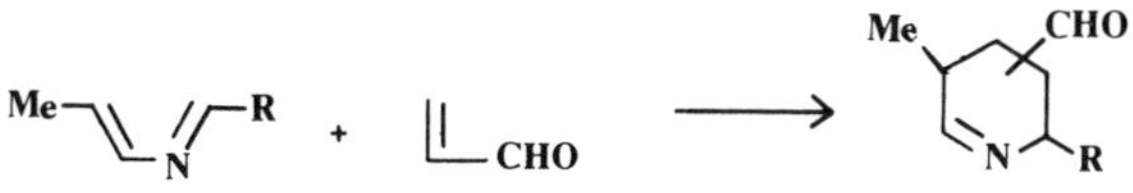

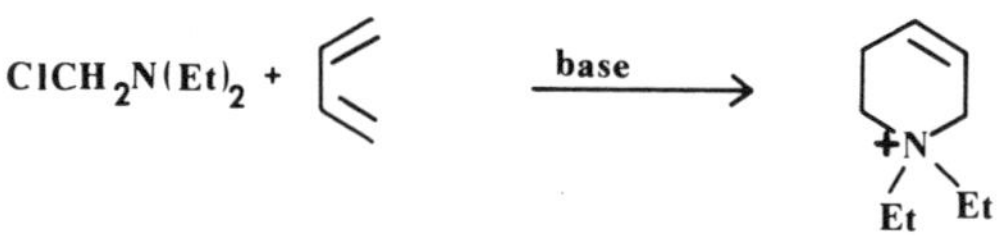

The interaction of 1,4-dicarbonyl compounds (**37**) with hydrazine, or the reaction of 1,2,4,5-tetrazines (**38**) with olefins, affords easy access to 1,4-dihydropyridazines. Both of these types of dihydropyridazines (**39** and **40**) are readily oxidized to the aromatic pyridazines (**41** and **42**, respectively).[2] Use of the Diels–Alder reaction generates tetrahydropyridazines (**44**).[2] Basic hydrolysis of the carbethoxy groups in **44** followed by acidification affords 1,2,3,6-tetrahydropyridazines (**43**).

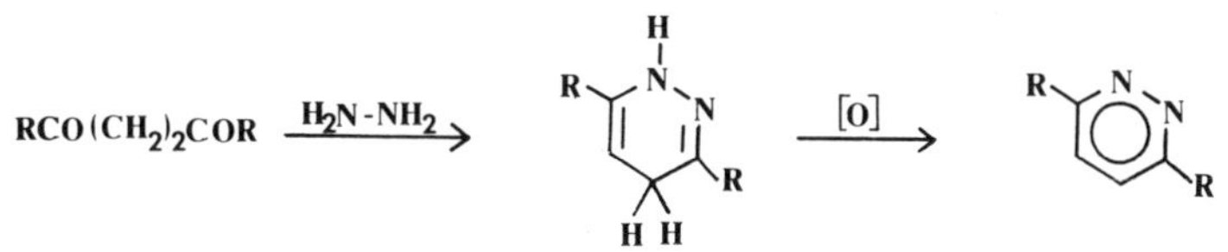

37 39 41

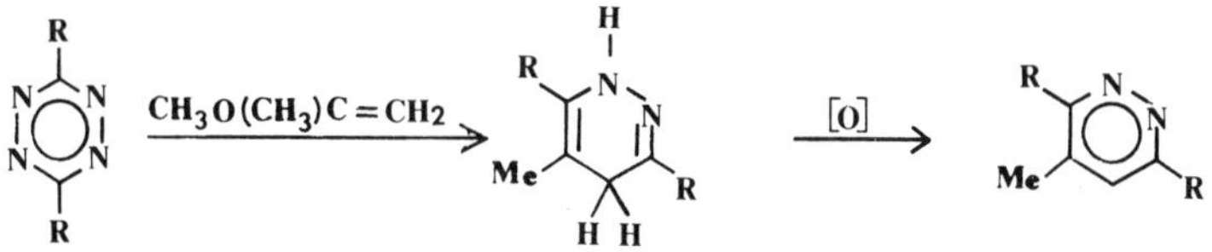

38 40 42

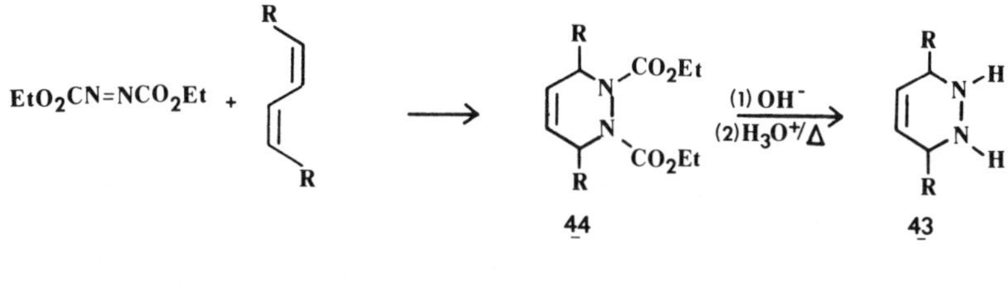

44 43

The reaction of urea with α, β-unsaturated carbonyl compounds yields 5,6-dihydropyrimidines (**45**).[20] When malondiamide reacts with appropriate malonic acid derivatives, 2,5-dihydropyrimidines (**46**) are obtained.[21] Acylation[22] of 1,3-diaminopropane also affords 1,4,5,6-tetrahydropyrimidines (**47**). Whenever the acyl halide is replaced by either an aldehyde or a ketone in the above reaction, hexahydropyrimidines (**48**) are obtained.[23] These 1,1-diaminoalkyl derivatives are not particularly stable and exist in equilibrium with the open-chain forms **49** and **50.**

O OH $\xrightarrow{H_2NCONH_2}$ O N H N O H

45

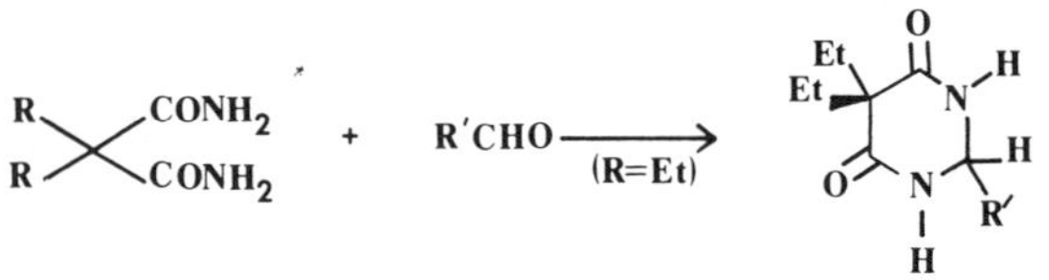

R R $CONH_2$ $CONH_2$ + $R'CHO \xrightarrow[(R=Et)]{}$ Et Et O N H H O N R' H

(R'=H, alkyl, aryl)

46

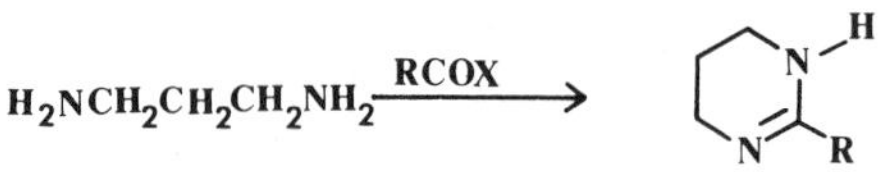

$H_2NCH_2CH_2CH_2NH_2 \xrightarrow{RCOX}$ N H N R

47

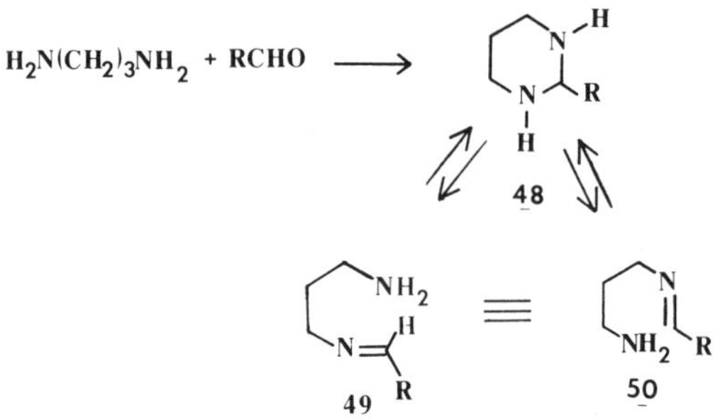

$H_2N(CH_2)_3NH_2$ + RCHO ⟶ N H N R H

48

NH_2 N H R ≡ N NH_2 R

49 50

The 5,6-dihydropyrazines (**51**) are the intermediates in pyrazine syntheses which normally result from the condensation of 1,2-ethylenediamine with α,β-dicarbonyl compounds.[24] These particular dihydro intermediates are amazingly stable. The facile dimerization of α-aminoketones gives rise to 3,6-dihydropyrazines (**52**) in a much employed reaction.[25]

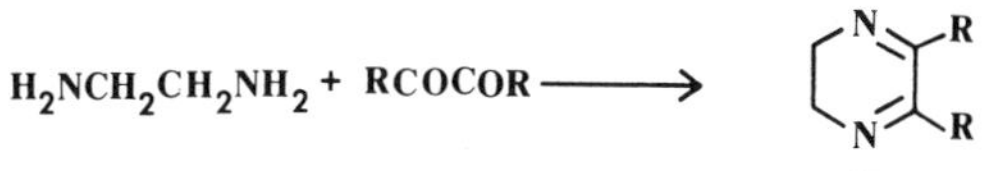

$$H_2NCH_2CH_2NH_2 + RCOCOR \longrightarrow$$

51

$$2\ H_2NCH_2COR \xrightarrow{\Delta}$$

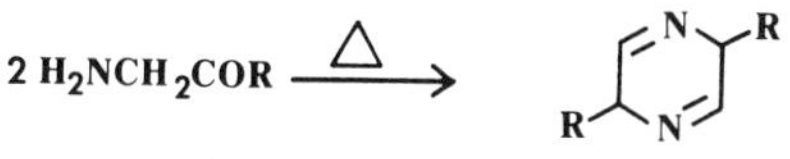

52

Tetrahydropyrazines are rather rare.[25] The 2-phenyl-3,3-dimethyl-3,4,5,6-tetrahydropyrazine (**53**) is formed by the following reaction:

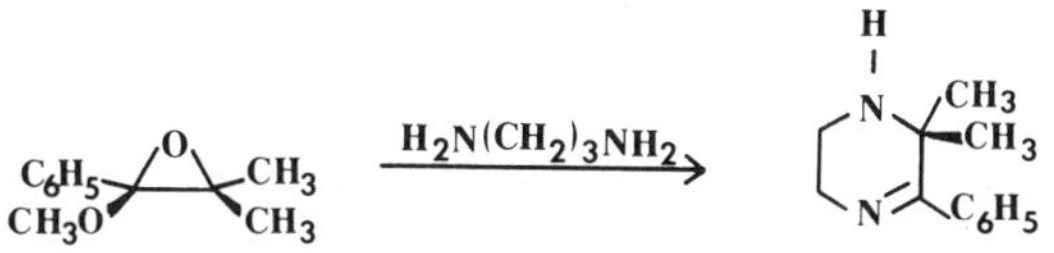

$$\xrightarrow{H_2N(CH_2)_3NH_2}$$

53 (91%)

Hexahydropyrazine (**25**), although directly available by reduction of pyrazine, is commercially prepared by the reaction of ethylene oxide with ethylenediamine.

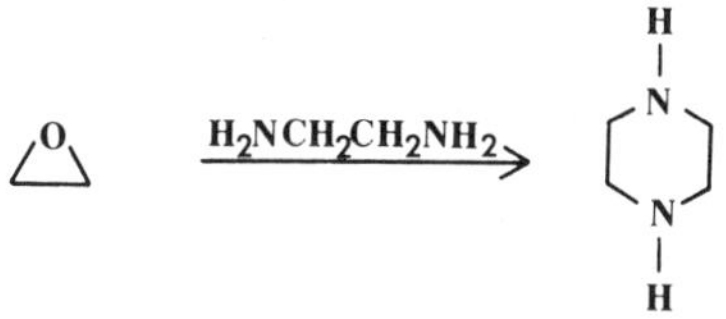

$$\xrightarrow{H_2NCH_2CH_2NH_2}$$

25

20.3. CHEMICAL PROPERTIES

The coenzyme dihydronicotinamide adenine dinucleotide (NADH) (**54**) contains a 1,4-dihydropyridine ring which can enzymatically reduce ketones and is converted to NAD^+ (containing a pyridinium ring). Much work has been directed toward the synthesis of chiral and achiral model systems that can achieve this goal. The anions **55** obtained during reduction of pyridine with $LiAlH_4$ or with sodium in liquid ammonia[3,31] readily undergo reaction with electrophiles, to afford 3-substituted products (**56**) via an enamine reaction. These compounds are also extremely reactive and are readily oxidized to the substituted pyridines (**57**). This substitution behavior does not apply to the anions of the 1,4-dihydropyridines (**58**). Furthermore, these anions are alkylated on nitrogen (**59**) rather than carbon.[32]

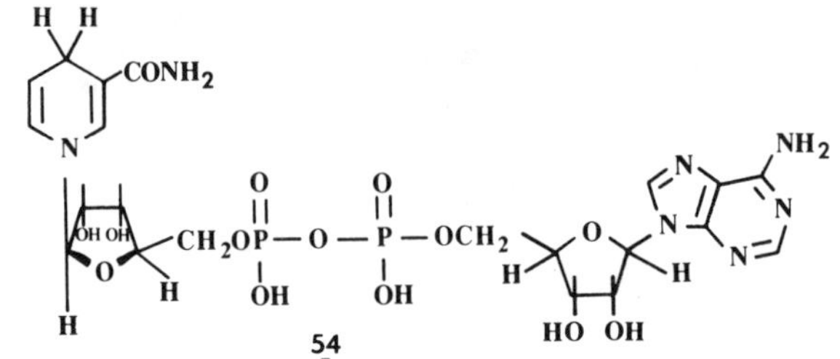

54

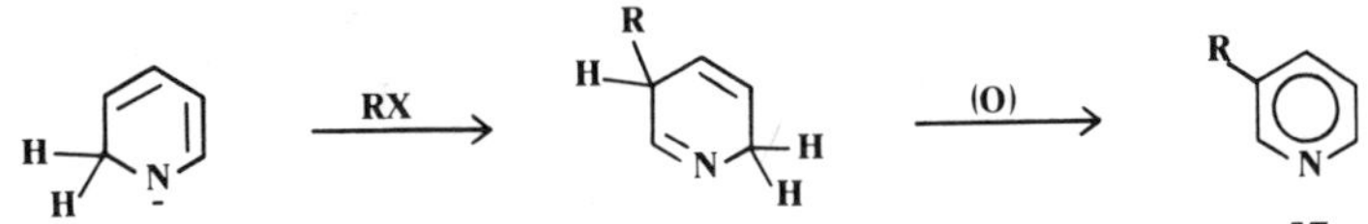

55 → RX → 56 → (O) → 57

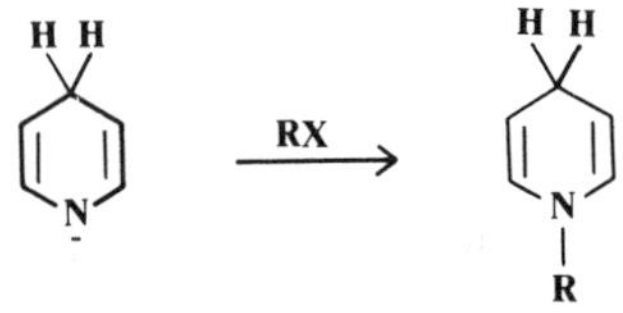

58 → RX → 59

N-Substituted 1,4-dihydropyridines (**59**) are much less reactive than their nonalkylated counterparts (**56**). One interesting example of a reaction of a *N*-substituted 1,4-dihydropyridine (**60**) is the following[33]:

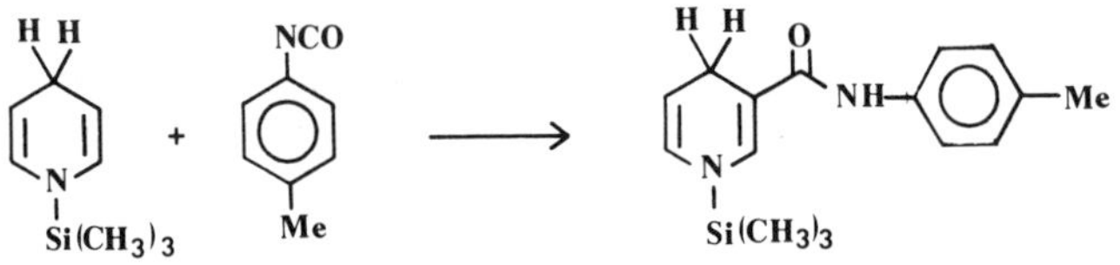

60

The reactivity of 1,4-dihydropyridines with Arrhenius or Lewis acids[34,35] is put to good use in the synthesis of tetrahydropyridines (**61**) by their reduction with sodium borohydride.

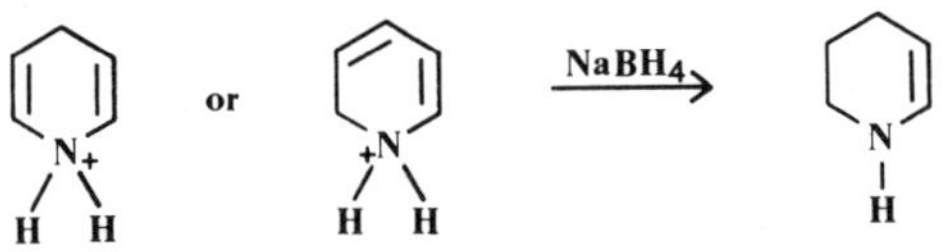

$\xrightarrow{NaBH_4}$

61

The diene moiety of 1,2-dihydropyridines (**62**) undergoes an intermolecular Diels–Alder reaction as long as no electron-donating substituents are present at the *terminal* points of the diene system.[36] A new synthesis of desethylcatharanthine (**63**) utilizes the cycloaddition reaction of **64** with **65.** Intramolecular Diels–Alder type cyclizations have been hypothesized (**66** → **63**).[37]

CN
H
H
N
$CH_2C_6H_5$
62
$CH_2{=}CH\ COCH_3$ →
NC
$N\text{-}CH_2C_6H_5$
$COCH_3$

$OSiMe_2(tbu)$
N
H
CO_2CH_3
64
+
N
$CO_2CH_2CCl_3$
65
→
$OSiMe_2(tbu)$
N
$CO_2CH_2CCl_3$
N
H
(30%) CO_2CH_3
(1) AcOH
(2) $MeSO_2Cl$
NEt_3
(3) Zn/CH_3CN

N
N
R
H
CO_2CH_3
66
Δ
(R = H)
N
N
H
CO_2CH_3
R
63

1,4-Dihydropyridines containing the structural moiety $-CH_2-$ halogen at C-4 rearrange with base to form azepines (**68**), which can rearrange further to form cyclopentadiene derivatives (**67**).[38]

Cl
R
H
H
NC
CN
H_3C
N
CH_3
CH_3
K_2CO_3
DMSO
→
CN
CH_3
R
$N-CH_3$
CH_3
CN
68
→
R
CN
CH_3
H_3COC
CN
67

3
N
→
N
N
N
69

Certain tetrahydropyridines are extremely reactive and readily trimerize (**69**).[10,39] A novel application of 1,2-dialkyl-1,2,3,6-tetrahydropyridines[40] has been devised in which they can be converted into (9*Z*, 11*E*)-tetradecadienyl acetate **70,** the sex pheromone of *Spodoptera littoralis*[41] and *S. litura.*[42] Compound **71** is alkylated to give the *N*-methylpyridinium salt which is then reduced with $NaBH_4$ to give **72** with high regioselectivity. Quaternization with methyl iodide and subsequent Hoffmann elimination gives the pure *N,N*-dimethyl (2*Z*, 4*E*)-alkadienamine **73,** which is again *N*-methylated. Coupling of **73** with a suitable Grignard reagent and subsequent minor alterations gives **70** (THP is tetrahydropyranyl ether).[43]

71 → (1) CH_3I/CH_3OH, (2) $NaBH_4$ → 72 → (1) CH_3I, CH_3OH, (2) 9 M KOH/Δ → 73 (45%)

→ (1) CH_3I, (2) Li_2CuCl_4 (5%), THF, (3) $ClMg(CH_2)_7OTHP$, (4) $TosOH/CH_3OH/\Delta$, (5) Ac_2O/Pyr → 70 (35%)

A novel procedure to generate the key intermediates in the Robinson annelation reaction has been devised by Danishefsky and co-workers[44]: the appropriately substituted pyridine (**74**) is reduced to the labile bis-enamine (**75**) by sodium in liquid ammonia. Subsequent hydrolysis of **75** results in the generation of a diketone which under the prevailing conditions cyclocondenses to give **76.**

→ (1) H_3O^+, (2) $HOCH_2CH_2OH/H^+$ → 74 → (1) Na/NH_3, (2) NaOH, (3) H_3O^+ → [75 → diketone] → → 76

In the quest for new syntheses of alkaloids, the palladium-induced hydrogenation of pyridines and their *N*-alkyl salts which possess a variety of β-acyl substituents has afforded a highly stable β-aminoacryl moiety: tetrahydropyridine **77.**[45] Application to the synthesis of lupinine **78** is shown by the reduction of **79** to give **80,** which upon treatment with acid in aprotic media yields **81.** Lithium aluminum hydride reduction, acid hydrolysis of the ketal, and Wolff–Kishner reduction gives lupinine **78** and epilupinine.[46]

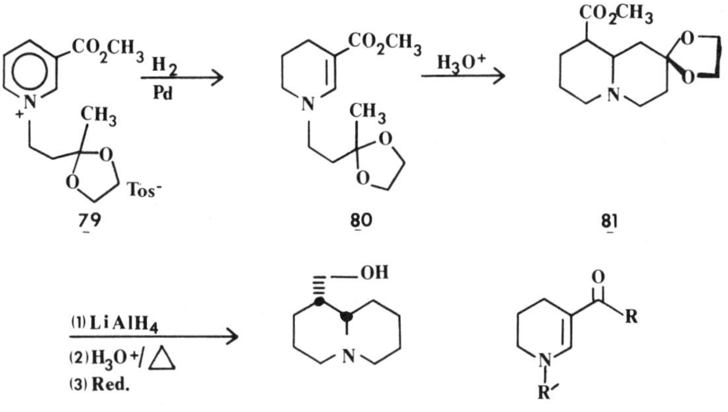

REFERENCES

1. R. E. Lyle and P. S. Anderson, *Adv. Heterocycl. Chem.* **6,** 45 (1966).
2. M. Tisler and B. Stanovnik, *Adv. Heterocycl. Chem.* **9,** 211 (1968).
3. F. Bohlmann, *Chem. Ber.* **85,** 390 (1952); also see A. J. De Koning, P. H. M. Budzelaar, J. Boerma, and G. J. van der Kerk, *J. Organomet. Chem.* **199,** 153 (1980).
4. M. Ferles, *Sb. Vys. Sk. Chem.—Technol. Praze Oddil Fak. Anorg. Org. Technol.,* **1960,** 519 [*Chem. Abstr.* **55,** 24740 (1961)].
5. U. Eisner and J. Kuthan, *Chem. Rev.* **72,** 1 (1972).
6. R. E. Lyle, in *Pyridine and Its Derivatives*, Suppl. 1, R. A. Abramovitch, Ed., Wiley, New York, 1974, p. 139.
7. E. Booker and U. Eisner, *J. Chem. Soc. Perkin I* **1975,** 929.
8. E. Bordignon, A. Signor, I. J. Fletcher, A. R. Katritzky, and J. R. Lee, *J. Chem. Soc.* (*B*) **1970,** 1567.
9. C. R. Smith, *J. Am. Chem. Soc.* **46,** 414 (1924).
10. E. Wenkert, K. G. Dave, and F. Haglid, *J. Am. Chem. Soc.* **87,** 5461 (1965).
11. P. A. Lavene and F. B. LaForge, *Chem. Ber.* **45,** 608 (1912).
12. C. S. Giam and J. L. Stout, *Chem. Commun.* **1970,** 478; C. S. Giam and E. E. Knaus, *Tetrahedron Lett.* **1971,** 4961.
13. C. S. Giam and S. D. Abbott, *J. Am. Chem. Soc.* **93,** 1294 (1971).
14. G. J. Dubsky and A. J. Guillarmod, *Helv. Chim. Acta* **52,** 1735 (1969).
15. R. E. Lyle and E. White, *J. Org. Chem.* **36,** 772 (1971).
16. G. Fraenkel and J. C. Cooper, *Tetrahedron Lett.* **1968,** 1825.

17. R. A. Sulzbach and A. F. M. Iqbal, *Angew. Chem. Int. Ed.* **10,** 127 (1971).
18. W. Traber and P. Karrer, *Helv. Chim. Acta* **41,** 2066 (1958).
19. A. Sims and P. Smith, *Proc. Chem. Soc.* **1958,** 282.
20. E. Fischer and G. Roeder, *Chem. Ber.* **34**, 3751 (1901).
21. H. Burrows and C. A. Keane, *J. Chem. Soc.* **1907,** 269.
22. A. W. Hoffmann, *Chem. Ber.* **21,** 2332 (1888).
23. G. E. K. Branch, *J. Am. Chem. Soc.* **38,** 2466 (1916).
24. J. P. Marion, *Chimia* **21,** 510 (1967).
25. C. L. Stevens, K. A. Taylor, and M. E. Mink, *J. Org. Chem.* **29,** 3574 (1964), and refs. cited therein.
26. K. Blaha and O. Cervinka, *Adv. Heterocycl. Chem.* **6,** 147 (1966).
27. R. Lukes and J. Kovar, *Coll. Czech. Chem. Commun.* **19,** 1227 (1954).
28. G. Y. Kondrateva and Y. S. Dolskaya, *Zh. Org. Khim.* **6,** 2200 (1970).
29. H. Böhme, K. Hartke, and A. Müller, *Chem. Ber.* **96,** 607 (1963).
30. W. J. Middleton and C. G. Krespan, *J. Org. Chem.* **33,** 3625 (1968).
31. R. A. Dommisse and F. C. Alderwireldt, *Bull. Soc. Chim. Belg.* **82,** 441 (1973).
32. F. J. Villani, C. A. Ellis, M. D. Yudis, and J. B. Morton, *J. Org. Chem.* **36,** 1709 (1971).
33. E. J. Moriconi and R. E. Misner, *J. Org. Chem.* **34,** 3672 (1969).
34. H. B. Charman and J. M. Rowe, *Chem. Commun.* **1971,** 476.
35. F. Liberatore, V. Carelli, and M. Cardellini, *Tetrahedron Lett.* **1968,** 4735.
36. D. Craig, A. Kuder, and J. Efroymson, *J. Am. Chem. Soc.* **72,** 5236 (1950); Y. Ban, T. Wakamatsu, Y. Fujimoto, and T. Oishi, *Tetrahedron Lett.* **1868,** 3383.
37. C. Marazano, J. -L. Fourrey, and B. C. Das, *J. Chem. Commun.* **1981,** 37.
38. P. J. Brignell, U. Eisner, and H. Williams, *J. Chem. Soc.* **1965,** 4226.
39. E. Wenkert, K. G. Dave, F. G. Haglid, R. G. Lewis, T. Oishi, R. V. Stevens, and M. Terashima, *J. Org. Chem.* **33,** 747 (1968).
40. M. Saunders and E. H. Gold, *J. Am. Chem. Soc.* **88,** 3376 (1966).
41. B. F. Nesbitt et al., *Nature (London), New Biol.* **244,** 208 (1973).
42. Y. Tamaki, H. Noguchi, and T. Yushima, *Appl. Entomol. Zool.* **8,** 200 (1973).
43. G. Decodts, G. Dressaire, and Y. Langlois, *Synthesis* **1979,** 510.
44. S. Danishefsky and R. Cavanaugh, *J. Am. Chem. Soc.* **90,** 520 (1968); S. Danishefsky and P. Cain, *J. Org. Chem.* **39,** 2925 (1974), and refs. cited therein.
45. E. Wenkert, *Accts. Chem. Res.* **1,** 78 (1968).
46. E. Wenkert, K. G. Dave, and R. V. Stevens, *J. Am. Chem. Soc.* **90,** 6177 (1968).

Chapter 21

ALKYL-SUBSTITUTED PYRIDINES AND DIAZINES

21.1. GENERAL COMMENTS

The removal of a proton from a methyl, methylene, or methine carbon that is substituted on a carbon atom either α or γ to the ring nitrogen in π-deficient heteroaromatics affords carbanions which are considerably stabilized by delocalization involving the ring nitrogen(s) (**1a–3a**):

N CH$_2^-$ **1** ⟷ N CH$_2$ **1a**   N CH$_2^-$ **2** ⟷ -N CH$_2$ **2a**

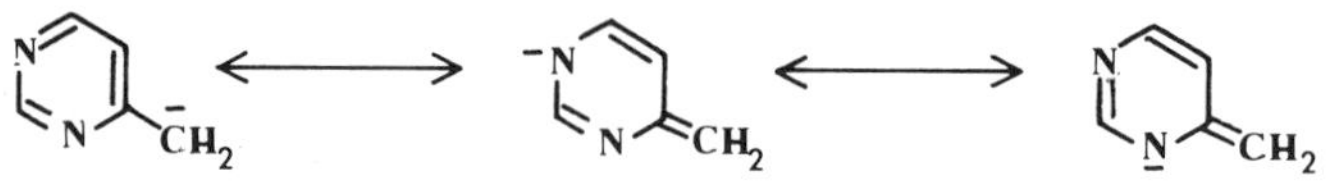

3 3a 3b

When the ring nitrogen(s) carry a positive charge, such as in *N*-methyl (**4**) and *N*-oxide (**5**) derivatives, this stabilization is expected to be even more significant than for the neutral analogues. The enhanced acidity of alkyl substituents in electron-deficient heteroaromatics account for their chemical reactivities.

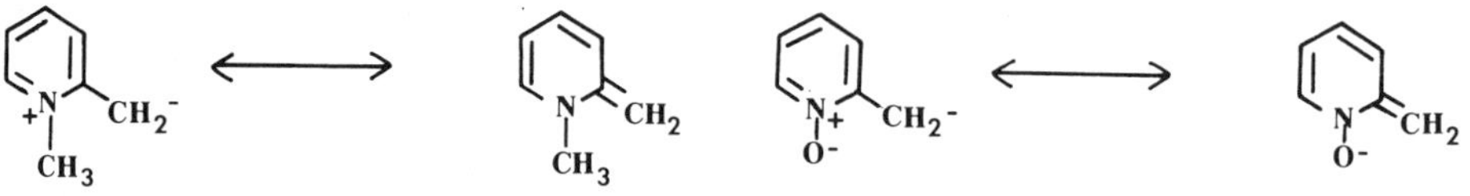

4 4a 5 5a

21.2. REACTIONS WITH CARBONYL COMPOUNDS

When either 2-methyl- (**6**) or 4-methylpyridine (**7**) (common names: α- and γ-picoline, respectively) is treated with acetic anhydride and zinc chloride, the corresponding condensation products **8** and **9** are obtained in reasonable yields.[1-5] Reaction with benzaldehyde, followed by treatment with acid, affords the corresponding stilbazoles (**10** and **11**, respectively).[2] In a similar manner, 2,5-dimethylpyrimidine (**12**) affords the benzylidene derivative (**13**).[6] Clearly, the formation of **13** to the exclusion of **14** (the latter involving a methyl group β to the ring nitrogen) is in agreement with the increased acidity of the α- as opposed to the β-alkyl hydrogens. 4-Methylpyridazine (**15**) reacts with chloral to afford the alcohol **16**.[7] Under more strenuous conditions, **16** could presumably be dehydrated to **17**.

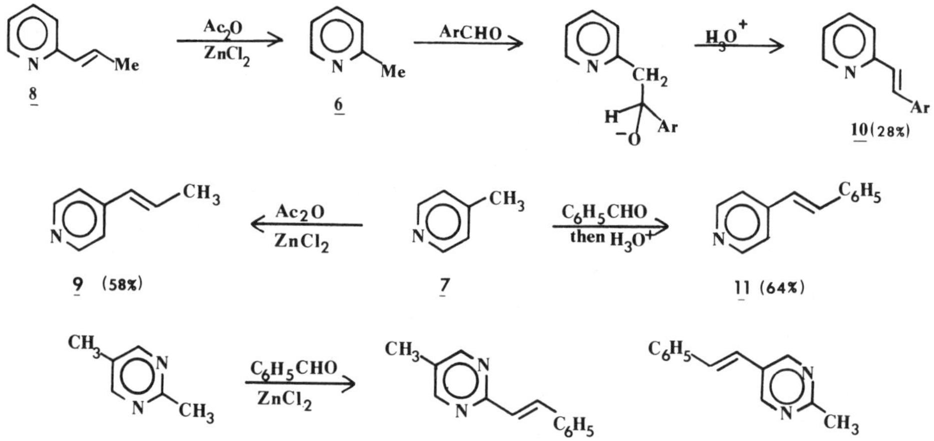

8 ← Ac_2O, $ZnCl_2$ — 6 — ArCHO → — H_3O^+ → 10 (28%)

9 (58%) ← Ac_2O, $ZnCl_2$ — 7 — C_6H_5CHO then H_3O^+ → 11 (64%)

12 — C_6H_5CHO, $ZnCl_2$ → 13; 14 (none)

15 → (Cl_3CCHO) → 16 (CH_2CHCCl_3, OH) → (?) → 17 (CCl_3)

The "Anil synthesis" also takes advantage of the activated methyl group, in that 2,6-dimethylpyridine (**18**) reacts with anils of aromatic aldehydes in *N,N*-dimethylformamide (DMF) and potassium *t*-butoxide to give distyryl derivatives (**19**).[8] Care must be exercised when DMF is used since it must be rigorously purified to remove traces of cyanide ion which result from photodecomposition.[9,10] If unpurified or "aged" DMF is used, alternative reactions, derived from cyanide catalysis, become competitive.

18 + Ar—N=CH—Ph → (KOt bu, DMF) → 19 (78%)

21.3. REACTIONS WITH STRONG BASES

When any of these electron-deficient compounds are treated with sodamide in liquid ammonia, or with phenyllithium, the corresponding carbanions (**20** and **21**) are obtained. Ethereal solutions of these carbanions are highly colored (red); for multiple anion formation, the concentration of anions is reflected by even darker coloration. Either one of these anionic intermediates readily reacts with alkylhalides to give **22,** with carbon dioxide to form **23,** or with any other suitable electrophile.[2,4,5] Thus pyrazinyl ketones (**24**) are readily available by this procedure.[11]

2-picoline (Me) → ($NaNH_2$, NH_3 (l)) → 20 (CH_2^- Na^+); → (ArLi) → 21 (CH_2^- Li^+); 20 or 21 → (CO_2) → 23 ($CH_2CO_2^-$); 20 or 21 → (RX) → 22 (CH_2R)

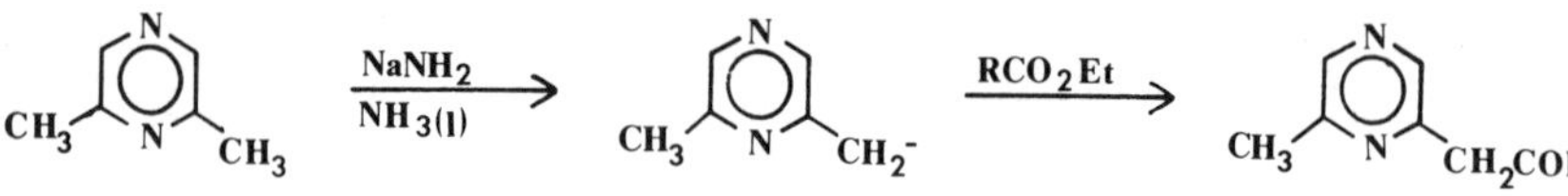

24 (50-85%)

When 2,4,6-trimethylpyrimidine (**25**) is treated with phenyllithium followed by methyl iodide, 2,6-dimethyl-4-ethylpyrimidine (**26**) is formed selectively.[12]

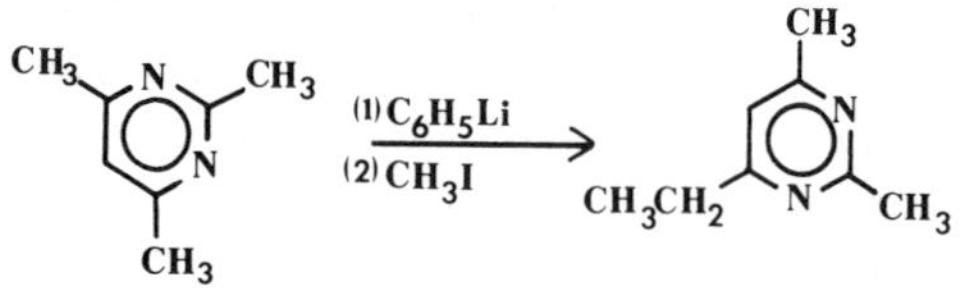

25 26

The Claisen condensation, when applied to 2-methylpyridine *N*-oxide (**27**), affords the expected product (**28**) in excellent yield.[13] In the absence of the *N*-oxide function, this reaction only occurs when an electron-withdrawing group, such as an NO_2 function, is present on the pyridine ring. An interesting application of the reactivity of the methyl group of 2-methylpyridine involves the following synthesis of pyrrocolines (**29**):[14]

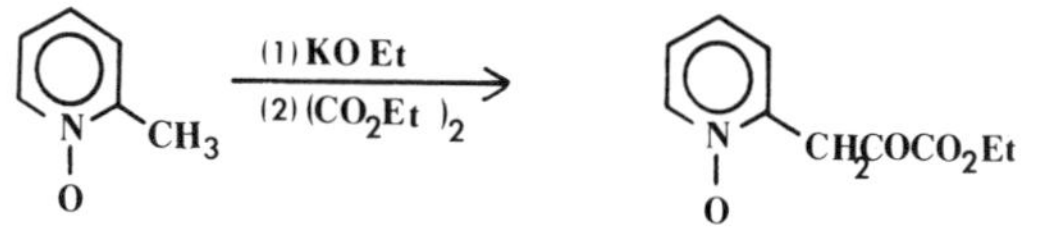

27 28

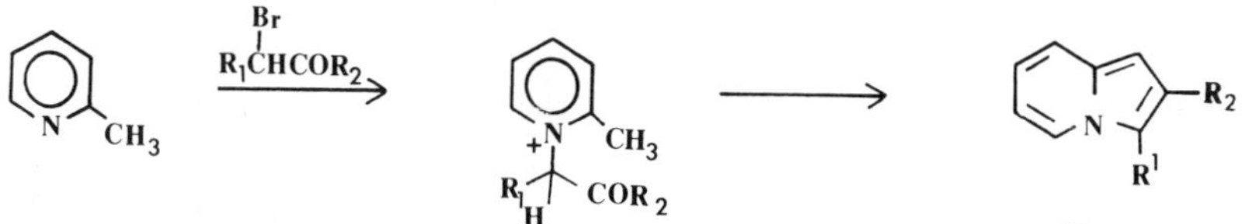

29

The activating effect of a ring nitrogen is also seen in the reactivity of 2-vinylpyridine (**30**) with sodium ethoxide. The intermediate anion (**31**), when quenched with aqueous acid, affords the *β*-ethoxy derivative **32.**[15]

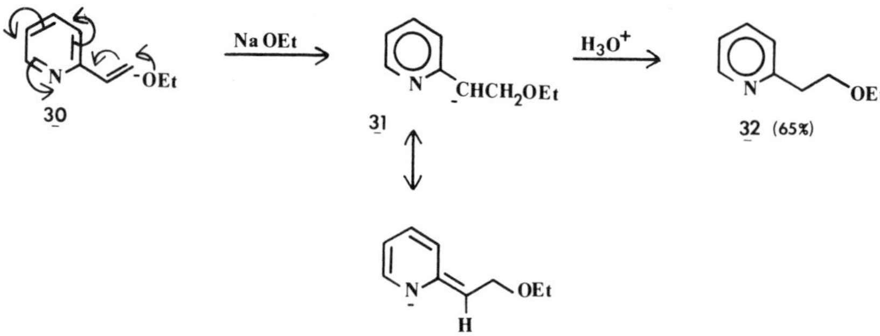

30 31 32 (65%)

Although one would not expect the methyl group in 3-methylpyridine (**33**) to be as acidic as those in the 2- and 4-positions, it is nevertheless more acidic than the methyl group in toluene. It does, for example, react with sodamide in liquid ammonia to afford the anion **34**, which when quenched with methyl iodide, gives 3-ethylpyridine (**35**).[16]

$$\mathbf{33}\ \xrightarrow[NH_3(l)]{NaNH_2}\ \mathbf{34}\ \xrightarrow{CH_3Cl}\ \mathbf{35}\ (54\%)$$

21.4. REACTIONS WITH NITROSO COMPOUNDS

The reactivity of active methylene carbaromatics with nitrous acid and its derivatives is well-known. Thus it is not surprising that these activated 2-alkyl positions on electron-deficient heteroaromatics react in a similar fashion.[17] It is noteworthy that the intermediate nitroso derivative **36** undergoes tautomeric conversion to the oximino derivative **37**. The *N*-methylpyridinium salt **38** can be condensed with nitrosoarenes (**39**) with equal ease, whereas 2-methylpyridine is *not* sufficiently activated to undergo this condensation.[18]

$$\xleftarrow[R_3N]{ArNO}\ \text{2-Me-pyridine N-oxide}\ \xrightarrow[NaNH_2]{RONO}\ \mathbf{36}\ \rightleftharpoons\ \mathbf{37}$$

$$\mathbf{38}\ \xrightarrow[R_3N]{ArNO\,(\mathbf{39})}$$

Although no precise data are yet available, it appears that the relative reactivity (acidities) of the methyl groups in 2-methylpyridine and in the corresponding methyldiazines is as follows:

2-methylpyridine < 2-methylpyrazine < 4-methylpyrimidine < 2-methylpyrimidine

21.5. OXIDATIONS

As is true for the methyl group in toluene, the methyl and homologous alkyl derivatives of pyridine (**40**) and the related azines can be oxidized to the corresponding carboxylic acids (**41**), which exist mainly in their zwitterionic forms (**41a**).[19] The α-, β-, and γ-carboxylic acids are frequently referred to as picolinic, nicotinic, and isonicotinic acid, respectively. Among these, the α- and γ-isomers (**42** and **43**) are unstable to heat and are decarboxylated to give the corresponding pyridine.[20]

40 (Py–CH_3) —$KMnO_4$→ 41 (Py–CO_2H) ⇌ 41a (PyH$^+$–CO_2^-)

42 (2-Py–CO_2H) —Δ→ pyridine ←Δ— 43 (4-Py–CO_2H)

Similar behavior is evidenced among the alkyldiazines. Oxidation of the methyl group on **6** can be stopped at the aldehyde stage by the use of selenium dioxide[21,22] or by reaction of either the α- or γ-methyl derivatives with a halogen and dimethyl sulfoxide.[21,23]

6 (2-Py–CH_3) —SeO_2→ 2-Py–CHO

6 —I_2→ 2-Py–CH_2I —Me_2SO (DMSO)→ 2-Py–CH(H)–O–$\overset{+}{S}Me_2$ I^- —Δ or base→ 2-Py–CHO

The most facile and widely used oxidative substitution procedure is the transformation of *N*-oxides (for example, **27**) to the α-acetoxyalkyl derivatives **44.** The mechanistic aspects have been considered.[24] A sequence of reactions ending with hydrolysis converts **44** to an aldehyde. Other reactions of this type that have been employed to functionalize the α-position use different chlorination agents: phosphoryl chloride,[25] trichloroacetyl chloride,[26] benzenesulfonyl chloride,[27] and *p*-toluenesulfonyl chloride.[28]

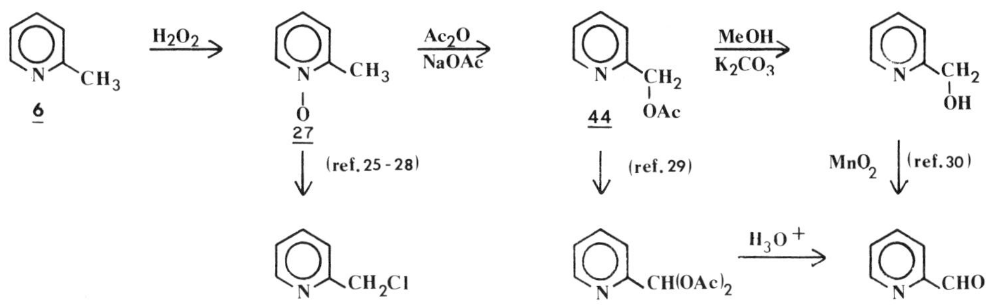

REFERENCES

1. C. E. Kaslow and R. D. Stayner, *J. Am. Chem. Soc.* **67,** 1716 (1945).
2. D. Jerchel and H. E. Heck, *Chem. Ann.* **613,** 171 (1958).
3. O. F. Beumel, W. N. Smith, and B. Rybalka, *Synthesis* **1974,** 43.
4. W. Baker, K. M. Buggle, J. F. W. McOmie, and D. A. M. Watkins, *J. Chem. Soc.* **1958,** 3594.
5. R. B. Woodward and E. C. Kornfeld, *Org. Synth. Coll. Vol.* **3,** 413.
6. A. Holland, *Chem. Ind.* (*London*) **73,** 786 (1954).
7. R. H. Mizzoni and P. E. Spoerri, *J. Am. Chem. Soc.* **76,** 2201 (1954).
8. A. E. Siegrist, H. R. Meyer, P. Gassmann, and S. Moss, *Helv. Chim. Acta* **63,** 1311 (1980).
9. G. R. Newkome and J. M. Robinson, *Tetrahedron Lett.* **1974,** 691.
10. J. C. Trisler, B. F. Freasier, and S. -M. Wu, *Tetrahedron Lett.* **1974,** 687.
11. M. R. Kamal and R. Levine, *J. Org. Chem.* **27,** 1355 (1962).
12. J. C. Roberts, *J. Chem. Soc.* **1952,** 3065.
13. R. Adams and S. Miyano, *J. Am. Chem. Soc.* **76,** 3168 (1954).
14. E. T. Borrows and D. O. Holland, *Chem. Revs.* **42,** 615 (1948).
15. W. E. Doering and R. A. N. Weil, *J. Am. Chem. Soc.* **69,** 2461 (1947).
16. H. C. Brown and W. A. Murphey, *J. Am. Chem. Soc.* **73,** 3308 (1951).
17. T. Kato and Y. Goto, *Chem. Pharm. Bull.* (*Japan*) **11,** 461 (1963) [*Chem. Abstr.* **63,** 5596 (1965)].
18. A. Kauffmann and L. G. Valette, *Chem. Ber.* **45,** 1736 (1912).
19. (a) R. W. Green and M. K. Tong, *J. Am. Chem. Soc.* **78,** 4896 (1956); (b) H. P. Stephenson and H. Sponer, *J. Am. Chem. Soc.* **79,** 2050 (1957).
20. S. Hoogewerff and W. A. van Dorp, *Chem. Ber.* **13,** 61 (1880).
21. T. Slebodinsky et al., *Przemysl. Chem.* **48,** 90 (1969) [*Chem. Abstr.* **71,** 38751 (1969)].
22. N. Rabjohn, *Org. Reactions* **24,** 261 (1976).
23. A. Markovac, C. L. Stevens, A. B. Ash, and B. H. Hackley, *J. Org. Chem.* **35,** 841 (1970).
24. K. Schofield, *Hetero-aromatic Nitrogen Compounds—Pyrroles and Pyridines*, Plenum Press, New York, 1967, pp. 340–343.
25. T. Kato, *J. Pharm. Soc.* (*Japan*) **75,** 1239 (1955) [*Chem. Abstr.* **50,** 8665 (1956)].
26. T. Koenig and J. S. Wieczorek, *J. Org. Chem.* **33,** 1530 (1968).
27. J. Vozza, *J. Org. Chem.* **27,** 3856 (1962).
28. E. Matsumura, *J. Chem. Soc. Japan* **74,** 363 (1953) [*Chem. Abstr.* **48,** 6442 (1954)].
29. V. Boekelheide and W. J. Linn, *J. Am. Chem. Soc.* **76,** 1286 (1954).
30. E. P. Papadopoulos, A. Jarrar, and C. H. Issidorides, *J. Org. Chem.* **31,** 615 (1966); A. J. Lin, K. C. Agrawal, and A. C. Sartorelli, *J. Med. Chem.* **15,** 615 (1972).

Chapter 22

SELECTED NATURALLY OCCURRING AND PHARMACEUTICALLY INTERESTING π-DEFICIENT HETEROAROMATIC COMPOUNDS

22.1. GENERAL COMMENTS

Although neither pyridine nor any of the diazines occur free in nature, a very large number of derivatives of pyridine, pyrimidine, and pyrazine have been isolated. To date no pyridazine derivatives of biological origin have been found. The numerous naturally occurring pyridine derivatives vary in complexity from nicotinamide (**1**), vitamin B_6 (pyridoxine) (**2**), and nicotine (**3**) to highly complex systems, such as nicotinamide adenine dinucleotide (NAD) (**4**), and alkaloids such as quinine (**5**).

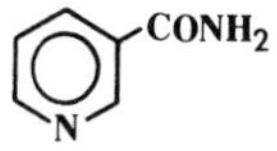

$CONH_2$ 1   HO, CH_2OH, CH_2OH, Me, N 2   N, CH_3 3

H, O, OH, HO, H, N, N, N, N, NH_2, $CH_2O-P(=O)(OH)-O-P(=O)(O^{\ominus})-OCH_2$, H, OH, OH, H, O, $N^{\oplus}$, $CONH_2$ 4

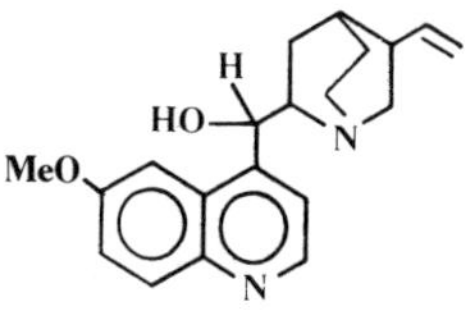

H, HO, MeO, N, N 5   N, OCH_3, CH_3, N, CH_3 6   N, n-bu, n-bu, N, O, OH 7

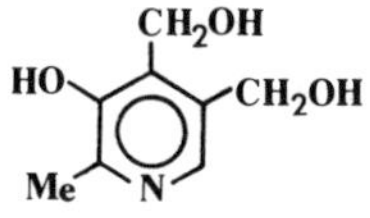

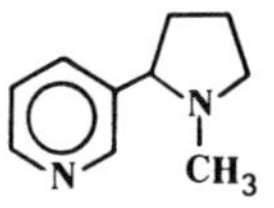

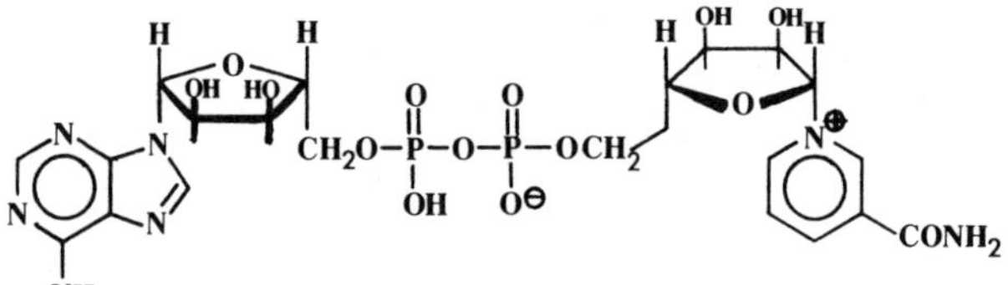

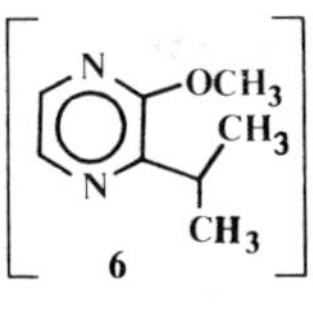

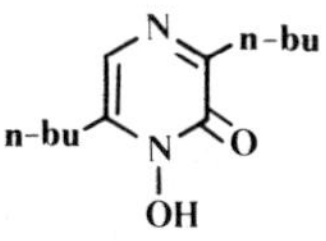

Many pyrazine derivatives occur naturally. Sweet corn and some legumes, for example, contain the methoxypyrazine **6.** Aspergillic acid (**7**) is a naturally occurring antibiotic agent. Within the pteridine family there are numerous important compounds such as tetrahydropteroylglutamic acid (**8;** folinic acid), a substance that is essential for mammalian cell division.

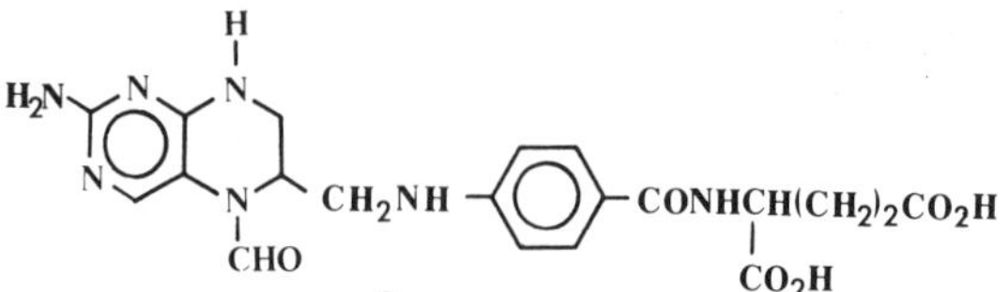

8

The number of substituted pyrimidines found in nature is legion. Among these are vitamins B_1 (**9**, thiamine) and B_2 (**10**, riboflavin), a number of nucleosides [for example, uridine riboside (**11**)], numerous alkaloids, and other related compounds. The syntheses of naturally occurring compounds to be discussed in this chapter have been selected as representative examples[1] which utilize some of the general reactions of π-deficient heteroaromatic compounds.

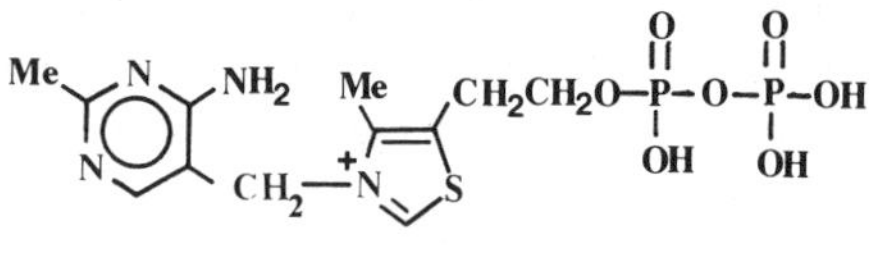

9

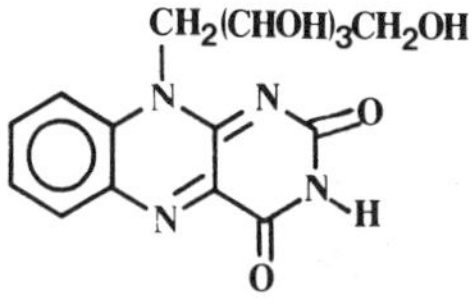

10

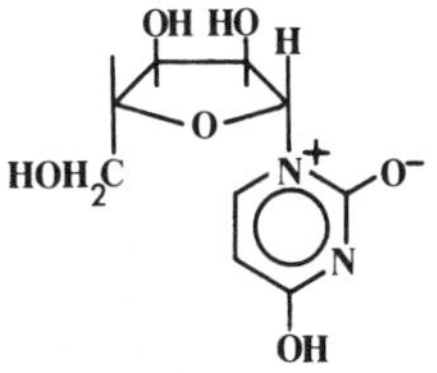

11

22.2. SYNTHESES OF SELECTED NATURALLY OCCURRING π-DEFICIENT HETEROAROMATICS

22.2.1. NICOTINE

Nicotine (**3**), a compound fatal to man at levels greater than 40 mg, has been synthesized by several different routes. Among these, the Claisen condensation (Scheme 22.1) of ethyl nicotinate (**12**) with *N*-methyl-2-pyrrolidone (**13**) affords a convenient starting material (**14**), which is hydrolyzed to give intermediate **15.** Under these conditions **15** readily decarboxylates to **16.** After reduction of the carbonyl group, the resulting alcohol is converted to the iodide (**17**), which spontaneously cyclizes upon warming to afford the *d,l*-nicotine **3.**[2]

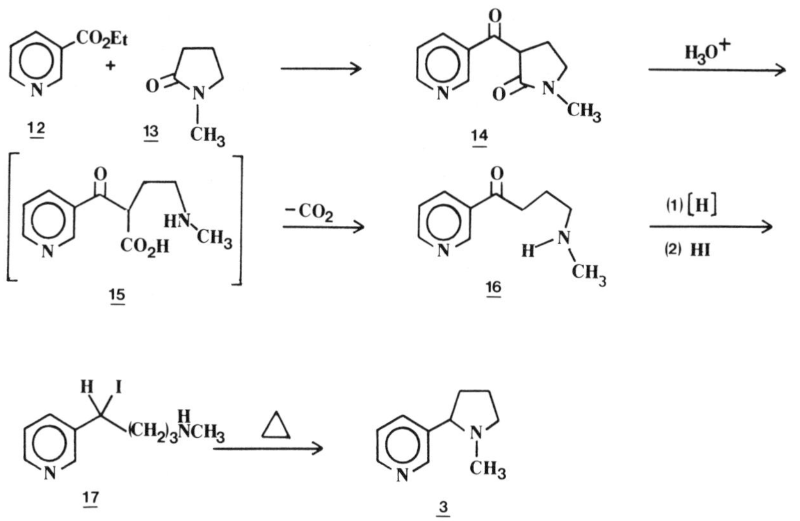

Scheme 22.1

22.2.2. CARPYRINIC ACID[3]

An elegant synthesis (Scheme 22.2) of methyl carpyrinate (**18**) starts with 2-acetylfuran (**19**).[4] Treatment of **19** with ammonia provides an excellent source of 3-hydroxy-2-methylpyridine (**20**). A Kolbe synthesis, followed by protection of the phenolic group by methylation with dimethyl sulfate, affords the carboxylic acid **21,** which undergoes a Hammick condensation with aldehyde **22** to give the hydroxyester **23.** Oxidation of **23** with manganese dioxide, subsequent Huang–Minlon reduction, and removal of the protecting group with pyridinium hydrochloride affords **18.** Other routes to **18** utilize the acid-

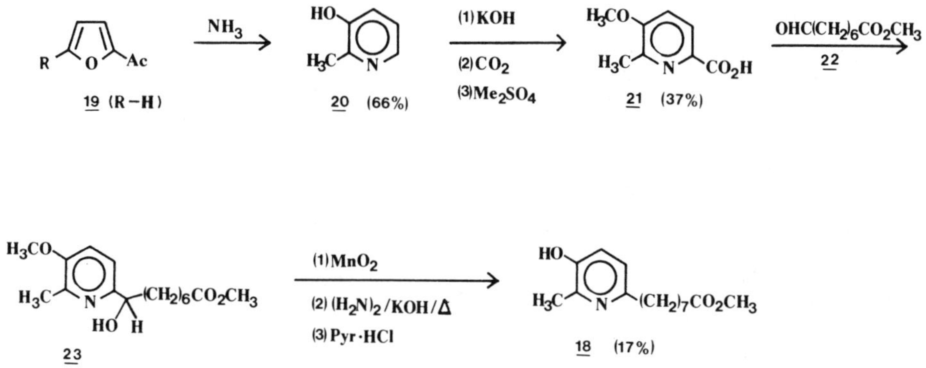

Scheme 22.2

ity of the α-methyl hydrogens (**24**) and the enhanced electron richness of phenols (**25**),[5] or the ability of furans to undergo electrophilic substitution (acylation of **26**), followed by facile rearrangement of substituted **19** [R = $(CH_2)_7CO_2CH_3$].[3]

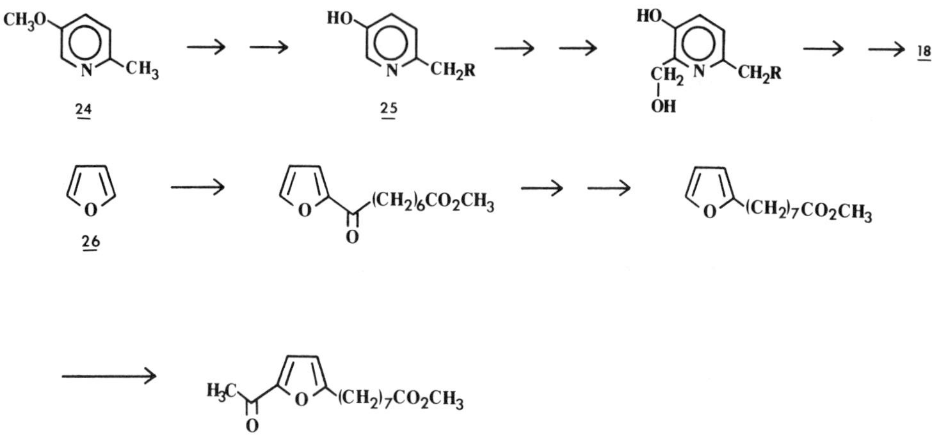

22.2.3. ELLIPTICINE

Because of the potential anticancer activity of ellipticine (6*H*-pyrido[4,3-*b*]carbazole, **27**) and its derivatives, numerous synthetic pathways to these compounds have recently been proposed. The first synthesis of **27** (Scheme 22.3) by Woodward in 1959 consists of but three steps from 3-acetylpyridine (**28**).[6] Condensation of **28** with indole in the presence of zinc chloride gives **29**, which is reduced with zinc and acetic anhydride at reflux; the resulting 1,4-dihydropyridine intermediate is pyrolyzed to give **27** in about 2% yield. More recently, a simple synthesis [7a] of **27** circumvents many of the obvious problems

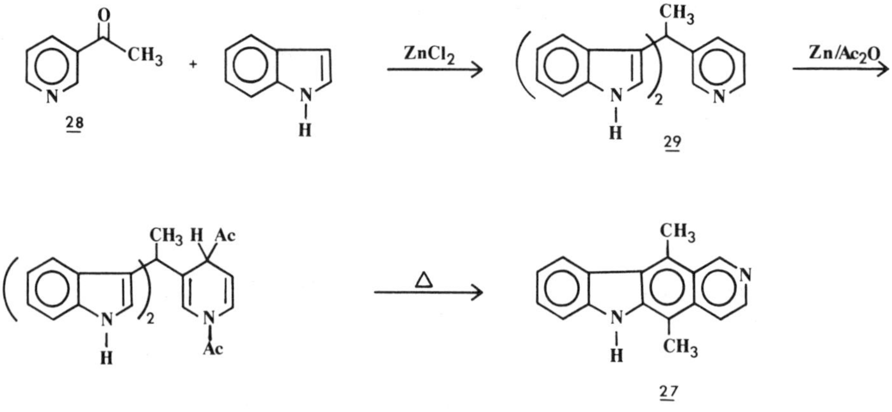

Scheme 22.3

associated with the earlier routes. Indolylmagnesium bromide is combined with **30,** then acetylated to give **31.** Subsequent treatment of **31** with *o*-mesitylsulfonylhydroxylamine, acetic anhydride, and methyl iodide gives a salt **32,** which is allowed to react with potassium cyanide to afford nitrile **33.** Ellipticine is isolated in 25–30% overall yield when **33** is treated with methyllithium and the imine intermediate is hydrolyzed with aqueous acetic acid. 8-Hydroxyellipticine, a metabolite of **27** from *Aspergillus allioceus,* was synthesized from 6-methoxyindolylmagnesium bromide by a similar procedure.[7a]

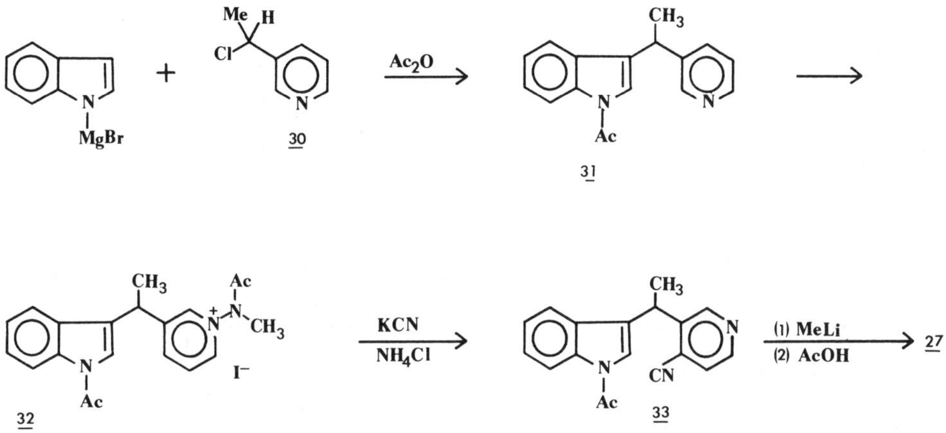

22.2.4. LYSERGIC ACID

Fresh ergot contains the extremely pharmacologically active lysergic acid alkaloids. When the dihydrolysergic acid (**35**) was synthesized, it lacked the $\Delta^{9,10}$-double bond. The introduction of which is shown in Scheme 22.4.[8] Condensa-

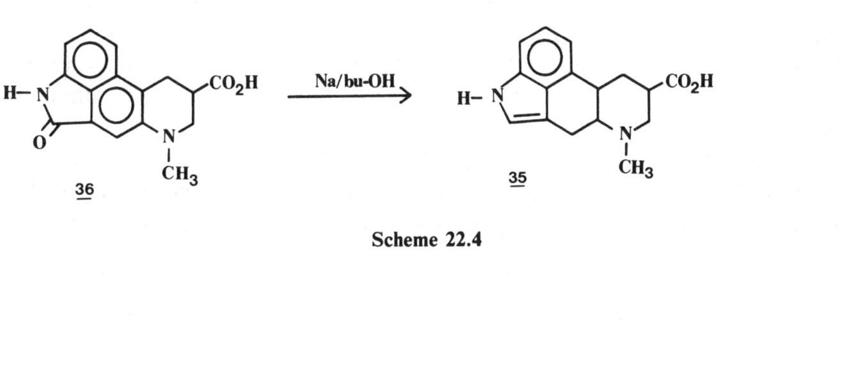

Scheme 22.4

tion of **34** with cyanomalonic dialdehyde, followed by cyclization, gives the lysergic acid ring system containing a carboxylic acid group at the appropriate position. *N*-Methylation, followed by catalytic reduction, affords the tetrahydro derivative (**36**), which can be chemically reduced with sodium and butanol to give limited quantities of (±)-dihydrolysergic acid (**33**). The Kornfeld synthesis[9] of (±)-lysergic acid (**37**) constructs the $\Delta^{3,4}$-unsaturated site in the pyridine ring, then functionalizes the remainder of the molecule (**38** → **37**).

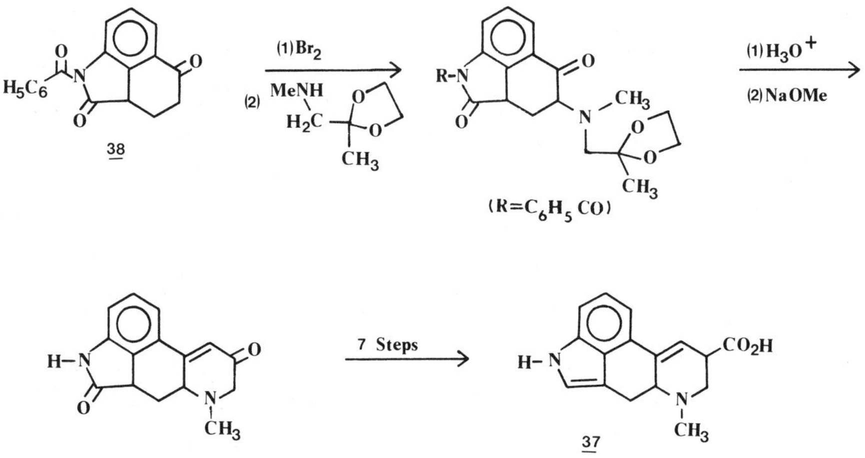

22.2.5. PYRIDOXINE

A procedure for the commercial scale synthesis of pyridoxine is shown in Scheme 22.5 (p. 313). Initial formation of the pyridine nucleus **39** employs one of the general pyridine syntheses already described. Nitration at C-5 and conversion of the oxo group to a 2-chloro function generates intermediate **40.** Two consecutive reduction reactions, followed by acid hydrolysis of the "benzylic" ether function, affords the primary amine **41,** which when diazotized gives pyridoxine.[10a] Shorter (three steps), more elegant routes to **2** are also available[10b] from simple starting materials, such as disubstituted oxazole and diethyl malonate.

22.2.6. QUININE

Malaria causes suffering on a global scale: almost one-third of the world's population is regularly exposed to the carriers of this debilitating disease. The oldest (seventeenth century) effective drug for the treatment of malaria is quinine (**5**),[11-13] the syntheses of which represent classic examples of brilliant synthetic work. One sequence of reactions which leads to its synthesis is outlined in Scheme 22.6 (p. 314). The ring expansion (**42** → **43**) followed by rearrangement of the seven-membered ring (via **44**) and its thermal rearrangement with concomitant loss of N_2 (**45**) represents an unique transformation which ulti-

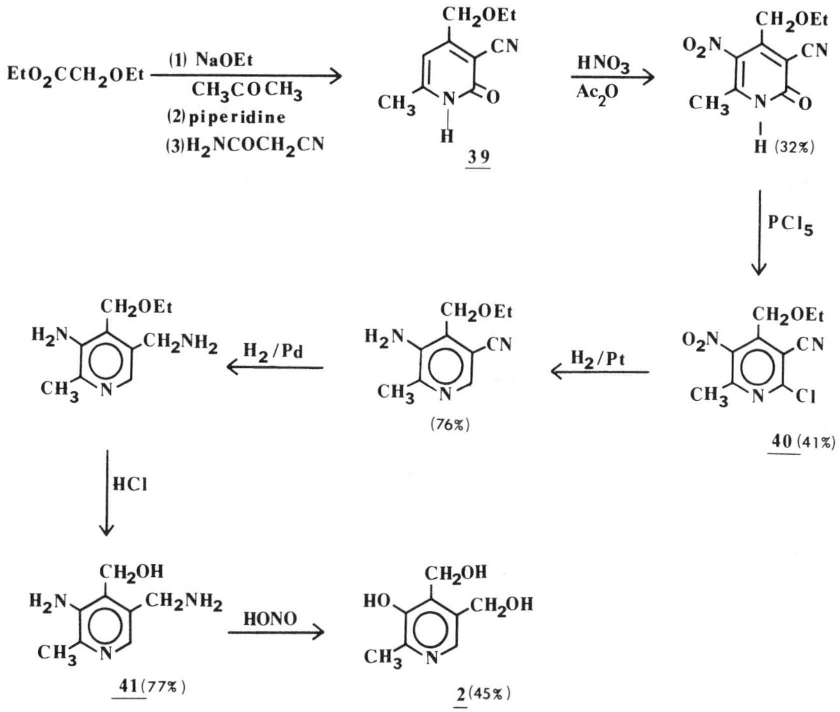

Scheme 22.5

mately affords the piperidine derivative **46,** the key compound in this synthesis of quinine. Condensation of carbanion **47** and **46,** followed by introduction of the double bond between the quinoline and piperidine rings (**48** → **49**), cyclization by a Michael-type reaction, and a novel oxidation at the α-methylenic position gives quinine (**5**) along with its isomers.

22.2.7. EPIIBOGAMINE

The preparation of epiibogamine (**50**) represents an excellent example (Scheme 22.7) of the use of a pyridine derivative in the synthesis of a very complex molecule.[14]

There are several unique features to this synthesis. The initial [4 + 2] cycloaddition depends on the use of the proper dihydro isomer (**51**). Conversion of the stereoisomers **52** and **53** to the exo-exo (**54**) and exo-endo (**55**) isomers, respectively, eliminates one reaction "by-product" (**53**). Subsequent conversion of **54** and **55** to **56** is accomplished by standard carbocyclic transformations. The final step in this synthesis (**56** → **50**) involves a clever application of the Fischer indole synthesis.

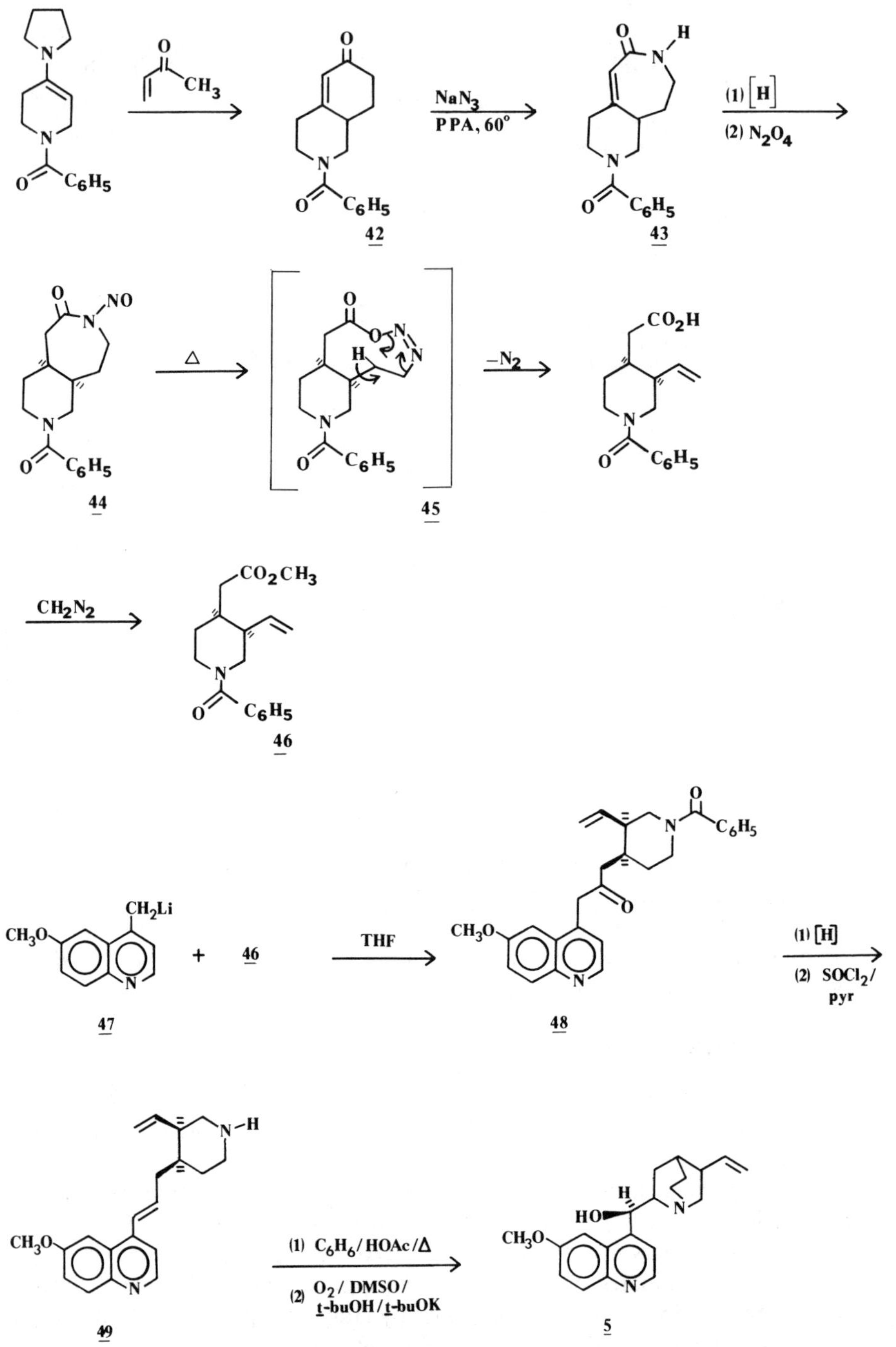

Scheme 22.6

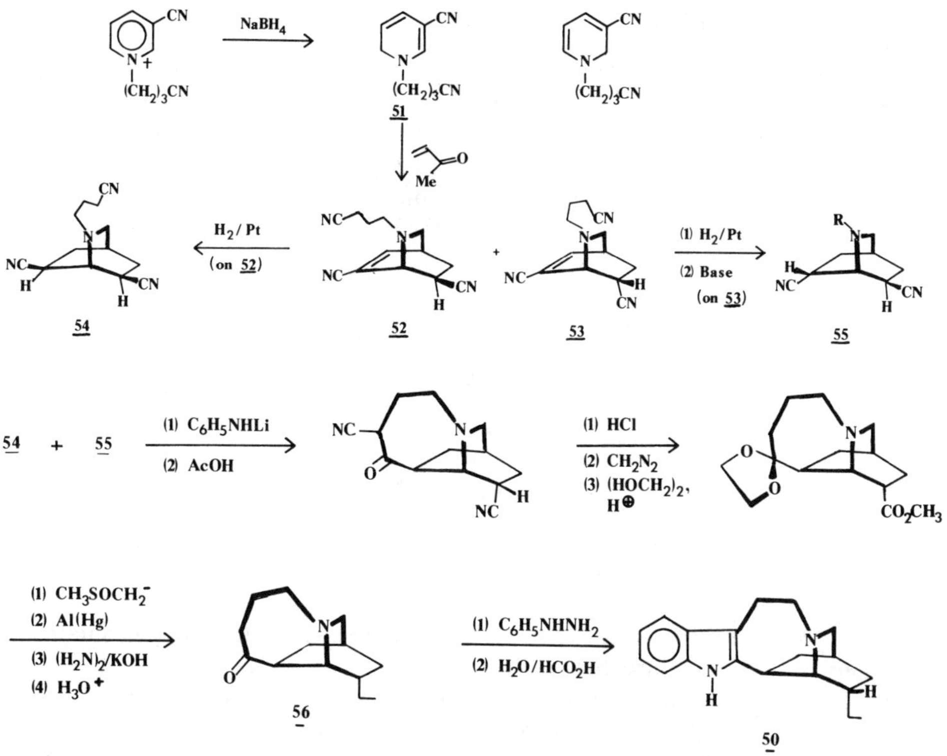

CN, N^+, $(CH_2)_3CN$ —$NaBH_4$→ CN, N, $(CH_2)_3CN$ **51**; CN, N, $(CH_2)_3CN$

51 → (methyl vinyl ketone, O, Me)

52: NC, N, NC, CN, H + **53**: CN, N, NC, H, CN

52 —H_2/Pt (on **52**)→ **54**: CN, N, NC, H, CN, H

53 —(1) H_2/Pt (2) Base (on **53**)→ **55**: R, N, H, NC, CN, H

54 + **55** —(1) C_6H_5NHLi (2) AcOH→ NC, N, O, NC, H —(1) HCl (2) CH_2N_2 (3) $(HOCH_2)_2$, $H^{\oplus}$→ O, O, N, CO_2CH_3

—(1) $CH_3SOCH_2^-$ (2) Al(Hg) (3) $(H_2N)_2$/KOH (4) H_3O^+→ **56** (N, O) —(1) $C_6H_5NHNH_2$ (2) H_2O/HCO_2H→ **50** (N, N–H, H)

Scheme 22.7

22.2.8. PTEROYLGLUTAMIC ACID

The antianemia factor pteroylglutamic acid (**57**) can be synthesized as in Scheme 22.8.[15] *N,N'*-Bis-alkylation of amine **58** with the α,β-dibromopropionaldehyde gives **59,** which after imine formation and reduction (reductive amination) generate **57.**

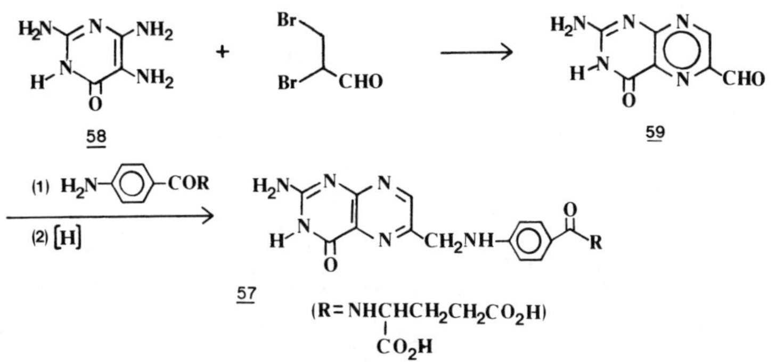

58 (H_2N, N, NH_2, H, N, NH_2, O) + Br, Br, CHO → **59** (H_2N, N, N, H, N, N, CHO, O)

—(1) H_2N–C_6H_4–COR (2) [H]→ **57** (H_2N, N, N, H, N, N, CH_2NH, O, R)

(R = $NHCHCH_2CH_2CO_2H$ with CO_2H)

Scheme 22.8

22.2.9. YOHIMBINE

Yohimbine (**60**) represents one of many examples of a naturally occurring substance which contains both a pyrrole and a pyridine ring system in its structure. Its synthesis by Stork and Guthikonda[16] exemplifies an elegant and unusually simple synthetic approach to a highly complex organic compound (Scheme 22.9). The diagonal line drawn on structure **60** indicates the antithetic approach to its synthesis, whereby the indole portion and the remaining part (including the piperidine ring) are synthesized separately.

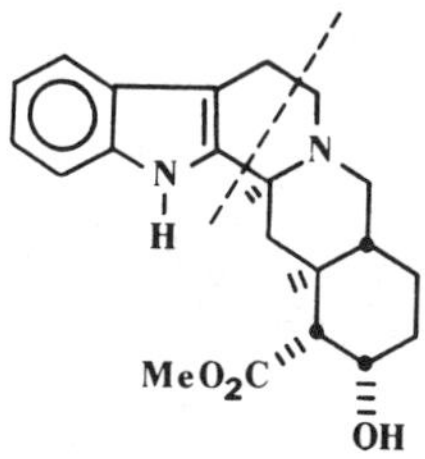

N
H
MeO_2C
OH

60

The synthesis of 3-oxo-4-pentenoic acid (**61**) in good yield greatly facilitates the annelation procedure that results in the formation of the piperidine derivative **62.** The close similarity between **62** and the previously considered **42** should be noticed. Conversion of **63** to **65** via the *N*-cyano derivative **64** is noteworthy (von Braun synthesis—demethylation), as is the transformation of **67** → **60.** This latter conversion involves prior formation of a —C=N— linkage, which reacts with the α-position of the indole ring.

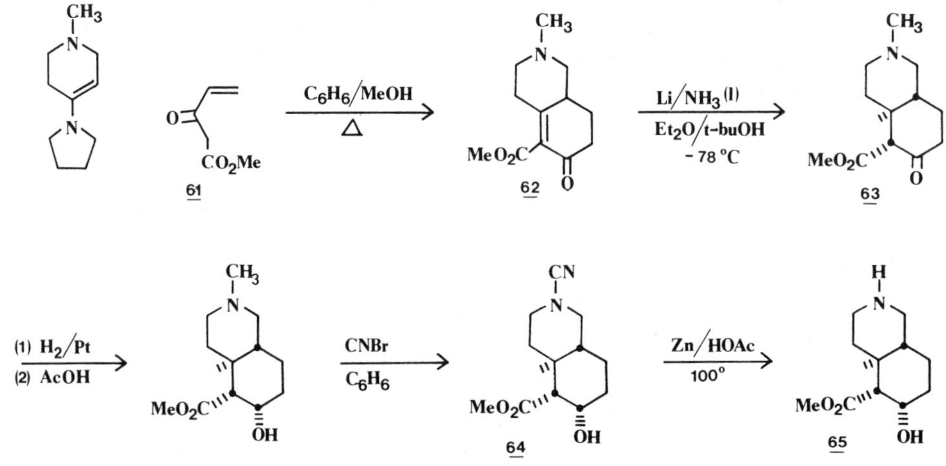

CH_3 N N; O CO_2Me 61 —(C_6H_6/MeOH, Δ)→ CH_3 N MeO_2C O 62 —(Li/NH_3 (l), Et_2O/t-buOH, - 78 °C)→ CH_3 N MeO_2C O 63

—((1) H_2/Pt, (2) AcOH)→ CH_3 N MeO_2C OH —(CNBr, C_6H_6)→ CN N MeO_2C OH 64 —(Zn/HOAc, 100°)→ H N MeO_2C OH 65

Scheme 22.9. Phase A.

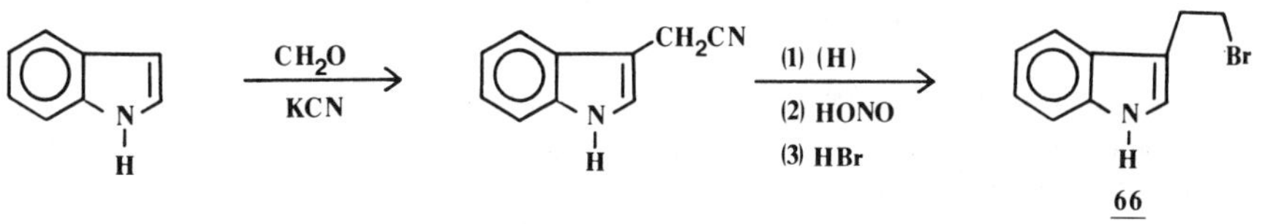

CH_2O / KCN; CH_2CN; (1) (H) (2) HONO (3) HBr; Br; N H; 66

Scheme 22.9. Phase B.

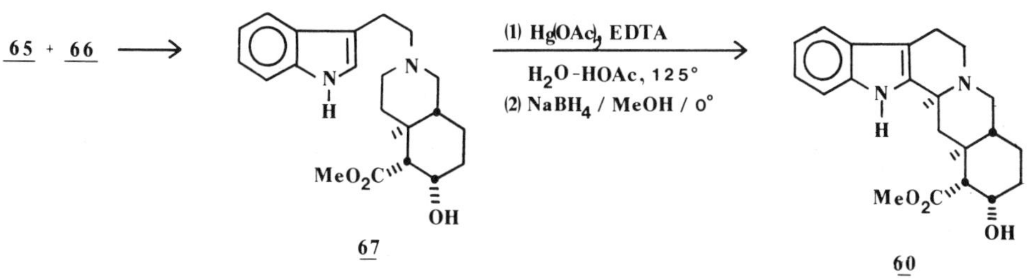

65 + 66 ⟶ 67; (1) $Hg(OAc)_2$, EDTA, H_2O-HOAc, 125° (2) $NaBH_4$ / MeOH / 0°; MeO_2C; OH; 60

Scheme 22.9. Phase C.

22.3. SYNTHESES OF SELECTED PHARMACEUTICAL AGENTS

22.3.1. PYRIMETHAMINE

The observation that 2,4-diaminopyrimidines inhibit the growth of some microorganisms led, ultimately, to the development of numerous antimalarial agents; of these, the synthesis (Scheme 22.10) of pyrimethamine (**68**) is but one example.[17]

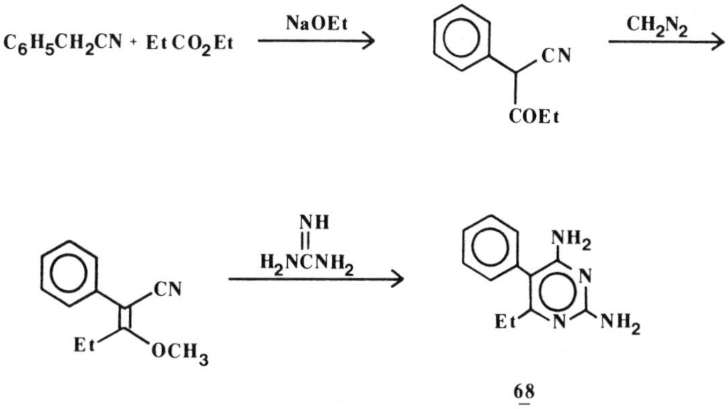

$C_6H_5CH_2CN$ + $EtCO_2Et$ —NaOEt→ (CN, COEt) —CH_2N_2→ (Et, CN, OCH_3) —$H_2NC(=NH)NH_2$→ (NH_2, N, Et, N, NH_2) 68

Scheme 22.10

22.3.2. MINOXIDAL

A very effective hypotensive agent, minoxidal (**69**) represents a compound in which a relatively minor structural change (compare to **68**) alters the biological activity drastically.[18]

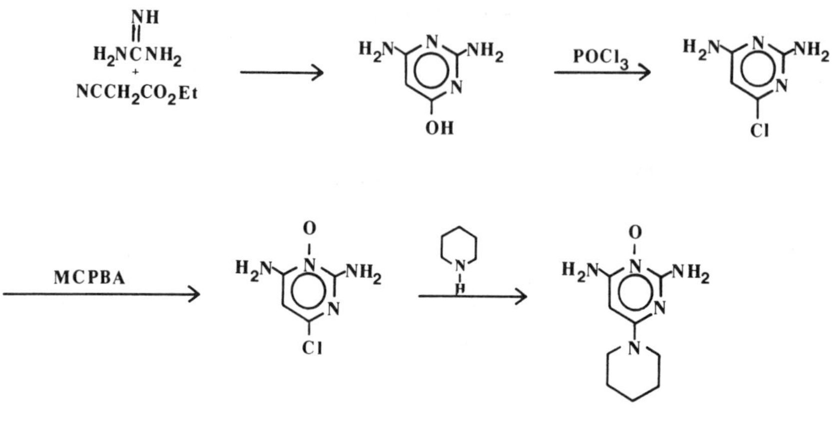

$H_2NC(=NH)NH_2$ + $NCCH_2CO_2Et$ → POCl$_3$ → MCPBA →

69

Scheme 22.11

22.3.3. NIFLUMINIC ACID

Substitution of pyridine for benzene in fenamic acid-type analgesics gives rise to nifluminic acid **70** (Scheme 22.12). Treatment of nicotinic acid *N*-oxide with phosphorus trichloride followed by hydrolysis affords **71,** which is highly susceptible to heteroaromatic nucleophilic substitution (**71** → **70**).[19]

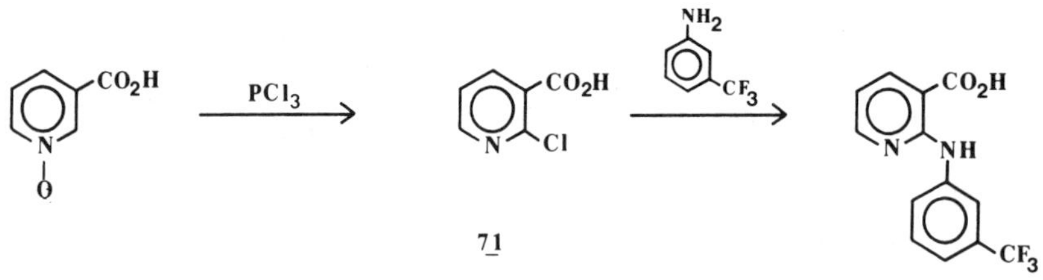

PCl$_3$ →

71

70

Scheme 22.12

22.3.4. MORPHAZINAMIDE

The syntheses (Scheme 23.13) of pyrazinamide (**72**) and morphazinamide (**73**) represent novel interconversions.[20] Quinoxaline (**74**) is prepared by standard bis-imine formation, then oxidation of the original aromatic nucleus generates a bis-carboxylic acid **75.** Facile thermal decarboxylation, esterification, and amminolysis give **72;** subsequently Mannich reaction of **72** yields **73.**

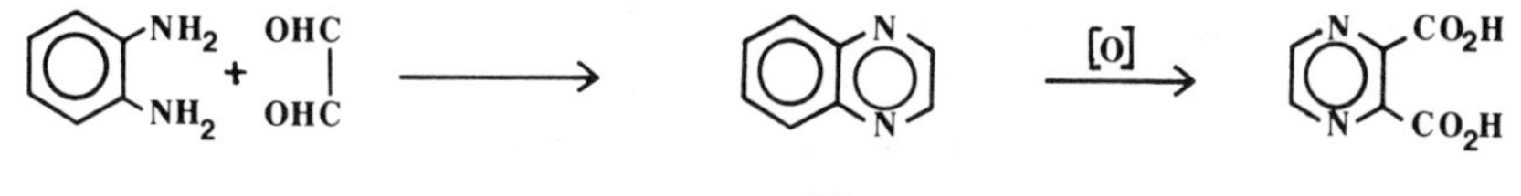

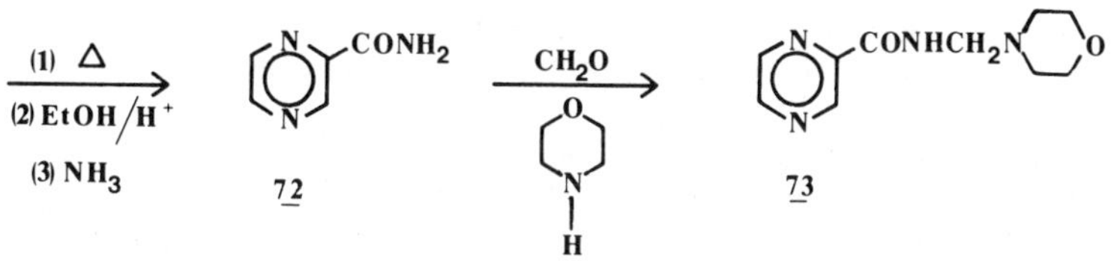

Scheme 22.13

22.3.5. BENZOMORPHAN SYSTEM

Nucleophilic 1,2-addition of a benzylmagnesium chloride to 3,4-dimethylpyrridinium iodide leads to the more hindered dihydro **76** (Scheme 22.14). Reduction of the enamine double bond affords the $\Delta^{3,4}$-olefin **77,** which upon treatment with acid gives the benzomorphan system (**78**).[21] De-*O*- and *N*-methylation and subsequent *N*-alkylation affords a potent series of analgesic agents.

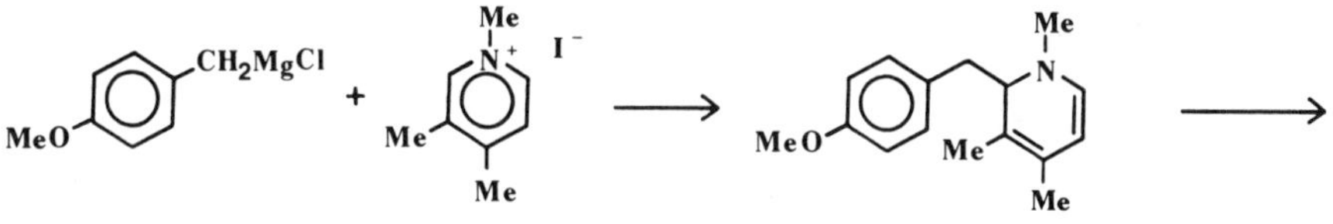

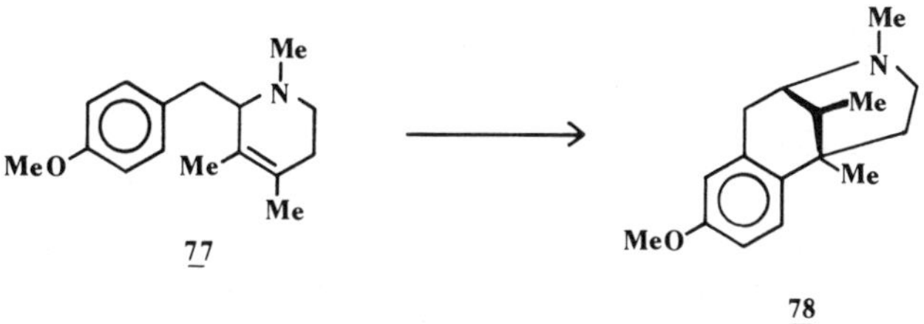

Scheme 22.14

REFERENCES

1. Reviews: S. W. Pelletier, *Chemistry of the Alkaloids,* Van Nostrand-Reinhold Company, 1970.
2. E. Spath and H. Bretschneider, *Chem. Ber.* **61,** 327 (1928).
3. G. Fador, J. -P. Fumeaux, and V. Sankaran, *Synthesis* **1972,** 464.
4. H. Rapoport and E. J. Volcheck, *J. Am. Chem. Soc.* **78,** 2451 (1956).
5. T. R. Govindachari, N. S. Narasimhan, and S. Rajadurai, *J. Chem. Soc.* **1957,** 560.
6. R. B. Woodward, G. A. Iacobucci, and F. A. Hochstein, *J. Am. Chem. Soc.* **81,** 4434 (1959).
7. (a) M. Sainsbury and R. F. Schinazi, *J. Chem. Soc. Chem. Commun.* **1975,** 540; (b) D. Dolman and M. Sainsbury, *Tetrahedron Lett.* **22,** 2119 (1980). Also see J. Bergman and H. Goonewardena, *Acta Chem. Scand.* **B34,** 763 (1980); R. Besselievre and H. -P. Husson, *Tetrahedron* **37,** Suppl. 1, 241 (1981). Review: R. Barone and M. Chanon, *Heterocycles* **16,** 1357 (1981).
8. F. C. Ukle and W. A. Jacobs, *J. Org. Chem.* **10,** 76 (1945); also see R. Ramage, V. M. Armstrong, and Si Coulton, *Tetrahedron* **37,** Suppl. 1, 157 (1981).
9. E. D. Kornfeld, E. J. Fornefeld, G. B. Kline, M. J. Mann, R. G. Jones, and R. B. Woodward, *J. Am. Chem. Soc.* **76,** 5236 (1954).
10. S. A. Harris and K. Folkers, *J. Am. Chem. Soc.* **61,** 1245, 3307 (1939); E. E. Harris et al., *J. Org. Chem.* **27,** 2705 (1962).
11. R. B. Woodward and W. E. Doering, *J. Am. Chem. Soc.* **67,** 860 (1945).
12. M. R. Uskokovic and G. Grethe, *The Alkaloids* **14,** 181 (1973).
13. M. R. Uskokovic, J. Gutzwiller, and T. Henderson, *J. Am. Chem. Soc.* **92,** 203 (1970).
14. Y. Ban, T. Wakamatsu, Y. Fujimoto, and T. Oishi, *Tetrahedron Lett.* **1968,** 3383.
15. C. W. Waller et al., *J. Am. Chem. Soc.* **70,** 19 (1948).
16. G. Stork and R. N. Guthikonda, *J. Am. Chem. Soc.* **94,** 5109 (1972).
17. P. B. Russell and G. H. Hitchings, *J. Am. Chem. Soc.* **73,** 3763 (1951).
18. W. C. Anthony and J. J. Ursprung, U. S. Pat. 3,461,461 [*Chem. Abstr.* **68,** 21947 (1968)].
19. C. Hoffmann and A. Faure, *Bull. Soc. Chim. Fr.* **1966,** 2316.
20. S. Kushner, H. Dalalian, J. L. Sanjurjo, F. L. Bach, S. R. Safir, V. K. Smith, and J. H. Williams, *J. Am. Chem. Soc.* **74,** 3617 (1952).
21. E. L. May and J. H. Ager, *J. Org. Chem.* **24,** 1432 (1959).

Chapter 23

SYNTHESES AND REACTIONS OF SELECTED FUSED, MIXED HETEROAROMATIC COMPOUNDS

23.1. GENERAL COMMENTS

Of the fused ring compounds that possess a single heteroatom per ring, there are six possible isomers, excluding those in which the heteroatom lies at a bridgehead position. The simplest examples in this series are represented by the pyrrolo- (X = NH), thieno- (X = S), and furano- (X = O) [*x, y, z*] pyridines (**1–6**). These can be grouped in two classes based on the benzopyridine model: [*b*]-fused (**1–3**) (quinoline isosteres) and [*c*]-fused (**4–6**) (isoquinoline isosteres). This chapter focuses on the influences in condensed systems of *N*-heteroaromatics which have opposite distributions of π-electron density within each ring.

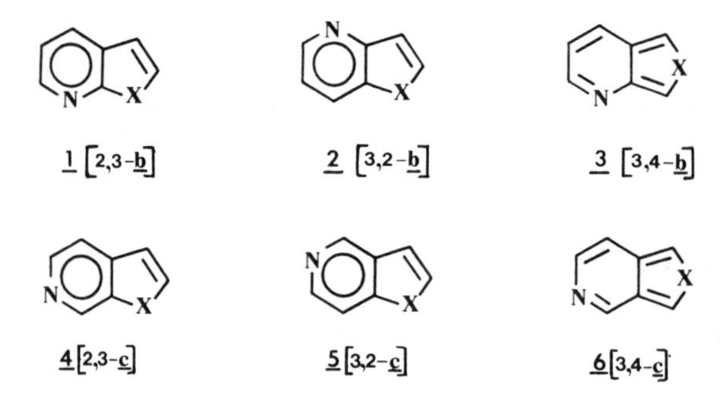

23.2. PYRROLOPYRIDINES (MONOAZAINDOLES)[1]

In view of the new antibiotics[2] (tubercidin, toyocamycin, and sangivamycin), the biochemical aspects of porphobilinogens, and the psychotropic and cardiomuscular activity of pyrrolopyridines, this class of compounds has been vigorously investigated over the past decade.

23.2.1. ⟨3 + 2⟩ COMBINATIONS

The *Fischer indole* reaction has been utilized to a limited degree, owing to the unfavorable electrophilic cyclization into a pyridine nucleus. The syntheses of 5- and 7-azatryptamines (**7, 8**) from pyridylhydrazines (**9**) and the lactam of an α-enamine also offer a convenient route to both the [2,3-*b*] and [3,2-*c*] ring systems.[2]

23.2.2. ⟨4 + 1⟩ COMBINATIONS

The *Reissert method* has also been adapted to the syntheses of azaindoles.[3] The drawback to this procedure is the decarboxylation step. However, treatment of the carboxylic acid (for example, of **10**) with copper bronze in refluxing biphenyl results in facile decarboxylation.[4]

Cyclization of 2- and 4-chloro-3-(β-chloroethyl)pyridines (**11**) with ammonia and primary or secondary amines is a general route to the azaindolines[5]; substituted pyrimidines (**12**) give rise to the related diazaindoles.[6]

23.2.3. ⟨3 + 3⟩ COMBINATIONS

Reaction of aminopyrroles with dicarbonyl compounds leads to the syntheses of a wide range of 7-azaindole derivatives.[7] Thus **13** undergoes cyclocondensation with β-dicarbonyl compounds or their acetals to generate **14.**

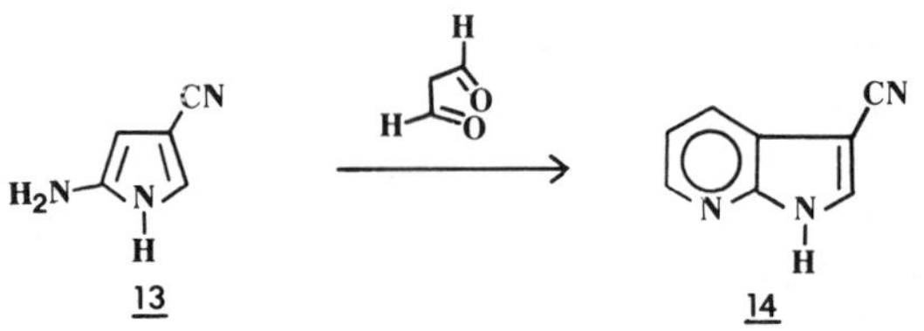

23.2.4. ⟨5⟩ CYCLIZATIONS

When *N*-(5-bromo-3-picolyl)-*N*-methylaminoacetate (**15**) is treated with potassium amide in liquid ammonia, the transient pyridyne intermediate (**16**) is trapped by intramolecular cyclization to give **17,** which can lead to several pyrrolo[3,4-*c*]pyridines (**18, 19**), as well as to a dihydro derivative **20** (the major product).[8] Cycloaddition reactions of ethyl 2-methyl-2*H*-pyrrolo[3,4-*c*]-pyridine-6-carboxylate have recently been reported.[9]

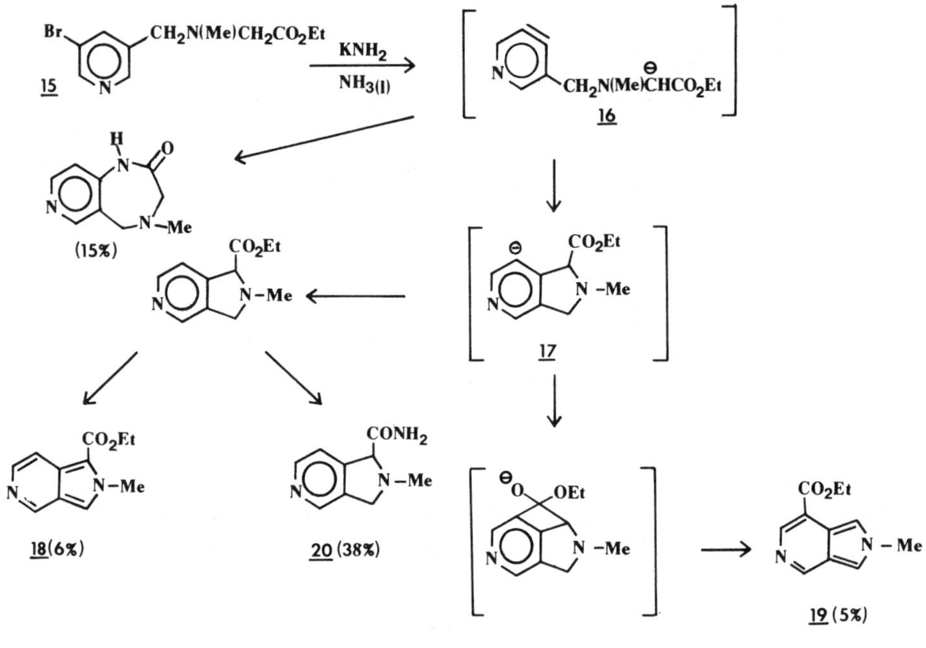

Other more exotic routes to these compounds are given in Reference 1.

23.3. THIENOPYRIDINES[10]

Currently, there is no known utility to this family of compounds. Should some use be discovered, however, shale oil of high sulfur content does contain selected methyl thieno[2,3-*b*]- and [3,2-*b*]pyridines.[11]

23.3.1. ⟨3 + 3⟩ COMBINATIONS

Unlike the syntheses of azaindoles, the common routes to the thienopyridines center around the construction of the pyridine ring; thus the appropriately substituted thiophenes are necessary. The *Skraup synthesis* is readily adapted to the syntheses of these compounds because the electron-rich thiophene ring acts similarly to that of benzene in the synthesis of quinolines. The salt of 2-aminothiophene (**21**) reacts with methyl vinyl ketone to give 4- (**22**) and some 6-methylthieno[2,3-*b*]pyridine (**23**)[12]; the 3-amino isomer of **21** also gives a mixture of thieno[3,2-*b*]pyridines. Schiff base chemistry has been successfully applied to the generation of a fused pyridine ring: treatment of **21** with acetylacetone gives an imine which in the presence of a Lewis acid cyclizes to give **24.**[12]

$FeCl_3/ZnCl_2$

(as salt)

21 22 23

(1) $CH_3COCH_2COCH_3$

(2) $-H_2O$ (acid)

24

23.3.2. ⟨3 + 2 + 1⟩ COMBINATIONS

The *Bischler–Napieralski* synthesis of "isoquinoline" again generates the pyridine ring, in this case from the appropriately substituted thiophene. Thieno[3,2-*c*]pyridine (**25**) is prepared in reasonable overall yield from **26** via a simple five-step sequence.[13] Other isoquinoline syntheses, for example, the *Pomeranz–Fritsch synthesis,*[13] the *Pictet–Spengler synthesis,*[14] and the *Pictet–Gams synthesis,*[15] have all been applied to the construction of this fused system.

$R'CH_2NO_2$, NaOH

$LiAlH_4$, $Ac_2O/NaOH$

26

$POCl_3/C_6H_6$, Pd/C/Xylene

25

The [3,4-*c*] series has been synthesized from **27** by the following attractive route:

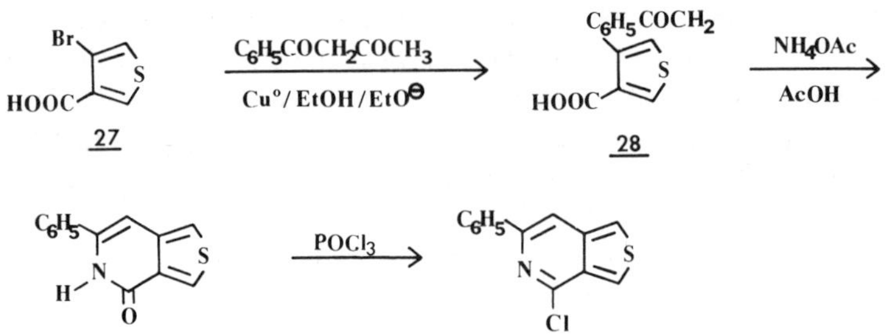

27 28 29 30

Facile deacylation occurs for certain carbanions to give **28,** which readily cyclizes to the pyridinone **29.** Treatment of **29** with phosphorus oxychloride affords the aromatized product **30.**[16]

23.3.3. ⟨5 + 1⟩ COMBINATIONS

From attempts to formylate 3-thienylacetonitrile (**31**) via the *Vilsmeier–Haack* reaction, thienopyridine **32** is isolated (34%); under the same conditions, 3-acetamidothiophene (**33**) gives five products, including **34a** and **34b.**[17] Isomeric thienopyridines are prepared from the appropriate acetamidothiophene.

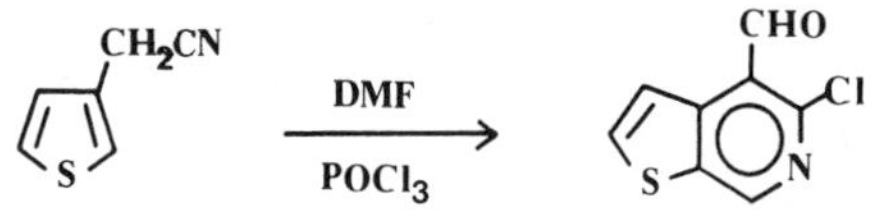

31 32 (34%)

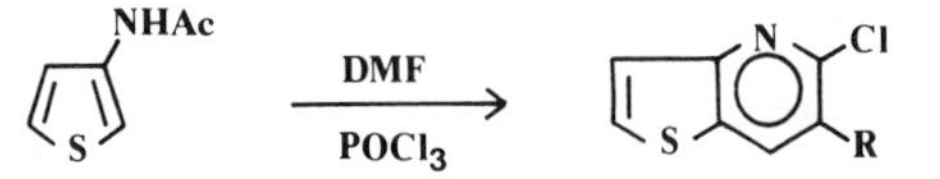

33

34a (R=H; 23%)
34b (R=CHO; 5%)

Limited chemistry of these thienopyridines has been reported. However, Meth-Cohn and Narine[17] have shown that **35** undergoes electrophilic substitution at C-5, nucleophilic substitution at C-2 (electron-deficient ring), and metal-halogen exchange at C-6, as one would predict.

23.4. FURANOPYRIDINES

Little synthetic work has been conducted on these ring systems, but routes to the thienopyridines should, in many cases, be applicable. The parent furano[3,4-*c*]pyridine (**36**) has recently been prepared for the first time by the flash vacuum thermolysis of 5,8-epoxy-5,6,7,8-tetrahydroisoquinoline (**37**), which is obtained by the catalytic hydrogenation of **38.**[18] Compound **36,** quantitatively formed by this procedure, is stable at room temperature, but quickly decomposes upon contact with air.

23.5. PURINES (7*H*- OR 9*H*-IMIDAZO[4,5-*d*]PYRIMIDINES)

Purine (**39**) chemistry dates back to the infancy of organic chemistry, as demonstrated by the isolation of uric acid (**40**) from gallstones in 1776[19]; the structural elucidation was completed nearly a century later by Liebig.[20] Caffeine (**41**) is a member of this family, as are guanine (**42**), xanthine (**43**), and adenine (**44**), three compounds of paramount importance in biochemistry. Numerous reviews[21] detail the diversified chemistry of this series, a result of the disparate electronic nature of the component rings: pyrimidine (an electron-deficient ring) and imidazole (an electron-excessive ring). Although illogical with respect to current rules, the ring numbering system of **39** dictated by Fischer in 1884[22] is nonetheless a matter of chemical convention.

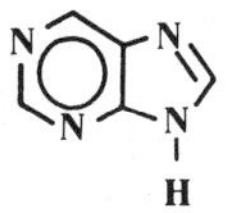

39 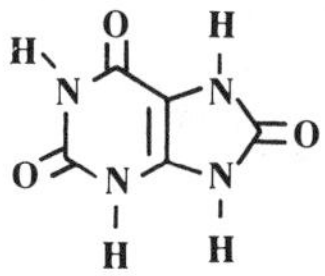 | uric acid 40 | caffeine 41

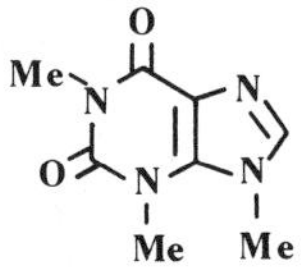

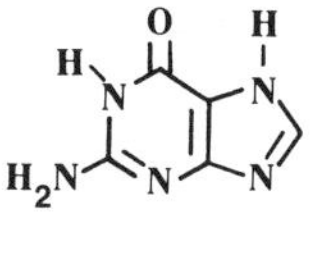

guanine 42 | xanthine 43 | adenine 44

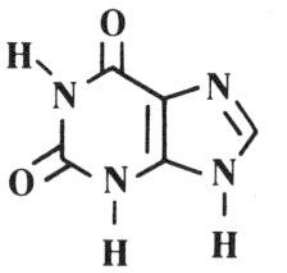

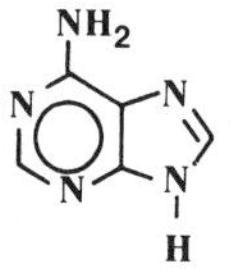

23.5.1. ⟨4 + 1⟩ COMBINATIONS

The majority of purine syntheses have been based on a pyrimidine precursor, and the *Traube synthesis*[23] is the most widely used procedure. This synthetic route generally parallels the *Isay reaction* in the synthesis of pteridines and can be described as the introduction of a one-carbon unit between the 4,5-diamino groups of a pyrimidine (**45**). Such one-carbon units (**46**) are carboxylic acids, amides, acyl halides, haloformates, ureas, thioureas, and amidines, from which water, ammonia, hydrohalic acid, or mixtures thereof are expelled. Xanthine (**43**) is prepared from **47** upon treatment with formamide.[24]

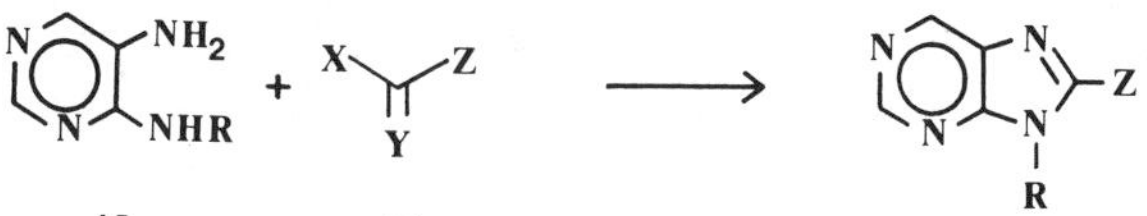

45 + 46 ⟶

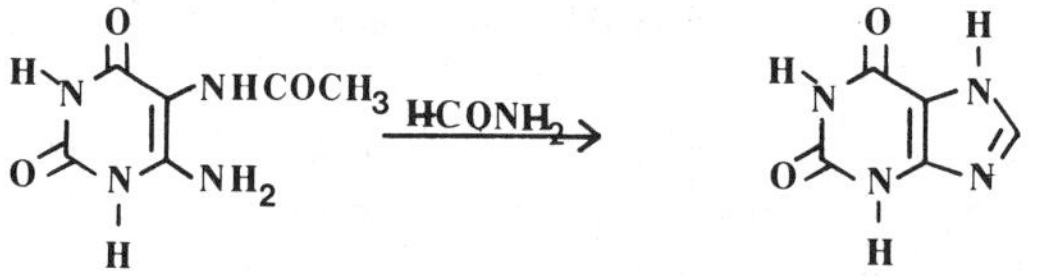

47 $\xrightarrow{HCONH_2}$ 43

The *in situ* reduction of a nitroso or nitro group with formic acid and zinc or Raney nickel affords a convenient route to purine (**39**).[25]

$$\text{4-nitroso-5-aminopyrimidine} \xrightarrow[\text{Zn}]{HCO_2H} \mathbf{39}$$

23.5.2. ⟨5 + 1⟩ COMBINATIONS (VIA IMIDAZOLES)

7-Methylxanthine (**48**) may be synthesized from an appropriate imidazole derivative upon treatment with an one-carbon reagent such as ethyl carbonate.[26] Other common one-carbon reagents are formic acid, esters and ortho esters, amides, urea, cyanates, isothiocyanates, chlorocarbonates, carbon disulfide, and thiophosgene.

$$\xrightarrow[\Delta]{(EtO)_2CO} \mathbf{48}$$

23.5.3. MISCELLANEOUS ROUTES

Amazingly, the fusing of two equivalents of urea with trichloroacetamide (**49**) gives uric acid (**40**)[27]! Similarly, purine (**39**) may be prepared by heating formamide with ammonia at 180°C,[28] and adenine (**44**) can be constructed (43%) from formamide and phosphoryl chloride[29] under pressure. Numerous examples of these magical combinations to give purine derivatives have been reported.[21] However, the ultimate combination is the reaction of ammonia and hydrogen cyanide to give adenine![30,31]

$$(H_2N)_2CO + \underset{\mathbf{49}}{Cl_3CCONH_2} \longrightarrow \mathbf{40}$$

$$HCN \xrightarrow{NH_3} \mathbf{44}$$

23.6. REACTIONS OF PYRROLOPYRIDINES

23.6.1. *N*-ALKYLATION AND ACYLATION

In the presence of strong base, pyrrolopyridines are alkylated on the pyrrole nitrogen; in the absence of strong base, a mixture of *N*-alkylated products results. Azaindole **50** reacts with ethylmagnesium iodide to give a salt, which is converted to the *N*-acetyl derivative with acetyl chloride.[32]

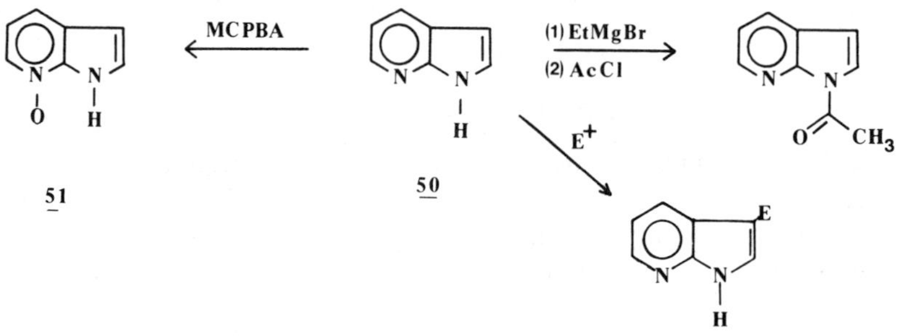

MCPBA (1) EtMgBr (2) AcCl E^+ O CH$_3$ E

51 50

23.6.2. *N*-OXIDATION

Treatment of **50** with *m*-chloroperbenzoic acid (MCPBA) gives the corresponding *N*-oxide, **51,** which is capable of typical *N* → *C* rearrangements with either acetic anhydride or phosphoryl chloride.[33]

23.6.3. ELECTROPHILIC SUBSTITUTION

The presence of the fused π-electron-deficient ring causes behavioral changes with respect to the indole nucleus. The Mannich, Vilsmeier, cyanomethylation, halogenation, and nitration reactions of **50** all give substitution patterns similar to the reactions on the isomeric 4-, 5-, and 6-azaindoles.[32–34]

23.6.4. NUCLEOPHILIC SUBSTITUTION

The standard Chichibabin reaction of electron-deficient systems does not occur with the azaindoles, probably due to the electron donation by the pyrrole ring; rather, the pyrrole ring is opened to give **52.**[33] Substitution of α- or γ-halogens on the pyridine ring (**53**) occurs readily by alkoxide ion,[5] but only under more severe conditions than for pyridine itself.

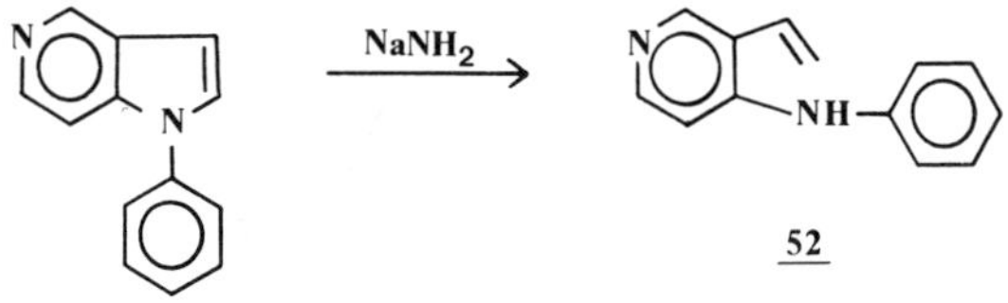

$NaNH_2$

52

RO^-

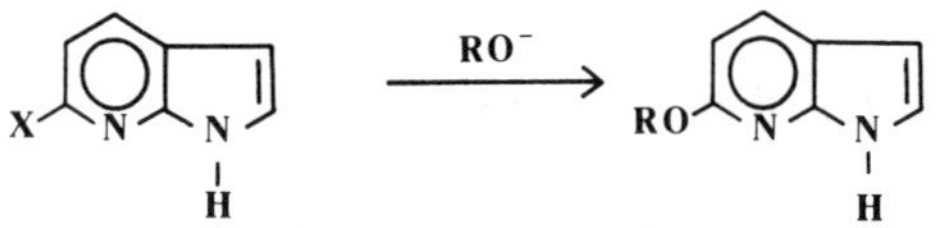

53

REFERENCES

1. Reviews: L. N. Yakhontov and A. A. Prokopov, *Russ. Chem. Rev.* **49,** 428 (1980); R. E. Willette, *Adv. Heterocycl. Chem.* **9,** 27 (1963); L. N. Yakhontov, *Russ. Chem. Rev.* **37,** 551 (1968).
2. L. N. Yakhontov, R. G. Glushkov, E. V. Pronina, and V. G. Smirnova, *Dokl. Akad. Nauk SSSR* **212,** 389 (1973).
3. M. N. Fisher and A. R. Matzuk, *J. Heterocycl. Chem.* **6,** 775 (1969); A. I. Scott, C. A. Townsend, K. Okada, and M. Kajiwara, *J. Am. Chem. Soc.* **96,** 8054 (1974).
4. L. N. Yakhontov, V. A. Azimov, and E. I. Lapan, *Tetrahedron Lett.* **1969,** 1909.
5. L. N. Yakhontov, E. I. Lapan, and M. V. Rubtsov, *Khim. Geterotsikl. Soed.* **5,** 550 (1969).
6. L. N. Yakhontov, M. S. Sokolova, N. I. Koretskaya, K. A. Chkhikvadze, O. Yu. Magidson and M. V. Rubtsov, *Khim. Geterotsikl. Soed.* **5,** 145 (1969).
7. A. Brodrick and D. G. Wibberley, *J. Chem. Soc. Perkin Trans. I* **1975,** 1910; W. Zimmermann, K. Eger, and H. J. Roth, *Arch. Pharm.* (Weinheim) **309,** 597 (1976).
8. I. Ahmed, G. W. H. Cheeseman, and B. Jaques, *Tetrahedron* **35,** 1145 (1979).
9. J. Duflos and G. Queguiner, *J. Org. Chem.* **46,** 1195 (1981).
10. J. M. Baker, in *Advances in Heterocyclic Chemistry,* Vol. 21, A. R. Katritzky and A. J. Boulton, Eds., Academic Press, New York, pp. 65–118; S. W. Schneller, *Int. J. Sulfur Chem.* **7,** 309 (1972).
11. M. Pailer and W. Jiresch, *Monatsch. Chem.* **100,** 121 (1969).
12. L. M. Klemm, C. E. Klopfenstein, R. Zell, and D. R. McCoy, *J. Org. Chem.* **34,** 347 (1969).
13. M. L. Dressler and M. M. Joullie, *J. Heterocycl. Chem.* **7,** 1257 (1970).
14. S. Gronowitz and E. Sandberg, *Ark. Kem.* **32,** 217 (1970).
15. W. Herz and L. Tsai, *J. Am. Chem. Soc.* **77,** 3529 (1955).
16. D. E. Ames and O. Ribeiro, *J. Chem. Soc. Perkin Trans. I* **1975,** 1390.
17. O. Meth-Cohn and B. Narine, *Tetrahedron Lett.* **1978,** 2045; O. Meth-Cohn, B. Narine, and B. Tarnowski, *J. Chem. Soc. Perkin I* **1981,** 1531.
18. U. E. Wiersum, C. D. Eldred, P. Vrijhof, and H. C. van der Plas, *Tetrahedron Lett.* **1977,** 1741.
19. Scheele, "Examen Chemicum Calculi Urinarii," *Opuscula* **2,** 73 (1776).
20. J. Liebig, *Ann. Chem.* **10,** 47 (1834); F. Wöhler and J. Liebig, *Ann. Chem.* **26,** 241 (1838).
21. J. H. Lister, in *Fused Pyrimidines.* Part II. *Purines,* D. J. Brown, Ed., Wiley-Interscience, New York, 1971, Chap. 1, ref. 11, 16–26; D. T. Hurst, in *An Introduction to the Chemistry and Biochemistry of Pyrimidines, Purines, and Pteridines,* Wiley, 1980.
22. E. Fischer, *Chem. Ber.* **17,** 329 (1884).
23. W. Traub, *Chem. Ber.* **33,** 1371, 3035 (1900).
24. D. S. Acker and J. E. Castle, *J. Org. Chem.* **23,** 2010 (1958).
25. C. E. Liau, K. Yamashita, and M. Matsui, *Agric. Biol. Chem.* (*Japan*) **26,** 624 (1962).
26. J. Sarasin and E. Wegmann, *Helv. Chim. Acta* **7,** 713 (1924).
27. J. Horbaczewski, *Monatsh. Chem.* **8,** 201, 584 (1887).
28. H. Bredereck, H. Ulmer, and H. Waldmann, *Chem. Ber.* **89,** 12 (1956).
29. M. Ochiai, R. Marumoto, S. Kobayashi, H. Shimazu, and K. Morita, *Tetrahedron* **24,** 5731 (1968).
30. J. Oro, *Nature* **191,** 1193 (1961).
31. C. U. Lowe, M. W. Rees, and R. Markham, *Nature* **199,** 219 (1963).

32. R. Herbert and D. G. Wibberley, *J. Chem. Soc. (C)* **1969,** 1505.

33. B. A. J. Clark, M. M. S. El-Bakoush, and J. Parrick, *J. Chem. Soc. Perkin Trans. I* **1974,** 1531; D. M. Krasnokut-skaya and L. N. Yakhontov, *Khim. Geterotsikl. Soedin.* **13,** 380 (1977).

34. A. A. Prokopov and L. N. Yakhontov, *Khim. Geterotsikl. Soedin.* **14,** 496 (1978).

Chapter 24

MESO-IONIC HETEROCYCLIC COMPOUNDS

24.1. GENERAL COMMENTS

As previously described, the reaction of pyridine with methyl iodide affords *N*-methylpyridinium iodide (**1**), in which the positive charge of the nitrogen is balanced by the negative iodide counterion. When the pyridinium ring contains a functional group which can sustain a negative charge, as is the case for pyridinium-3-olates (**2**), the compound is an "internal" salt rather than an "external" one, as in the instance of pyridinium ion **1.** These "internal" salts are known as *mesomeric betaines.*[1,2]

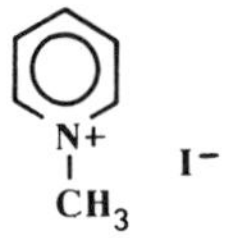

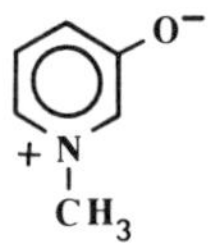

Earl and Mackney[3] dehydrated *N*-nitroso-*N*-phenylglycine (**3**) and obtained a heterocycle which is best described as a resonance hybrid of the canonical structures **4a** ↔ **4c.** The Australian chemists (University of Sydney) coined the name *N*-phenyl-"sydnone" for this compound, and this sydnone can be represented by general structure **5.**[2]

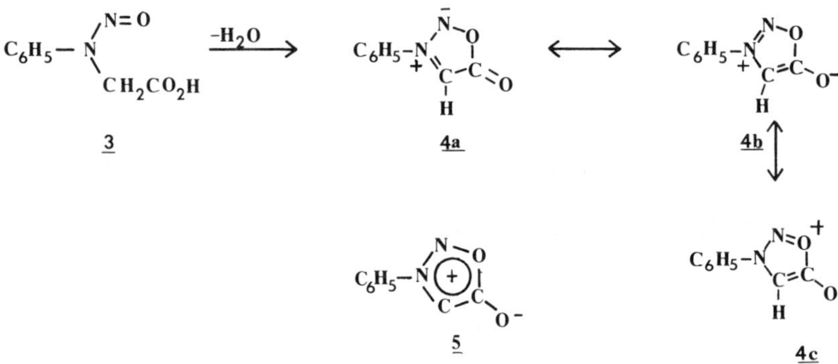

This type of ionic heterocycle is referred to as "*meso-ionic*" and represents a subdivision of the general class of mesomeric betaines. The most recently proposed definition of a meso-ionic compound is as follows[3]: *A five-membered heterocycle which cannot be represented satisfactorily by any one covalent or polar structure and possesses a sextet of electrons in association with the five atoms comprising the ring.*

Meso-ionic compounds can be further subdivided into two distinct types (A and B), depending upon whether each of two adjacent atoms of the five-membered ring contributes two π-electrons (type B) or whether each of two nonadjacent atoms (type A) contributes two π-electrons to the aromatic sextet. In the most general sense, meso-ionic compounds are symbolized as in **6.**

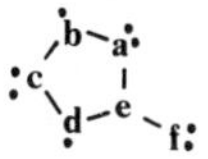

Type A 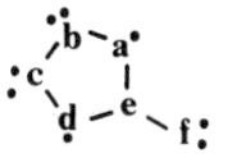 Type B 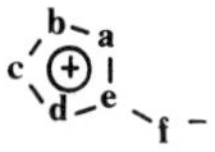 6

The various positions in meso-ionic compounds can be occupied by suitably substituted carbon, nitrogen, oxygen, or sulfur atoms. When only these are considered (selenium, etc., are also possible ring members), 144 different meso-ionic structures of type A can be drawn, of which 48 different systems have been prepared.[6] Of the 84 different type B meso-ionic systems that are possible, only eight have so far been reported.

Many of the type A meso-ionic compounds participate in 1,3-dipolar cycloaddition reactions where canonical forms of structure **7** must make a major contribution. In fact, this structure can be thought of as being formed from the union of the 1,3-dipolar structure **8** and the heterocumulene **9.** Among the known type A meso-ionic compounds, most are associated with one of the nitrogen-containing fragments **10, 11,** or **12.**

7 8 9

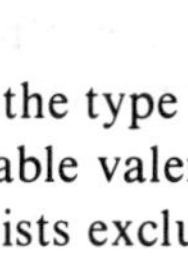

10 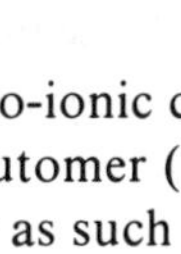 11 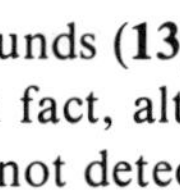 12

Consideration of the type B meso-ionic compounds (**13**) suggests the possible existence of a stable valence tautomer (**14**). In fact, although the meso-ionic compound **15** exists exclusively as such (**16** is not detectable), triketone **17** is stable and none of its possible meso-ionic tautomer **18** has been observed.[6]

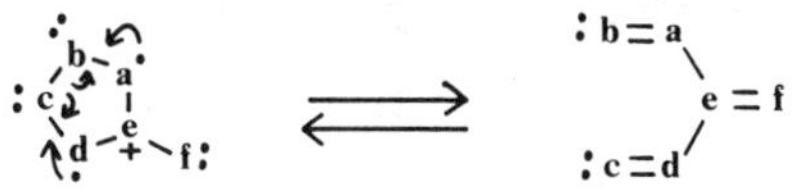

13 14

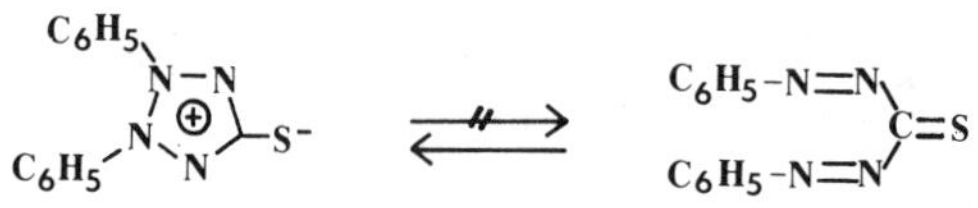

15 16

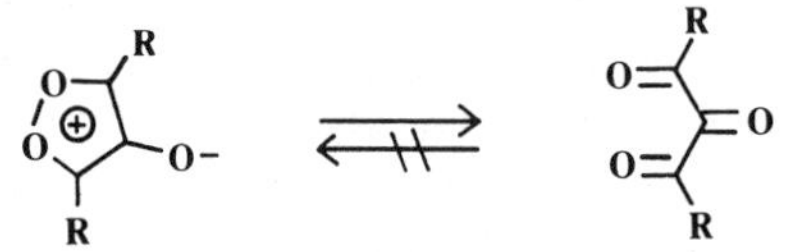

18 17

There is still some ambiguity in the naming of meso-ionic compounds; for example, *N*-phenylsydnone (**5**) has been described as meso-ionic 1,2,3-oxadiazole-5-one[6] and as meso-ionic 1,2,3-oxadiazolium-5-olate. The former name emphasizes the double bond character of the exocyclic group, whereas the latter emphasizes the charge on the exocyclic group. Since the latter more clearly describes the ground state of meso-ionic compounds, such nomenclature is followed in this text.

24.2. SYNTHESES AND PROPERTIES OF SOME TYPE A MESO-IONIC COMPOUNDS

24.2.1. 1,3-OXAZOLIUM-4-OLATES

The thermal decomposition of diazoketones of general structure **19** in the presence of cupric acetylacetonate[7,8] affords **21,** presumably via the intermediacy of carbene **20.** These red, highly crystalline compounds are reactive 1,3-dipoles. For example, reaction of **21** with acetylenes affords the bicyclic intermediate **22,** which upon heating, is transformed to the furan **23.**

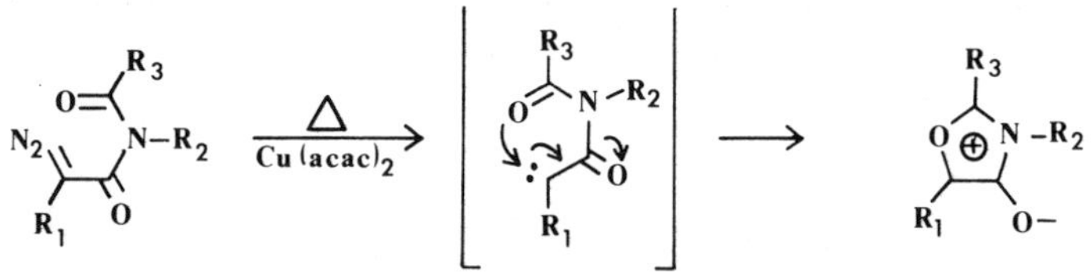

Δ Cu (acac)$_2$

19 20 21

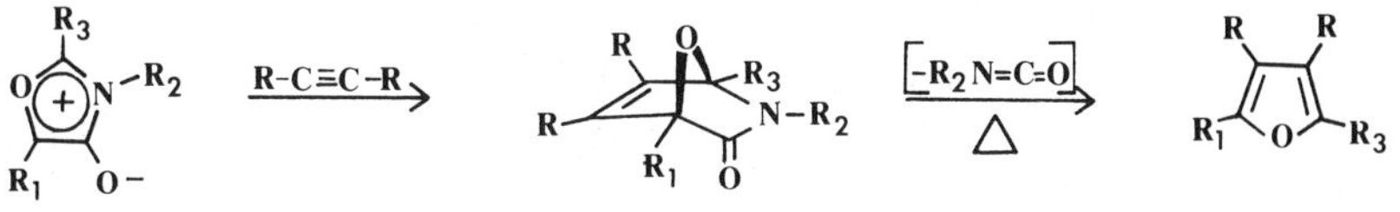

21 → (R-C≡C-R) → 22 → ([-R₂N=C=O], Δ) → 23

24.2.2. 1,3-OXAZOLIUM-5-OLATES

These compounds (**24**), which are isomeric with the 1,3-oxazolium-4-olates, have been named "münchnones" by Huisgen (University of München), and are prepared by the cyclodehydration of α-acylamino acids (**25**), usually by means of acetic anhydride under moderate conditions. The products **24** react with acetic anhydride at elevated temperatures to form ring-opened amides **26.**

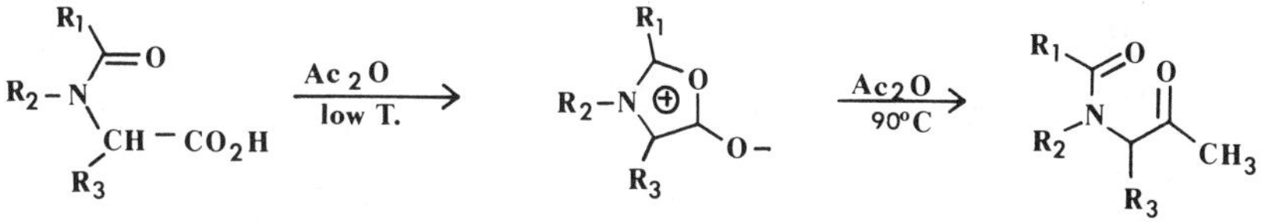

25 → (Ac_2O, low T.) → 24 → (Ac_2O, 90°C) → 26

The yellow trisubstituted 1,3-oxazolium-5-olates (**24**) are moisture sensitive, and the disubstituted analogues [(R_2 = H) **24**] are too unstable to be isolated. Münchnones (**24**) are very susceptible to attack by water, alcohols, and primary amines and afford the corresponding nucleophilic reaction products **27, 28,** and **29,** respectively.[9] These 1,3-oxazolium-5-olates (**24**) are also very reactive as 1,3-dipoles. Examples of the many reactions studied[9,10] are shown below and on the next page:

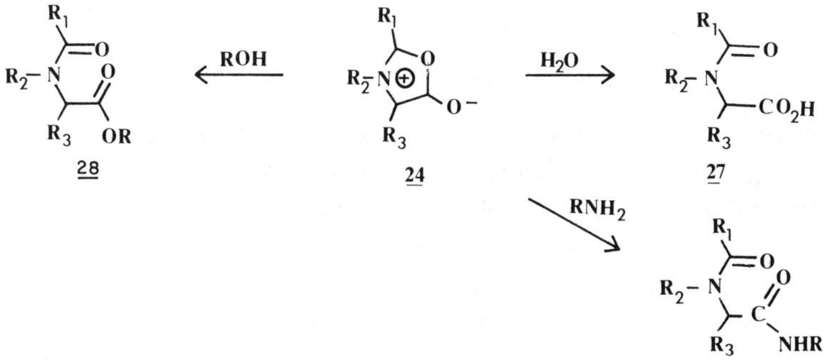

28 ← (ROH) ← 24 → (H_2O) → 27

24 → (RNH_2) → 29

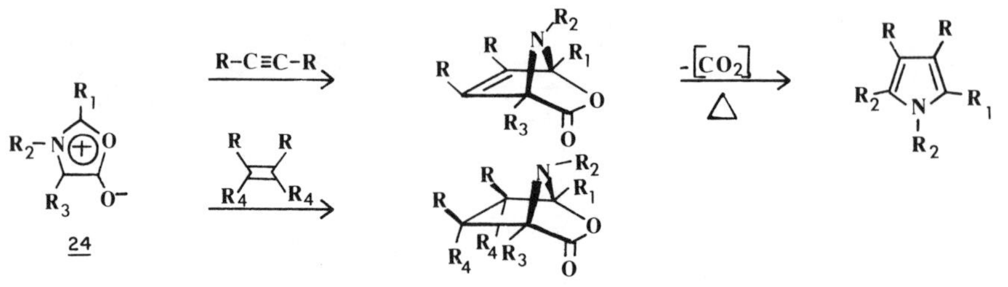

The β-lactam **31,** derived from some münchnones (for example, **30**) with carbodiimides (**32**), suggests that a small equilibrium concentration of the valence-tautomeric ketone **30A** exists in solution.[11]

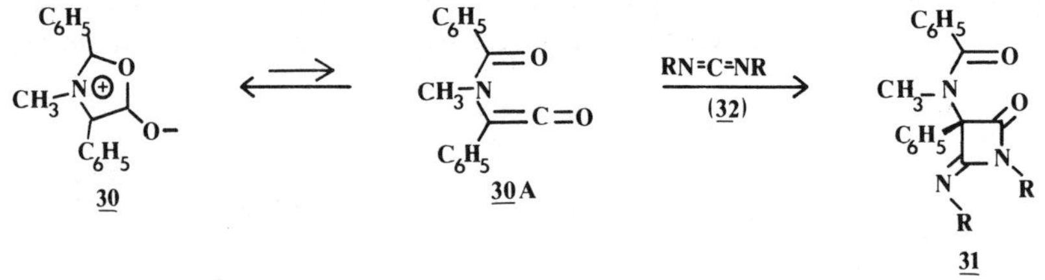

24.2.3. 1,3-OXATHIOLIUM-4-OLATES

The sulfur analogues (**33**) of the 1,3-oxazolium-4-olates, formed by cyclodehydration of the acid derivatives **34,** are too unstable to be isolated,[12] but react *in situ* with acetylenes to afford substituted furans (**35**).

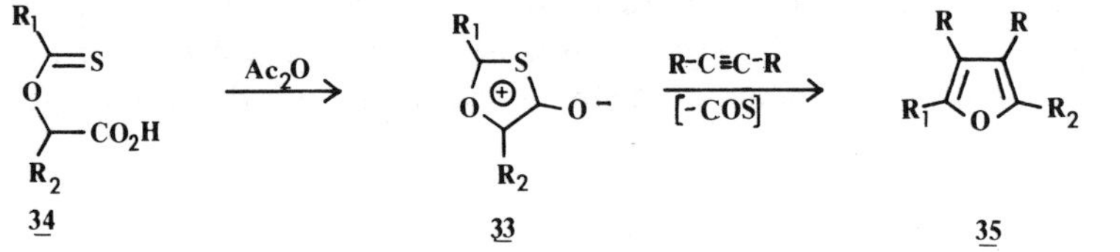

24.2.4. 1,3-OXATHIOLIUM-5-OLATES

Interestingly, when the sulfur and ether oxygen in **34** are interchanged, and the resulting acid (**36**) is dehydrated with trifluoroacetic anhydride (TFAA), the stable, red, crystalline, meso-ionic **37**[13] results.

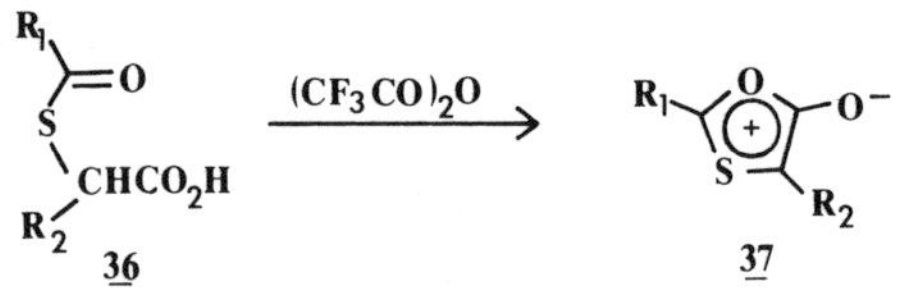

24.2.5. 1,3-DIAZOLIUM-4-OLATES

These imidazole-related meso-ionic compounds[14-16] can be prepared in a number of different ways, but the transformation **38** to **39** affords the best yields. These compounds (**39**) are amazingly stable toward acid and alkali.

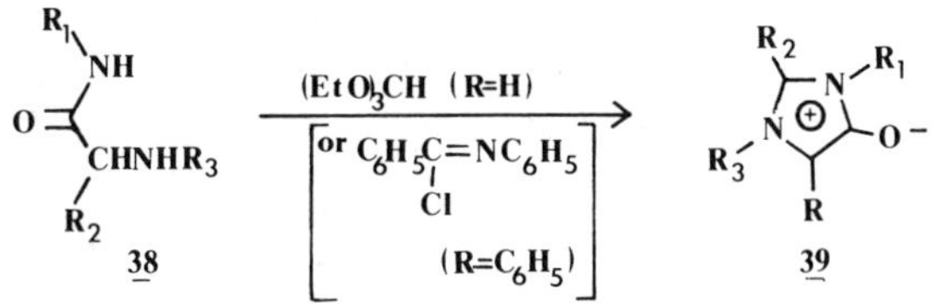

The tetraphenyl derivative **40** undergoes facile 1,3-dipolar additions.[16] The bicyclic intermediates formed by reaction with alkynes are thermally unstable and are converted to the expected pyrroles (**41**). Treatment of **40** with dimethyl fumarate affords the predicted addition product (**42**), which loses methanol to give the bicyclic imide **43.**

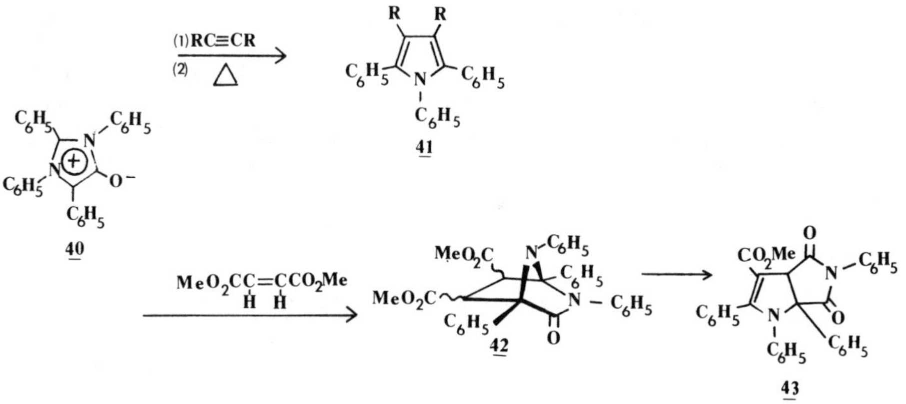

24.2.6. 1,3-DIAZOLIUM-4-AMIDES

In this class of compounds (for example, **46**) the negative charge is concentrated on an exocyclic nitrogen rather than an oxygen atom. The synthesis of **46** is unique in that it involves the intermediacy of the stable aminoimidazolium salt **45,** which has been prepared from nitrile (**44**) and an acid chloride. When **45** is treated with sodium bicarbonate, it is converted to the yellow meso-ionic **46.** As expected, these compounds also react as 1,3-dipoles. Thus, when treated with dimethyl acetylenedicarboxylate, **47** affords pyrrole **48** without intermediate isolation of the addition product **49.**[17] When **47** is treated with diethyl azodicarboxylate (**50**), the bicyclic adduct **51** is obtained.

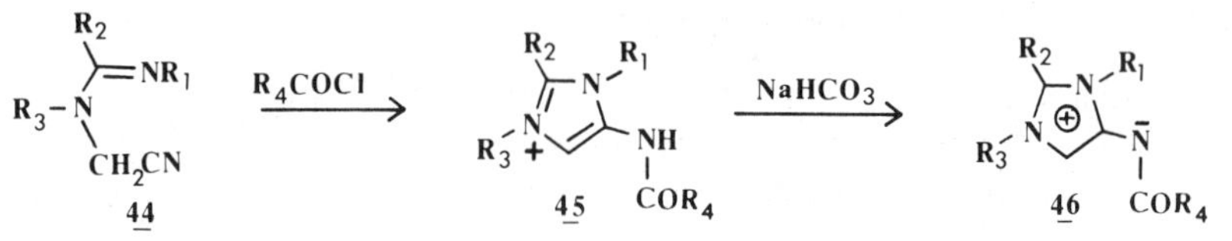

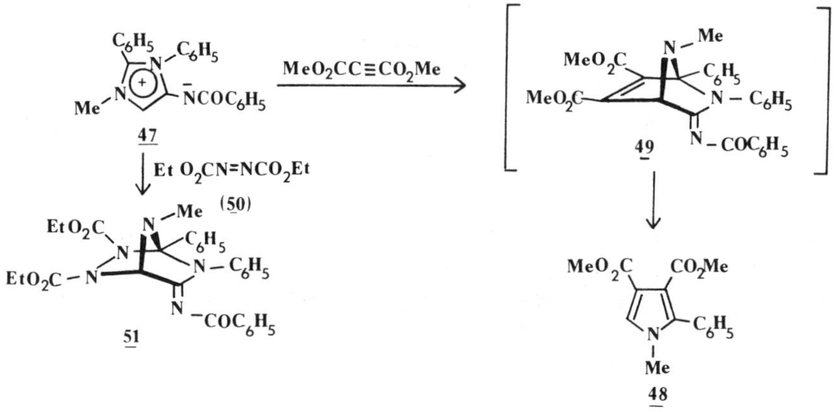

24.2.7. 1,3-DITHIOLIUM-4-OLATES

Cyclodehydration of thioglycolic acid (**52**) generates the highly colored sulfur analogues (**53**) of the 1,3-diazolium meso-ionic compounds.[18–20] Again, these compounds react as 1,3-dipolarophiles, affording thiophenes (**54**) upon reaction with alkynes and a bicyclic 1 : 1 addition product (**55**) when treated with alkenes.

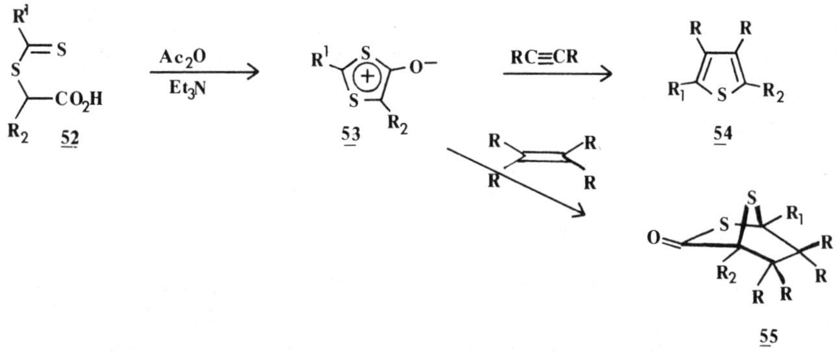

24.2.8. 1,2,3-OXADIAZOLIUM-5-OLATES

These sydnones were the first meso-ionic compounds which were studied in great detail,[2,21,22] although only one synthetic route is available (**56** → **58**). This general synthesis fails only when R_1 in **56** is a hydrogen. This is not surprising, since the resulting meso-ionic compound would immediately tautomerize to the stable oxadiazole derivative **59.**

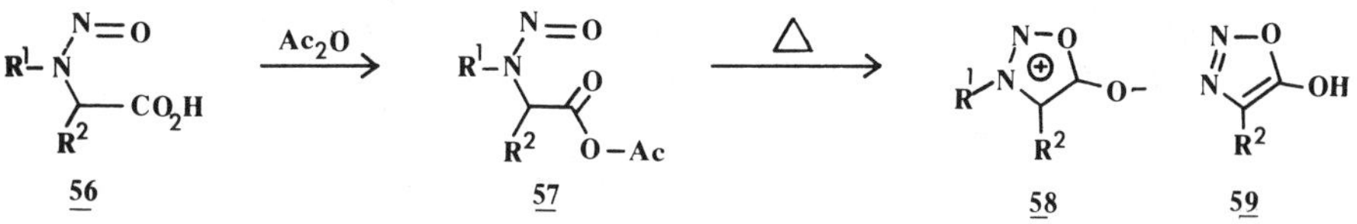

The physical properties of sydnones have been carefully examined and are consistent with their formulation as meso-ionic compounds. For example, the "carbonyl" bond absorption is in the region 1718–1770 cm^{-1}, the dipole moment of the *N*-phenyl derivative is 6.48D, and ESCA spectra show a considerable difference between the two nitrogen atoms.[2,21,22] When *N*-phenylsydnone (**60**) is dissolved in "super acid" ($HFSO_3$–SbF_5), protonation occurs on the exocyclic oxygen (**61**).[23] Alkylation with Meerwein's reagent (triethyloxonium tetrafluoroborate) results in *O*-alkylation (**62**) on the same site.[24]

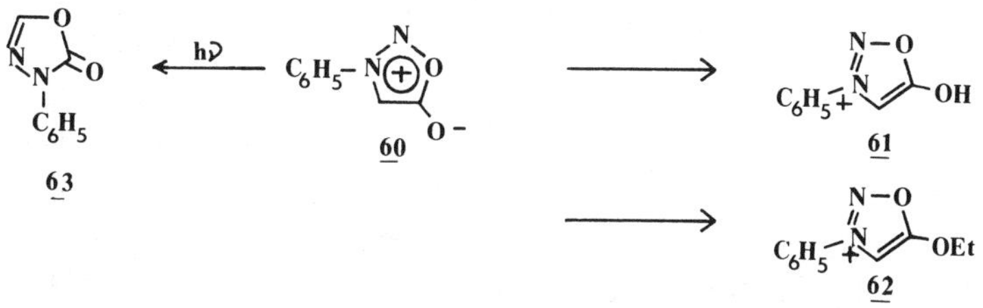

Many photochemical transformations of sydnones have been described.[6] Among these, the formation of 4-phenyl-Δ^2-1,3,4-oxadiazol-5-one (**63**) from *N*-phenylsydnone (**60**) is of some interest. Although the mechanism of this reaction has not been conclusively established, it may well involve the following steps:

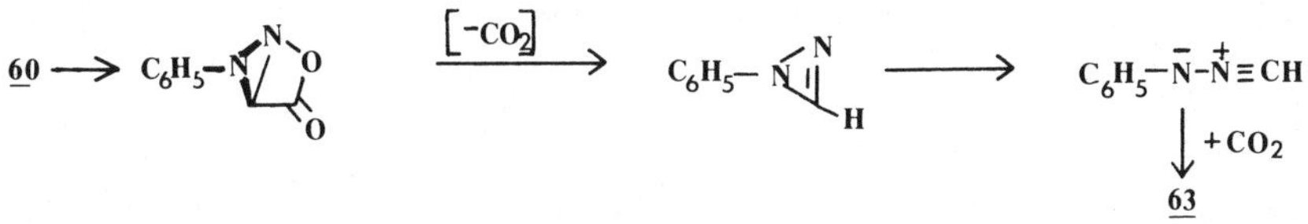

Sydnones that have no substituent at the 4-position can be brominated with bromine or *N*-bromosuccinimide and undergo a number of other typical electrophilic substitution reactions at C-4.[21] 4-Lithio derivatives (**64**)[25,26] are available from either 4-unsubstituted sydnones or their 4-halo derivatives.

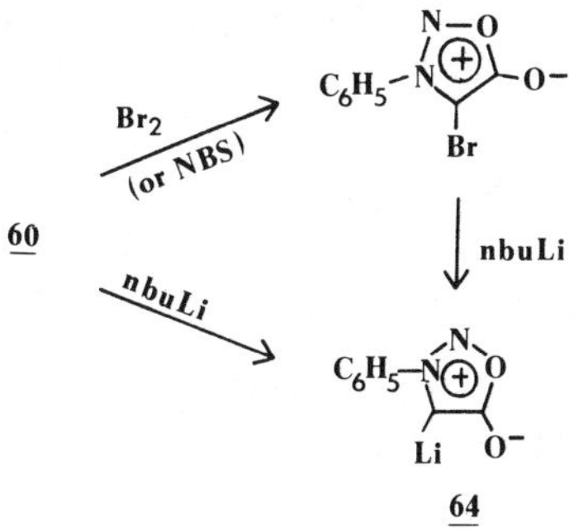

The 1,3-dipolar cycloaddition reactions of sydnones have received considerable attention.[9] Some examples of these reactions follow:

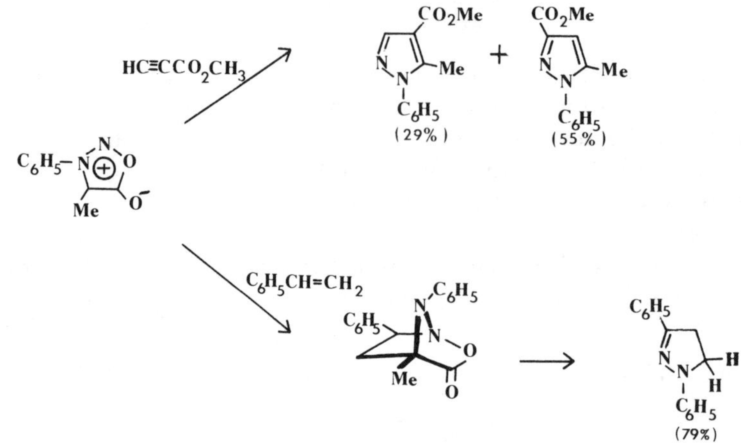

24.3. SYNTHESES AND PROPERTIES OF SOME TYPE B MESO-IONIC COMPOUNDS

24.3.1. 1,2-DIAZOLIUM-4-AMIDES

Treatment of **65** with methyl fluorosulfonate, followed by deprotonation of the diazolium perchlorate (**66**) with aqueous potassium hydroxide,[27] affords this meso-ionic ring **67.** The tetramethyl derivative **68,** when heated in benzonitrile, rearranges to the isomeric pyrazole **69.**

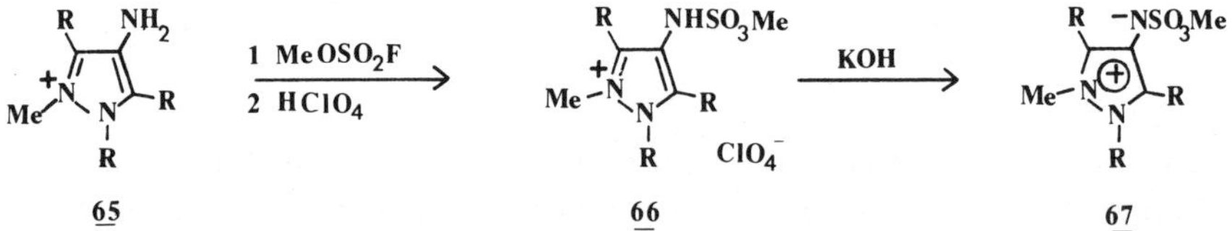

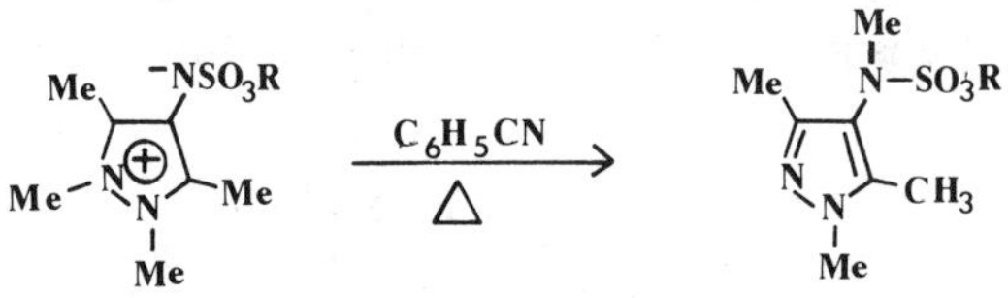

24.3.2. 1,2-DITHIOLIUM-4-OLATES

Diaryl derivatives of this meso-ionic ring system (**70**) are also available by any one of the following sequences[28–30]:

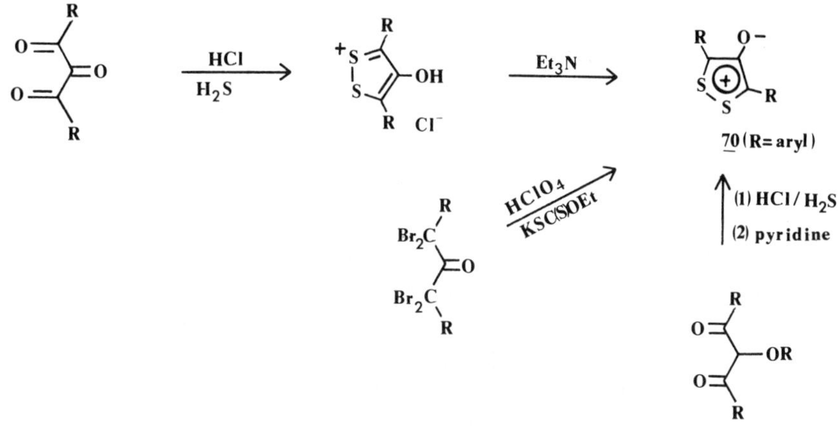

The 3,5-diaryl-1,2-dithiolium-4-olates (**71**) are high-melting crystalline solids.[28–30] Some of the chemical reactions of these meso-ionic compounds are indicative of their relative chemical stability.

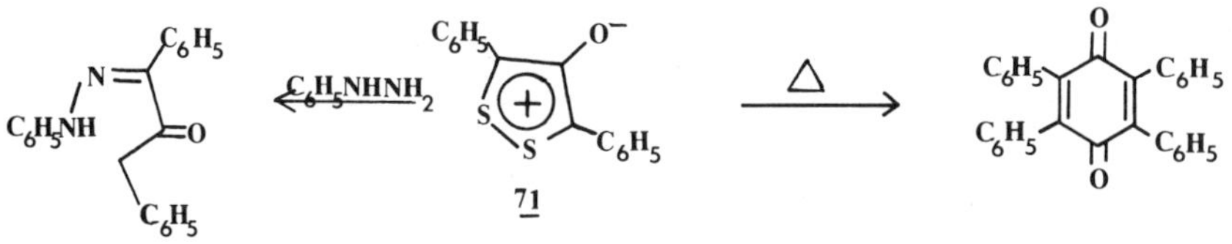

24.3.3. 1,2,3,4-TETRAZOLIUM-5-THIOLATES

Although the only known example of this meso-ionic system is the diphenyl derivative **72,** its unique properties warrant inclusion in this text. The common name of this derivative is dehydrodithizone. The red material, originally obtained in 1882, is prepared by manganese dioxide oxidation of dithizone (**73**).[31] The structure of **72** has only recently been confirmed by X-ray crystallography.[32] The compound rearranges, thermally, to the isomeric type A meso-ionic compound **74.**

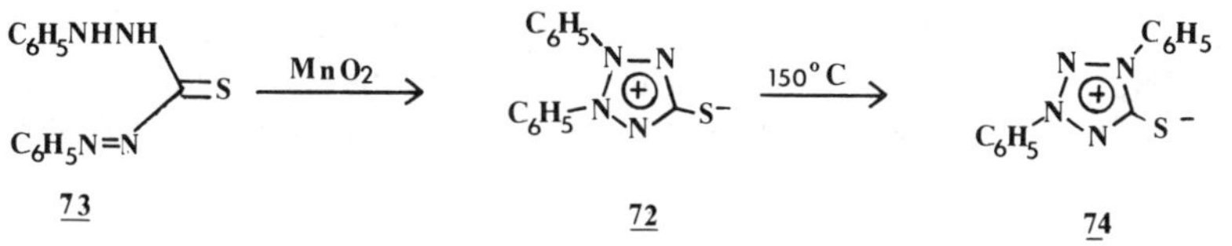

This system is unusual in that many of its reactions are best explained by invoking its ring-opened valence-bond tautomer **75,** as admirably demonstrated by the following[33–36]:

$N{=}N{-}C_6H_5$, Me, NEt_2, C_6H_5

S, $N = N - C_6H_5$, N, N, H

$MeC{\equiv}CNEt_2$ AcOH Δ

$C_6H_5N = N$, $N = N - C_6H_5$, S

75

C_6H_5, C_6H_5, N, N, N, S⁻

72

$MeO_2CC{\equiv}CCO_2Me$

CO_2Me, CO_2Me, C_6H_5, C_6H_5, N, N, N, N, S

NC, CN, NC, CN

CN CN, C_6H_5, C_6H_5, N, N, N, N, S, CN, CN

Reaction with electron-deficient alkynes and alkenes, however, gives products that are best explained by using the cyclic tautomer **74.**[35]

24.4. SYNTHETIC APPLICATIONS OF MESO-IONIC COMPOUNDS

The 1,3-dipolar cycloaddition of alkynes to meso-ionic compounds represents a valuable synthetic tool since the intermediate bicyclic derivatives can often be converted to five-membered heterocyclic rings of different types. The other novelty is found in the observation that these meso-ionic compounds rarely have to be isolated from their acyclic precursors and can, in fact, be prepared *in situ*. The overall reaction sequence is the following: when X = O and Z = RN, pyrroles are obtained; furans are synthesized when X = S and Z = O. Thiophenes become available when X = S or C_6H_5N and Z = S.

R_2, C = X, Z, CO_2H, R_1, H $\xrightarrow{-H_2O}$ R_2, X, Z, ⊕, O^-, R_1 $\xrightarrow{R_3C{\equiv}CR_4}$ R_3, R_4, R_1, Z, R_2

REFERENCES

1. W. Baker, W. D. Ollis, and V. D. Poole, *J. Chem. Soc.* **1949,** 307; **1950,** 1542.
2. W. Baker and W. D. Ollis, *Quart. Rev.* **11,** 15 (1957).
3. J. C. Earl and A. W. Mackney, *J. Chem. Soc.* **1935,** 899.
4. R. A. Eade and J. C. Earl, *J. Chem. Soc.* **1946,** 591.
5. Y. Noël, *Bull. Soc. Chim. Fr.* **1964,** 173.
6. W. D. Ollis and C. A. Ramsden, *Adv. Heterocycl. Chem.* **19,** 1 (1976).
7. M. Hamaguchi and T. Ibata, *Tetrahedron Lett.* **1974,** 4475.
8. T. Ibata, M. Hamaguchi, and H. Kiyohara, *Chem. Lett.* **1975,** 21.
9. R. Huisgen, *Chem. Soc. Spec. Publ.* **21,** 51 (1967).
10. R. Knorr and R. Huisgen, *Chem. Ber.* **103,** 2598 (1970).
11. R. Huisgen, E. Funke, F. C. Schaefer, and R. Knorr, *Angew. Chem. Int. Ed.* **6,** 367 (1967).
12. H. Gotthardt, M. C. Weisshuhn, and K. Dörhöfer, *Angew. Chem. Int. Ed.* **14,** 422 (1975).
13. K. T. Potts, J. Kane, E. Carnahan, and U. P. Singh, *J. Chem. Soc. Chem. Commun.* **1975,** 417.
14. V. R. Grashey, E. Jänchen, and J. Litzke, *Chem. Z.* **97,** 657 (1973).
15. A. Lawson and D. H. Miles, *J. Chem. Soc.* **1959,** 2865.
16. T. Shiba and H. Kato, *Bull. Chem. Soc. Japan* **43,** 3941 (1970).
17. K. T. Potts and S. Husain, *J. Org. Chem.* **36,** 3368 (1971).
18. K. T. Potts, D. R. Choudhury, A. J. Elliott, and U. P. Singh, *J. Org. Chem.* **41,** 1724 (1976).
19. H. Gotthardt and B. Christl, *Tetrahedron Lett.* **1968,** 4743, 4747, 4751.
20. H. Kato, M. Kawamura, T. Shiba, and M. Ohta, *Chem. Commun.* **1970,** 959.
21. F. H. C. Stewart, *Chem. Rev.* **64,** 129 (1964).
22. L. B. Kier and E. B. Roche, *J. Pharm. Sci.* **56,** 149 (1967).
23. G. A. Olah, D. P. Kelly, and N. Suciu, *J. Am. Chem. Soc.* **92,** 3133 (1970).
24. K. T. Potts, E. Houghton, and S. Husain, *Chem. Commun.* **1970,** 1025.
25. C. V. Greco, M. Pesce, and J. M. Franco, *J. Heterocycl. Chem.* **3,** 391 (1966).
26. H. Kato and M. Ohta, *Bull. Chem. Soc. Japan* **32,** 282 (1959).
27. G. V. Boyd and T. Norris, *J. Chem. Soc. Perkin I* **1974,** 1028.
28. A. Chinone, K. Inouye, and M. Ohta, *Bull. Chem. Soc. Japan* **45,** 213 (1972).
29. D. Barillier, P. Rioult, and J. Vialle, *Bull. Soc. Chim. Fr.* **1976,** 444.
30. A. Schönberg and E. Frese, *Chem. Ber.* **103,** 3885 (1970).
31. E. Fischer and E. Besthorn, *Ann. Chem.* **212,** 316 (1882).
32. Y. Kushi and Q. Fernando, *J. Am. Chem. Soc.* **92,** 1965 (1970).
33. J. W. Ogilvie and A. H. Corwin, *J. Am. Chem. Soc.* **83,** 5023 (1961).
34. W. S. McDonald, H. M. N. H. Irving, G. Raper, and D. C. Rupainwar, *Chem. Commun.* **1969,** 392.
35. G. V. Boyd, T. Norris, and P. F. Lindley, *J. Chem. Soc. Perkin I* **1976,** 1673.
36. P. Rajagopalan and P. Renev, *Chem. Commun.* **1971,** 490.

Chapter 25

THREE- AND FOUR-MEMBERED HETEROCYCLES

25.1. GENERAL COMMENTS

Three-membered heterocycles are usually prepared by modifications of the general reaction:

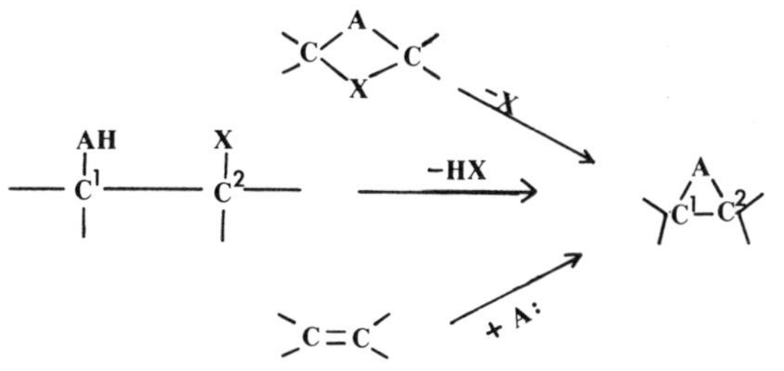

where A is O, NR, or S. The dehydrohalogenation is normally accomplished by alkali and proceeds by backside attack of the A^- species at the carbon bearing the X group (halogen, —OSO_3R, or other good leaving group). Thus inversion of configuration at C-2 occurs via this intramolecular *SN*2 reaction. There are two additional routes to this family; a [2 + 1] cycloaddition (two bonds being generated simultaneously) and contraction of a larger ring.[1a]

25.2. THREE-MEMBERED AZA-HETEROCYCLES

25.2.1. AZIRIDINES (AZACYCLOPROPANES)

25.2.1.1. General Comments. Of the three saturated three-membered ring heterocycles to be discussed, aziridine is unique in that the substituent on the nitrogen atom does not lie in the plane of the ring. Thus a properly substituted aziridine is a potentially chiral molecule (for example, **1a** and **1b**). At room temperature the inversion process between **1a** and **1b** is

too rapid to allow isolation of the enantiomers. However, at temperatures below about −40°C, the inversion rate is slowed sufficiently to permit the determination of the inversion frequency by NMR.

25.2.1.2. Syntheses. The *Gabriel method* affords aziridines (**2**) through cyclodehydration of β-chloroamines (**3**). The β-chloroamines are generally available from β-hydroxyamines, which can be generated from the reaction of ammonia or primary amines with epoxides (**4**).[1-3]

RCH—CHR (O) **4** —$R'NH_2$→ RCH(OH)—CHR(NHR') —HCl→ RCH(Cl)—CHR(NHR') **3** —Base→ RCH—CHR (N—R') **2**

In the *Wenkert method,* β-amino alcohols are treated with sulfuric acid to afford the β-amino hydrogen sulfate derivatives (**5**) which, when treated with base, are readily converted to the corresponding aziridines.

RCH(OH)—CHR(NHR') —H_2SO_4→ RCH(HO_3SO)—CHR(NHR') **5** ⟶ RCH—CHR (N—R') **2**

Neither of these procedures is applicable to the preparation of 2,2,3,3-tetraalkylaziridines because the corresponding β-haloamine precursors cannot be readily obtained.[3]

A new method for the syntheses of aziridines, applicable to all substituted aziridines, involves the treatment of olefins (**6**) with iodoisocyanate (**7**) and proceeds via the intermediacy of the iodonium ion **8.** The resulting *trans*-iodoisocyanate **9,** when treated with methanol, affords the corresponding β-iodocarbamate (**10**), which is converted to the desired aziridine (**11**) by treatment with alcoholic base.[4] The stereospecific nature of these cyclization procedures is exemplified by the following sequence in which **12** gives chiral **13.**

6 —INCO (**7**)→ [**8**] —NCO^-→ **9** (N=C=O, I) —MeOH→

10 ($NHCO_2CH_3$, I) —KOH, EtOH→ **11** (N—H)

(−)-erythro-**12** (NH_2, H, C_6H_5, C_6H_5, H, Cl) —KOH, EtOH→ **13** (80% optically active) (NH, H, C_6H_5, C_6H_5, H)

The direct insertion of nitrenes into an olefinic linkage represents another synthetic route to the aziridines (for example, **14**).[3b] When the nitrene is either photochemically or thermally prepared, the yields of azetidine are not very impressive. However, if ethyl azidoacetate reacts under mild conditions (room temperature) with an olefin, the intermediate 1,2,3-triazole (**15**) can readily be isolated. Photolysis of the triazole provides **16** in excellent yield.[5]

N_3CO_2Et —hν (or Δ)→ [$:\ddot{N}-CO_2Et$] → **14** (N-CO₂Et aziridine)

$O_2N-C_6H_4-SO_2ONHCO_2Et$ —Et_3N→ [$:\ddot{N}-CO_2Et$]

norbornene + N_3CO_2Et —25°→ **15** (94%) —hν→ **16** (95%)

25.2.2. AZIRINES (AZACYCLOPROPENES): SYNTHESES

Two isomeric azirines (**17** and **18**) are clearly possible. Of these, only the 2*H*-azirines (**18**) are stable; the 1*H*-azirines (**17**) have merely been postulated as intermediates in some transformations.[6] It has been suggested that the instability of the 1*H*-azirines (**17**) is due to the potential overlap of the nitrogen lone pair of electrons and the olefinic π-electrons (**19**). In fact, *ab initio* calculations[6] indicate that the 2*H*-azirine is more stable than the 1*H* isomer by 169 kJ mole^{-1}. The first general synthesis of 2*H*-azirines (for example, **20**) is due to Smolinsky[7] and involves the vapor-phase pyrolysis of α-vinylazides (**21**). Vinylazides (**22**) can be prepared by the following sequence of reactions.[8]

17 **18** **19**

21 —100–105°C [or hν (λ = 365 nm) <−30°C, N_2]→ **20**

$$RCH{=}CH_2 \xrightarrow[NaN_3]{ICl} RCH(N_3)CH_2I \xrightarrow[Et_2O,\ 0^\circ C]{KOtbu} R(N_3)C{=}CH_2 \ (\mathbf{22}) \xrightarrow{h\nu} [\text{nitrene}] \longrightarrow \text{2H-azirine}$$

The vinyl azides are best transformed photochemically via the nitrene to the 2*H*-azirines (**23**). A remarkable wavelength-dependent photoinduced isoxazole–oxazole rearrangement has been described.[6]

$$\text{isoxazole} \underset{(\lambda = >300)}{\overset{h\nu\ (\lambda = 253{\cdot}7)}{\rightleftharpoons}} \mathbf{23} \xrightarrow[(\lambda = 253{\cdot}7)]{h\nu} \text{oxazole}$$

25.2.3. REACTIONS

25.2.3.1. Aziridines. Ring-opening reactions initiated by nucleophilic reagents occur largely with inversion of configuration (**24** to **25**) at the site of nucleophilic attack.[9] When unsymmetrical aziridines (**26**) are involved in these reactions, there are two possible sites of attack, and ring-opened products (**27**, **28**) arising from attack at either or both locations are observed. Generally, the less hindered carbon is attacked, although the product ratios are often reagent and solvent dependent and are difficult to predict (some specific examples in oxirane chemistry are described later).

$$\mathbf{24}\ (\text{chiral}) \xrightarrow{CH_3CH_2NH_2} \mathbf{25}\ (\text{meso})$$

$$\mathbf{26} \xrightarrow{Nu^-} R_1CH(NH_2)CHR_2(Nu)\ (\mathbf{27}) + R_1CH(Nu)CHR_2(NH_2)\ (\mathbf{28})$$

The ring-opening reactions of aziridines occur with great ease in *acidic* media. In fact, aziridine itself reacts, often explosively, with hydrohalic acids to form a polymer (**29**).

$$\text{CH}_2\text{CH}_2\text{NH (aziridine)} \xrightarrow{\text{HX}} [\text{X}-\text{CH}_2\text{CH}_2\text{NH}_2] \longrightarrow -[\text{CH}_2\text{CH}_2\text{NH}]_n- \quad \mathbf{29}$$

The stereoselective nature of these electrophilic ring-opening reactions is exemplified by the conversion of **30** to the *ax–ax* ring-opened product, in which the predicted antiperiplanar orientation is realized in these rigid systems.[10]

$$\mathbf{30} \xrightarrow[\text{(2) Ac}_2\text{O, Pyr}]{\text{(1) CH}_3\text{CO}_2\text{H}} \text{AcNH, OAc product}$$

The difference in the site of attack by hydrochloric acid on 2-methyl- and 2-phenylaziridine (**31** and **32**) is noteworthy.[11,12]

$$\text{C}_6\text{H}_5\text{CH(Cl)CH}_2\text{NH}_2 \xleftarrow[\mathbf{32};\ \text{R=C}_6\text{H}_5]{\text{HCl}} \text{RCH}-\text{CH}_2\text{(NH)} \xrightarrow[\mathbf{31};\ \text{R=CH}_3]{\text{HCl}} \text{CH}_3\text{CH(NH}_2\text{)CH}_2\text{Cl}$$

Attempts to quaternize *N*-alkylaziridines generally result in ring-opening unless nucleophiles of very low nucleophilicity (for example: perchlorate, *p*-toluenesulfonate) are present.[13-15] These quaternary aziridinium salts (**33**) react *very* readily with various nucleophiles.[14]

$$\mathbf{33}\ (\text{ClO}_4^-) \xrightarrow{\text{CH}_3\text{OH}} \text{CH}_3\text{O-cyclohexyl-CH}_2\text{-N}^+\text{H(pyrrolidinium)}\ \text{ClO}_4^-$$

Kinetic studies indicate that during the solvolysis of *β*-chloroamines, such as **34,** the aziridinium ion **35** acts as an intermediate.[16]

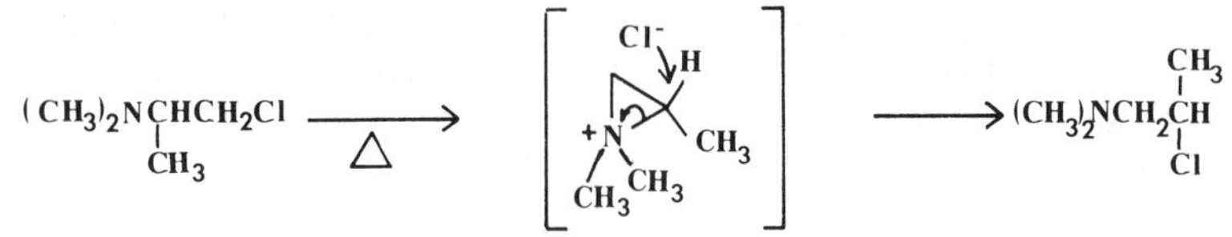

$$(CH_3)_2NCH(CH_3)CH_2Cl \xrightarrow{\Delta} \left[Cl^- \; H, \; {}^+N(CH_3)(CH_3), \; CH_3 \right] \longrightarrow (CH_3)_2NCH_2CH(CH_3)Cl$$

34 35

Aziridines, when treated with carbonyl compounds, afford excellent yields of oxazolidines (**36**).[17]

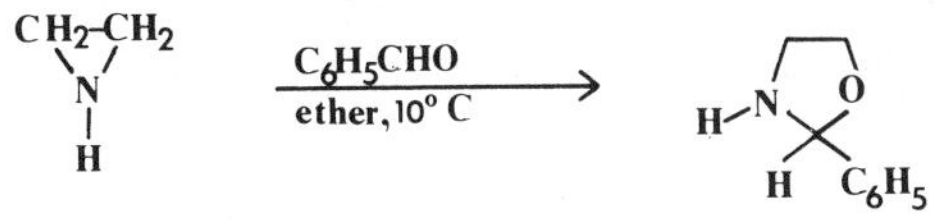

$$CH_2{-}CH_2{-}N{-}H \xrightarrow[\text{ether, } 10°\,C]{C_6H_5CHO} H{-}N,\; O,\; H,\; C_6H_5$$

36

An extremely useful synthetic technique involves the stereoselective deamination of aziridines by nitrosating agents such as nitrosyl chloride or methyl nitrite.[18,19] The *N*-nitroso derivative **37** has been isolated in this reaction at −20°C.

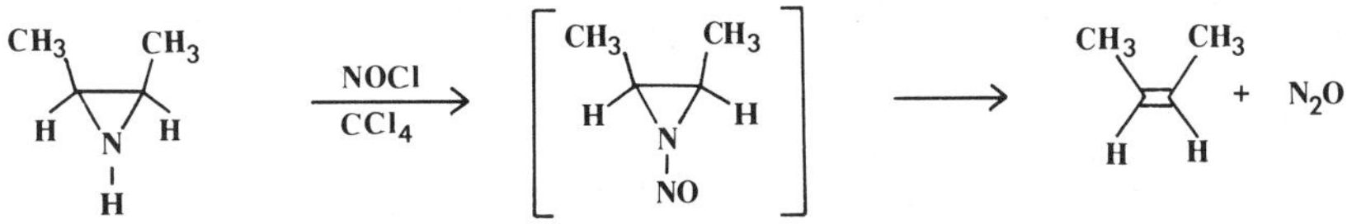

$$CH_3,\ CH_3,\ H,\ H,\ N{-}H \xrightarrow[CCl_4]{NOCl} \left[CH_3,\ CH_3,\ H,\ H,\ N{-}NO \right] \longrightarrow CH_3,\ CH_3,\ H,\ H\ (\text{alkene}) + N_2O$$

37

N-Acyl (**38**) derivatives of ethylenimine can be rearranged to oxazolidones (**39**), either thermally or by acid catalysis.[20] The corresponding thio derivatives react similarly.

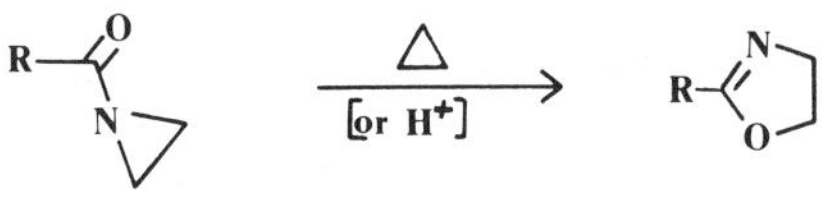

$$R{-}C(=O){-}N\text{(aziridine)} \xrightarrow[\text{[or } H^+]]{\Delta} R{-}C(=N)(O)\text{(ring)}$$

38 39

Iodide ion is an effective catalyst for the isomerization of all aziridines; the process presumably occurs by attack of I^- at the least hindered carbon of the aziridine ring.[21] Alkyl aziridinium derivatives (for example, **40**), rather than undergoing a thermal ring expansion, undergo a *stereospecific cis* elimination via the process depicted in structure **41.**[22]

R—C(=O)—N (aziridine), CH_3 CH_3 **40** —NaI→ R—C(O^-)=N—C(CH_3)(CH_3)—CH_2—I → R—oxazoline—CH_3, CH_3

40 —160–220 °C→ CH_2=C(CH_3)CH_2NHCOR

CH_2R, CH_2Cl, N, CH_3, O

41

25.2.3.2. Azirines. Although the high reactivity of 2*H*-azirines makes them very useful synthetic reagents, their skin-irritating properties demand care in their handling. It is desirable to prepare the 2*H*-azirines *in situ* when they are used as synthons.[6,7,24]

The strained nature of 2*H*-azirines is reflected in their proton and ^{13}C NMR spectra. The imine hydrogen (C_3—H) has a chemical shift close to 0 ppm, that is, *extremely shielded.* The coupling constant between C_2 and H_2 is 186 Hz, a value consistent with a high degree (37%) of *s*-character in the C_2—H_2 bond.[25] Structure **42** indicates the ^{13}C chemical shifts for 2,3-diphenyl-2*H*-azirine.[25]

C_6H_5, C_6H_5, H, N

δ 163.8 (ppm)

δ 22.7 (ppm)

42

The nonbasic 2*H*-azirines, when dissolved in anhydrous perchloric acid, behave as ring-opened cations (**43**) and react with acetone or acetonitrile to form the cycloadducts **44** and **45,** respectively.[8] When 2*H*-azirines (**46**) are treated with aqueous acids they are converted to α-amino ketones (**47**) in a manner analogous to the Neber reaction (**48** → **47**).[26]

C_6H_5, N, CH_3, CH_3 —$HClO_4$→ [C_6H_5, CH_3, +, H—N, CH_3] **43** —CH_3COCH_3→ CH_3 CH_3 CH_3, C_6H_5, N—O, CH_3 **44**

43 —CH_3CN→ CH_3 CH_3, C_6H_5, CH_3, N—N **45**

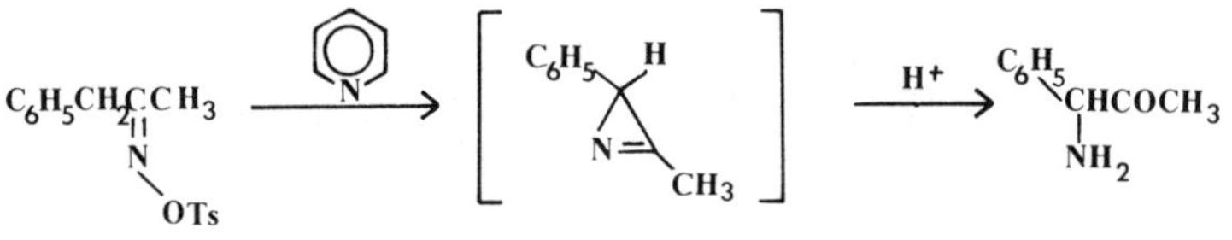

Addition of nucleophiles to the C=N bond of 2*H*-azirines takes place stereospecifically on the least hindered side of the molecule.[27] For example, lithium aluminum hydride reduction of the 2*H*-azirine **49** affords the *cis*-aziridine **50** in almost quantitative yield. Grignard reagents react in a similar stereoselective manner. The reactions of various carbanions (**51**) with 2*H*-azirines have also been studied, and in all instances, the carbanion attacks the imine carbon (formation of **52**). The resulting intermediate ring-cleaves, then frequently recyclizes to form five-membered rings (**53**).[27,28]

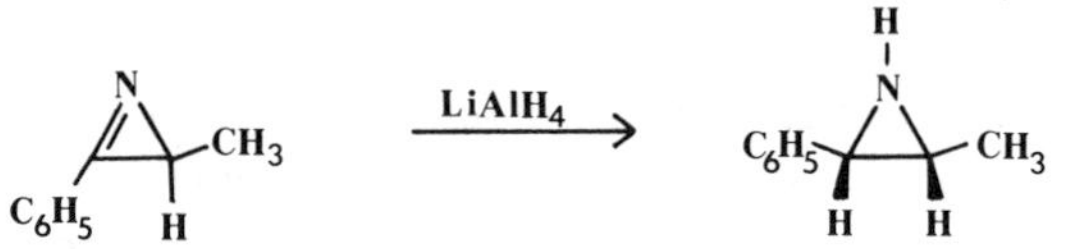

The trichloromethide anion [$^{-}CCl_3$] adds stereoselectively to 2*H*-azirines to form aziridines (**54**) which, when treated with nitrous acid, afford olefins (**55**) via the same *N*-nitroso compounds (**56**) as previously described for aziridines.[27] The trichloromethylaziridine **54,** when treated with base, yields the azetidine **57.** This compound is envisioned to form via the bicyclobutane derivative **58.** A stable analogue of **58** (that is, **59**) is obtained by the following reaction[29]:

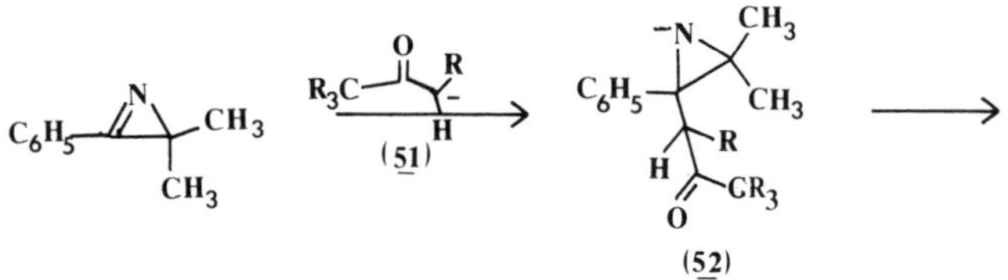

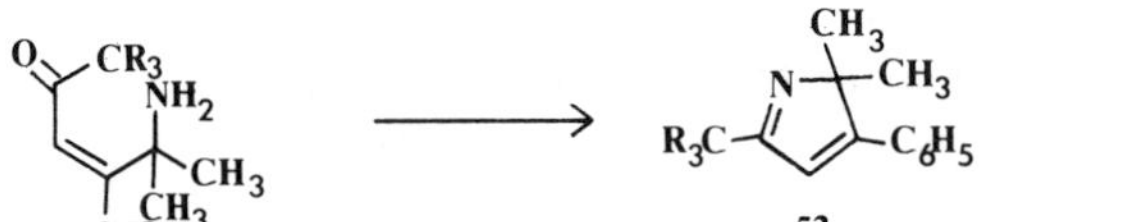

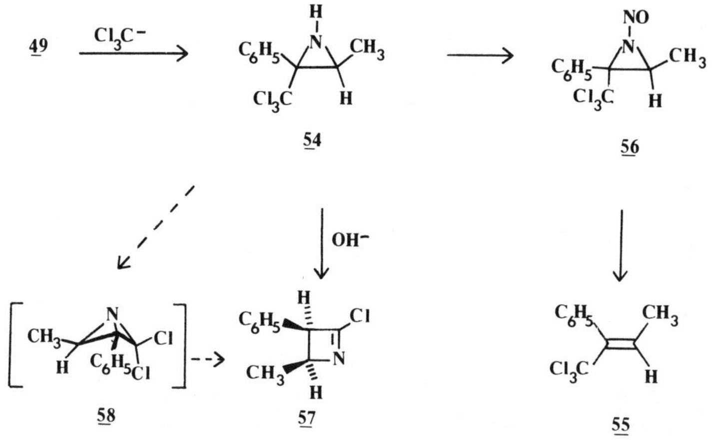

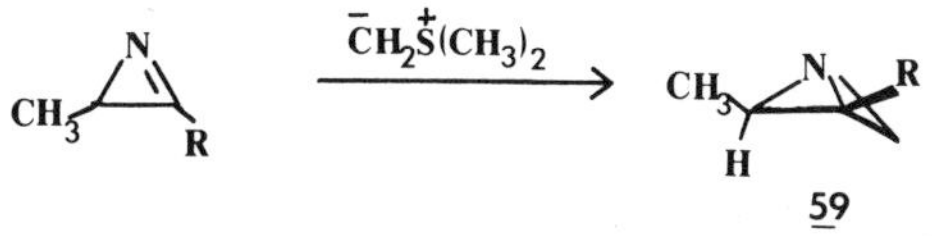

Photolysis of 2*H*-azirines produces nitrilium ylide intermediates (**60**) which can be trapped by a number of different dipolarophiles (**60** → **61**).[21]

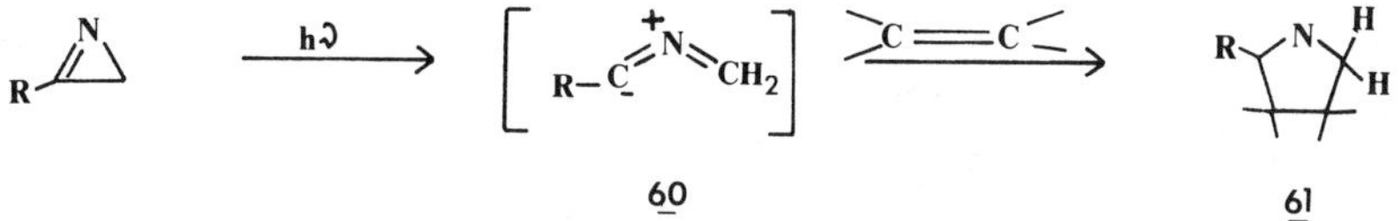

The reactive double bond in 2*H*-azirines (for example, **62**) participates readily in Diels–Alder type reactions. Some of the numerous examples are presented below.[25]

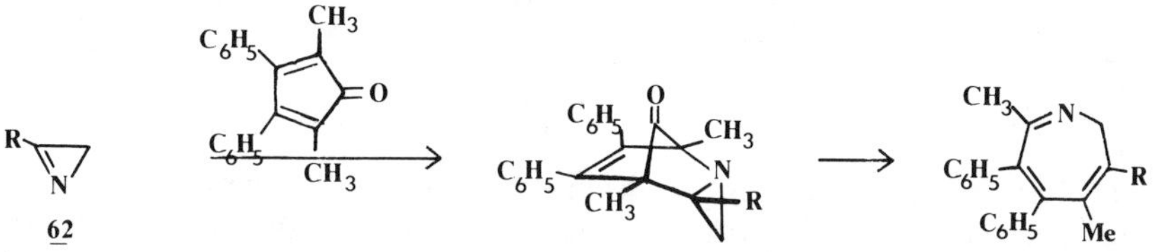

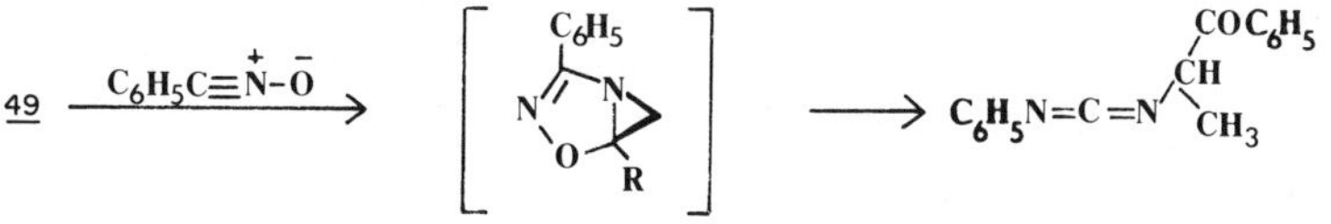

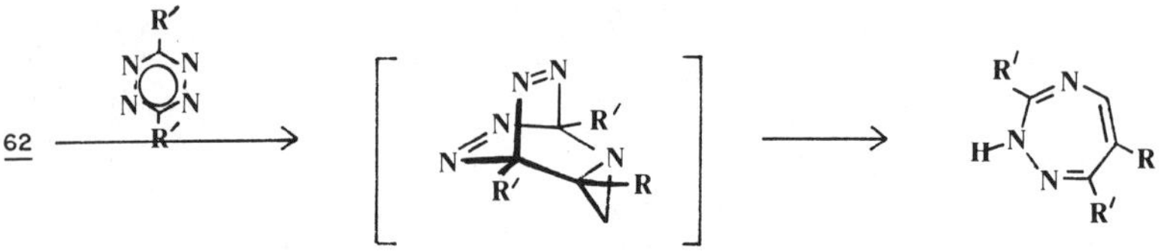

25.3. THREE-MEMBERED OXA-HETEROCYCLES

25.3.1. EPOXIDES: SYNTHESES

These compounds (**63**) are generally prepared by peracid oxidation of the appropriate olefins.[30] The most convenient of the peracids available for this reaction is *m*-chloroperbenzoic acid.[31] As shown, the reaction proceeds by an electrophilic attack of the peracid on the double bond.[32] Thus the site of epoxidation is highly dependent upon the electron density of the double bond: the more electron-rich the double bond, the more readily it reacts (**64** → **65**).[33]

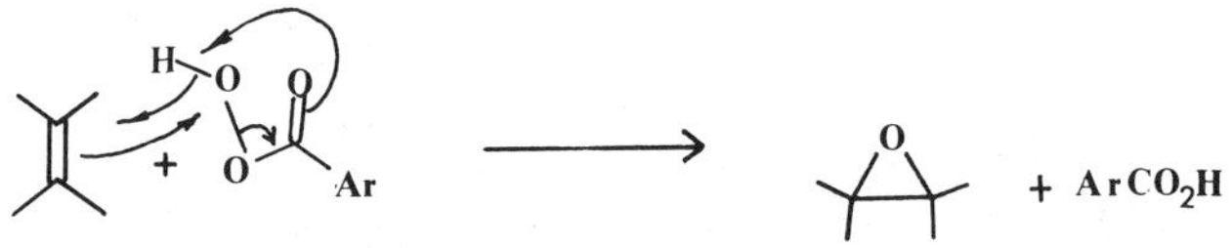

+ $ArCO_2H$

63

64

65

The stereoselective nature of these epoxidations is reflected in the following two reactions[30]:

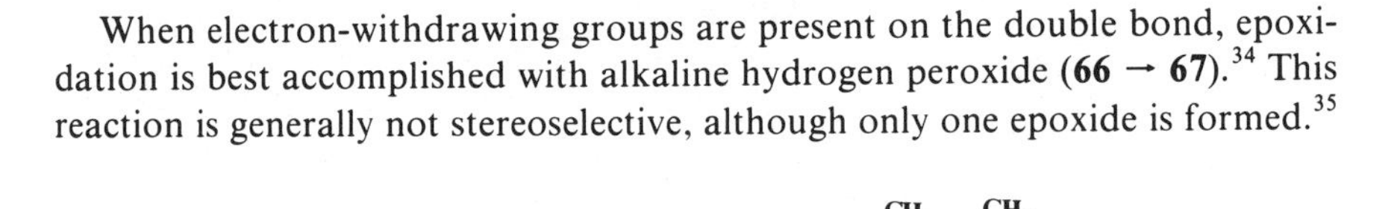

When electron-withdrawing groups are present on the double bond, epoxidation is best accomplished with alkaline hydrogen peroxide (**66** → **67**).[34] This reaction is generally not stereoselective, although only one epoxide is formed.[35]

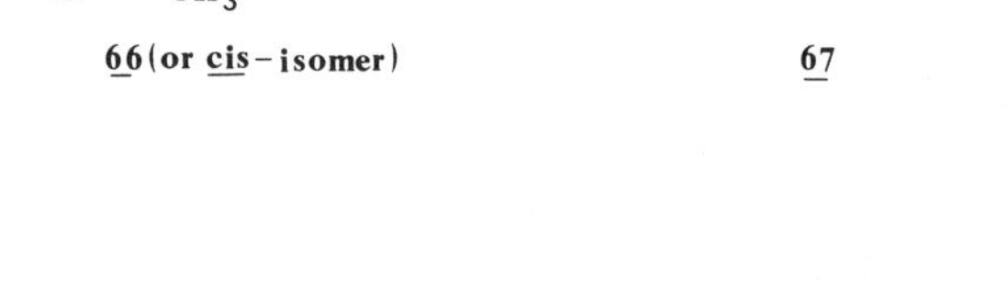

66 (or cis-isomer)

67

The use of alkaline *tert*-butyl hydroperoxide to effect epoxidation prevents hydrolysis of reactive functional groups such as the cyano group in **68.**[35]

Dimethyloxosulfonium methylide (**69**) reacts with carbonyl compounds to afford epoxides by a cleverly devised nonoxidative process.[36] Dimethylsulfonium methylide (**70**) also acts as a methylene group source, although the sites of attack are different from those observed for **69.** The latter attacks the carbonyl group from the less hindered side, whereas the former does not.[36,37]

Although diazomethane and its derivatives react with many aldehydes and ketones to yield epoxides, the reaction often produces complex mixtures.[38] The well-known reaction of olefins with hypohalous acids (*N*-haloacetamides and aqueous perchloric acid are excellent choices for the preparation) gives rise to *trans*-halohydrins (**71** → **72**).[39] When treated with base these compounds are readily dehydrohalogenated in a trans manner to afford epoxides (**73**).

The *Darzens reaction* is a means of preparing α,β-epoxycarbonyl compounds (**74**) and is delineated below.[40] The reaction is stereoselective and produces *trans*-epoxides.[40]

Phosphorus triamides (**75**) react with aromatic aldehydes and, when an electron-withdrawing group is present on the aromatic ring, afford epoxides.[42]

$C_6H_5CHO + ClCH_2COC_6H_5 \xrightarrow{NaOH}$ **74**

$(R_2N)_3P + C_6H_5CHO \longrightarrow$ (cis and trans)
75

25.3.2. OXIRENE (OXACYCLOPROPENE): PROPERTIES

Oxirene (**76**) is a four-π-electron heterocycle and should be anti-aromatic. In fact, molecular orbital calculations suggest that it should be 50 kJ mole^{-1} less stable than the isomeric formylcarbene.[43] Thus it is not surprising that oxirene is not known. A recent attempt at its synthesis from an epoxybarrelene (**77**)[44] failed. At 300–400°C only cycloheptatriene aldehyde derivatives were obtained; however, under flash thermolytic conditions at 600–800°C, some ketone was isolated. Thus, it is possible that oxirene is an intermediate in the ketone formation. Oxirene derivatives (**78**) have been postulated as reaction intermediates in the peracid oxidation of acetylenes and in the photochemical Wolff rearrangement of α-diazocarbonyl compounds.[44,45] Oxirene has also been shown to be an intermediate in a photochemically induced rearrangement of ketones.[47]

77 $\xrightarrow{(300-400°C)}$ (not formed) **76**

77 $\xrightarrow{600-800°C}$ [oxirene] $\longrightarrow CH_2{=}C{=}O$

$RC(=O)C(=N_2)R \xrightarrow{h\nu}$ [**78**] $\longrightarrow O{=}C{=}CR_2$

25.3.3. EPOXIDES: REACTIONS

Nucelophilic attack on epoxides is accompanied by essentially complete inversion of configuration at the site of attack. For example, the *trans*-epoxide **79** when treated with ammonium hydroxide gives the *trans*-**80**.[48] Although nucleophilic attack generally occurs at the less-hindered carbon of the epoxide ring, solvent and reagent differences often exert considerable influence. Thus the styrene epoxide **81** gives the following product ratios under the indicated conditions.[49]

79 $\xrightarrow[25°C]{NH_4OH}$ 80

$C_6H_5CH\text{—}CH_2$ (epoxide, O bridge) **81** $\xrightarrow{C_6H_5OH}$ $C_6H_5CH(OH)CH_2OC_6H_5$ + $C_6H_5CH(OC_6H_5)CH_2OH$

	$C_6H_5CH(OH)CH_2OC_6H_5$	$C_6H_5CH(OC_6H_5)CH_2OH$
NaOH / H_2O	20%	65%
NaOEt / dioxane	41%	36%

Lithium aluminum hydride reduction of unsymmetrical epoxides (for example, **81**) generates the more highly substituted alcohols (**82**), whereas in the presence of aluminum trichloride the less substituted alcohols (**83**) are formed.[50]

81 $\xrightarrow{LiAlH_4}$ $C_6H_5CHOHCH_3$ (82)

81 $\xrightarrow[AlCl_3]{LiAlH_4}$ $C_6H_5CH_2CH_2OH$ (83)

Phosphonate carbanions (**84**) convert epoxides to cyclopropanes (**85**) under moderate reaction conditions (85°C)[51]; other phosphorus derivatives have been employed in this transformation as well.[52–54] The fact that chiral epoxides afford optically active cyclopropanes in these reactions is of great synthetic utility,[55] as illustrated by the conversion of optically active **81** to (1*S*, 2*S*)-**85**.

81 $\xrightarrow[(Na^+)]{(EtO)_2P(O)\bar{C}HCO_2Et}$ 85 (84)

81 $\xrightarrow{(C_6H_5)_3\overset{+}{P}\overset{-}{C}HCO_2Et}$ [oxaphospholane, $P(C_6H_5)_3$] $\longrightarrow$ (1s, 2s)-85

The treatment of epoxides with Grignard reagents serves to generate alcohols.[56]

$$\underset{\underline{4}}{RCH\underset{O}{-}CHR} \xrightarrow{CH_3CH_2MgBr} \underset{OH}{RCH}-\underset{CH_2CH_3}{CHR}$$

Electrophilic reagents react readily with epoxides (for example, **86**), primarily by cleavage of the bond between oxygen and the least substituted carbon in the epoxide ring. Again, solvent and reagents have a profound effect on the product distribution.[57] The direction of ring cleavage in styryl epoxide (**81** and **87**) is strongly dependent on the nature of the substituent present on the benzene ring.[58]

$$\underset{\underline{86}}{CH_3CH\underset{O}{-}CH_2} \xrightarrow{HCl} CH_3\underset{OH}{CH}CH_2Cl + CH_3\underset{Cl}{CH}CH_2OH$$

ether, − 50°C	90%	10%
H_2O, 65-83°C	56-82%	44-18%

$$\underline{81}\ (R = H),\ \underline{87}\ (R = NO_2) \xrightarrow[HI]{(R = H)} C_6H_5\underset{I}{CH}CH_2OH$$

$$\xrightarrow[HI]{(R = NO_2)} o\text{-}NO_2C_6H_4\underset{OH}{CH}CH_2I$$

In a manner analogous to that of the aziridines, epoxides react with carbonyl compounds to form dioxolanes (**88**).[59]

$$\underline{4} \xrightarrow[SnCl_4]{C_6H_5CHO} \underset{\underline{88}}{\text{RCH–O, RCH–O ring; C(H)(}C_6H_5\text{)}}$$

Extrusion of the heteroatom from epoxides is readily accomplished by trialkylphosphines and yields olefins with high stereoselectivity,[60] the major olefinic product having the opposite stereochemistry from the starting epoxide. This behavior can be accounted for by the intermediacy of **89**, its rotation to **90**, and subsequent syn elimination of the phosphine oxide.

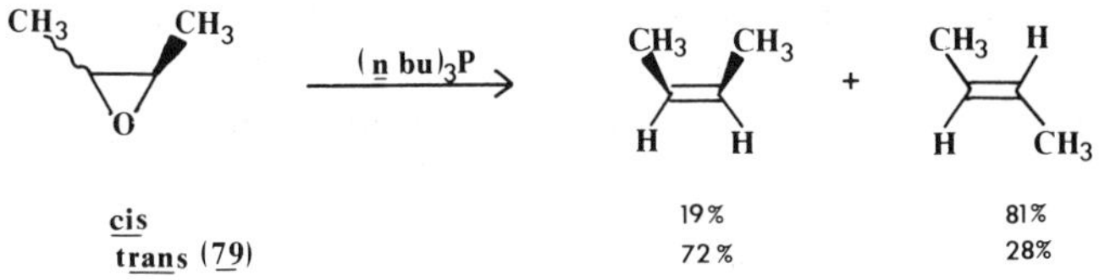

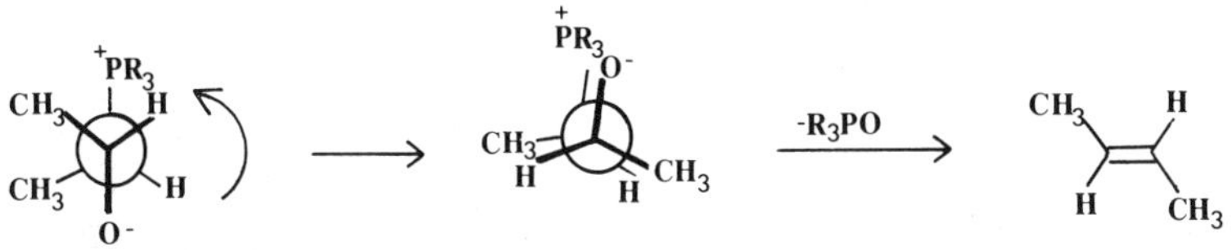

cis | 19% | 81%
trans (79) | 72% | 28%

89 90

Epoxides (for example, **91** and **92**) undergo thermal isomerization to carbonyl compounds. The same transformation can be more readily accomplished by means of catalytic amounts of mineral acids, boron trifluoride etherate in benzene, or anhydrous magnesium bromide in ether or benzene.[3]

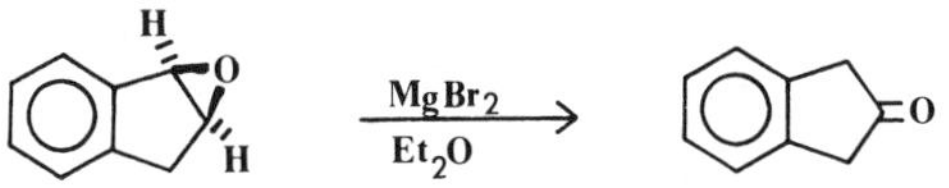

91 (100%)

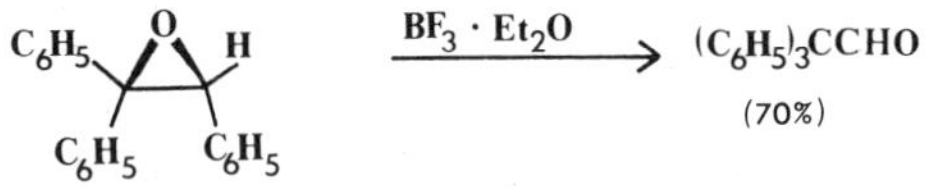

92 $(C_6H_5)_3CCHO$ (70%)

25.4. THREE-MEMBERED THIA-HETEROCYCLES

25.4.1. EPISULFIDES (THIACYCLOPROPANES): SYNTHESES

The most efficacious syntheses of episulfides involve the conversion of the corresponding, readily available epoxides, as shown by the transformation of **93** to **94**. Thiourea reacts with epoxides in a similar fashion.[63]

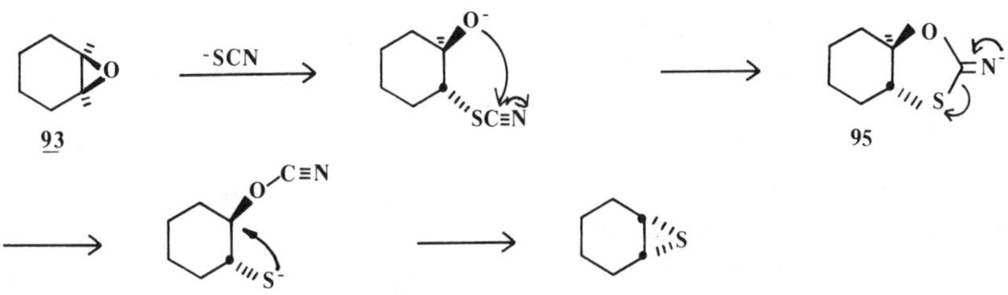

93 95

94 (73%)

That the epoxide of cyclopentene does not react to form the episulfide in this manner lends strong support to the proposed mechanism, since the cyclopentane analogue of **95** cannot exist in the required *trans*-ring fused fashion (**96**).

96

The reaction of thioketones (**97**) with dimethyloxosulfonium methylide affords episulfides (**98**) in relatively good yields.[36] Treatment of aromatic thioketones (**97**) with diazomethane also affords episulfides.[64] Sulfenes (**99**) react with diazomethane in a similar manner and afford episulfones (**100**).[65,66]

$(C_6H_5)_2C{=}S + (CH_3)_2\overset{+}{S}(O)CH_2^- \longrightarrow$ 98; $(C_6H_5)_2C{=}S$ (97) $\xrightarrow{CH_2N_2}$ 98

$C_6H_5CH{=}SO_2 \leftrightarrow C_6H_5\overset{-}{C}H{-}\overset{+}{S}O_2$ (99) $\xrightarrow{CH_2N_2}$ $C_6H_5CH{-}CH_2$ (SO_2 ring) (100)

25.4.2. THIIRENES (THIACYCLOPROPENES): SYNTHESES AND PROPERTIES

Thiirene (**101**) belongs, as does the oxygen analogue, to the class of four-π-electron antiaromatic heterocycles, and is therefore not expected to be endowed with much stability. In fact, **101** has now been obtained from the photochemical decomposition of 1,2,3-thiadiazole (**102**) in an argon matrix at 89 K.[67]

102 $\xrightarrow[(89°K, Argon)]{h\nu\ (\lambda = 235\text{-}280\ nm)}$ 101 + $CH_2{=}C{=}S$ (103) + $HC{\equiv}C{-}SH$ (104)

When irradiated with light at 3300–3700 Å, thiirene is converted to **103** and **104,** the acetylene derivative (**104**) being the major product. The dimethyl, mono-, and dideuteriothiirenes were prepared as well and their infrared spectra obtained.[67]

The instability of the thiirene can be reduced by formation of thiirenium salts (**105**), thiirene 1-oxides (**106**), and thiirene 1,1-dioxide (**107**). Because the free pair of electrons is no longer present in these systems, the compounds should resemble cyclopropene, a two-π-electron ring system. The *S*-methylthiirene derivative **108** has been prepared from methanesulfenyl chloride and di-*tert*-butylacetylene.[68] The proton resonances for **108** are at δ 2.62 and 1.54 (ppm) for the S—CH_3 and *tert*-butyl methyl groups, respectively. The carbon resonances are reported to be at δ 113.42 and 33.12 ppm, respectively. The compound is stable at room temperature for several days.

105 106 107

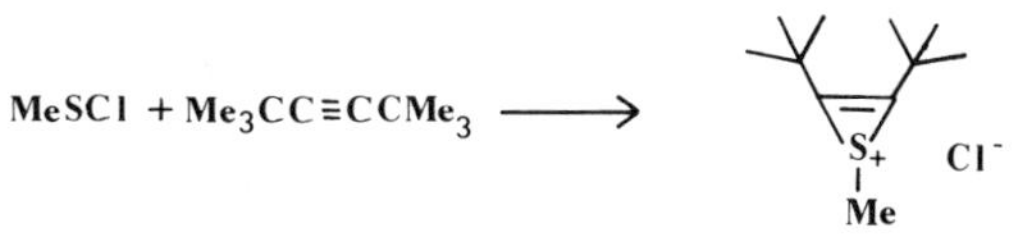

$MeSCl + Me_3CC{\equiv}CCMe_3 \longrightarrow$ 108 Cl^-

The 2,3-diphenylthiirene 1-oxide (**109**) is available from α,α'-dibromodibenzyl sulfoxide (**110**) when it is treated with triethylamine.[69] Some of the reactions of this compound are indicated. Based on CNDO/2 calculations,[70] the thiirene 1,1-dioxide (**111**) is expected to be more stable than the sulfoxide **109,** as the result of a stabilizing π(carbon)–π(oxygen) interaction. These compounds have also been prepared by the dehydrohalogenation of 2-bromothiirane 1,1-dioxides (**112**).[71,72]

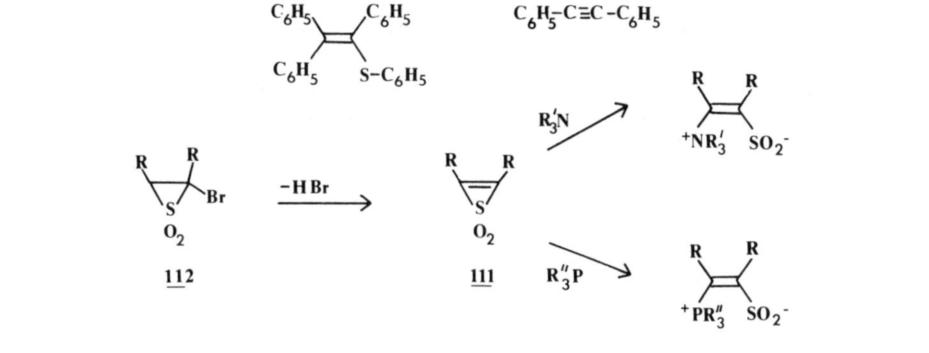

110 $\xrightarrow{Et_3N}$ 109 $\xrightarrow{C_6H_5CHN_2}$

109 $\xrightarrow{C_6H_5MgBr}$

109 $\xrightarrow{h\nu}$ $C_6H_5{-}C{\equiv}C{-}C_6H_5$

112 $\xrightarrow{-HBr}$ 111

111 $\xrightarrow{R'_3N}$ $^+NR'_3$ SO_2^-

111 $\xrightarrow{R''_3P}$ $^+PR''_3$ SO_2^-

The starting material for the dehydrohalogenation can be obtained either by base-catalyzed ring closure of dibromosulfones or by carbene addition to an α-bromosulfene.[71,72] Among the various, facile, ring-opening reactions known for these type of compounds, their reaction with *tert*-amines and *tert*-phosphines is of some interest.[73]

25.4.3. EPISULFIDES: REACTIONS

As is true for the epoxides and aziridines, episulfides react with nucleophiles with inversion of configuration at the site of nucleophilic attack. Thus treatment of the *trans*-2,3-dimethylepisulfide (**113**) with lithium aluminum deuteride affords the mercaptan **114.**[74]

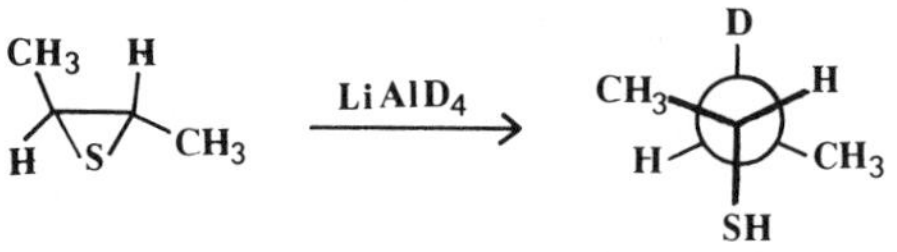

CH_3 H / H S CH_3 —$LiAlD_4$→ D, CH_3, H, H, CH_3, SH

113 114

Electrophilic attack by hydrohalic acids or acid chlorides on episulfides (**115**) follows the same pattern as observed for the epoxides.[75]

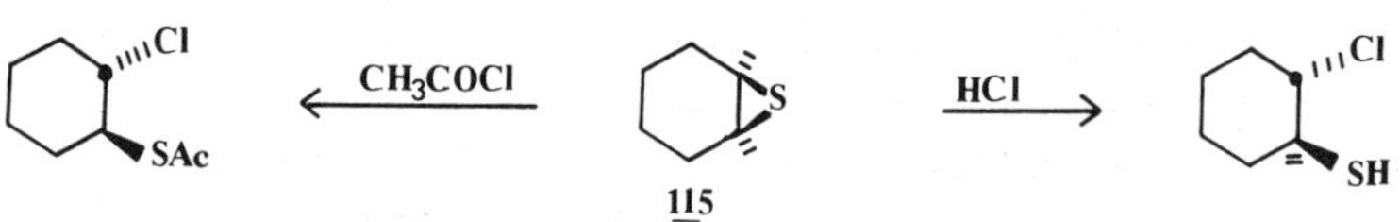

Cl, SAc ←CH_3COCl— S —HCl→ Cl, SH

115

The treatment of episulfides with tertiary phosphines or phosphites removes the sulfur stereospecifically, giving the corresponding olefin (for example, **116**) in very high yields.[60]

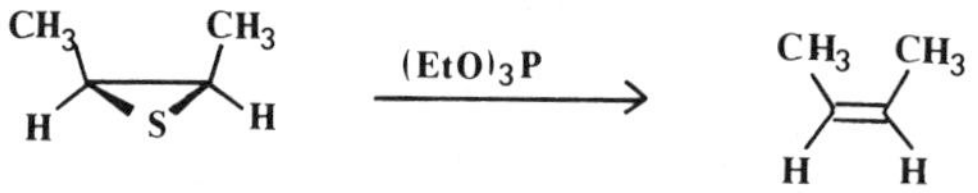

CH_3 CH_3 / H S H —$(EtO)_3P$→ CH_3 CH_3 / H H

116 (100%)

25.5. FOUR-MEMBERED AZA-HETEROCYCLES

The mononitrogen analogue of cyclobutadiene is known as azete (**117**), and its two dihydro derivatives (**118, 119**) are referred to as 1- and 2-azetine, respectively.[76]

N N H N

117 118 119

25.5.1. AZETE

Molecular orbital calculations[77,78] on this antiaromatic cyclobutadiene derivative predict that the antiaromatic destabilization of azete **117** is 10 kJ mole^{-1} less than that of cyclobutadiene.[77] Suggestions have been made that application of the "push–pull" effect (the presence of electron-donating and/or electron-withdrawing substituents) should stabilize this molecule.[79] A similar stabilization should be caused by the presence of bulky substituent groups. Seybold and co-workers[80] prepared the stable azete **120,** which is a red solid derived from pyrolysis of the 1,2,3-triazine **121** and trapped on a cold (−196°C) surface. The possibility that **120** could actually exist as the betaine **122** has been considered, although calculations suggest a planar four-membered ring, such as **120,** with strongly distorted diagonals.

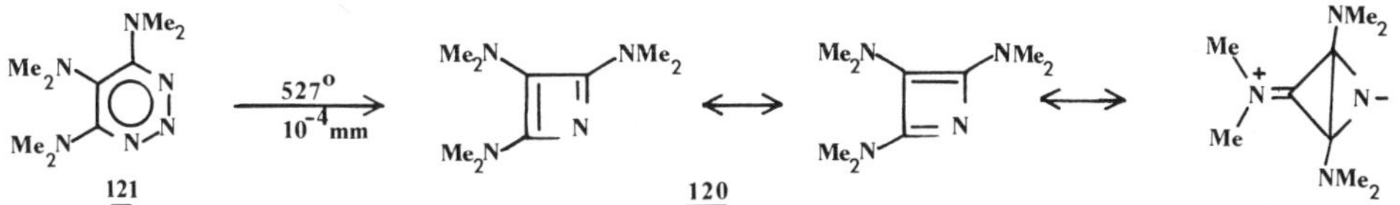

121 —(527°, 10^{-4} mm)→ 120 ⟷ ⟷ 122

The benzazetes **123** are generally more stable than azetes and are prepared by thermolysis of benzo-1,2,3-triazines **124.**[81] The antiaromatic azete ring is somewhat stabilized by fusion to the benzene ring, and a *p*-methoxyphenyl group at C-2 of the benzazete **126** markedly increases the stability in compounds of this type. Thus **127** is more stable than **126.** The compounds, trapped at −80°C as dark-red materials, are thermally labile, and, on warming, readily dimerize to form compounds of general structure **125.**[82,83]

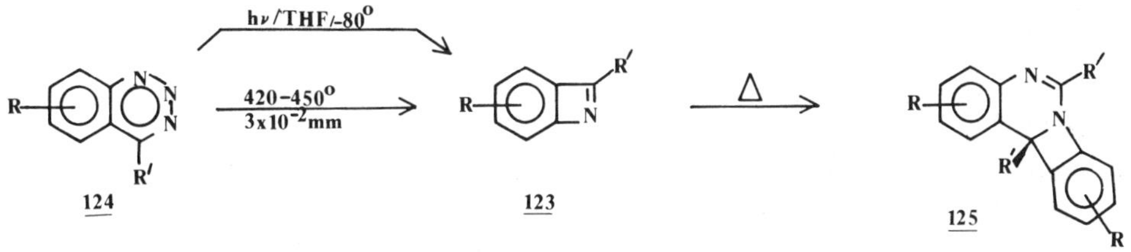

124 —(420–450°, 3×10^{-2} mm; hν/THF/−80°)→ 123 —Δ→ 125

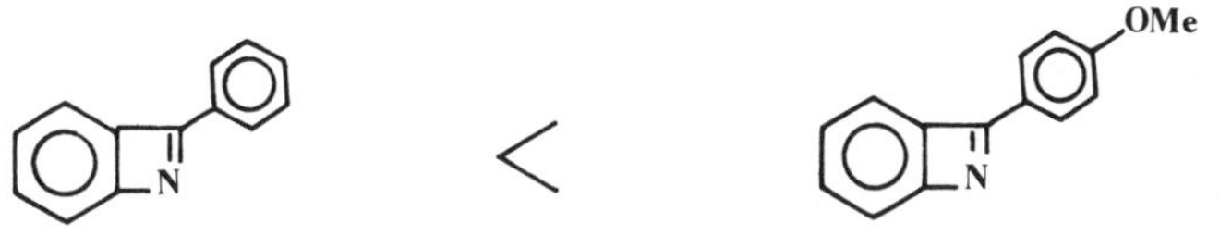

126 < 127

2-Phenylbenzazetes (**128**) behave as strained imines[82] and are hydrolyzed to *o*-aminobenzophenones **129** by aqueous acid. The C=N reactivity in these compounds reflects the ease with which they undergo cycloaddition reactions with a variety of 1,3-dipoles or dienes (for example, **130**). Nitrile oxide adduct **131** rearranges, interestingly, to the 1,3,5-isomer **132.**[84]

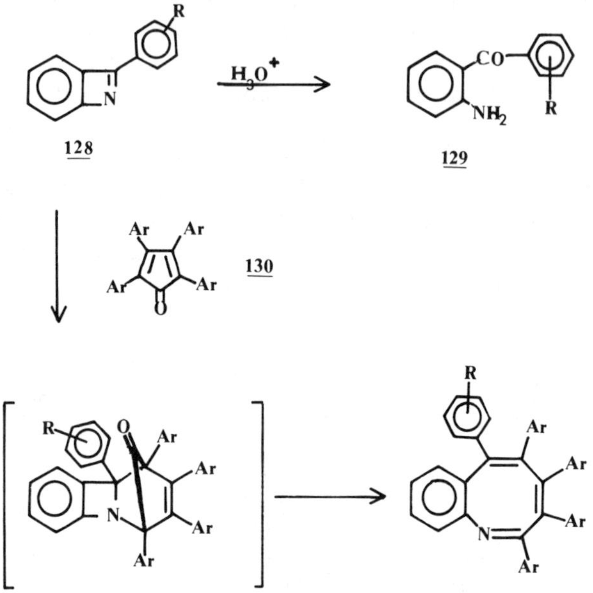

128 129 130

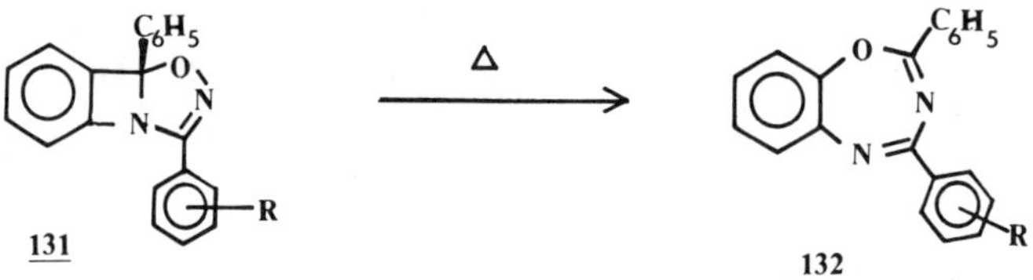

131 132

The first isolable benzoazetidinone (**133;** R = *tert*-butyl) was prepared in 1971.[85] Anthranilium salts **134** when treated with triethylamine afford **133,** in which stability increases as the size of R increases (Me < Et < *i*-Pr < *t*-Bu). However, only the *tert*-butyl derivative is sufficiently stable to allow purification by vacuum distillation. When R = 1-adamantyl and the benzo group is expanded to the *b*-naphtho group, the remarkably stable azetidinone **135** is isolated.

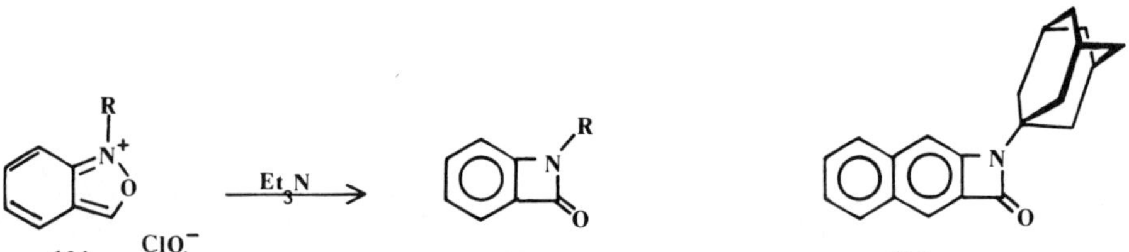

134 133 135

25.5.2. 1- AND 2-AZETIDINES

A general method for the synthesis of 1-azetidines (**136**) involves the following sequence of reactions[86]:

$ClSO_2NCO$ → $(Et_3O)BF_4$ → **136**

The pyrolysis of cyclopropylazides **137** affords a convenient alternative route to azetidines **138**[87,88] via a nitrene intermediate.

137 —105–125°→ [nitrene] → **138**

The relatively uncommon 2-azetines (**139**) can be obtained from the thermal [2 + 2] cycloaddition of *N*-sulfonylimines **140** and *N,N*-ketene-acetals **141**.[89] The first totally characterized azetinone **142** was obtained from the triazene **143** upon treatment with BF_3·etherate.[90] Spectral data (NMR, MS) are available.[89,91]

$C_6H_5CH{=}NSO_2Ar$ (**140**) + $CH_2{=}C(NR_2)_2$ (**141**) → **139**

143 —$BF_3 \cdot Et_2O$→ **142**

REFERENCES

1. (a) H. Quast, *Heterocycles* **14,** 1677 (1980); (b) S. Garbriel, *Chem. Ber.* **21,** 1049 (1888).
2. (a) P. E. Fanta in *The Chemistry of Heterocyclic Compounds,* Vol. 19, Part I, A. Weissberger, Ed., Interscience, New York, 1964, Chapter 2; (b) K. Hafner and C. Koenig, *Angew. Chem. Int. Ed.* **2,** 96 (1963), and refs. cited therein.

3. H. Wenker, *J. Am. Chem. Soc.* **57,** 2328 (1935).
4. A. Hassner and C. Heathcock, *J. Org. Chem.* **29,** 3640 (1964).
5. P. L. Levins and Z. B. Papanastassiou, *J. Am. Chem. Soc.* **87,** 826 (1965).
6. F. W. Fowler, *Adv. Heterocycl. Chem.* **13,** 45 (1971).
7. G. Smolinsky and C. A. Pryde, *J. Org. Chem.* **33,** 2411 (1968), and refs. cited therein.
8. F. W. Fowler, A. Hassner, and L. A. Levy, *J. Am. Chem. Soc.* **89,** 2077 (1967).
9. R. Ghirardelli and H. J. Lucas, *J. Am. Chem. Soc.* **79,** 734 (1957).
10. A. Hassner and C. Heathcock, *J. Org. Chem.* **30,** 1748 (1965).
11. S. Gabriel and H. Ohle, *Chem. Ber.* **50,** 804 (1917).
12. F. Wolfheium, *Chem. Ber.* **47,** 1440 (1914).
13. J. G. Allen and N. B. Chapman, *J. Chem. Soc.* **1960,** 1482, and refs. cited therein.
14. N. J. Leonard, J. V. Paukstelis, and L. E. Brady, *J. Org. Chem.* **29,** 3383 (1964).
15. N. P. Neureites and F. G. Bordwell, *J. Am. Chem. Soc.* **85,** 1209 (1963).
16. E. M. Schultz and J. M. Sprague, *J. Am. Chem. Soc.* **70,** 48 (1948).
17. J. S. Doughty, C. L. Lazell, and A. R. Collett, *J. Am. Chem. Soc.* **72,** 2866 (1950).
18. R. D. Clark and G. K. Helmkamp, *J. Org. Chem.* **29,** 1310 (1964).
19. C. L. Bumgardner, K. S. McCallum, and J. P. Freeman, *J. Am. Chem. Soc.* **83,** 4417 (1961).
20. S. Gabriel and R. Stelzner, *Chem. Ber.* **28,** 2929 (1895).
21. H. W. Whitlock, Jr. and G. L. Smith, *Tetrahedron Lett.* **1965,** 1389, and refs. cited therein.
22. H. W. Heine, *Angew. Chem. Int. Ed.* **1,** 528 (1962).
23. A. Loewenstein, J. F. Neumer, and J. D. Roberts, *J. Am. Chem. Soc.* **82,** 3599 (1960).
24. K. Isomura, M. Okada, and M. Taniguchi, *Tetrahedron Lett.* **1969,** 4073.
25. V. Nair, *Org. Magn. Res.* **6,** 483 (1974).
26. A. Hassner and F. W. Fowler, *J. Am. Chem. Soc.* **90,** 2869 (1968).
27. A. Hassner, *Heterocycles* **14,** 1517 (1980).
28. A. Laurent, P. Mison, A. Nafti, and N. Pellissier, *Tetrahedron Lett.* **1979,** 3955.
29. A. G. Hortmann and D. A. Robertson, *J. Am. Chem. Soc.* **94,** 2758 (1972).
30. H. O. House, *Modern Synthetic Reactions,* 2nd ed. Benjamin, New York, 1972, Chap. 6.
31. N. N. Schwartz and J. H. Blumbergs, *J. Org. Chem.* **29,** 1976 (1964).
32. B. M. Lynch and K. H. Pausacker, *J. Chem. Soc.* **1955,** 1525.
33. W. Hückel and V. Wörffl, *Chem. Ber.* **88,** 338 (1955).
34. H. O. House and R. S. Ro, *J. Am. Chem. Soc.* **80,** 2428 (1958).
35. Y. Ogata and Y. Sawaki, *Tetrahedron* **20,** 2065 (1964).
36. E. J. Corey and M. Chaykousky, *J. Am. Chem. Soc.* **87,** 1353 (1965), and refs. cited therein.
37. C. E. Cook, R. C. Corley, and M. E. Wall, *Tetrahedron Lett.* **1965,** 891.
38. C. D. Gutsche, *Org. Reactions* **8,** 364 (1954).
39. R. Filler, *Chem. Rev.* **63,** 21 (1963).
40. M. Ballester, *Chem. Rev.* **55,** 283 (1955).
41. C. C. Tung, A. J. Speziale, and H. W. Frazier, *J. Org. Chem.* **28,** 1514 (1963).
42. V. Mark, *J. Am. Chem. Soc.* **85,** 1884 (1963).
43. O. P. Strausz, R. K. Gosavi, A. S. Denes, and I. G. Czizmadia, *J. Am. Chem. Soc.* **98,** 4784 (1976).
44. E. Lewars and G. Morrison, *Tetrahedron Lett.* **1977,** 501.
45. K. P. Zeller, *Tetrahedron Lett.* **1977,** 707.

46. I. G. Czizmadia, J. Font, and O. P. Strausz, *J. Am. Chem. Soc.* **90,** 7360 (1968).
47. R. L. Russell and F. S. Rowland, *J. Am. Chem. Soc.* **92,** 7510 (1970).
48. F. H. Dickey, W. Fickett, and H. J. Lucas, *J. Am. Chem. Soc.* **74,** 944 (1952).
49. R. Fuchs and C. A. Vanderwerf, *J. Am. Chem. Soc.* **76,** 1631 (1954).
50. M. N. Rerick and E. L. Eliel, *J. Am. Chem. Soc.* **84,** 2356 (1962).
51. W. S. Wadsworth, Jr. and W. D. Emmons, *J. Am. Chem. Soc.* **83,** 1733 (1961).
52. D. B. Denney and M. J. Boskin, *J. Am. Chem. Soc.* **84,** 3944 (1962).
53. L. Horner, H. Hoffmann, and V. G. Toscano, *Chem. Ber.* **95,** 536 (1962).
54. I. Tömösközi, *Chem. Ind.* (*London*) **1965,** 689.
55. Y. Inouye, T. Sugita, and H. M. Walborsky, *Tetrahedron* **20,** 1695 (1964).
56. N. G. Gaylord and E. I. Becker, *Chem. Rev.* **49,** 413 (1951).
57. C. A. Stewart and C. A. Vanderwerf, *J. Am. Chem. Soc.* **76,** 1259 (1954).
58. F. Arndt, J. Amende, and W. Ender, *Monatsh. Chem.* **59,** 202 (1932), and refs. cited therein.
59. M. T. Bogert and R. O. Roblin, Jr., *J. Am. Chem. Soc.* **55,** 3741 (1933).
60. D. E. Bissing and A. J. Speziale, *J. Am. Chem. Soc.* **87,** 2683 (1965).
61. A. C. Cope, P. A. Trumbull, and E. R. Trumbull, *J. Am. Chem. Soc.* **80,** 2844 (1958).
62. E. E. van Tamelen, *J. Am. Chem. Soc.* **73,** 3444 (1951).
63. C. C. J. Culvenor, W. Davies, and W. E. Savige, *J. Chem. Soc.* **1952,** 4480, and refs. cited therein.
64. H. Staudinger and J. Siegwart, *Helv. Chim. Acta* **3,** 833 (1970).
65. L. A. Paquette, *J. Org. Chem.* **29,** 2851 (1964).
66. G. Opitz and K. Fischer, *Angew. Chem.* **77,** 41 (1965).
67. A. Krantz and J. Laureni, *J. Am. Chem. Soc.* **103,** 486 (1981), and refs. cited therein.
68. G. Capozzi, V. Luccnini, G. Modena, and P. Scrimin, *Tetrahedron Lett.* **1977,** 911.
69. H. -W. Chen, Ph. D. Thesis, Univ. of Massachusetts, 1972, *Diss. Abs. Int.* (*B*) **33,** 2525 (1972).
70. D. Clark, *Int. J. Sulfur. Chem.* (*C*) **7,** 11 (1972).
71. J. Philips, J. Swisher, D. Haidukewych, and I. Morales, *Chem. Commun.* **1971,** 22.
72. M. Rosen and G. Bonet, *J. Org. Chem.* **39,** 3805 (1974).
73. B. Jarvis, W. Tong, and H. Ammon, *J. Org. Chem.* **40,** 3189 (1975).
74. G. W. Helmkamp and N. Schnautz, *Tetrahedron* **2,** 304 (1958).
75. C. C. J. Culvenor, W. Davies, and N. S. Heath, *J. Chem. Soc.* **1949,** 282.
76. J. A. Moore, in *Chemistry of Heterocyclic Compounds,* Vol. 19, Part II, A. Weissberger, Ed., Interscience, New York, 1964, p. 916.
77. B. A. Hess, Jr., L. J. Schaad, and C. W. Holyoke, Jr., *Tetrahedron* **31,** 295 (1975).
78. M. J. S. Dewar and N. Trinajshic, *Theor. Chim. Acta* **17,** 235 (1970).
79. B. A. Hess, Jr. and L. J. Schaad, *J. Org. Chem.* **41,** 3058 (1976).
80. G. Seybold, U. Jersak, and R. Gompper, *Angew Chem. Int. Ed.* **12,** 847 (1973).
81. R. J. Kobylecki and A. McKillop, *Adv. Heterocycl. Chem.* **19,** 215 (1976).
82. B. M. Adger, C. W. Rees, and R. C. Storr, *J. Chem. Soc. Perkin I* **1975,** 45.
83. C. W. Rees, R. C. Storr, and P. J. Whittle, *J. Chem. Soc. Chem. Commun.* **1976,** 411.
84. C. W. Rees, R. Somanathan, R. C. Storr, and A. D. Woolhouse, *J. Chem. Soc. Chem. Commun.* **1975,** 740.
85. R. A. Olofson, R. K. Van der Meer, and S. Stournas, *J. Am. Chem. Soc.* **93,** 1543 (1971).
86. G. Pifferi, P. Consomi, G. Pelizza, and E. Testa, *J. Heterocycl. Chem.* **4,** 619 (1967).

87. A. B. Levy and A. Messner, *J. Am. Chem. Soc.* **93,** 2051 (1971).
88. G. Szeimies, U. Siefken, and R. Rinck, *Angew. Chem. Int. Ed.* **12,** 161 (1973).
89. F. Effenberger and R. Maier, *Angew. Chem. Int. Ed.* **5,** 416 (1966).
90. K. R. Henery-Logan and J. V. Rodricks, *J. Am. Chem. Soc.* **85,** 3524 (1963).
91. A. Hassner, J. O. Currie, Jr., A. S. Steinfeld, and R. F. Atkinson, *J. Am. Chem. Soc.* **95,** 2982 (1973).

Chapter 26

SYNTHESES AND REACTIONS OF SIMPLE SEVEN-MEMBERED UNSATURATED HETEROCYCLES

26.1. GENERAL COMMENTS

The simplest examples of seven-membered unsaturated heterocycles[1] are 1*H*-azepine (**1**), oxepin (**2**), and thiepin (**3**). These monocyclic 8π-heterocycles are termed antiaromatic, if planar, and are isoelectronic with the cycloheptatrienide anion (**4**). Since each of these $4n\pi$ heterocycles can potentially exist as a mixture of valence tautomers (**5** ⇄ **6**), the value of the equilibrium constant for the substituted parents has been obtained, but only recently. Unsubstituted **1** and **2** have been synthesized; **3** has defied preparation.

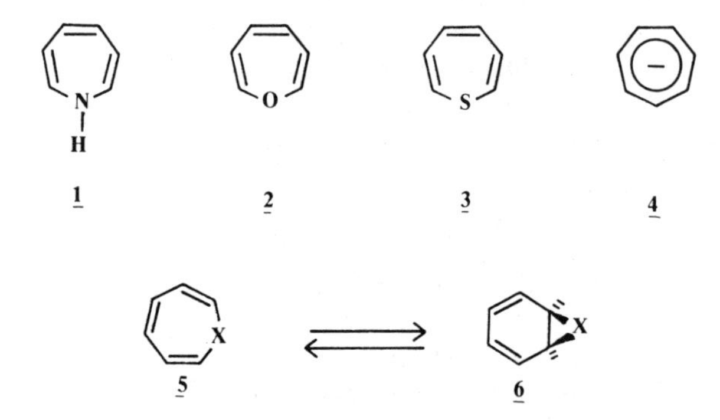

26.2. SYNTHESES AND REACTIONS OF 1*H*-AZEPINES

26.2.1. ⟨6 + 1⟩ COMBINATIONS

The first synthesis of an *N*-derivative of **1**, compound **7**, is by addition to benzene of a carbethoxynitrene, generated by either pyrolysis or photolysis of ethyl azidoformate.[2] This cycloaddition process is concerted and probably involves the singlet nitrene.[3] As envisioned, such a nitrene insertion route would add to substituted benzenes to give a mixture of products. Mild transesterification of **7** with iodotrimethylsilane gives **8**, which is hydrolyzed at 78°C to afford the thermally unstable carbonic acid **9**.[4] Facile decarboxylation of **9** gives the unsubstituted 1*H*-azepine **1**, which is stable in solution at −78°C for only a few hours.[4] The location of the equilibrium between azepine and benzene imine [**6** (X = NH)] has been determined: 1*H*-azepine exists exclusively (>99%) as the uniform seven-membered ring valence tautomer **1**. Catalytic amounts of acid or base cause **1** to isomerize to the more stable 3*H*-azepine (**10**).

N_3CO_2R / hν or Δ → [N-CO_2R] ⇌ N-CO_2R **7**

7 —Me_3SiI→ N-CO_2SiMe_3 **8** (71%) —MeOH, −78°C→ N-CO_2H **9** (74%)

9 —Δ (−60°C to 25°C) (<1 mm)→ N−H **1** —NMe_3→ H H, N **10**

A more versatile route to **7** has been devised by Paquette and co-workers,[3,4] in which 1,4-cyclohexadiene is treated with silver cyanate and iodine in ether, then methanol, to give a stable carbamate **11**. Cyclization of **11** by treatment with methoxide affords aziridine **12**. Subsequent bromination-dehydrobromination of **12** generates the desired substituted azepine.[5] This procedure permits the incorporation of nonreactive substituents on the azepine nucleus.

(1) INCO, (2) CH_3OH → I, $NHCO_2CH_3$ **11** (54%) —NaOMe, THF→ N-CO_2CH_3 **12** (79%)

—(1) Br_2, (2) −HBr→ **7** (R = CH_3; 45%)

26.2.2. ⟨5 + 2⟩ COMBINATIONS

A standard route to these seven-membered heterocycles is by application of the Diels–Alder reaction. The reaction of pyrrole derivatives with dimethyl acetylenedicarboxylate gives the desired [4 + 2] cycloadduct **13.** Photolysis of **13** generates the thermally labile azaquadricyclane **14,** which smoothly ring-opens at 20–40°C to yield **15.**[6]

N–R $\xrightarrow[\Delta]{MeO_2CC\equiv CCO_2Me}$ **13** (N–R, CO_2Me, CO_2Me) $\underset{\Delta}{\overset{h\nu}{\rightleftarrows}}$ **14** (N–R, CO_2Me, CO_2Me) $\xrightarrow{\Delta}$ **15** (MeO_2C, MeO_2C, N–R)

26.2.3. REACTIONS OF *N*-SUBSTITUTED AZEPINES

26.2.3.1. Thermal Dimerization. *N*-Carbalkoxy-1*H*-azepines (**7**) thermally dimerize to give a [6 + 6]-fused product (**17**).[7] A thermal [6 + 6] cycloaddition is disallowed on the basis of orbital symmetry[8]; it has been shown that the allowed [6 + 4] dimer (**16**) is initially formed and at higher temperatures rearranges to the thermodynamically stable product. The *N*-methyl-1*H*-azepine readily dimerizes at 0°C in ether,[9] supportive of enhanced antiaromatic character (with electron-donating groups) relative to the *N*-carbalkoxy derivatives.

7 (N–CO_2R) $\xrightarrow{\Delta}$ [**16** (RO_2C–N, N–CO_2R)] $\longrightarrow$ **17** (N–CO_2R, N–CO_2R)

26.2.3.2. Thermal Cycloadditions. 1*H*-Azepines (**7**) undergo a smooth cycloaddition with the dienophile tetracyanoethylene to give **18.**[10] It is surprising that **7** does not react via its benzene imine tautomer, as do other members of this class, a fact that is indicative of a significant energy barrier between **5** and **6** (X = NR). The olefinic nature of **7** is further demonstrated when it is treated with active dienes, for example, **19.**[11]

(22%; R=Et) adduct (O, C_6H_5, C_6H_5, H, H, N–CO_2R) $\xleftarrow[C_6H_5CH_3/\Delta]{(19)}$ **7** $\xrightarrow[0°/ether]{(NC)_2C=C(CN)_2}$ **18** (N–CO_2R, CN, CN, CN, CN)

26.2.3.3. Thermal Aromatization. The energy of activation required to transform **7** to the benzene imine tautomer can be supplied thermally at reasonable temperatures. When **7** has alkyl substituents at the 2-, 4-, and/or 7-positions, the rate of dimerization is decreased with respect to the competing reaction sequence: tautomerization, homolytic C—N bond rupture, and hydrogen transfer. Thermolysis of **20** gives **21.**[12]

CH_3 CO_2CH_3 20 ⟶ [H $N\text{-}CO_2CH_3$ CH_3] ⟶ $NHCO_2CH_3$ CH_3 21

26.2.3.4. Photochemical Cyclizations. Irradiation of **7** leads to the photochemically allowed [3.2.0]-bicyclic product **22.**[13] This electrocyclic reaction is in accord with those of carbocyclic analogues.

$N\text{-}CO_2R$ 7 ⟶ N CO_2R 22

26.2.3.5. Reductions. Since these azepines are polyenic in character, catalytic reduction gives the fully reduced perhydro derivatives **23** (azepanes). Because such a route is not useful on a large scale, a better procedure is to reduce lactam **24** with lithium aluminum hydride. In view of the industrial importance of (capro) lactams, as in the preparation of nylon 6, standard ring expansion routes [e.g., Beckmann[14] (**25** → **24**) or Schmidt[15] reactions] on the readily available cyclohexanone offer more convenient methods to prepare **24.**

N R $\xrightarrow[\text{Cat.}]{H_2}$ N R 23 $\xleftarrow[\text{ether}]{LiAlH_4}$ N R O 24 $\xleftarrow{}$ or (1) NH_2OH/PPA (2) HN_3/PPA O 25

26.3. SYNTHESES OF SELECTED DIAZEPINES[16] AND BENZODIAZEPINES[17]

This topic, albeit limited, is included in view of the important physiological activity of several members of this seven-membered heterocyclic ring system. Librium (introduced in 1961; **26**), valium (diazepam; **27**), and anthramycin (**28**) are but a few examples of this diverse group of diazepines. The unique features associated with the tranquilizing effects of the 1,4-benzodiazepine family are an electron-withdrawing group at position 7, a N-1 methyl group, an *o*-halogen on the 5-phenyl ring, and a two- (or three-) carbon side chain with an *ω*-dialkylmethyl moiety. Thus many of the recent compounds in this and related series possess one or more of these functional groups.

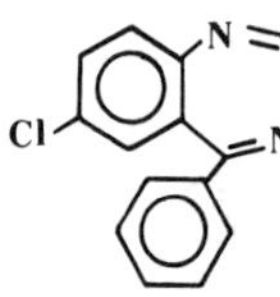

26 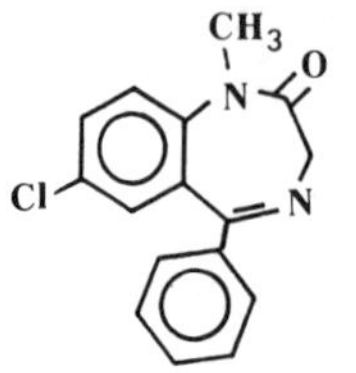 27 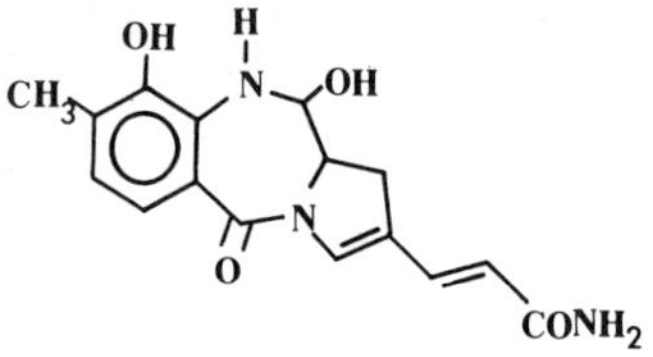 28

26.3.1. 1,2-DIAZEPINES

Currently, no unsubstituted 1*H*-1,2-diazepines are known; however, *N*-substituted members are conveniently prepared by the photolysis of **29.** The rearrangement proceeds via intermediate **30.** The tautomeric equilibrium greatly favors the diazepine **31,** for which the shallow boat conformation reduces the antiaromatic character of the ring.[18] Unlike the azepines and 1*H*-diazepines, the 5*H*-diazepine **32** exists preferentially as the bicyclic diazanorcaradiene **33.**[19]

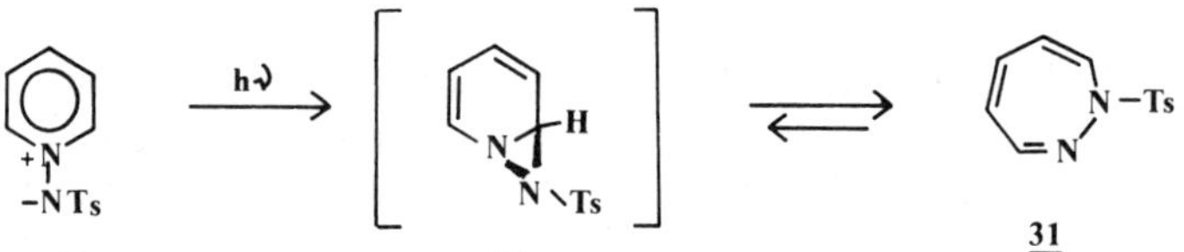

29 30 31

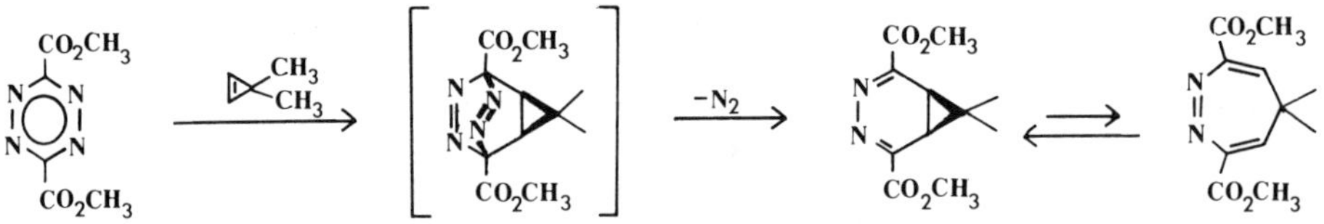

33 32

26.3.2. 1,4-BENZODIAZEPINES

Numerous routes to the valium and librium family have been devised. It is impossible to expound on all possibilities. Instead, several general routes are considered.

26.3.2.1. ⟨6 + 1⟩ Combination. Treatment of 2-chloromethylquinazoline 3-oxide (**34**) with methylamine gives librium (**26**)[20] directly via a ring-expansion sequence. The expected 2-methylamino derivative of **34** is not observed.

34 —MeNH2→ [ ] → 26

26.3.2.2. ⟨4 + 3⟩ **Combination.** 2-Aminobenzophenone or its *N*-methyl derivative is a convenient starting point for the synthesis of valium. Treatment of **35** with bromoacetyl bromide gives **36,** which readily cyclizes in ethanolic ammonia. Amide **37** is isolable and can be cyclized upon heating.[21] *N*-oxides at N-4 are readily obtained by treatment of the diazepinones with peracids.

35 —$BrCH_2COBr$→ 36 —NH_3→ 37 —Δ→ 27

35 —$H_3\overset{+}{N}CH_2CO_2Et$ (15–75%) 21b→ 27

26.3.2.3. ⟨8 → 7⟩ **Contraction.** Benzoxodiazocine **38,** prepared from oxime **39,** undergoes a ring contraction to generate **40.**[22]

39 —[OH]⁻→ 38 (34%) —OH⁻→ [ ] → 40

26.3.2.4. ⟨5 + 2⟩ **Ring Expansion.** Amine **35** (R = Me) is treated with paraformaldehyde and potassium cyanide to give excellent yields of nitrile **41,** which undergoes a facile cyclization to indole **42** in the presence of base or acid catalysts.[23a] Catalytic hydrogenation of **42** affords the pivotal amine, which under oxidative conditions suffers cleavage of the indole 2,3-bonds, thus generating the labile aminoketone **37.** Cyclization of **37** gives valium.[23b]

35 (R = Me) —HCHO, KCN / AcOH→ 41 (89%) —Na / MeOH→ 42 (92%) —(1) H_2/cat. (2) O_3 or CrO_3→ 37

26.4. SYNTHESES AND REACTIONS OF OXEPINS

26.4.1. ⟨6 → 7⟩ REARRANGEMENT

The first synthesis of the parent oxepin (**2**) was reported in 1964 by Vogel and co-workers.[25] Bromination of cyclohexene oxide **43** produces the dibromide **44;** subsequent treatment of **44** with sodium methoxide gives the desired unsubstituted oxepin **2.** Spectral (NMR) evidence suggests that **2** exists as a mixture of tautomeric isomers.[25b] Further proof of structure is its thermal rearrangement to phenol at temperatures >70°C. Numerous substituted oxepins have been synthesized by this procedure.[26]

43 —Br_2→ 44 —NaOMe or DBN→ ⇌ 2 —Δ→ phenol (OH)

26.4.2. ⟨5 + 2⟩ COMBINATION

Application of the Diels–Alder reaction to this problem, a route described for the synthesis of azepines, is quite successful. Upon irradiation, adduct **45** is transformed to oxaquadricyclane **46,** which when heated above 100°C gives oxepin **47.**[27] This compound (**47**) is also available by a modified procedure, in which **48** undergoes free radical substitution and is then debrominated with sodium iodide.[27]

furan —$MeO_2CC{\equiv}CCO_2Me$, Δ→ 45 (70–75%) (CO_2Me, CO_2Me) —hν→ 46 (55–60%) (CO_2Me, CO_2Me)

46 —100°C→ MeO_2C, MeO_2C (oxepin) 47 (70–75%) ←NaI, $(CH_3)_2CO$— MeO_2C, MeO_2C, Br, Br, O ←NBS— MeO_2C, MeO_2C, O 48

26.4.3. ⟨6 + 1⟩ COMBINATION

Epoxidation of Dewar benzene (**49**) with *m*-chloroperbenzoic acid gives epoxide **50**, which upon either photolysis or pyrolysis rearranges to the parent oxepin **2.**[28]

49 —$ArCO_3H$→ 50 —hν or Δ→ 2

26.4.4. REACTIONS OF OXEPINS

26.4.4.1. Reductions. Since these compounds are olefinic in character, they are generally reduced catalytically to the perhydro derivatives **51** or rearrange efficiently to the "phenolic" derived products, such as cyclohexanol.[25b] In the presence of strong nucleophiles, ring contraction is the likely mode.[26]

(benzene) ←$-H_2O$— OH ←$LiAlH_4$ (R=H)— 2 (R, R, O) —H_2/Pd-C (R=H)→ 51 + OH

26.4.4.2. Cycloaddition Reactions. The benzene oxide tautomer of **2** plays an important role in the cycloaddition course, in that **2** typically reacts with dienophiles, such as dimethyl acetylenedicarboxylate, to give the [4 + 2] adduct (**52**).[26]

2 ⇌ (benzene oxide, O) —$MeO_2CC{\equiv}CCO_2Me$, Δ→ 52 (O, CO_2Me, CO_2Me)

26.4.4.3. Photochemical Rearrangements. As shown for the azepines, substituted oxepins undergo a photochemically induced rearrangement to the corresponding *cis*-2-oxabicyclo[3.2.0]hepta-3,6-dienes (**53**).[29] By alteration of the irradiation conditions, exclusive formation of phenol can be realized.[29]

(R=H)
-80°C / hν
2537 Å
(acetone)

hν
λ > 310 mm
(Et_2O)

53

26.4.5. BENZOXEPINS

The synthetic route to 1-benzoxepin **54** is quite interesting because two rearrangements are applicable in the final stage. Vogel et al.[30] used the very successful double dehydrobromination procedure in the synthesis of benzoxepin from the tetrabromide **55,** which was in turn derived from bromination of diene **56.** The "benzene oxide" intermediate **57** can undergo a 1,2-O → C migration to give the aromatic ring in **54,** or can simply tautomerize to generate the novel 10 π-electron 1,6-oxa[10]annulene (**58**). The latter has been proposed to be the main reaction product, which subsequently isomerizes to **54** during purification.

Br_2/CCl_4

56 55 57 58

H_2 / Pd

54

26.5. SYNTHESES AND REACTIONS OF THIEPINS[31]

To date the unsubstituted thiepin (**3**) has not been synthesized.

26.5.1. THIEPIN 1,1-DIOXIDE

Mock successfully synthesized the bis-oxide of **3** by a convenient three-step sequence. *cis*-Hexatriene upon treatment with sulfur dioxide gives the allowed 1,6-addition product **59.** Subsequent bromination-debromination generates the bis-oxide **60.**[32] Thermolysis of **60** leads to extrusion of sulfur dioxide via the episulfone **61.**

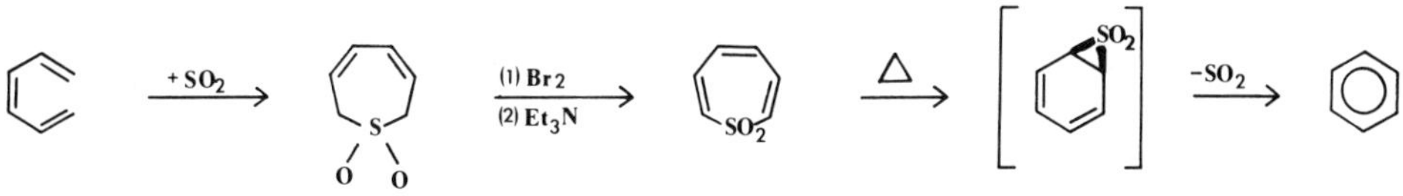

+ SO_2 → 59 —(1) Br_2 (2) Et_3N→ 60 —Δ→ [61] —$-SO_2$→

26.5.2. FUSED THIEPINS

26.5.2.1. *d*-Fused Thiepins. A convenient dicondensation of *o*-phthaldehyde with diethyl thioacetate affords the stable diester and diacid (**62**) after hydrolysis.[33] Upon mild warming, **62** eliminates sulfur to give the dicarboxylic acid.

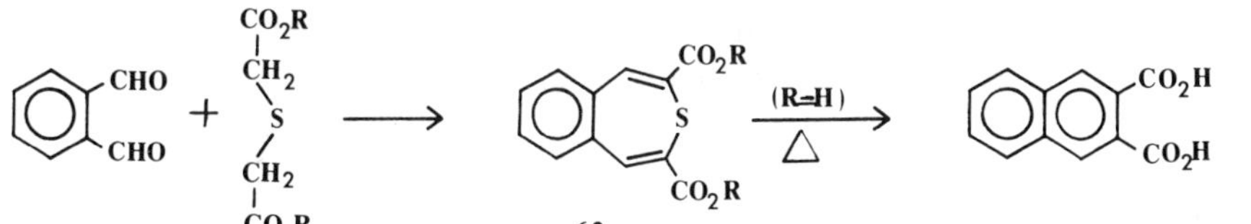

CHO, CHO + CO_2R, CH_2, S, CH_2, CO_2R → 62 (CO_2R, S, CO_2R) —(R=H) Δ→ CO_2H, CO_2H

Mild oxidation of **63** gives the vinyl sulfoxide **64** and after brief treatment with acetic anhydride, **64** is converted to the thiophene-fused thiepin **65**.[34a] Upon warming, **65** decomposes to a complex mixture. When warmed in the presence of dienophiles, **65** undergoes an addition reaction followed by expulsion of sulfur. Compound **65** is stable to oxidation, as shown by the isolation of sulfoxide **66** and sulfone **67**.[34b]

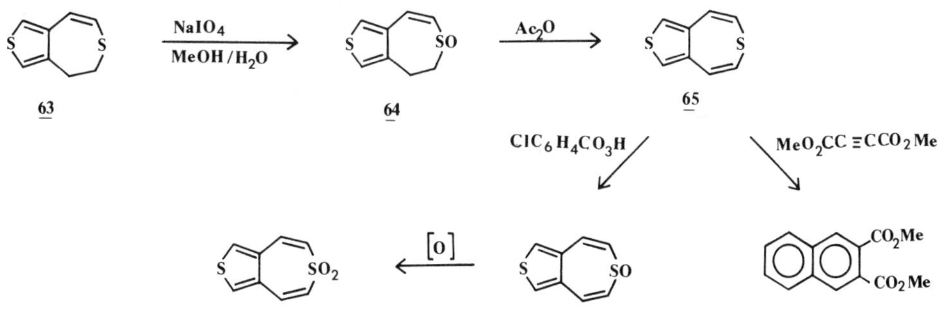

63 —$NaIO_4$, MeOH/H_2O→ 64 —Ac_2O→ 65

65 —$ClC_6H_4CO_3H$→ 66 —[O]→ 67

65 —$MeO_2CC\equiv CCO_2Me$→ CO_2Me, CO_2Me

26.5.2.2. *b*-Fused Thiepins. Attempts to prepare **68** have resulted most often in the isolation of naphthalene; however, the desired benzothiepin (**68**) is probably an intermediate.[35] Very few ether-substituted fused thiophenes have been prepared[31]; the stabilized sulfones or sulfoxides are more common.

SO_2Cl_2 68 −S

26.6. CONCLUSION

Contrary to Paquette's concluding remarks[1] in 1969, much work remains to be done in this series of antiaromatic compounds.

REFERENCES

1. L. A. Paquette in *Nonbenzenoid Aromatics,* J. P. Snyder, Ed., Academic Press, New York, 1969, Chap. 5, pp. 249–310.
2. K. Hafner and C. König, *Angew. Chem.* **75,** 89 (1963); W. Lwowski, T. J. Maricich, and T. W. Mattingly, Jr., *J. Am. Chem. Soc.* **85,** 1200 (1963); *ibid.,* **87,** 3630 (1965); R. J. Cotter and W. F. Beach, *J. Org. Chem.* **29,** 751 (1964).
3. W. Lwowski and R. L. Johnson, *Tetrahedron Lett.* **1967,** 891.
4. E. Vogel, H.-J. Altenbach, J.-M. Drossard, H. Schmickler, and H. Stegelmeier, *Angew. Chem. Int. Ed.* **19,** 1016 (1980).
5. L. A. Paquette, D. E. Kuhla, J. H. Barrett, and R. J. Haluska, *J. Org. Chem.* **34,** 2866 (1969).
6. H. Prinzbach, R. Fuchs, and R. Kitzing, *Angew. Chem. Int. Ed.* **7,** 67 (1968).
7. L. A. Paquette and J. H. Barrett, *J. Am. Chem. Soc.* **88,** 2590 (1966); L. A. Paquette, J. H. Barrett, and D. E. Kuhla, *J. Am. Chem. Soc.* **91,** 3616 (1969).
8. R. Hoffmann and R. B. Woodward, *J. Am. Chem. Soc.* **87,** 4388 (1965).
9. K. Hafner and J. Mondt, *Angew. Chem. Int. Ed.* **5,** 839 (1966).
10. A. S. Kende, P. T. Izzo, and J. E. Lancaster, *J. Am. Chem. Soc.* **87,** 5044 (1965).
11. L. A. Paquette, D. E. Kuhla, J. H. Barrett, and L. M. Leichter, *J. Org. Chem.* **34,** 2888 (1969).
12. L. A. Paquette, D. E. Kuhla, and J. H. Barrett, *J. Org. Chem.* **34,** 2879 (1969).
13. L. A. Paquette and J. H. Barrett, *J. Am. Chem. Soc.* **88,** 1718 (1966).
14. E. C. Horning and V. L. Stromberg, *J. Am. Chem. Soc.* **74,** 2680 (1952); also see, for an alternate procedure, D. H. R. Barton, M. J. Day, R. H. Hesse, and M. M. Pechet *Chem. Commun.* **1971,** 945.
15. H. Wolff, *Org. Reactions* **3,** 307 (1946).
16. F. D. Popp and A. C. Noble, *Adv. Heterocycl. Chem.* **8,** 21 (1967); also see: D. Lloyd, H. P. Cleghorn, and D. R. Marshall, *Adv. Heterocycl. Chem.* **17,** 1 (1974).
17. D. Lloyd and H. P. Cleghorn, *Adv. Heterocycl. Chem.* **17,** 27 (1974).

18. R. Allmann, A. Frankowski, and J. Streith, *Tetrahedron* **28,** 581 (1972).

19. A. Steigel, J. Sauer, D. A. Kleier, and G. Binsch, *J. Am. Chem. Soc.* **94,** 2770 (1972).

20. L. H. Sternbach, E. Reeder, O. Keller, and W. Metlesics, *J. Org. Chem.* **26,** 4488 (1961).

21. (a) S. C. Bell, T. S. Sulkowski, C. Gochman, and S. J. Childress, *J. Org. Chem.* **27,** 562 (1962); (b) L. H. Sternbach, R. I. Fryer, W. Metlesics, E. Reeder, G. Sach, G. Saucy, and A. Stempel, *J. Org. Chem.* **27,** 3788 (1962).

22. A. Stempel, I. Douvan, E. Reeder, and L. H. Sternbach, *J. Org. Chem.* **32,** 2417 (1967).

23. (a) G. N. Walker, A. R. Engle, and R. J. Kempton, *J. Org. Chem.* **37,** 3755 (1972); (b) see Ref. 34 therein.

24. A. Rosowsky, Ed., *Seven-Membered Heterocyclic Compounds Containing Oxygen and Sulfur,* Wiley-Interscience, New York, 1972, pp. 1–570.

25. (a) E. Vogel, R. Schubart, and W. A. Böll, *Angew. Chem. Int. Ed.* **3,** 510 (1964); (b) E. Vogel, W. A. Böll, and H. Günther, *Tetrahedron Lett.* **1965,** 609.

26. E. Vogel and H. Günther, *Angew. Chem. Int. Ed.* **6,** 385 (1967).

27. H. Prinzbach, M. Arguëlles, and E. Druckrey, *Angew. Chem. Int. Ed.* **5,** 1039 (1966).

28. E. E. van Tamelen and D. Carty, *J. Am. Chem. Soc.* **89,** 3922 (1967).

29. J. M. Holovka and P. D. Gardner, *J. Am. Chem. Soc.* **89,** 6390 (1967).

30. E. Vogel, M. Bishop, W. Pretzer, and W. A. Böll, *Angew. Chem. Int. Ed.* **2,** 642 (1964); F. Sondheimer and A. Shani, *J. Am. Chem. Soc.* **86,** 3168 (1964).

31. Ref. 24, pp. 573–896.

32. W. L. Mock, *J. Am. Chem. Soc.* **89,** 1281 (1967).

33. G. P. Scott, *J. Am. Chem. Soc.* **75,** 6332 (1953); K. Dimroth and G. Lenke, *Chem. Ber.* **89,** 2008 (1956).

34. (a) R. H. Schlessinger and G. S. Ponticello, *J. Am. Chem. Soc.* **89,** 7138 (1967); (b) *Tetrahedron Lett.* **1968,** 3017.

35. V. J. Traynelis and J. R. Livingston, Jr., *J. Org. Chem.* **29,** 1092 (1964).

Chapter 27

LARGE-MEMBERED HETEROCYCLES

27.1. EIGHT-MEMBERED RINGS

27.1.1. AZOCINE (1)

27.1.1.1. General Comments. The ultraviolet spectrum[1] of 2-methoxyazocine (**2**) suggests that the members of the azocine family possess a tublike conformation, similar to that found in cyclooctatetraene, and that they should thus have chemical properties analogous to those of a cyclopolyolefin. Azocines undergo a facile two-electron reduction,[2] either polarographically or chemically (potassium in THF or liquid ammonia), to generate a planar dianion (**3**) which possesses the π-electron delocalization of an aromatic 10π system. The diamagnetic ring current in **3** is easily ascertained by NMR spectroscopy.[3] When dianion **3** is quenched with water, a mixture of dihydroazocines is produced. The conjugated isomer **4** is the more stable, as demonstrated by the base-catalyzed isomerization of **5** to **4**, which on further treatment with base, ring-contracts and expels methanol to generate a cyclobuteno[2,3]pyridine (**6**).[4]

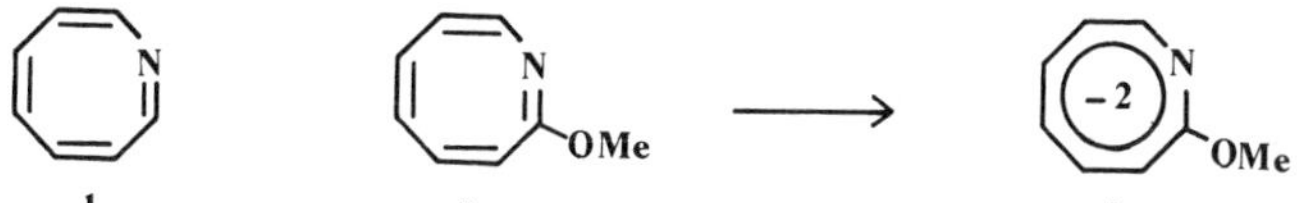

N OMe → (−2) N OMe
1 2 3

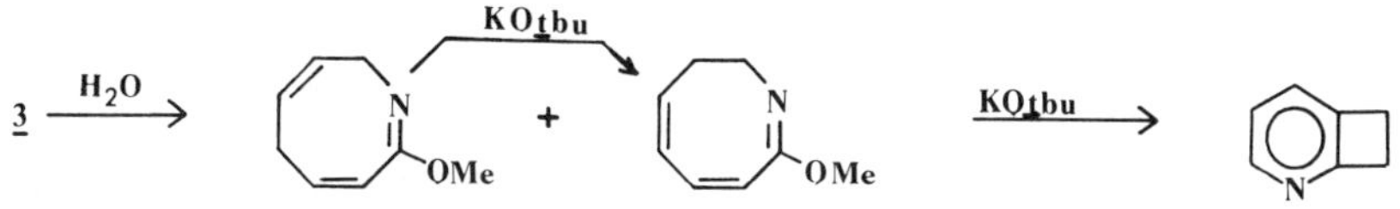

3 $\xrightarrow{H_2O}$ N OMe + N OMe $\xrightarrow{KOtbu}$ N
KOtbu
5 4 6

27.1.1.2. Syntheses and Reactions of Azocines. *(a) ⟨10 → 8⟩ Method.* Azocine (**1**) was initially produced when diazabaskette (**7**)[5] was pyrolyzed in an attempt to expel nitrogen and generate cubane (**8**). However, only intractable materials were obtained in certain cases[6]. Under photolytic conditions, **7** gives only low yields of cyclooctatetraene.[6] When **7** is pyrolyzed at low pressures and short contact times (i.e., mass spectrometric conditions), azocine (**1**) is obtained in 60% yield[6]! The formation of **1** from **7** is envisioned as proceeding through a retro Diels–Alder reaction, rearrangement, and loss of HCN.

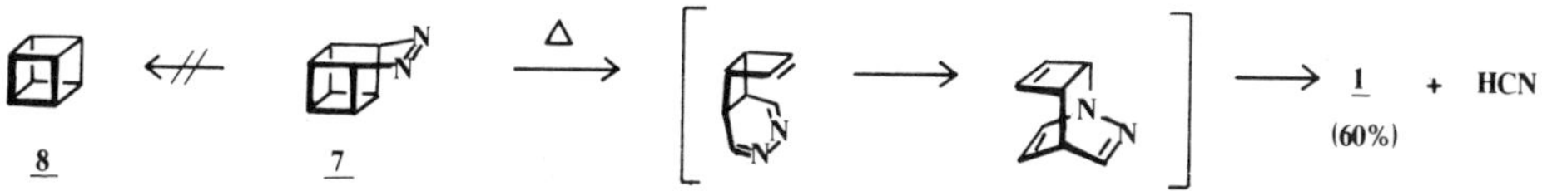

8 ⇷ 7 N N $\xrightarrow{\Delta}$ [N N → N N] → 1 + HCN
(60%)

(b) ⟨6 + 2⟩ Route. Circumvention of the rapid decomposition of **1** is easily accomplished by the preparation of **2** via a simple four-step sequence.[1] Cycloaddition of chlorosulfonyl isocyanate to 1,4-cyclohexadiene **9,** followed by reduction, gives the β-lactam **10.** *O*-Methylation of **10** with Meerwein's reagent, trimethyloxonium fluoroborate, gives **11,** which undergoes allylic bromination by *N*-bromosuccinimide and base-induced dehydrohalogenation to afford ether **2.** Tautomer **2** is highly favored over the four isomeric bicyclic tautomers.

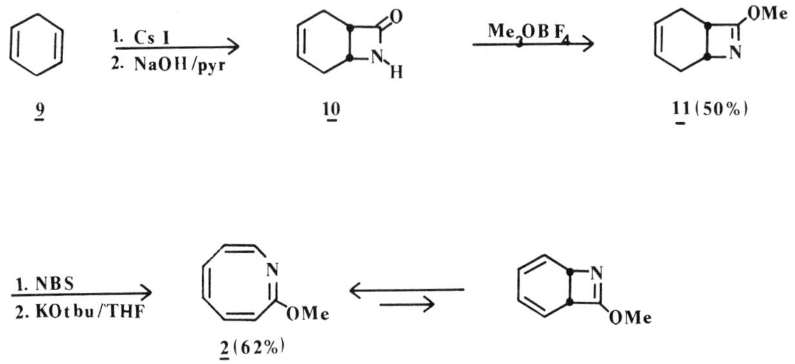

1. Cs I
2. NaOH/pyr

Me_3OBF_4

9 10 11 (50%)

1. NBS
2. KOtbu/THF

OMe

2 (62%)

(*c*) ⟨*4* + *4*⟩ *Route.* Substituted azocines (e.g., **14**) can be conveniently prepared by a Diels–Alder/retro-Diels–Alder sequence between substituted triazines **12** and **13.**[7] In this more substituted compound, no direct evidence for the bicyclic tautomers is obtained.

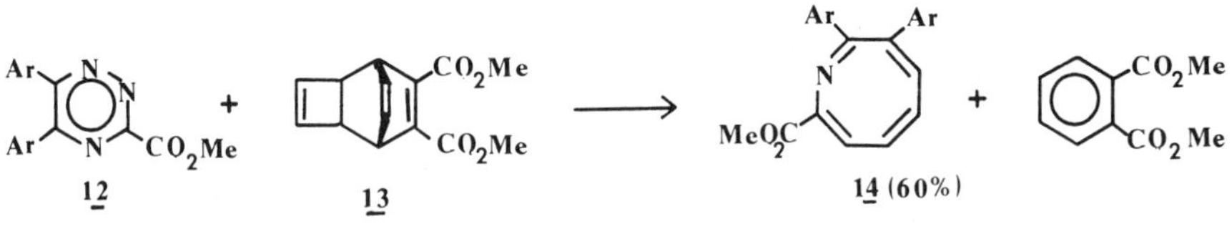

12 13 14 (60%)

(*d*) *Reactions of Azocines.* Although only limited chemical investigations have been conducted on this family, there is ample evidence for the polyene character of the compounds. When **2** is reduced with 10% palladium on carbon, ether **15** is isolated.[1] This same compound can be readily synthesized from lactam **16** by treatment with Meerwein's reagent; lactam **16** is obtained by ring expansion of cycloheptanone under Schmidt reaction conditions[8] or from the corresponding oxime via the Beckmann rearrangement.

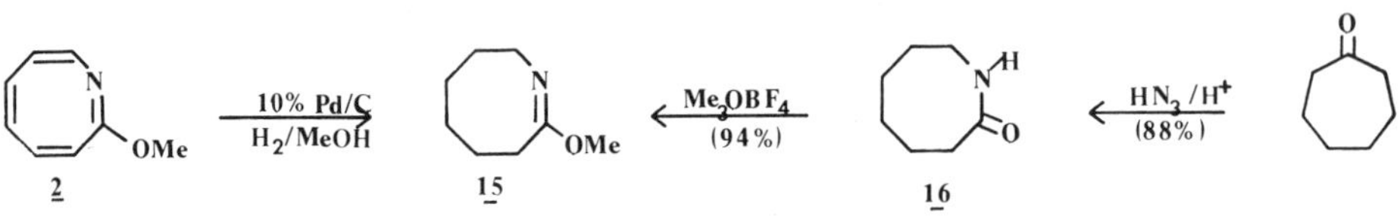

10% Pd/C, H_2/MeOH; Me_3OBF_4 (94%); HN_3/H^+ (88%)

2 15 16

Two reactions of **2** strongly suggest that bicyclic tautomer **17** is the reacting species. When **17** (or **2**) is treated with base, benzonitrile is isolated; and when a Diels–Alder reaction with *N*-phenylmaleimide is done, a [4 + 2] adduct **18** with the appropriate stereochemistry is obtained.[1]

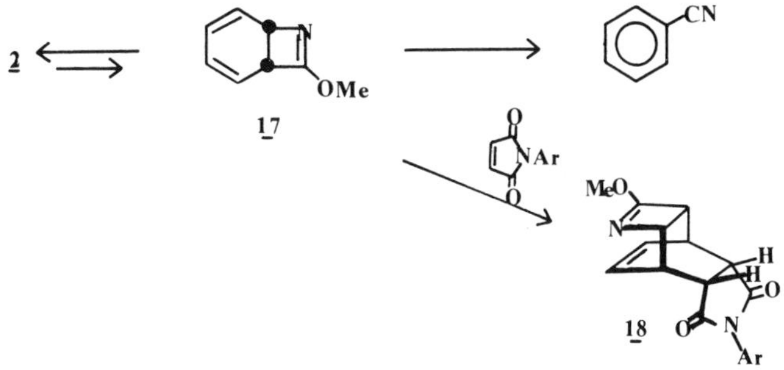

27.1.2. DIHETEROCINES

The 10π-electron "aromatic" character of the parent diheterocines was recognized in a review by Vol'pin,[9] in which the prediction was made that a 1,4-diheteroatom relationship would afford the most likely orientation for "aromatic" character.

27.1.2.1. Syntheses and Reactions of Diazocine. The parent 1,2-diazocine (**19**) is synthesized by the photolytic decomposition of the diazatetracycle **20**,[10] which is elegantly prepared in seven steps from sodium cyclopentadienide.[11] The major mechanism for this transformation is proposed to be a [2 + 2] photoreversion to give the diazabicyclooctatriene **21**, which rearranges to diazocine **19** in 70% yield. Continued photolysis of **19** gives benzene in quantitative yield.

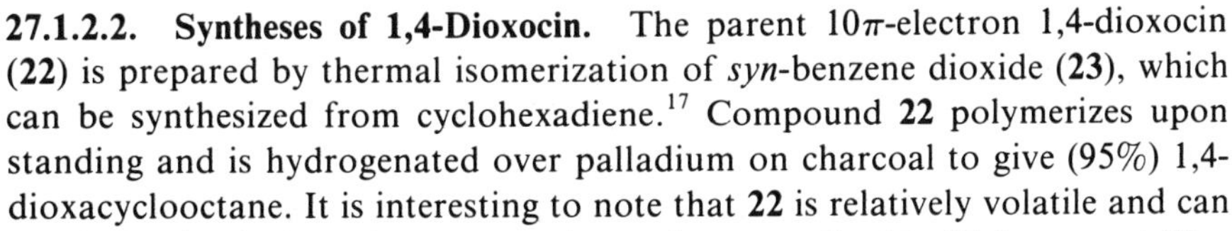

27.1.2.2. Syntheses of 1,4-Dioxocin. The parent 10π-electron 1,4-dioxocin (**22**) is prepared by thermal isomerization of *syn*-benzene dioxide (**23**), which can be synthesized from cyclohexadiene.[17] Compound **22** polymerizes upon standing and is hydrogenated over palladium on charcoal to give (95%) 1,4-dioxacyclooctane. It is interesting to note that **22** is relatively volatile and can be distilled, whereas the tautomeric *syn*-benzene dioxide **23** is a crystalline compound that has been synthesized independently.[13] At 60°C in benzene, the tautomeric equilibrium is 5:95 (**23**:**22**).

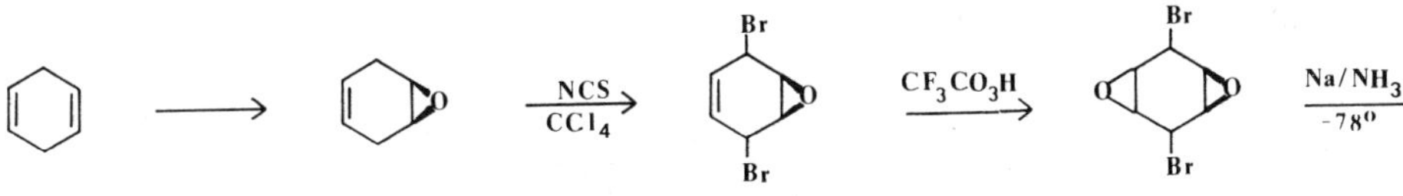

Br 63% 35:65 Br HSn(bu)$_3$ 22(32%) 23

Since arene oxides are metabolites of aromatic compounds and are generally carcinogenic, the isolation of antibiotic **24** from an unspecified fungus is most remarkable.[14]

24

27.1.2.3. Syntheses of 1,4-Dithiocins. In a synthesis similar to that of **22,** the preparation of the parent 1,4-dithiocin **25** was attempted. Treatment of **23** with thioacetic acid affords **26,** which is smoothly transformed to the bis-tosylate **27.** Cyclization of **27** to the desired precursor **28** is conducted at −20°C, since at temperatures in excess of 20°C, facile loss of sulfur occurs. The conclusion is that due to the thermal lability of **28** and in view of theoretical predictions, **25** is not a 10π-aromatic compound.[15]

23 CH_3COSH / pyr/0° → 26(13%) TsCl / pyr/0° → 27(60%) KOH / MeOH / −20° →

28(80%) >20° → benzene + S; 25

A simple synthetic scheme has been developed to produce the benzo-fused 1,4-dithiocin family.[16] Condensation of *cis*-1,2-dimercaptoethenes with *cis*-3,4-dichlorocyclobutene gives the desired bicyclic precursor **29.** Pyrolysis of **29** affords **30.** Irradiation of **30** transforms it back to **29** and upon vacuum pyrolysis **30** produces naphthalene. In most of these larger thio cycles, a facile cheleotropic expulsion of sulfur is realized under thermal and sometimes under photolytic conditions. This degradative process curtails many attempts to form related novel sulfur-containing systems. This 10π-electron system is deemed nonaromatic because of the absence of a diamagnetic ring current.[16]

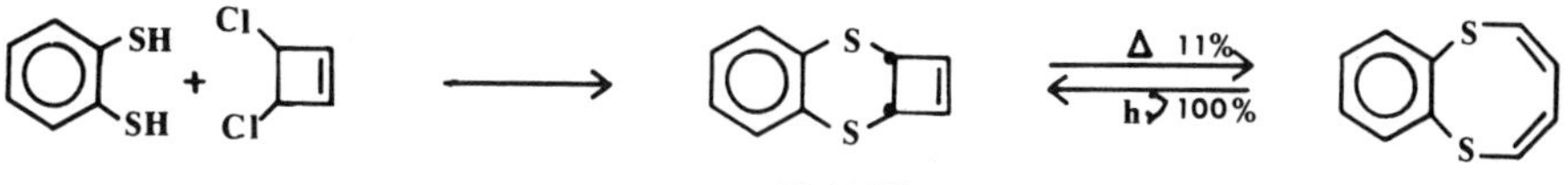

27.1.2.4. Syntheses of 1,4-Diazocine. *cis*-"Benzene triimine" (**31**) is converted, albeit poorly, to the difunctionalized derivative **32,** which can be mono-*N*-nitrosated with nitrosyl tetrafluoroborate at low temperatures. Upon warming, nitrous oxide gas is evolved to give the expected *cis*-"benzene diimine" (**33**), which thermally tautomerizes to the 1,4-diazocine (**34**).[17] When $R = CH_3$, **32** is readily transformed to **34** at $-30°C$; the half-life of **33** is about 2 minutes at 20°C in chloroform. Based on NMR spectral data, **34** is more "aromatic" than **25** or **22;** this phenomenon is attributed to a flattening of a twist conformation.

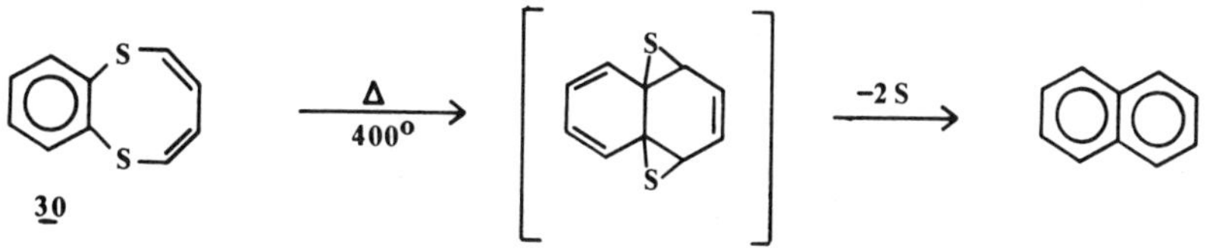

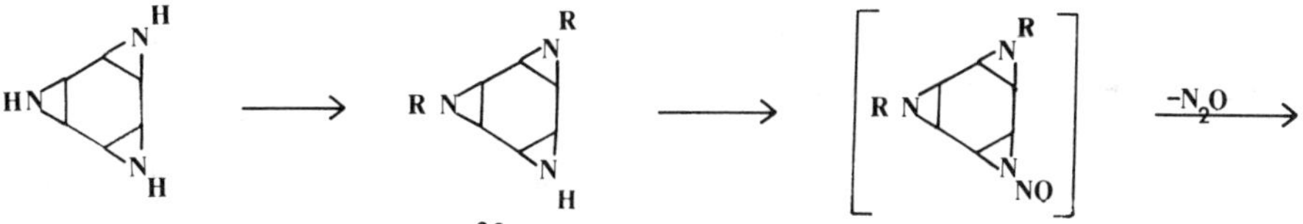

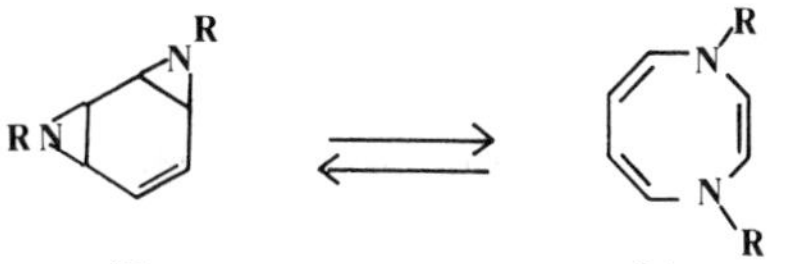

27.2. NINE-MEMBERED RINGS

The nine-membered ring system is unique in that the molecular geometry offers insight to the ability of the heteroatom to contribute to the overall π system.[18]

27.2.1. MONOHETEROATOM RINGS

The parent compounds, azonine (**35**; X = NR)[19] and oxonin (**36**, X = O),[20] were both synthesized by the same procedure, a photoinduced retro-electrocyclization of the corresponding cyclooctatriene derivative **37.** Oxonin is slightly antiaromatic because the oxygen atom is reluctant to contribute to the π-system; azonine (**35**; R = H) appears to be planar and possesses a highly delocalized 10π-electron system which has a *greater* diamagnetic ring current than either benzene or pyrrole. *N*-substituents (**35**, R ≠ H) diminish the aromaticity and the molecule takes on polyene characteristics. Thionin (**35**; X = S) has thus far eluded synthesis.[21]

37 — hν → 35 X=NR, 36 X=O — X=O, hν ⇄ — <−15° → 38

X=O, 30° → benzofuran product; X=NCO₂Et → 39 (N-CO₂Et)

These compounds behave as polyenes in chemical reactions. Isomerization of the double bond from cis to trans is effected photochemically; subsequent thermolysis affords cyclized products (**38**). Azonine (**35**; X = NCO_2Et) undergoes a thermal [4 + 2] cycloaddition with dienophiles and at 55°C for 15 minutes cyclizes to give *N*-carbethoxy-*cis*-8,9-dihydroindole (**39**).

27.2.2. TRIHETEROATOM RINGS

When either *syn*-benzene trioxide (**40**)[22] or triimine (**41**)[23] is heated, there occurs a facile transformation to the *cis,cis,cis*-1,4,7-trioxonine (**42**) or -triazonine (**43**), respectively.

40 X=O, 41 X=NR — Δ → 42 X=O, 43 X=NR — X=NMe, 150° → (N,N'-dimethyl bicyclic product) + $MeNH_2$

27.3. TEN- OR LARGER-MEMBERED RINGS

A review of the π-excessive heteromacrocycles is available.[18] Other compounds of this description include the macrocyclic polyethers ("crown ethers"), the polysulfides ("thiacrown ethers"), and the tetraazamacrocycles. During the last decade each of these subjects has grown to an extent that defies simple summation in this text; therefore, several reviews are referenced.[24]

Although the chemical investigation of small- and medium-sized heterocyclic systems is far from complete, the inherent diversity of the larger systems presages an explosion of research activity. An increase in ring size affords the possibility of an immense number of monocyclic compounds possessing different heteroatom combinations. If subcarbocyclic or subheterocyclic rings are incorporated in the macrocycles, the number of compounds that can be imagined is myriad.

That a simple cyclic molecule which contains a heteroatom can be elaborated to provide a multitude of new compounds is not a matter of great surprise. What is surprising is that so many of the substances have proved to be useful. These extended annular systems have already found application as catalysts and as agents for selective metal ion solubilization. They have also been employed for the detoxification of heavy metals, for the stabilization of energetic transition states, and for ion-selective electrodes. Some of the compounds offer promise as mimics of complex biological molecules such as enzymes and as a means of transporting small neutral molecules across cell membranes. The future will no doubt demonstrate that current estimations of the utility of these substances have been extremely conservative.

REFERENCES

1. L. A. Paquette, T. Kakihana, J. F. Hansen, and J. C. Philips, *J. Am. Chem. Soc.* **93,** 152 (1971).
2. L. B. Anderson, J. F. Hansen, T. Kakihana, and L. A. Paquette, *J. Am. Chem. Soc.* **93,** 161 (1971).
3. L. A. Paquette, J. F. Hansen, and T. Kakihana, *J. Am. Chem. Soc.* **93,** 168 (1971).
4. L. A. Paquette and T. Kakihana, *J. Am. Chem. Soc.* **93,** 174 (1971).
5. R. Askani, *Chem. Ber.* **102,** 3304 (1969).
6. D. W. McNeil, M. E. Kent, E. Hedaya, P. F. D'Angelo, and P. O. Schissel, *J. Am. Chem. Soc.* **93,** 3817 (1971), and refs. cited therein.
7. J. A. Elix, W. S. Wilson, and R. N. Warrener, *Tetrahedron Lett.* **1970,** 1837.
8. F. F. Blicke and N. J. Doorenbos, *J. Am. Chem. Soc.* **76,** 2317 (1954).
9. M. E. Vol'pin, *Russ. Chem. Rev.* **29,** 129 (1960).
10. B. M. Trost and R. M. Cory, *J. Am. Chem. Soc.* **93,** 5573 (1971).
11. B. M. Trost and R. M. Cory, *J. Am. Chem. Soc.* **93,** 5572 (1971).
12. E. Vogel, H. -J. Altenbach, and D. Cremer, *Angew. Chem. Int. Ed.* **11,** 935 (1972).
13. H. -J. Altenbach and E. Vogel, *Angew. Chem. Int. Ed.* **11,** 937 (1972).

14. D. B. Borders, P. Shu, and J. E. Lancaster, *J. Am. Chem. Soc.* **94,** 2540 (1972).

15. E. Vogel, E. Schmidbauer, and H. -J. Altenbach, *Angew. Chem. Int. Ed.* **13,** 736 (1974).

16. D. L. Coffin, Y. C. Poon, and M. L. Lee, *J. Am. Chem. Soc.* **93,** 4627 (1971).

17. H. Prinzbach, M. Breuninger, B. Gallenkamp, R. Schwesinger, and D. Hunkler, *Angew. Chem. Int. Ed.* **14,** 348 (1975); also see E. Vogel et al., *Angew. Chem. Int. Ed.* **18,** 962 (1979); H. Prinzbach et al., *Angew. Chem. Int. Ed.* **18,** 964 (1979).

18. A. G. Anastassiou and H. S. Kasmai, *Adv. Heterocycl. Chem.* **23,** 55 (1978).

19. A. G. Anastassiou and J. H. Gebrian, *J. Am. Chem. Soc.* **91,** 4011 (1969); *Tetrahedron Lett.* **1970,** 825; S. Masamune, K. Hojo, and S. Takada, *Chem. Commun.* **1969,** 1204; A. G. Anastassiou, S. W. Eachus, R. P. Cellura, and J. H. Gebrian, *Chem. Commun.* **1970,** 1133; A. G. Anastassiou and S. W. Eachus, *J. Am. Chem. Soc.* **94,** 2537 (1972).

20. A. G. Anastassiou and R. P. Cellura, *Chem. Commun.* **1969,** 903, 1521; S. Masamune, S. Takada, and R. T. Seidner, *J. Am. Chem. Soc.* **91,** 7769 (1969).

21. A. G. Anastassiou and B. Chao, *Chem. Commun.* **1971,** 979.

22. E. Vogel, H. -J. Altenbach, and C. -D. Sommerfeld, *Angew. Chem. Int. Ed.* **11,** 939 (1972).

23. H. Prinzbach, R. Schwesinger, M. Breuninger, B. Gallenkamp, and D. Hunkler, *Angew. Chem. Int. Ed.* **14,** 347 (1975).

24. *Coordination Chemistry of Macrocyclic Compounds,* G. A. Melson, Ed., Plenum Press, New York, 1979; *Progress in Macrocyclic Chemistry,* R. M. Izatt and J. J. Christensen, Eds., Wiley-Interscience, New York; *Synthetic Multidentate Macrocyclic Compounds,* R. M. Izatt and J. J. Christensen, Eds., Academic Press, New York, 1978.

Appendix A

TABLE A.1. Atomic Electron Populations for Selected π-Excessive Heteroaromatics[a,b]

	1	2	3	4	5	Dipole Moment[c] (Debye)
O-1	8.3104	5.7715	6.1083	6.1083	5.7717	0.19
S-1	16.1276	5.9119	6.0355	6.0356	5.9119	2.22
N-1	6.9197	6.0297	6.0637	6.0637	6.0297	1.97
P-1	15.0194	5.9818	6.0513	6.0513	5.9818	1.89
N-1, N-2[d]	6.8551	7.1310	5.9624	6.1166	6.0280	1.52
	(6.9965)	(6.9965)	(5.9892)	(6.1166)	(5.9892)	
N-1, N-3[d]	6.9563	5.9154	7.1714	5.9872	6.0702	3.01
	(7.0639)	(5.9154)	(7.0639)	(6.0287)	(6.0287)	
O-1, N-3, N-4	8.3637	5.6952	7.1310	7.1310	5.6952	2.12

[a] These densities are derived from MINDO/3 calculations with the addition of inner shell electrons. All structures were optimized until Hebert's test was satisfied. Calculations were performed by G. Baker and K. Theriot (LSU).

[b] The excess or deficit π-electron density is equal to the atomic electron population less the atomic number of the atom.

[c] Calculated dipole moments.

[d] Numbers in parentheses are averages which account for the equivalence of the nitrogens due to rapid proton transfer.

TABLE A.2. Atomic Electron Populations for Selected π-Deficient Heteroaromatics[a,b]

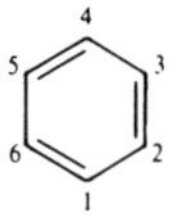	1	2	3	4	5	6	Dipole Moment[c]
N-1	7.1662	5.8653	6.0638	5.9303	6.0639	5.8650	1.31
N-1, N-2	7.0222	7.0222	5.9732	5.9857	5.9857	5.9732	2.27
N-1, N-3	7.2252	5.7493	7.2252	5.8032	6.1368	5.8033	1.20
N-1, N-4	7.1219	5.9292	5.9292	7.1219	5.9293	5.9292	0.00
N-1, N-2, N-3	7.1239	6.8809	7.1239	5.8824	6.0534	5.8824	2.61
N-1, N-2, N-4	6.9792	7.0821	5.8380	7.1737	5.8478	6.0415	1.21
N-1, N-3, N-5	7.2802	5.6941	7.2802	5.6941	7.2802	5.6941	0.00

[a] These densities are derived from MINDO/3 calculations with the addition of inner shell electrons. All structures were optimized until Hebert's test was satisfied. Calculations were performed by G. Baker and K. Theriot (LSU).

[b] The excess or deficit π-electron density is equal to the atomic electron population less the atomic number of the atom.

[c] Calculated dipole moments.

TABLE A.3. Atomic Electron Populations for Selected Fused π-Deficient Heteroaromatics[a,b]

	1	2	3	4	10	5	6	7	8	9	Dipole Moment[c]
N-1	7.1839	5.8422	6.0696	5.9374	6.0458	5.9771	6.0138	5.9706	6.0354	5.8643	1.31
N-2	5.8537	7.1639	5.8812	6.0702	5.9286	6.0242	5.9707	6.0082	5.9778	6.0533	1.34
N-1, N-2	7.0182	6.9964	6.0056	5.9854	5.9816	6.0026	5.9891	5.9894	6.0015	5.9801	2.09
N-1, N-3	7.2340	5.7459	7.2299	5.7932	6.1157	5.9548	6.0284	5.9529	6.0548	5.8199	1.33
N-1, N-4	7.1384	5.9113	5.9112	7.1386	5.9253	6.0111	5.9926	5.9927	6.0110	5.9256	0.13
N-1, N-5	7.1644	5.8621	6.0484	5.9708	5.9296	7.1644	5.8621	6.0484	5.9708	5.9296	0.00
N-1, N-6	7.1985	5.8240	6.0860	5.9133	6.1187	5.8314	5.1791	5.8631	6.1012	5.8161	1.60
N-1, N-7	7.1672	5.8561	6.0600	5.9584	5.9959	6.0438	5.9032	7.1489	5.8845	5.9353	2.24
N-1, N-8	5.1960	5.8267	6.0900	5.9153	6.1065	5.9153	6.0900	5.8267	7.1960	5.7531	2.70
N-2, N-3	5.9461	7.0446	7.0447	5.8458	5.9925	6.0022	5.9867	5.9867	6.0022	5.9925	2.67
N-1, N-3, N-5, N-8	7.2297	5.7491	7.2292	5.8043	6.0562	7.1031	5.9439	5.8759	7.1682	5.7766	

[a] These densities are derived from MINDO/3 calculations with the addition of inner shell electrons. All structures were optimized until Hebert's test was satisfied. Calculations were performed by G. Baker and K. Theriot (LSU).

[b] The excess or deficit π-electron density is equal to the atomic electron population less the atomic number of the atom.

[c] Calculated dipole moments.

TABLE A.4. Atomic Electron Populations for Selected Fused π-Deficient Heteroaromatics[a,b]

	1	2	3	4	5	6	7	8	9	Dipole Moment[c]
N-1, O-9	7.2177	5.8415	6.1164	5.9099	6.1083	5.6092	6.1326	5.7267	8.3549	1.88
N-1, N-9	7.2080	5.8350	6.1136	5.9081	6.0837	5.8289	6.0818	5.9667	7.0201	1.14
N-1, N-7, N-9	7.1920	5.8541	6.1000	5.9304	6.0144	5.8583	7.1875	5.8402	7.0585	1.72
N-1, N-3, N-7, N-9	7.2492	5.7491	7.2464	5.8070	6.0810	5.8099	7.1739	5.8455	7.0675	2.83

[a] These densities are derived from MINDO/3 calculations with the addition of inner shell electrons. All structures were optimized until Hebert's test was satisfied. Calculations were performed by G. Baker and K. Theriot (LSU).

[b] The excess or deficit π-electron density is equal to the atomic electron population less the atomic number of the atom.

[c] Calculated dipole moments.

Appendix B

TABLE B.1. ^{1}H, ^{13}C, and ^{15}N Chemical Shifts of Some π-Deficient Heteroaromatic Compounds

	^{1}H Chemical Shifts (δ, ppm)[a]					^{13}C Chemical Shifts (δ, ppm)[a,c]					^{15}N Chemical Shifts (δ, ppm)[b]
	H-2	H-3	H-4	H-5	H-6	C-2	C-3	C-4	C-5	C-6	
	7.27	7.27	7.27	7.27	7.27	128.5	128.5	128.5	128.5	128.5	—
	8.50	7.06	7.46	7.06	8.50	150.6	124.5	136.4	124.5	150.6	317
	—	9.17	7.68	7.68	9.17	—	152.8	127.6	127.6	152.8	400

	9.15	—	8.60	7.09	8.60	159.5	—	157.5	122.1	157.5	298
	8.50	8.50	—	8.50	8.50	145.6	145.6	—	145.6	145.6	338
	—	9.88	—	8.84	9.48	—	163.0	—	142.0	150.2	$N_1 = 442$ $N_2 = 382$ $N_4 = 282$
	9.18	—	9.18	—	9.18	167.5	—	167.5	—	167.5	282

[a] In $CDCl_3$ solutions using TMS as internal reference.

[b] With reference to $NH_3 = 0$ ppm[7]; all in DMSO or DMF, except pyrazole.

[c] See footnote *b* in Table B.2 and R. J. Radel, B. T. Keen, C. Wong, and W. W. Paudler, *J. Org. Chem.* **42,** 546 (1977), and references therein.

TABLE B.2. ^{1}H, ^{13}C, and ^{15}N Chemical Shifts of Some π-Excessive Heteroaromatic Compounds

	^{1}H Chemical Shifts (δ, ppm)[a,b]				^{13}C Chemical Shifts (δ, ppm)[a]				^{15}N Chemical Shifts (δ, ppm)[c,d]
	H-2	H-3	H-4	H-5	C-2	C-3	C-4	C-5	N
Cyclopentadienide anion	7.27	7.27	7.27	7.27	103.0	103.0	103.0	103.0	—
Pyrrole (N–H)	6.62	6.05	6.05	6.62	118.5	108.2	108.2	118.5	149
Furan (O)	7.40	6.30	6.30	7.40	142.6	109.6	109.6	142.6	—
Thiophene (S)	7.19	7.04	7.04	7.19	125.4	127.2	127.2	125.4	—
Selenophene (Se)	7.88	7.23	7.23	7.88	127.3	129.8	129.8	127.3	—
Tellurophene (Te)	8.85	7.72	7.72	8.85	127.3	138.0	138.0	127.3	—

N O	—	—	—	—	—	—	—	—	381
N S	—	—	—	—	—	—	—	—	298
3 4 5 N N–H	—	7.55	6.25	7.55	—	134.3	105.2	135.3	247.3 ($CHCl_3$)
2 4 5 N N–H	7.70	—	7.14	7.14	136.2	—	122.3	122.3	209
2 4 5 N O	7.95[e]	—	7.09	7.69	—	—	—	—	—
2 4 5 N S	8.88	—	7.98	7.41	—	—	—	—	—

[a] In $CDCl_3$ solutions using TMS as internal reference.

[b] G. C. Levy and G. L. Nelson, *Carbon-13 NMR for Organic Chemists,* Wiley-Interscience, New York, 1972.

[c] With reference to $NH_3 = 0$ ppm; all in DMSO or DMF, except pyrazole.

[d] G. C. Levy, R. L. Lichtes, *N-15 Nuclear Magnetic Resonance,* Wiley, New York, 1979, and refs. cited therein.

[e] In CCl_4 solution; D. J. Brown and P. P. Ghosh, *J. Chem. Soc. B* **1969,** 270.

TABLE B.3. ^{1}H, ^{13}C, and ^{15}N Chemical Shifts of Some Bicyclic Heteroaromatic Compounds

Position of	^{1}H Chemical Shifts (δ, ppm)								^{13}C Chemical Shifts (δ, ppm)										^{15}N Chemical Shifts
Heteroatom(s)	H-1	H-2	H-3	H-4	H-5	H-6	H-7	H-8	C-1	C-2	C-3	C-4	C-5	C-6	C-7	C-8	C-9	C-10	(δ, ppm)
1[a]	—	8.81	7.26	8.00	7.68	7.43	7.61	8.05[b]	—	150.9	121.5	136.0	128.3	126.8	129.7	130.1	149.0	128.7	316.2
2[a]	9.13	—	8.45	7.50	7.71	7.57	7.49	7.86[b]	153.1	—	143.8	120.8	126.8	130.5	127.5	127.9	129.0	136.0	334.0
1,4[a]	—	8.73	8.73	—	8.06	7.67	7.67	8.06	—	145.5	145.5	—	129.8	129.9	129.9	129.8	143.2	143.2	336.0
2,3[a,c]	9.60	—	—	9.60	8.13	8.00	8.00	8.13[b]	152.0	—	—	152.0	126.7	133.2	133.2	126.7	126.7	126.7	398.0
1,2[a,c]	—	—	9.22	7.76	7.76	7.76	7.76	8.48	—	—	146.1	124.6	127.9	132.3	132.1	129.5	151.0	126.8	—
1,3[a]	—	9.57	—	9.58	7.70	7.50	7.60	8.03	—	160.5	—	155.7	127.4	127.9	134.1	128.6	150.1	125.2	—
1,8[c,e]	—	9.22	7.48	8.22	8.22	7.48	9.22	—	—	153.0	122.8	138.9	138.9	122.8	153.0	—	156.4	123.4	—
1,6[c,e]	—	9.91	7.52	8.28	9.28	—	8.76	7.93	—	143.0	126.0	138.0	150.0	—	147.0	125.0	150.0	136.0	—
1,7[c,e]	—	9.14	7.67	8.26	7.72	8.73	—	9.66	—	152.0	125.5	137.0	120.0	145.6	—	153.0	150.0	139.0	—
1,5[c,e]	—	8.97	7.58	8.40	—	8.97	7.58	8.40	—	151.7	125.0	140.4	—	151.7	125.0	140.4	144.2	144.2	—
2,7[c,e]	9.28	—	8.76	7.50	7.50	8.76	—	9.28	153.4	—	147.2	119.6	119.6	147.2	—	153.4	124.3	138.5	—
2,6[c,e]	9.39	—	8.75	7.78	9.39	—	8.75	7.78	152.6	—	145.4	119.9	119.9	—	145.4	119.9	131.0	131.0	—
1,4,5	—	9.50	9.50	—	—	9.50	8.58	9.50	—	—	—	—	—	—	—	—	—	—	—
1,3,8	—	9.57	—	9.58	8.44	7.72	9.33	—	—	—	—	—	—	—	—	—	—	—	—
1,4,5,8	—	9.49	9.49	—	—	9.49	9.49	—	—	—	—	—	—	—	—	—	—	—	—
1,3,5,8	—	9.66	—	9.19	—	9.16	9.16	—	—	—	—	—	—	—	—	—	—	—	—

[a] R. J. Qugmise, D. M. Grand, R. K. Robins, and R. K. Robins, Jr., *J. Am. Chem. Soc.* **91,** 6381 (1969).

[b] In CCl_4.

[c] W. W. Paudler, T. J. Kress, *Adv. Heterocycl. Chem.* **11,** 123 (1970).

[d] P. J. Black and M. L. Heffernan, *Aust. J. Chem.* **18** 707 (1965).

[e] W. W. Paudler, *Adv. Heterocycl. Chem.* **24,** in press (1982)

Appendix C

TABLE C.1. pK_a Data of Some Representative Heteroaromatic Compounds

	pK_a		pK_a
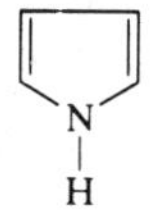 N, H	−0.27	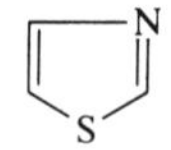 N, S	2.53
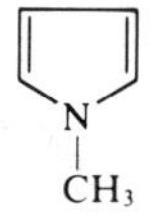 N, CH_3	−1.80	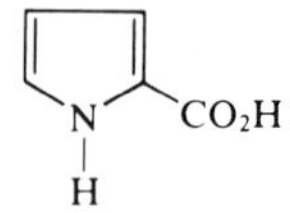 N, H, CO_2H	4.45
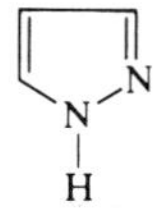 N, N, H	2.47	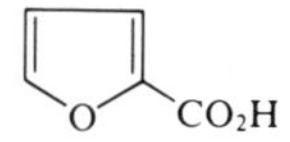 O, CO_2H	3.16
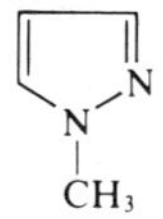 N, N, CH_3	2.04	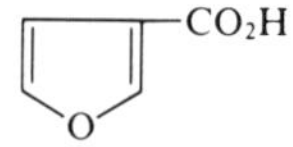 CO_2H, O	3.95
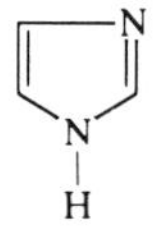 N, N, H	6.95	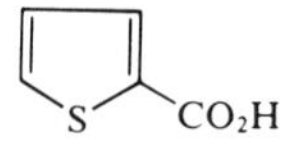 S, CO_2H	3.53
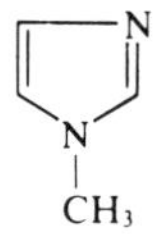 N, N, CH_3	7.33	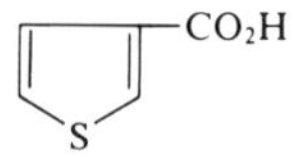 CO_2H, S	4.10
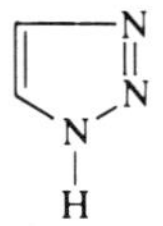 N, N, N, H	1.17	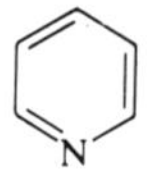 N	5.23
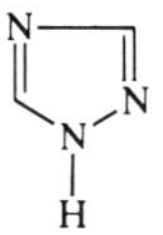 N, N, N, H	2.30	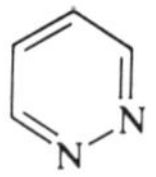 N, N	2.33
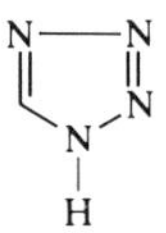 N, N, N, N, H	4.89	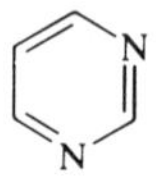 N, N	1.30
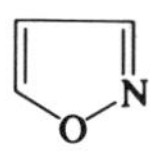 N, O	1.30	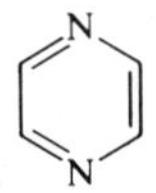 N, N	0.65

Source. A. Albert, *Physical Methods in Heterocyclic Chemistry,* Vol. I, A. R. Katritzky, Ed., Academic Press, 1963, pp. 44.

Appendix D

TABLE D.1. Ultraviolet Spectral Data of Some Representative Heteroaromatic Compounds [λ_{max}^{m}(log E)] (Solvent)[a]

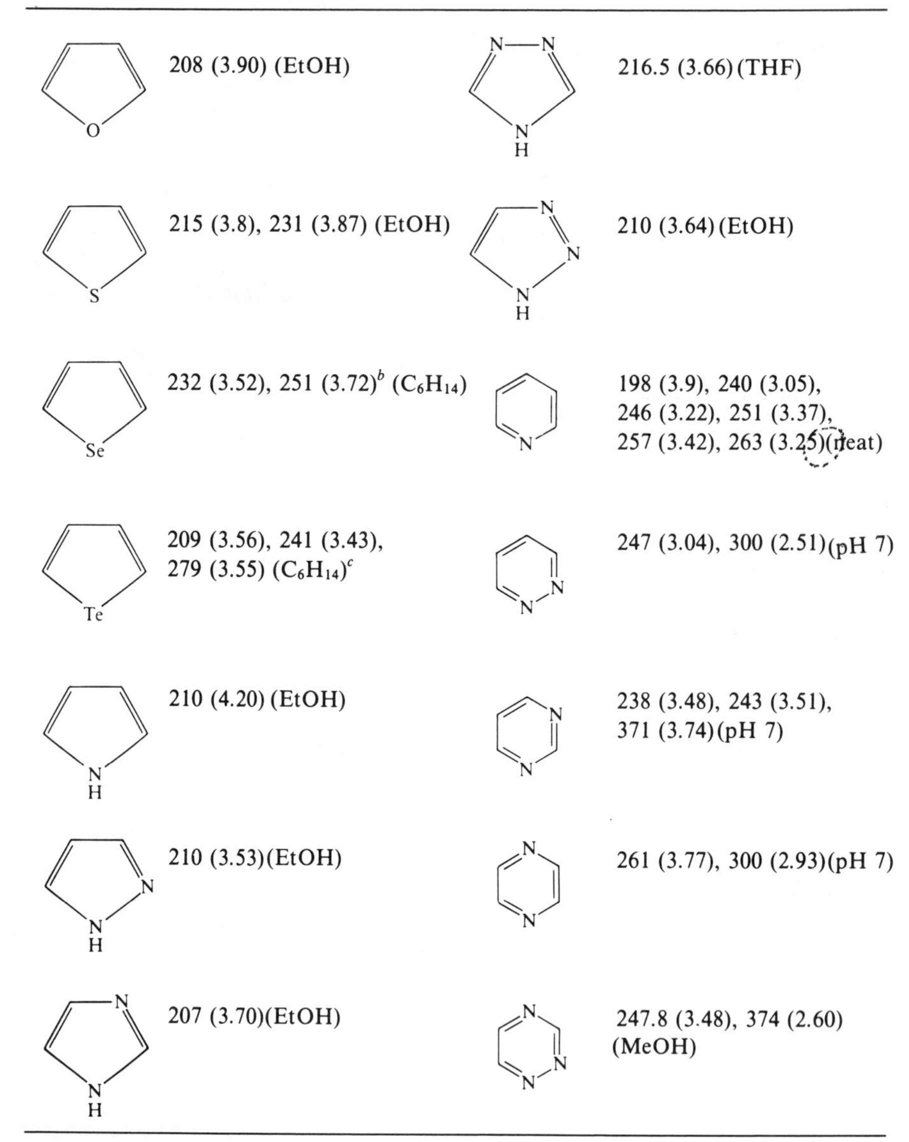

208 (3.90) (EtOH)	216.5 (3.66) (THF)
215 (3.8), 231 (3.87) (EtOH)	210 (3.64) (EtOH)
232 (3.52), 251 (3.72)[b] (C_6H_{14})	198 (3.9), 240 (3.05), 246 (3.22), 251 (3.37), 257 (3.42), 263 (3.25)(neat)
209 (3.56), 241 (3.43), 279 (3.55) (C_6H_{14})[c]	247 (3.04), 300 (2.51)(pH 7)
210 (4.20) (EtOH)	238 (3.48), 243 (3.51), 371 (3.74)(pH 7)
210 (3.53)(EtOH)	261 (3.77), 300 (2.93)(pH 7)
207 (3.70)(EtOH)	247.8 (3.48), 374 (2.60) (MeOH)

[a] W. L. F. Armarego, in "Physical Methods in Heterocyclic Chemistry," Vol. III., A. R. Katritzky, Ed., Academic Press, New York, 1971, Chapter 4.

[b] A. Bellotti and L. Chierici, *Gazz. Chim. Ital.* **90,** 1125 (1960); L. Chierici and G. Pappalardo, *ibid.* **88,** 453 (1958).

[c] F. Fringuelli and A. Taticchi, *J. Chem. Soc. Perkin* I **1972,** 199.

Appendix E

TABLE E.1. Molecular Dimensions of Some π-Excessive Heteroaromatic Compounds[a]

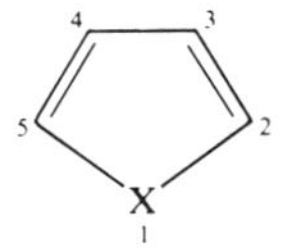

X	Bond Lengths (Å)						Bond Angles				
X	X—C_2	C_2—C_3	C_3—C_4	C_2—H_2	C_3—H_3	N—H	X_1—C_2—C_3	C_2—C_3—C_4	C_5—X_1—C_2	X_1—C_2—H	C_4—C_3—H
O	1.362	1.361	1.431	1.075	1.077		114°41′	106°31′	106°33′	115°55′	127°57′
S	1.714	1.370	1.423	1.080	1.080		111°28′	112°27′	92°10′	119°51′	124°00′
Se	1.855	1.370	1.433	1.070	1.079		111°34′	114°33′	87°46′	121°44′	122°52′
Te[b]	2.055	1.375	1.423	1.078	1.081		110°81′	117°93′	82°53′	124°59′	121°04
N—H	1.370	1.382	1.417	1.070	1.077	0.996	107°42′	107°24′	109°48′	121°30′	127°06′

[a] J. B. Auliniu, A. A. Shapkin, and N. N. Maztesieva, *Dokl. Akad. Nauk. S.S.S.R.* **185**, 384 (1968); also see footnote *b*.

TABLE E.2. Molecular Dimensions of Some π-Deficient Heteroaromatic Compounds

Ring System	Bond Lengths (Å)									Bond Angles					
	1—2	2—3	3—4	4—5	5—6	6—1	2—H	3—H	4—H	1—2—3	2—3—4	3—4—5	1—2—H	2—3—H	3—4—H
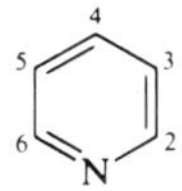	1.340	1.395	1.394	—	—	—	1.084	1.081	1.077	123.9	118.5	118.3	115.9 2—1—6: 116.8	121.3	120.8
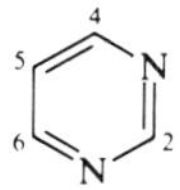	1.317	—	1.344	1.358	—	—	—	—	—	128.2	115.2	122.7	4—5—6: 116.3	—	—
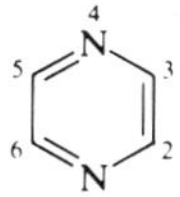	1.314	1.358	—	—	—	—	—	—	—	122.4	—	—	2—1—6: 115.1	—	—
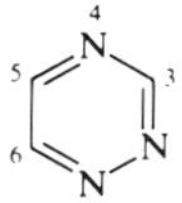	1.335	1.314	1.339	1.317	1.401	1.317	—	—	—	—	—	—	—	—	—
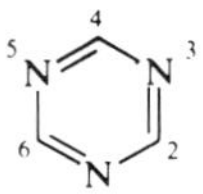	1.319	—	—	—	—	—	0.998	—	—	126.8	—	—	—	—	—

Source. W. Cohran, *Physical Methods in Heterocyclic Chemistry,* Vol. I, A. R. Katritzky, Ed., Academic Press, 1963, pp. 164ff.

Appendix F

TABLE F.1. Major Mass Spectral Fragmentation Patterns of Some Simple Heteroaromatic Compounds

O $\oplus\cdot$ $\longrightarrow$ $C_3H_3^{\oplus}$

O $CH_3\cdot$ $\oplus\cdot$ $\longrightarrow$ $CH_3CO^{\oplus}$, O $\oplus\cdot$

S $\oplus\cdot$ $\longrightarrow$ $C_2H_2S^{\oplus}$

CH_3 S $\oplus\cdot$ $\longrightarrow$ S $\oplus$

$\oplus$ N H $\xrightarrow{-HCN}$ $C_3H_4^{\oplus}$

CH_3 $\oplus$ N H $\longrightarrow$ N $\oplus$

TABLE F.1. (*Continued*)

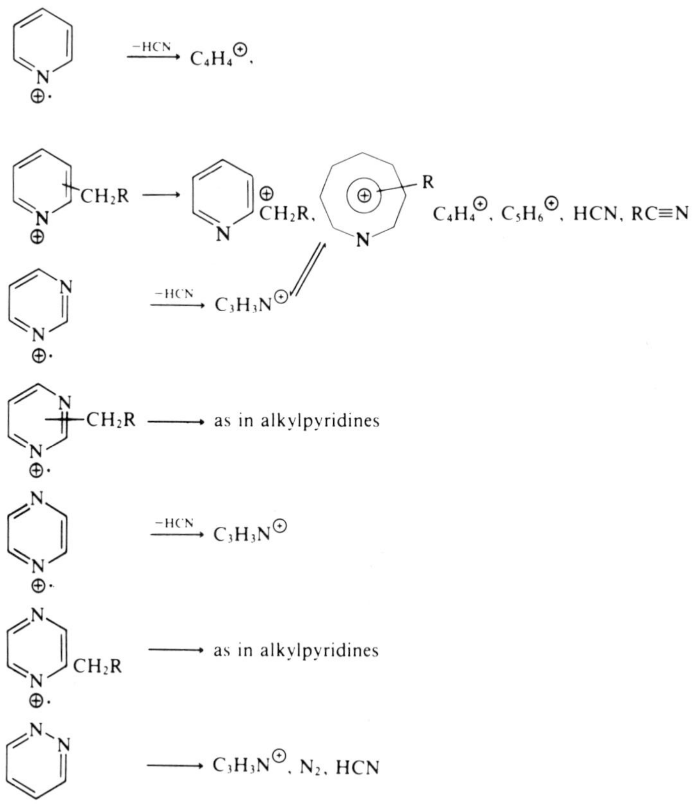

Pyridine radical cation (N⊕·)	$\xrightarrow{-HCN}$ $C_4H_4^{\oplus}$,
Alkylpyridine radical cation (—CH_2R, N⊕)	⟶ pyridine ⊕CH_2R, azatropylium (⊕—R) ion, $C_4H_4^{\oplus}$, $C_5H_6^{\oplus}$, HCN, RC≡N
Pyrimidine radical cation (N⊕·)	$\xrightarrow{-HCN}$ $C_3H_3N^{\oplus}$
Alkylpyrimidine radical cation (—CH_2R, N⊕·)	⟶ as in alkylpyridines
Pyrazine radical cation (N⊕·)	$\xrightarrow{-HCN}$ $C_3H_3N^{\oplus}$
Alkylpyrazine radical cation (CH_2R, N⊕·)	⟶ as in alkylpyridines
Pyridazine	⟶ $C_3H_3N^{\oplus}$, N_2, HCN

Source. A. Spiteller, *Physical Methods in Heterocyclic Chemistry,* Vol. III, A. R. Katritzky, Ed., Academic Press, 1971, p. 223.

INDEX